Extreme Natural Events

A. S. Unnikrishnan · Fredolin Tangang ·
Raymond J. Durrheim
Editors

Extreme Natural Events

Sustainable Solutions for Developing
Countries

 Springer

Editors
A. S. Unnikrishnan
Physical Oceanography Division
CSIR-National Institute of Oceanography
Dona Paula, Goa, India

Fredolin Tangang
Department of Earth Sciences
and Environment
Universiti Kebangsaan Malaysia
Bangi, Selangor, Malaysia

Raymond J. Durrheim
School of Geosciences
University of the Witwatersrand
Johannesburg, South Africa

ISBN 978-981-19-2513-9   ISBN 978-981-19-2511-5 (eBook)
https://doi.org/10.1007/978-981-19-2511-5

This Springer imprint is published by the registered company Springer Nature Singapore Pte Ltd.
The registered company address is: 152 Beach Road, #21-01/04 Gateway East, Singapore 189721, Singapore

Foreword by Philip L. Woodworth

Two recent events have underlined the importance of extreme natural events to our planet's populations. The first event was the publication in 2021 of *The Physical Science Basis* by Working Group I (WG1) of the Intergovernmental Panel on Climate Change (IPCC). Two further reports concerned with *Impacts, Adaptation and Vulnerability* and *Mitigation of Climate Change* will be published by Working Groups II and III, respectively in 2022, thereby completing the overall IPCC's Sixth Assessment Reports (AR6). These reports provide irrefutable evidence for anthropogenic climate change during the past century and show that extreme events connected with climate change are now occurring more frequently than previously, with the largest events often impacting on developing countries. The reports provide projections of changes in many climate parameters towards 2100 and beyond, and thereby changes in the occurrence of extreme events, from which one concludes that, without mitigation and adaptation measures being taken, impacts on developing countries will be even more severe.

The second event was the 26th United Nations Climate Change Conference of the Parties (COP26) held in Glasgow during October–November 2021. Discussions at that conference underlined the need for urgency in tackling climate change, so that its impacts might be constrained as far as possible. It also demonstrated how difficult and costly that effective action will be.

For example, within my own field of sea-level science, the AR6 WG1 suggested that global average sea level could rise by 0.28–0.55 m by 2100 (relative to 1995–2014), assuming a very low greenhouse gas emission scenario, or 0.63–1.01 m, assuming a very high emission scenario. Such a rise in sea level, combined with possible changes in the frequency and intensity of storms and their associated storm surges, will require major expenditure in raising coastal defences, with sometimes consequent undesirable modifications to coastal environments. Moreover, in some cases, such as small island states, sea-level rise will represent a major threat to the people.

The AR6 and COP26 have made clear that climate change will impact every country in the world in different ways. In particular, rising temperatures and sea levels and modifications to rainfall patterns will have major impacts on natural environments and agriculture and have many consequences for the built environment and infrastructure. A large number of these topics are discussed in this volume.

The editors are to be congratulated on assembling an excellent set of chapters which together underline the importance of the above topics. The first chapters in the volume are concerned with the provision of information on, and systems for addressing, extreme climate events. There then follow two sets of chapters dealing with extreme rainfall and thunderstorm events, which often lead to river flooding, and extreme wave and sea-level events, which can lead to coastal flooding. Further chapters discuss the extreme events associated with earthquakes and landslides, a reminder that extreme natural events are not confined to those associated with climate change. For example, recent years have demonstrated the threat posed to coastal populations by undersea earthquakes and tsunamis. Some of the chapters in the book discuss impacts of climate change on agriculture and integrated methods to reduce the impacts of disastrous events.

The chapters in this volume show that although the importance of individual types of extreme event varies between countries, they all require addressing worldwide. The chapters also demonstrate that major investment is required in infrastructure, including regional and global monitoring networks for a range of climate parameters. In addition, development is needed in new analysis and modelling techniques for the forecasting of extreme events (e.g. downscaling of global and regional climate projections to the more practically useful short spatial scales). International collaboration will be essential in network development and the use of the resulting data sets. The ultimate aim must be to provide the best possible information and advice to decision makers and the public, so that systems can be constructed by which the impacts of extreme events might be mitigated.

Therefore, it is gratifying that the chapters in this volume are written by authors from so many developing countries. I can recommend the volume as an important contribution to research into the extreme events which have such scientific and practical importance for us all.

Philip L. Woodworth, Ph. D.
Former Director of the Permanent
Service for Mean Sea Level
Emeritus Fellow
National Oceanography Centre
Liverpool, UK

Introduction by Amitava Bandopadhyay

Extreme natural events such as avalanches, earthquakes, tsunamis, wildfires, floods, droughts, cyclones, volcanoes, thunderstorms and intense rainfall events occur across the world. Many of these phenomena are affected by the climate change. Various climate assessment reports have indicated that their frequency or intensity has increased, and projections show further increase in the future. Developing countries are far more vulnerable to these extreme events due to their inadequate technological, financial and logistic capabilities.

Extreme events are likely to push millions in the developing world into poverty and reduce the prospects for sustainable development. Such events are likely to have a significant impact both on human growth and economics. Variations in the frequency, severity and duration of some extreme weather events increase risks to children's mental health and impact their development from infancy to adolescence. Similarly, these extremes will primarily impact economic growth of a nation through damage to property and infrastructure, loss in productivity, mass migration and security threats.

Solutions developed in technologically-advanced countries often cannot be simply applied in the developing world. It is thus imperative that the natural and social scientists and engineers in the developing world improve their ability to forecast and manage these extreme events in order to reduce the economic, social (human) and environmental impacts. Adaptation to these extreme natural events must be an integral part of the national policy of the developing countries dealing with disaster management.

This book in its seventeen chapters intends to explore the challenges of the developing countries to understand and manage the risks of extreme natural events. The book brings together scientific communities from Ghana, India, Indonesia, Malaysia, Philippines, Sri Lanka, South Africa and Venezuela to share their experience and expertise in different aspects of managing extreme natural events, particularly those related to climate.

In this connection, I am proud to mention that in the past the NAM S&T Centre has made significant contributions in capacity building and exchange of knowledge among its member countries on the subject of extreme natural events in partnership with various S&T institutions and agencies by organizing international workshops, roundtables, symposiums and training programmes and bringing out relevant scientific publications that are of significant interest to the global south.

I am extremely delighted that the NAM S&T Centre has reached another milestone by publishing its next scientific monograph titled *Extreme Natural Events: Sustainable Solutions for Developing Countries*. I gratefully acknowledge the contributions made by eminent experts from various countries on different themes on extreme natural events including climate extremes, *such as extreme rainfall events and thunderstorms, extreme waves, extreme sea-level changes, storm surges and coastal inundation, earthquakes and landslides, impact assessment and integrated disaster risk reduction*. I may mention here that the subject of "Lightning" as an important component of extreme natural events has not been included in the scope of this monograph as NAM S&T Centre has published a separate monograph titled *Lightning: Science, Engineering, and Economic Implications for Developing Countries* in August 2021 through Springer Nature, Singapore.

I am thankful to the editorial team of this book: Dr. A. S. Unnikrishnan (Former Chief Scientist, Physical Oceanography Division, CSIR-National Institute of Oceanography, Goa, India), Prof. Fredolin Tangang (Chairman and Professor, Department of Earth Sciences and Environment, Faculty of Science and Technology, Universiti Kebangsaan Malaysia, Kuala Lumpur) and Prof. Raymond J. Durrheim (South African Research Chair in Exploration, Earthquake and Mining Seismology, University of the Witwatersrand, Johannesburg, South Africa) for the scientific evaluation of the manuscripts and ensuring the best selection of the contents for dissemination of scientific knowledge on the subject in the developing world.

In addition, I am also grateful to some of the invited reviewers from India : (i) Dr. Umesh Chandra Kulshrestha (Professor, School of Environmental Sciences, Jawaharlal Nehru University, New Delhi); (ii) Dr. Someshwar Das (Former Adviser/Scientist 'G', Ministry of Earth Sciences, Government of India, New Delhi); (iii) Dr. M. R. Ramesh Kumar (Former Chief Scientist, Physical Oceanography Division, National Institute of Oceanography, Dona Paula, Goa); (iv) Dr. Kalachand Sain (Director, Wadia Institute of Himalayan Geology, Dehradun, Uttarakhand); (v) Dr. Smitha V. Thampi (Scientist F, Space Physics Laboratory, Vikram Sarabhai Space Centre, Indian Space Research Organisation, Trivandrum, Kerala); (vi) Prof. Bimal Kumar Roy (Head, R. C. Bose Centre for Cryptology and Security, and Former Director, Indian Statistical Institute, Kolkata) and (vii) Prof. Subimal Ghosh (Institute Chair Professor, Department of Civil Engineering and Convener, Interdisciplinary Program in Climate Studies, Indian Institute of Technology Bombay, Mumbai) for extending their support to the Centre in bringing out this valuable publication.

I express my sincere gratitude to Dr. Philip L. Woodworth, Former Director of the Permanent Service for Mean Sea Level at the National Oceanography Centre, Liverpool, UK, for kindly agreeing to write the "Foreword" of the monograph.

I am thankful to Dr. Loyola D'Silva, Executive Editor, Springer Nature, Singapore, for considering this book for publication through the reputed publishing house Springer Nature, and Ms. Vinothini Elango, Project Coordinator, Springer Nature, for monitoring and streamlining the publication process. I am confident that our association with "Springer" would lead to many more such valuable collaborative endeavours in future.

My sincere thanks are also due to the entire team of the NAM S&T Centre, especially to Mr. M. Bandyopadhyay, Senior Adviser and Ms. Jasmeet Kaur Baweja, Programme Officer and Contributing Staff Editor, NAM S&T Centre for facilitating this project. I am also thankful to Dr. Ranadhir Mukhopadhyay, Former Chief Scientist, CSIR-National Institute of Oceanography (NIO), Goa, for helping to bring out this publication. I also record my appreciation for the assistance rendered by my colleagues Mr. Rahul Kumra and Mr. Pankaj Button towards bringing out this publication.

I am sure that this book would be a valuable reference material for the scientists, researchers and other professionals working in the areas of extreme natural events, particularly those related to climate.

Amitava Bandopadhyay, Ph.D.
Director General, NAM S&T Centre
New Delhi, India

Preface

Extreme events of atmospheric, coastal, hydrological and geological origins (e.g. intense rainfall events, droughts, hurricanes, storm surges, floods, earthquakes, landslides) cause fatalities, property damage and socio-environmental disruption. The disasters occurring all over the globe have affected millions of people, causing loss of life and inflicting huge financial loss. Natural disasters the world over have been increasing due to global warming.

This state of affairs is particularly daunting for low-income developing countries, since they lack the resources to prepare, respond and mitigate. These extreme events represent major development impediments for low-income countries, as they hamper access to shelters, clean water, sanitation, cause food/nutritional shortage and increase the threat of communicable diseases.

Extreme events inflict considerable economic burden. The additional funding required for reconstruction and economic recovery efforts, particularly for developing countries, demonstrates the need to link the efforts for emergency disaster response to long-term development projects to sustain recovery. Hence, adapting and mitigating extreme events become a priority for governments the world over. The first step for achieving this is to understand these events, their occurrences in the past and changes that have been happening in recent years. It is also important to understand how climate-related disasters can be influenced by climate change in future periods.

The present monograph provides a comprehensive description of some of the major extreme events in atmosphere, ocean and land. Though the list may not be complete, a good attempt has been made to cover a wide spectrum of events covering different countries in many regions. The contributed articles are from authors belonging to various countries in the Non-Aligned Movement (NAM). The articles, in general, describe extreme events in the regions surrounding these countries. Developing countries are more vulnerable to the impacts of these events, because of lack of preparedness, lack of adequate adaptation and mitigation practises. Moreover, the high cost of mitigation also adds to the slow progress in implementation. We hope that this volume will be useful for improving the understanding of the extreme events in region-wise basis so that various governments and policymakers can develop long-term strategies for tackling these events.

Some articles in the present monograph focus on the scientific understanding of extreme events, while some are on adaptation and mitigation strategies to tackle these extremes. *The seventeen articles are divided into five parts.* Part I deals with a description of climate extremes. Part II covers extremes atmospheric events, such as intense rainfall and thunderstorms. Part III deals with a description of extremes waves, sea level and coastal flooding. Part IV deals with articles on earthquakes and landslides. Part V has only one article that deals with impact assessment, and Part VI deals with adaptation and mitigation approaches dealing in particular with hydroclimatic extreme events.

We believe that this volume will enhance public awareness and promote educational efforts to increase understanding of these events and their management, which will be the first step for planning, adaptation and mitigation. The Intergovernmental Panel on Climate Change (IPCC) published the Sixth Assessment Report (AR6) of WG I in 2021 and the reports of WG II and WG III are expected to be published in 2022. These reports provide comprehensive descriptions of the past changes and future projections of extreme events due to climate change and their impacts, vulnerability and mitigation. The present monograph provides a description of some of these events and adaptation practises followed in selected regions. A unique feature of the monograph is that it covers many regions in different continents such as Asia, Africa, North and South America. This will provide information on extreme events in different regions for various governments and policy makers. We hope that this monograph will generate a lot of public awareness and be useful for educational purposes. It is hoped that the monograph will complement AR6 reports of IPCC.

Goa, India   A. S. Unnikrishnan
Selangor, Malaysia   Fredolin Tangang
Johannesburg, South Africa   Raymond J. Durrheim

Contents

Part V   Impact Assessment

Part VI   Integrated Disaster Risk Reduction

Part I
Climate Extremes

Chapter 1
CORDEX Southeast Asia: Providing Regional Climate Change Information for Enabling Adaptation

Fredolin Tangang, Jing Xiang Chung, Faye Cruz, Supari, Jerasorn Santisirisomboon, Thanh Ngo-Duc, Liew Juneng, Ester Salimun, Gemma Narisma, Julie Dado, Tan Phan-Van, Mohd Syazwan Faisal Mohd, Patama Singhruck, John L. McGregor, Edvin Aldrian, Dodo Gunawan, and Ardhasena Spaheluwakan

Abstract This chapter describes the activities, progress and relevance of the Coordinated Regional Climate Downscaling Experiment—Southeast Asia (CORDEX-SEA) in providing regional climate change information for enabling adaptation in

F. Tangang (✉) · L. Juneng · E. Salimun
Department of Earth Sciences and Environment, Faculty of Science and Technology, Universiti Kebangsaan Malaysia, Bangi, Selangor, Malaysia
e-mail: tangang@ukm.edu.my

J. X. Chung
Faculty of Science and Marine Environment, Universiti Malaysia Terengganu, Kuala Nerus, Terengganu, Malaysia

F. Cruz · G. Narisma · J. Dado
Regional Climate Systems Laboratory, Manila Observatory, Quezon City, Philippines

Supari · D. Gunawan
Center for Climate Change Information, Agency for Meteorology Climatology and Geophysics (BMKG), Jakarta, Indonesia

F. Tangang · J. Santisirisomboon
Ramkhamhaeng University Center of Regional Climate Change and Renewable Energy (RU-CORE), Bangkok, Thailand

T. Ngo-Duc
REMOSAT Laboratory, University of Science and Technology of Hanoi, Academy of Science and Technology, Hanoi, Vietnam

G. Narisma
Atmospheric Science Program, Physics Department, Ateneo de Manila University, Quezon City, Philippines

T. Phan-Van
Department of Meteorology and Climate Change, VNU University of Science, Hanoi, Vietnam

M. S. F. Mohd
National Hydraulic Research Institute of Malaysia (NAHRIM), Seri Kembangan, Selangor, Malaysia

3

A. S. Unnikrishnan et al. (eds.), *Extreme Natural Events*,
https://doi.org/10.1007/978-981-19-2511-5_1

Southeast Asia. Southeast Asia is a region that has been affected by climate change, particularly by the weather and climate extremes, and will likely to be impacted more in the future decades as global temperature continues to increase. The number of climate-related disasters, particularly floods, has increased since the last few decades. For climate resilience, countries in the region need to mitigate and adapt, which both require regional future climate change information. However, for robust formulation of adaptation measures, future climate change information at local scales is required. CORDEX-SEA was established to provide multi-model and high-resolution climate projections and fulfil climate model data requirements in the region. CORDEX-SEA has generated a reasonably good model ensemble of regional climate downscaling from 11 global climate models (GCMs) using 7 regional climate models (RCMs) at a spatial resolution of 25 km. Based on these projections, depending on region and seasons, significant changes in mean and extreme precipitation are projected to occur in the future decades. The generated downscaled data can now be used for climate risk assessment in the Southeast Asia region. Furthermore, the establishment of the Southeast Asia Regional Climate Change Information System (SARCCIS), a data node of the Earth System Grid Federation (ESGF), facilitated the data archiving and dissemination of CORDEX-SEA data to end users and scientists involved in the assessment of vulnerability, impacts and adaptation (VIA).

1.1 Introduction

The world is at a critical juncture with respect to its current annual CO_2 emission of over 50 Gt. The Intergovernmental Panel on Climate Change (IPCC) Special Report on 1.5 °C concluded that the annual CO_2 emission would need to be reduced to 20–30 Gt by 2030 for ensuring the temperature increase is capped below 1.5 °C (Rogelj et al. 2018). Under the Paris Agreement, zero net emission would need to be achieved by the end of the century in order to cap global warming below 2.0 °C. However, the current Nationally Determined Contribution (NDC) pledges of the agreement are projected to elevate the global mean temperature by 2.5–2.8 °C relative to pre-industrial levels

P. Singhruck
Department of Marine Science, Faculty of Science, Chulalongkorn University, Bangkok, Thailand

J. L. McGregor
CSIRO Oceans and Atmosphere, Aspendale, Australia

E. Aldrian
Agency for the Assessment and Application of Technology (BPPT), Jakarta, Indonesia

A. Spaheluwakan
Center for Applied Climate Services, Agency for Meteorology Climatology and Geophysics (BMKG), Jakarta, Indonesia

F. Tangang
Natural Disaster Research Centre, Universiti Malaysia Sabah, Kota Kinabalu, Sabah, Malaysia

(OurWorldinData.org 2020). With current policies, the level of warming is projected to be even higher, i.e. around 2.8–3.2 °C (OurWorldinData.org 2020). In fact, to meet the 2 °C target, the emission reductions should increase by 80% beyond NDC (Liu and Raftery 2021). These uncertain prospects of climate change mitigations and magnitude of future warming emphasise the need for adaptation measures to increase climate resilience, especially in the developing and least-developed countries. In Southeast Asia, where the level of exposure and vulnerability to climate change impacts are considered high, adaptation measures are required for climate resilience.

The latest Sixth Assessment Report (AR6) of the Intergovernmental Panel on Climate Change (IPCC) indicated that observed hot extremes and heavy precipitation events in Southeast Asia have increased significantly over a period from the 1950s to the present (IPCC 2021; Seneviratne et al. 2021). Despite contributing minimally to the accumulated concentration of atmospheric greenhouse gases (GHG) thus far, many developing and least-developed countries have been impacted significantly and may face increasing risks of climate change impacts in the future decades, especially from weather and climate extremes. In Southeast Asia, where most countries are either developing or least developed (Fig. 1.1), many countries have experienced increased occurrences in climate-related disasters since the last few decades. In fact, from 1980 to 2017, Southeast Asia recorded the highest number of disasters compared to other regions in Asia (Dagli and Ferrarini 2019). The losses associated

Fig. 1.1 The map of Southeast Asia and the geographical locations of 11 countries within the region

with climate-related disasters, such as floods, droughts, forest fires and transboundary haze episodes and typhoons, have been steadily increasing over the last two decades (Hijioka et al. 2014).

In 2011, Thailand was struck by a massive flood event with staggering economic losses of USD46.5 billion and 815 deaths (Supharatid et al. 2016). Vietnam was hit by the worst-ever typhoon Linda in 1997 affecting more than 500,000 people with 3000 deaths (Anh et al. 2019). Typhoon Haiyan, which was the strongest ever recorded in the Philippines, struck the country in 2013 killing more than 6300 people and resulted in USD1.8 billion in economic losses (Hernandez et al. 2015). Typhoon Nargis that hit Myanmar in 2008 was considered the deadliest in the region killing more than 138,000 people (Firtz et al. 2009). Malaysia, despite being located outside the typhoon belt, had been experiencing recurrences of major flood events like the one in December 2014 that affected more than 500,000 people (Hai et al. 2017). Droughts are also becoming more frequent across the region. In 2015, Indonesia experienced a severe drought event that resulted in widespread forest fires and a prolonged period of haze episode that affected not only Indonesia but the surrounding countries including Malaysia, Singapore and Brunei (Lohberger et al. 2018). These are some examples of the major climate-related disasters that have occurred in the region in the past decades. As global temperature increases, extreme weather and climate events are likely to shift towards higher frequency, longer duration and greater intensity, which is projected to be the case in Southeast Asia in the future decades (e.g. Tangang et al. 2018; Supari et al. 2020).

As the world faces a real prospect of warming beyond 2 °C by the end of the twenty-first century, developing and least-developed countries, especially in Southeast Asia, need to increase their climate resilience through adaptation efforts. These countries should embrace sustainable development agendas as unsustainable practices would often exacerbate the risk of climate change impacts in the future. Climate risk assessments on key sectors often require future climate information, which together with the information on exposure and vulnerability will determine the level of risk (IPCC 2012; Lavell et al. 2012). Because such an assessment is often implemented at local scales, climate information needs to be tailored at similar scales. However, generating such future climate information, ideally using multiple global climate models (GCMs) and regional climate models (RCMs), can be highly technical and expensive for most least-developed and developing countries to implement individually unless conducted in a collaborative manner involving multiple institutions or countries. Top-down initiatives such as the Coordinated Regional Climate Downscaling Experiment (CORDEX), a programme developed under the World Climate Research Programme (WCRP) (Giorgi et al. 2009), can be a feasible platform for least-developed and developing countries to participate, either in running climate models or directly using the model's outputs. This chapter highlights how the CORDEX Southeast Asia, a CORDEX initiative of this region, can be a solution and a provider of climate information in the region.

1.2 Climate Information for Enabling Adaptation

Understanding the risk of future climate-related impacts to socio-economic sectors and the environment is crucial to the formulation of climate resilience policies. Managing a changing climate requires dual approaches, i.e. mitigation and adaptation (Fig. 1.2). Mitigation works in avoiding the risk, whereas adaptation reduces the risks of climate changes already locked in and possible future changes. While mitigation requires a globally coordinated effort to reduce the GHG emission to predetermined levels, adaptation is implemented at the local scales and can be sector or location dependent. Nonetheless, both approaches require future climate information. The levels of risk of climate-related impacts on sectors, assets, livelihoods and communities are shaped by the interactions of climate-related hazards and the levels of exposure and vulnerability of these sectors to the hazard concerned (Fig. 1.2). Hence, information on hazards, exposure and vulnerability needs to be quantified. In fact, these requirements have been outlined in the National Adaptation Plan (NAP) Process of the United Nations Framework Convention for Climate Change (UNFCCC LDC Expert Group 2012). In such a process, analysing current and future climate scenarios and assessing climate exposure and vulnerabilities are key factors in determining adaptation options.

Climate information should encompass both natural variability and anthropogenic climate change (Fig. 1.2). In future warmer periods, extreme events, e.g. floods, droughts and heatwaves, could be influenced both by anthropogenic climate change and climate variability. Modelling of climate change requires a GCM, a complex tool that is best described as a mathematical representation of the climate system and its processes. In addition, GCMs need to have the ability to simulate major

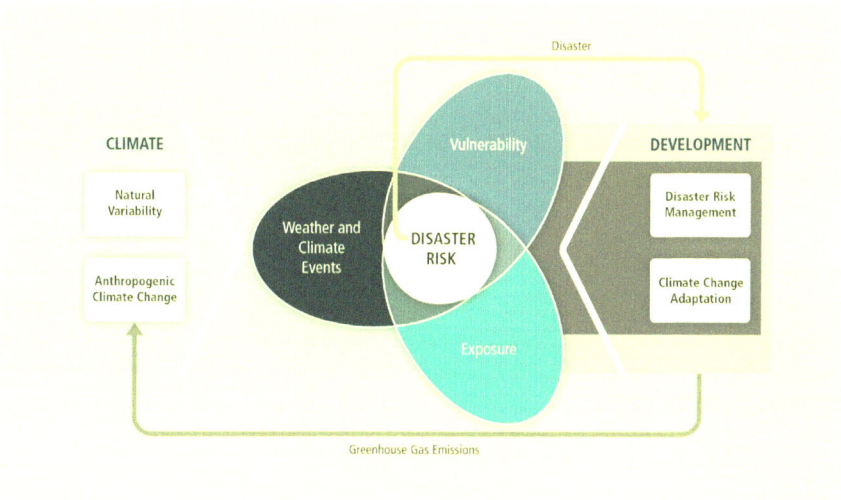

Fig. 1.2 Illustration highlighting the key concepts of the risk of climate impacts (IPCC 2012)

modes of natural variability in the climate system such as the El Niño–Southern Oscillation (ENSO), Indian Ocean Dipole (IOD) and Madden–Julian Oscillation (MJO). An RCM, which is nested within a GCM, should also have the ability to simulate information related to climate change and climate variability.

Relevant climate information can directly come from the model's output variables such as mean, minimum and maximum temperatures, rainfall, humidity, winds and solar radiation, usually presented in the form of changes over a particular future period relative to the historical period. For example, Tangang et al. (2020) described the changes in mean precipitation over Southeast Asia for early, middle and late periods of the twenty-first century. Such information on projected changes is relevant in assessing the risk of impacts in a particular sector. Climate indices are also often used to characterise future climatic conditions, e.g. standard precipitation index (SPI) for drought. For climate extremes, indices such as the extreme indices of the Expert Team on Climate Change Detection and Indices (ETCCDI) (Zhang et al. 2011) can be used. These indices can be computed from the basic climate model's variables, and future changes relative to historical periods can be quantified. For example, Supari et al. (2020) used ETCCDI indices in projecting changes of precipitation extremes in Southeast Asia. Such climate information can be useful in assessing the level of risk of climate and extremes impacts in relevant sectors. Supari et al. (2020) projected the changes in the consecutive dry days (CDD) over Indonesia under the highest-emission scenario (RCP8.5) exceeding 50% at the end of the twenty-first century (2081–2100) for the months of June to November relative to the historical period (1986–2005). These projected changes in CDD are comparable to those of past El Niño events, which were attributed to causing prolonged drought, forest fires and haze episodes (Supari et al. 2018). Hence, in future warmer periods, there is a greater likelihood for annual dry conditions in Indonesia from June to November even without the presence of an El Niño event. The intensity of dry conditions could further increase during El Niño periods. Changes in the intensity and frequency of El Niño could also pose additional risks to the region. Cai et al. (2014) indicated that extreme El Niño such as those of the 1997/98 and 2015/2016 could be twice more frequent in future warmer periods. Furthermore, relevant climate model output can also provide input to impact models, which in turn provide information on climate change impacts in specific sectors. For example, hydrological models such as the Soil and Water Assessment Tool (SWAT) model use future climate inputs to determine how climate change can impact water resources including hydrological droughts in future periods (e.g. Tan et al. 2019). Other examples are crop models such as the Decision Support System for Agro-technology Transfer (DSSAT), which use climate model output to predict yields of certain crops, which can be used to estimate the risk of impacts of climate change on certain crops, agriculture or food security (Boonwichai et al. 2018).

1.3 Climate Change Modelling: GCM and Regional Climate Downscaling

Assessment of the level of risks of climate change impacts requires climate model outputs for future periods. Forced with emission scenarios, GCMs can be numerically integrated to produce climate information in future periods. However, GCMs operate globally with coarse spatial resolutions, typically 100–300 km, ignoring local features such as local topography and coastlines. Hence, the simulated climate may not adequately represent that of a particular locality. To enable adaptation to future climate change impacts, GCM outputs need to be further refined through a process called regional climate downscaling (RCD). Dynamical downscaling (DD) is the most common approach in RCD where another model, called RCM, is nested within the GCM covering a particular region of interest (Fig. 1.3). The RCM is similar to the GCM but operates at much higher resolution typically < 50 km. At such resolutions, the RCM sees local features much better than the GCM. This is particularly important in a complex and unique region such as Southeast Asia that comprises thousands of islands that require high-resolution RCM to resolve (Fig. 1.1). However, the RCM cannot be integrated independently to simulate the future climate over the region of interest. The model is forced at the boundaries and at the surface by GCM outputs.

Over six decades since the first GCM was built in the 1960s (Manabe et al. 1965), tremendous advancement has been achieved with the development of many GCMs by various countries and institutions. Complete reviews of GCM development are available in Randall et al. (2007), Flato et al. (2013) and Stocker et al. (2014). GCMs can be different in terms of how the processes in the climate system are mathematically represented. For inter-comparisons of different GCMs, a well-coordinated community effort has been organised on a regular cycle, following the cycle of the IPCC, under the framework of the Coupled Model Inter-comparison Project (CMIP), e.g.

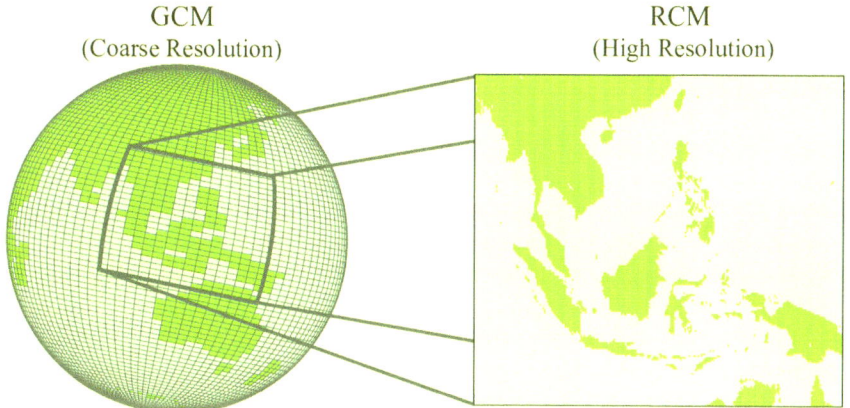

GCM
(Coarse Resolution)

RCM
(High Resolution)

Fig. 1.3 Illustration depicting nesting of an RCM over Southeast Asia region with a GCM

CMIP3, CMIP5, CMIP6 (Meehl et al. 2000; Eyring et al. 2016). Due to differences in the models, different GCMs often simulate different responses to the same forcings. This is also the case for RCMs. These different responses reflect uncertainties in the simulated climate (Latif 2011). Hence, a simulated climate response based on a combination of a particular GCM–RCM and an emission scenario is merely representing one realisation of many possible outcomes. For robust climate information, adequate sampling of many possible outcomes is needed. This is achieved by the application of multiple GCMs, RCMs and emission scenarios (Valle et al. 2009). This requirement makes multi-model and multi-emission scenarios by RCD very expensive and time consuming to implement and can be a daunting task, especially for countries in Southeast Asia.

1.4 Regional Climate Downscaling Activities in Southeast Asia

Not until very recently when CORDEX-SEA delivered its high-resolution multi-model climate simulation outputs, RCD in Southeast Asia has been limited mostly to a single GCM and a single RCM. Only a few countries have been able to conduct multi-model simulations over a domain best suited for them. Countries such as Vietnam (MONRE 2016), Malaysia (NAHRIM 2006) and the Philippines (PAGASA 2011) were able to conduct regional climate downscaling of CMIP3 GCMs but limited to a single RCM through collaboration between local and international institutions. At the regional level, the Centre for Climate Research Singapore (CCRS) of Meteorological Service Singapore and Hadley Centre, UK Met Office conducted the Southeast Asia Climate Analysis and Modeling (SEACAM) Project (Rahmat et al. 2014). Although this initiative is a step in the right direction for a coordinated effort, scientists in the region were mainly involved in analysing the model outputs. Another initiative on simulations at the regional scale was carried out by the Singapore-MIT Alliance for Research and Technology (SMART) and Center for Environmental Sensing and Modeling (CENSAM), Singapore (Kang et al. 2019). However, the application of a single RCM in both initiatives limits the assessment of uncertainty in regional climate downscaling (Rahmat et al. 2014; Kang et al. 2019).

There have been a number of other regional climate downscaling simulations in Southeast Asia. However, since these simulations are conducted with specific scientific objectives with limited temporal and domain coverage, it limits their usage in assessing the impacts of climate change. Overall, lack of coordinated effort in high-resolution multi-model regional climate downscaling was one of the key issues faced by the region. As a consequence, climate data end users and scientists from the vulnerability, impact and adaptation (VIA) assessment community had limited or no access to such climate information, which has had a negative impact on the progress in the climate change impact assessments for critical sectors in the region. Indeed, the assessment carried out by the Working Group II of the IPCC revealed significant knowledge gaps on impacts of climate change on critical sectors in Southeast Asia in the future (Hijioka et al. 2014). This eventually led to slow progress in addressing the issues related to climate change, including implementing adaptation in the region. These were the prime motivations that led to the establishment of CORDEX-SEA in 2013 (Tangang et al. 2020).

1.5 CORDEX Southeast Asia: Origin, Progress and Key Findings

CORDEX, a programme of the WCRP, was established to facilitate a coordination of activities in regional climate downscaling across the globe (Giorgi et al. 2009; www.cordex.org). Among the fourteen simulation domains under CORDEX, CORDEX-SEA covers Southeast Asia encompassing a region of 12.95° S to 25.64° N and 91.29° E to 145.15° E (Fig. 1.2). CORDEX can be a real opportunity for developing and least-developed countries to overcome barriers in generating and assessing climate change information for enabling adaptation. This has been the case for countries in Southeast Asia where climate data end users and VIA scientists can now have access to the high-resolution multi-model regional climate simulation outputs from CORDEX-SEA (Tangang et al. 2020).

Prior to being recognised as CORDEX-SEA, this initiative existed as the Southeast Asia Regional Climate Downscaling (SEACLID), a project funded by the Asia–Pacific Network for Global Change Research (APN; Tangang et al. 2018). SEACLID was then streamlined in CORDEX after its inception in 2013 and subsequently renamed as SEACLID/CORDEX Southeast Asia (http://www.ukm.edu.my/seaclid-cordex). Lately, it is referred to as CORDEX-SEA. The involvement of modelling groups from Thailand, Vietnam, Malaysia, Indonesia and the Philippines provided opportunity for capacity building in regional climate modelling through research and development activities and climate modelling training workshops. Many young scientists have been trained and secured their MSc and PhD degrees for their works in CORDEX-SEA. The scientists in these countries also had the opportunity to build their networks in regional climate modelling with the involvement of other modelling groups from countries outside the region including Australia, Germany, China, South Korea, Japan, Sweden and the United Kingdom. Tremendous progress has been achieved in scientific understanding of regional modelling and climate change in the region through various published articles in international journals (https://www.apn-gcr.org/project/southeast-asia-regional-climate-downscaling-project-seaclid/). In the earlier stages of CORDEX-SEA, the modelling groups in the region coordinated sensitivity experiments using the Regional Climate Model System (RegCM), a model developed by the Abdus Salam International Centre for Theoretical Physics (ICTP). This work enhanced understanding of various climate processes in the region and parameterisation schemes in the model, which led to the publications of three important publications, namely Juneng et al. (2016), Ngo-Duc et al. (2017) and Cruz et al. (2017). Further works on evaluating land surface schemes in the model led to the publication of Chung et al. (2018).

The first round of RCD of GCMs into 25 km spatial resolution over the CORDEX-SEA domain involved 11 CMIP5 models using 7 RCMs, representing a reasonably good size of ensemble members (Tangang et al. 2020). These simulations cover the historical period of 1976–2005 and future periods up to 2100, following two emission scenarios, i.e. the Representative Concentration Pathway 8.5 (RCP8.5) and RCP4.5.

Analysis of this high-resolution multi-model climate ensemble or its subset has yielded significant findings on future changes in climate and climate extremes in Southeast Asia. Tangang et al. (2018) found that duration, intensity and frequency of precipitation extremes were projected to increase and intensify over this region when global mean temperature increase reaches 2 °C above pre-industrial level. In Thailand, rainfall is projected to increase (decrease) over central and northern (southern) regions during boreal winter for the rest of the twenty-first century, but is projected to decrease throughout the country in boreal summer (Tangang et al. 2019). With the application of quantile mapping bias correction, Trinh-Tuan et al. (2019) projected a tendency for a drier condition over central and northern Vietnam during the wet season in the mid-twenty-first century.

The comprehensive analysis of Tangang et al. (2020) highlighted the added values of RCMs compared to those of GCMs in Southeast Asia. Other key findings include a projected increase in rainfall over most of Indo-China countries during boreal winter but with a tendency for drier conditions over Maritime Continent, especially Indonesia, during boreal summer (Tangang et al. 2020). While some countries in Indo-China, e.g. Thailand, Cambodia, Vietnam and Laos, were projected to experience a drier condition in boreal summer, Myanmar was projected to be in a wetter condition for both seasons (Fig. 1.4). Supari et al. (2020) extended the analysis and found significant and robust projected changes of extreme precipitation in most areas in Southeast Asia by the end of the twenty-first century if climate is not mitigated. Notably, consistent with the significant decrease of mean rainfall at the end of the twenty-first century under RCP8.5, the Maritime Continent, especially Indonesia, is projected to have significant and robust increase in consecutive dry days (CDD) during June to August (JJA) and September to November (SON) (Fig. 1.5). In a recent paper, Nguyen-Thi et al. (2021) conducted an interesting analysis of a climate analogue that presented the projection of novel climate in about 20% of the region by the end of the twenty-first century if climate change is not mitigated. CORDEX-SEA simulations have also been analysed to detect time emergence of climate change signals in Vietnam (Nguyen-Thi et al. 2021), and in the characterisation of typhoons affecting the Philippines (Tibay et al. 2021).

Other scientists or groups that were not involved in the CORDEX-SEA regional climate simulations have started using the model outputs. Ge et al. (2019) used a subset of CORDEX-SEA simulations for an analysis related to the extreme precipitation risks under global warming 1.5 and 2.0 °C over Southeast Asia. Some VIA scientists or groups have started to use CORDEX-SEA data in assessing climate change impacts of critical sectors (e.g. Tan et al. 2019, 2020). Furthermore, CORDEX-SEA simulation outputs have also been used by authorities in Indonesia, Vietnam and the

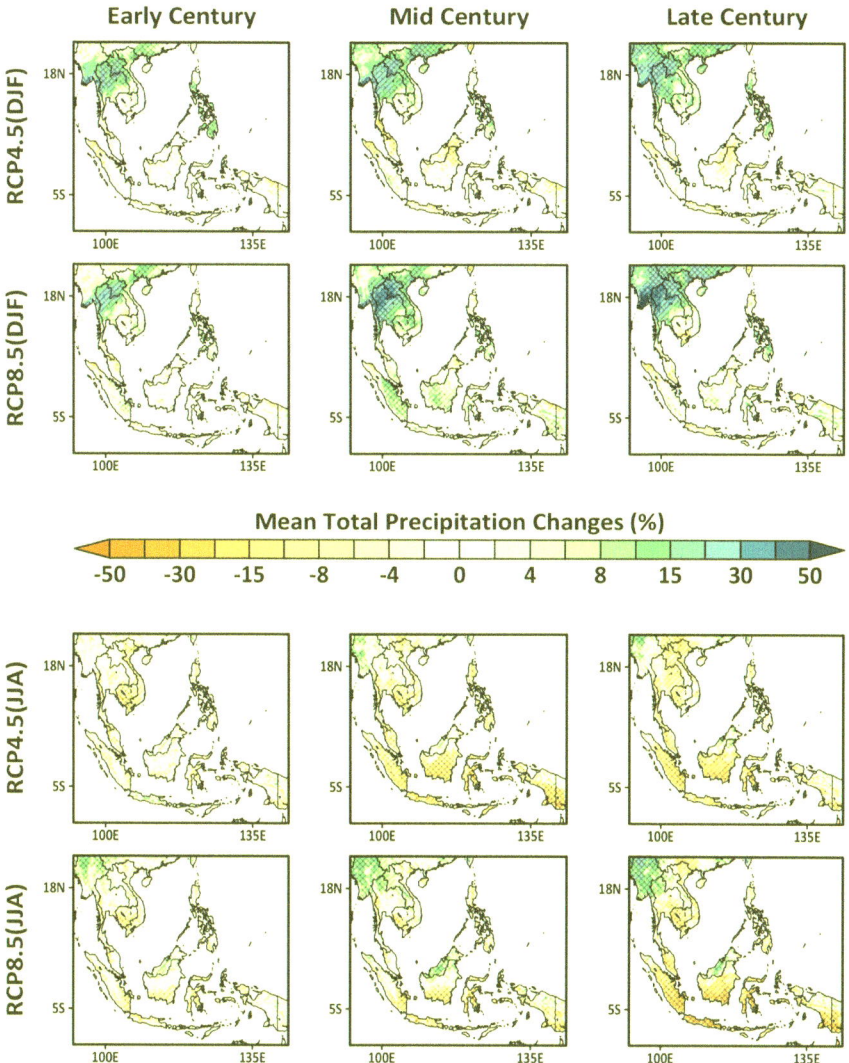

Fig. 1.4 Projected changes of rainfall based on CORDEX-SEA simulations during early century (2011–2040), mid-century (2041–2070) and late century (2071–2099) relative to the historical period (1976–2005). Forward slashes indicate changes are significant at 90% level, while backward slashes signify 75% agreement among models (adapted from Tangang et al. 2020)

Philippines for their national climate change assessments and analyses. It was also used in the Interactive Atlas for Southeast Asia of Working Group 1 of the IPCC Sixth Assessment Report (AR6) (https://interactive-atlas.ipcc.ch/).

F. Tangang et al.

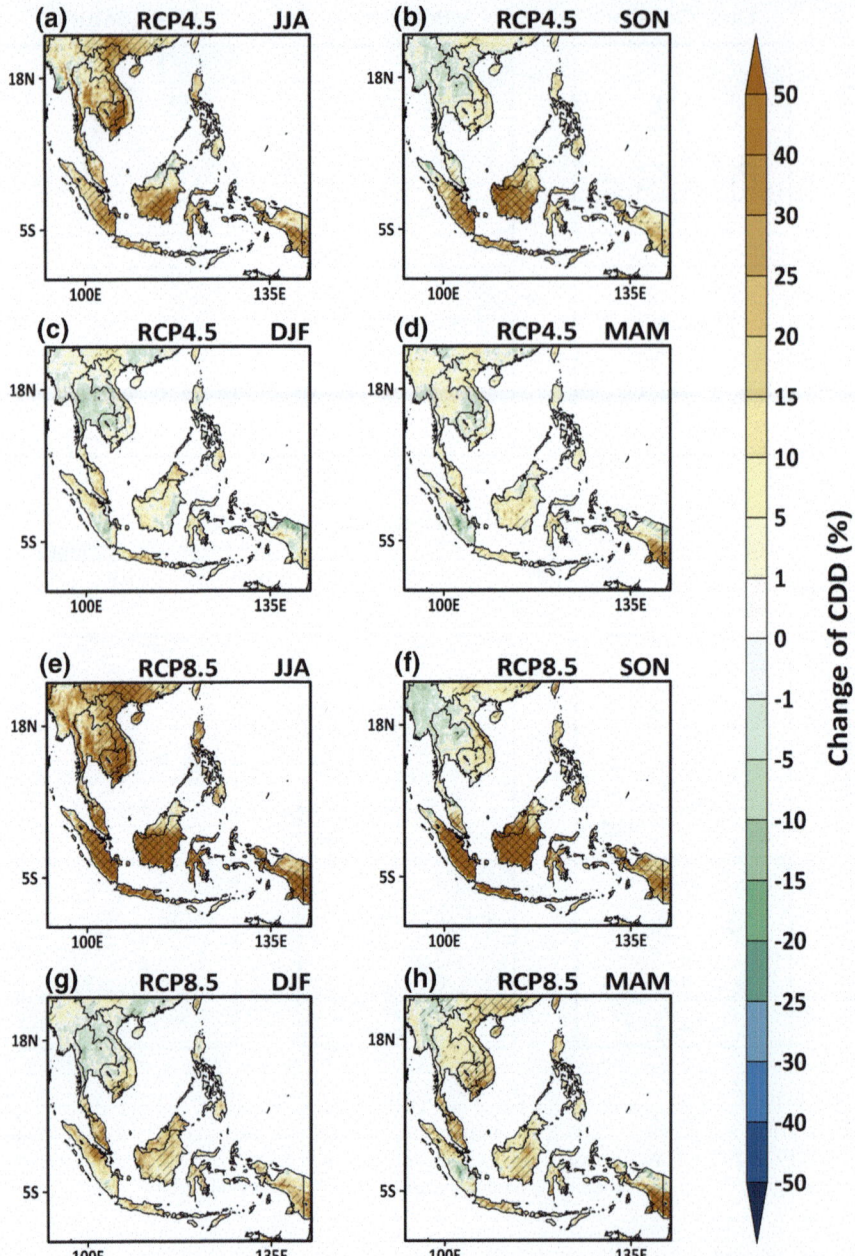

Fig. 1.5 Projected seasonal changes in CDD by the end of the twenty-first century (2081–2100), relative to the historical period (1986–2005). Forward slashes indicate significance backward slashes signify agreement among models (adapted from Supari et al. 2020)

1.6 SARCCIS: CORDEX-SEA Data Sharing Portal

It is crucially important that CORDEX regional climate simulation outputs be accessible to users and VIA scientific communities to generate climate information for enabling adaptation. With hundreds of terabytes of model output from CORDEX simulations, a systematic data sharing platform is needed for high accessibility and efficient data dissemination. CORDEX utilises the Earth System Grid Federation (ESGF) for data archiving and dissemination. The ESGF networks a global system of federated data centres located around the globe that archive and disseminate the world's largest collections of climate model data (Williams et al. 2013).

CORDEX-SEA archives and disseminates its climate model outputs through a data centre called the Southeast Asia Regional Climate Change Information System (SARCCIS; http://www.rucore.ru.ac.th/SARCCIS), which is an ESGF data node hosted by the Ramkhanghaeng University Center of Regional Climate Change and Renewable Energy (RU-CORE) in Bangkok, Thailand. SARCCIS is jointly managed by the Ramkhamhaeng University, Thailand and the National University of Malaysia, Malaysia. SARCCIS was launched on May 7, 2018, at the National University of Malaysia, Bangi, Selangor, Malaysia, jointly by the President of Ramkhamhaeng University and the Deputy Vice-Chancellor of the National University of Malaysia, reiterating the commitment of these universities in supporting SARCCIS and enabling climate change adaptation. SARCCIS can also archive in-house data in addition to those indexed under ESGF. It also provides spaces for archiving relevant information of case studies using CORDEX-SEA data. Additional features such as online analysis tools and visualisation can also be added. Hence, SARCCIS will strengthen initiatives in Southeast Asia in addressing climate change adaptation, disaster risk reduction and sustainable development goals, especially Goal 13 of Climate Action.

1.7 Challenges and the Way Forward

Regional climate simulation activities in Southeast Asia are relatively recent compared to countries in developed regions. Despite the progress and advancement made in regional climate downscaling within CORDEX-SEA, many challenges remain. First, representing the region's climate processes in a climate model can be very challenging, given its physical geography of sparsely distributed landmasses surrounded by many regional seas. Despite the climate model sensitivity experiments conducted in CORDEX-SEA (Juneng et al. 2016; Ngo-Duc et al. 2017; Cruz et al. 2017), many climate processes remain poorly understood. In fact, the 25 km model resolution adopted may not be optimum as large biases tended to be present in areas comprising many small islands, e.g. the Philippines and eastern parts of Indonesia (Tangang et al. 2020). This prompted some further downscaling at a resolution of 5 km over a number of subdomains, which is currently carried out under the second phase of CORDEX-SEA. Preliminary results showed that the 5 km simulations provided added values to the 25 km in a complex region such as Southeast Asia, especially in capturing extremes (e.g. Ngai et al. 2020). However, resolution may not be the only factor that determines the quality of the simulations. Various parameterisation schemes in the model can be potential sources of biases if not properly tuned. One particular scheme, the cumulus parameterisation scheme, is used to parameterise the precipitation process that occurs at a sub-grid scale. A simulation without such a parameterisation scheme can be done at a much high resolution (e.g. < 4 km) using a convective-permitting model (Prein et al. 2015). This could be the way forward for CORDEX-SEA in future simulations as it addresses the needs to resolve complex topography and precipitation processes. However, long-period climate simulations at such very high-resolution come at the expense of huge computing resources, which most modelling centres in the region do not have.

Regional climate downscaling exercises need to be repeated to follow the cycle of CMIP experiments. CORDEX-SEA simulations thus far have been based on the CMIP5 GCMs. With the availability of CMIP6 GCMs simulations, another round of CORDEX-SEA simulations would need to be carried out. This is also the requirement to be consistent with the AR6 cycle of the IPCC. While this can be another challenge to coordinate, this is also an opportunity to address some of the issues in the previous round of CMIP5 GCMs downscaling. For example, domain expansion may be needed and can only be ascertained through proper numerical experiments. Tibay et al. (2021) indicated that the existing simulations of CORDEX-SEA were not able to simulate the typhoons affecting the Philippines adequately, possibly due to the domain not extending far enough to the east to cover the cyclogenesis areas in the Pacific Ocean. Extension of the western boundary of the domain may also be needed to adequately cover the summer monsoonal wind, which is important for countries in Indo-China. Furthermore, the northern boundary needs to be extended as the previous domain excluded the northern part of Myanmar.

CORDEX-SEA also faces challenges in communicating the climate data to VIA communities. While the data are freely accessible, understanding the format and familiarisation with the data can be a real challenge to some scientists in VIA communities. Issues related to model uncertainty in the projection would need to be communicated to data users and VIA communities (Daron et al. 2018). Regular technical workshops may be needed to better communicate the technical aspects of climate data in SARCCIS or ESGF data nodes. Within CORDEX-SEA, this has been implemented a number of times. However, conducting technical workshops on a regular basis can be a real challenge without proper funding. Moreover, sustained engagement between climate scientists and end users may be necessary, especially if the climate information needs to be customised to fit the needs of the end users, which would require significant time and effort.

1.8 Conclusions

The establishment and successful implementation of CORDEX-SEA are a game changer for Southeast Asia in the generation of regional climate change information for enabling adaptation. From mostly relying on coarse GCM data or single model regional simulations, climate data users and VIA scientists in Southeast Asia now have free access to multi-model and high-resolution climate model data, which can be input to impact models to assess the risk of impacts of climate change and extremes on critical sectors and enable adaptation for climate resilience. With the current and future efforts in continuing the production of high-resolution multi-model simulations for CORDEX-SEA, especially for CMIP6 GCMs, and in making these available through SARCCIS, some of the gaps identified in Hijioka et al. (2014) can eventually be addressed. Southeast Asia will eventually be equipped with robust climate information to enable adaptation for climate resilience of the region.

Acknowledgements Fredolin Tangang was funded by the Malaysian Ministry of Higher Education (MOHE) Long-term Research Grant Scheme (LRGS/1/2020/UKM/01/6/1 Project under LRGS/1/2020/UKM/01/6 Programme). CORDEX-SEA was funded by the Asia-Pacific for Global Change Research (APN) (ARCP2015-04CMY-Tangang, ARCP2014-07CMY-Tangang, ARCP2013-17NMY-Tangang). We dedicate this chapter to the memory of late Prof. Gemma T. Narisma for her immense contribution to CORDEX-SEA.

References

Anh, L.T., H. Takagi and N.D. Thao. 2019. Storm surge and high waves due to 1997 Typhoon Linda: Uninvestigated worst storm event in southern Vietnam. *Journal of Japan Society of Civil Engineers, Ser. B3 (Ocean Engineering)* 75: I_73–I_78.

Boonwichai, S., S. Shrestha, M.S. Babel, S. Weesakul, and A. Datta. 2018. Climate change impacts on irrigation water requirement, crop water productivity and rice yield in the Songkhram River Basin, Thailand. *Journal of Cleaner Production* 198: 1157–1164.

Cai, W., S. Borlace, M. Lengaigne, P. Van Rensch, M. Collins, G. Vecchi, A. Timmermann, A. Santoso, M.J. Mcphaden, L. Wu, M.H. England, G. Wang, E. Guilyardi, and F.-F. Jin. 2014. Increasing frequency of extreme El Niño events due to greenhouse warming. *Nature Climate Change* 4: 111–116.

Chung, J.X., L. Juneng, F. Tangang, and A.F. Jamaluddin. 2018. Performances of BATS and CLM land-surface schemes in RegCM4 in simulating precipitation over CORDEX Southeast Asia domain. *International Journal of Climatology* 38: 794–810. https://doi.org/10.1002/joc.5211.

Cruz, F.T., G.T. Narisma, J.B. Dado, P. Singhruck, F. Tangang, U.A. Linarka, T. Wati, L. Juneng, T. Phan-Van, T. Ngo-Duc, and J. Santisirisomboon. 2017. Sensitivity of temperature to physical parameterization schemes of RegCM4 over the CORDEX-Southeast Asia region. *International Journal of Climatology* 37 (15): 5139–5153.

Dagli, S., and B. Ferrarini. 2019. The growth impact of disasters in developing Asia. ADB Economics Working Paper Series No. 585. Manila: ADB. https://doi.org/10.22617/WPS190 224-2.

Daron, J., I. Macadam, H. Kanamaru, T. Cinco, J. Katzfey, C. Scannell, R. Jones, M. Villafuerte, F. Cruz, G. Narisma, R.J. Delfino, R. Lasco, J. Manalo, E. Ares, A.L. Solis, R. De Guzman, J.

Basconcillo, and F. Tangang. 2018. Providing future climate projections using multiple models and methods: Insights from the Philippines. *Climatic Change* 148: 187–203.

Eyring, V., S. Bony, G.A. Meehl, C.A. Senior, B. Stevens, R.J. Stouffer, and K.E. Taylor. 2016. Overview of the coupled model intercomparison project phase 6 (CMIP6) experimental design and organization. *Geoscientific Model Development* 9: 1937–1958.

Flato, G., J. Marotzke, B. Abiodun, P. Braconnot, S.C. Chou, W. Collins, P. Cox, F. Driouech, S. Emori, V. Eyring, C. Forest, P. Gleckler, E. Guilyardi, C. Jakob, V. Kattsov, C. Reason and M. Rummukainen. 2013. Evaluation of climate models. In *Climate change 2013: The physical science basis. Contribution of working group I to the fifth assessment report of the intergovernmental panel on climate change*, eds. T.F. Stocker, D. Qin, G.-K. Plattner, M. Tignor, S.K. Allen, J. Doschung, A. Nauels, Y. Xia, V. Bex, and P.M. Midgley, 741–882. Cambridge University Press. https://doi.org/10.1017/CBO9781107415324.020.

Fritz, H.M., C.D. Blount, S. Thwin, M.K. Thu, and N. Chan. 2009. Cyclone Nargis storm surge in Myanmar. *Nature Geoscience* 2: 448–449.

Ge, F., S. Zhu, T. Peng, Y. Zhao, F. Sielmann, K. Fraedrich, X. Zhi, X. Liu, W. Tang and L. Ji. 2019. Risks of precipitation extremes over Southeast Asia: Does 1.5 °C or 2 °C global warming make a difference? *Environmental Research Letters* 14 (4): 044015.

Giorgi, F., C. Jones and G.R. Asrar. 2009. Addressing climate information needs at the regional level: The CORDEX framework. *World Meteorological Organization (WMO) Bulletin* 58 (3): 175.

Hai, O.S., A.A. Samah, S.N. Chenoli, K. Subramaniam, and M.Y. Ahmad Mazuki. 2017. Extreme rainstorms that caused devastating flooding across the east coast of peninsular Malaysia during November and December 2014. *Weather and Forecasting* 32: 849–872.

Hernandez, J.Y., Jr., R.E.R. Aquino, B.M. Pacheco, and E.C. Cruz. 2015. Damage caused by Typhoon Haiyan in the Philippines, review of structural regulations and practice, and research developments in wind engineering. *Wind Engineers, JAWE* 40: 270–274. https://doi.org/10.5359/jawe.40.270.

Hijioka, Y., E. Lin, J. Pereira, R. Corlett, X. Cui, G. Insarov, R. Lasco, E. Lindgren and A. Surjan. 2014. *Asia. Climate change 2014: Impacts, adaptation, and vulnerability. Part B: Regional aspects. Contribution of working group II to the fifth assessment report of the intergovernmental panel on climate change*, 1327–1370. Cambridge, United Kingdom and New York: Cambridge University Press.

IPCC. 2012. Managing the risks of extreme events and disasters to advance climate change adaptation. In *A special report of working groups I and II of the intergovernmental panel on climate change*, eds. C.B. Field, V. Barros, T.F. Stocker, D. Qin, D.J. Dokken, K.L. Ebi, M.D. Mastrandrea, K.J. Mach, G.-K. Plattner, S.K. Allen, M. Tignor, and P.M. Midgley. Cambridge, United Kingdom and New York, NY, USA: Cambridge University Press. Available at https://www.ipcc.ch/report/managing-the-risks-of-extreme-events-and-disasters-to-advance-climate-change-adaptation/.

IPCC. 2021. Summary for policymakers. In *Climate change 2021: The physical science basis. Contribution of working group I to the sixth assessment report of the intergovernmental panel on climate change*, eds. V. Masson Delmotte, P. Zhai, A. Pirani, S.L. Connors, C. Péan, S. Berger, N. Caud, Y. Chen, L. Goldfarb, M.I. Gomis, M. Huang, K. Leitzell, E. Lonnoy, J.B.R. Matthews, T.K. Maycock, T. Waterfield, O. Yelekçi, R. Yu, and B. Zhou. Cambridge University Press (in press).

Juneng, L., F. Tangang, J.X. Chung, S.T. Ngai, T.W. Tay, G. Narisma, F. Cruz, T. Phan-Van, T. Ngo-Duc, J. Santisirisomboon, and P. Singhruck. 2016. Sensitivity of Southeast Asia rainfall simulations to cumulus and air-sea flux parameterizations in RegCM4. *Climate Research* 69 (1): 59–77.

Kang, S., E-S. Im and E.A.B. Eltahir. 2019. Future climate change enhances rainfall seasonality in a regional model of western Maritime Continent. *Climate Dynamics* 52: 747–764. https://doi.org/10.1007/s00382-018-4164-9.

Latif, M. 2011. Uncertainty in climate change projections. *Journal of Geochemical Exploration* 110: 1–7. https://doi.org/10.1016/j.gexplo.2010.09.011.

Lavell, A., M. Oppenheimer, C. Diop, J. Hess, R. Lempert, J. Li, R. Muir-Wood and S. Myeong. 2012. Climate change: new dimensions in disaster risk, exposure, vulnerability, and resilience. In *Managing the risks of extreme events and disasters to advance climate change adaptation*, eds. C.B. Field, V. Barros, T.F. Stocker, D. Qin, D.J. Dokken, K.L. Ebi, M.D. Mastrandrea, K.J. Mach, G.-K. Plattner, S.K. Allen, M. Tignor, and P.M. Midgley. A special report of working groups I and II of the intergovernmental panel on climate change, 25–64. Cambridge, United Kingdom and New York, NY, USA: Cambridge University Press.

Liu, P.R., and A.E. Raftery. 2021. Country-based rate of emissions reductions should increase by 80% beyond nationally determined contributions to meet the 2 °C target. *Communications Earth and Environment* 2: 29. https://doi.org/10.1038/s43247-021-00097-8.

Lohberger, S., M. Stängel, E.C. Atwood, and F. Siegert. 2018. Spatial evaluation of Indonesia's 2015 fire-affected area and estimated carbon emissions using Sentinel-1. *Global Change Biology* 24: 644–654. https://doi.org/10.1111/gcb.13841.

Manabe, S., J. Smagorinsky, and R. Strickler. 1965. Physical climatology of a general circulation model with a hydrologic cycle. *Monthly Weather Review* 93: 769–798.

Meehl, G.A., G.J. Boer, C. Covey, M. Latif, and R.J. Stouffer. 2000. The coupled model intercomparison project (CMIP). *Bulletin of the American Meteorological Society* 81: 313–318.

MONRE. 2016. *Updates on the Climate change and sea level rise scenarios for Vietnam*. Hanoi.

NAHRIM. 2006. *Study of the impact of climate change on the hydrologic regime and water resources of Peninsular Malaysia*.

Ngai, S.T., H. Sasaki, A. Murata, M. Nosaka, J.X. Chung, L. Juneng, S.E. Supari and F. Tangang. 2020. Extreme rainfall projections for Malaysia at the end of 21st century using the high resolution non-hydrostatic regional climate model (NHRCM). *SOLA* 16: 132–139.https://doi.org/10.2151/sola.2020-023.

Ngo-Duc, T., F.T. Tangang, J. Santisirisomboon, F. Cruz, L. Trinh-Tuan, T. Nguyen-Xuan, T. Phan-Van, L. Juneng, G. Narisma, P. Singhruck, D. Gunawan, and E. Aldrian. 2017. Performance evaluation of RegCM4 in simulating extreme rainfall and temperature indices over the CORDEX-Southeast Asia region: Performance evaluation of RegCM4 over the CORDEX-Southeast Asia Region. *International Journal of Climatology* 37: 1634–1647.

Nguyen-Thi, T., T. Ngo-Duc, F. Tangang, F. Cruz, L. Juneng, J. Santisirisomboon, E. Aldrian, T. Phan-Van, and G. Narisma. 2021. Climate analogue and future appearance of novel climate in Southeast Asia. *International Journal of Climatology* 41 (S1): E392–E409. https://doi.org/10.1002/joc.6693.

OurWorldinData.org. 2020. Current climate policies will reduce emissions, but not quickly enough to reach international targets. https://ourworldindata.org/co2-and-other-greenhouse-gas-emissions. Accessed in April 2021.

PAGASA. 2011. Climate change in the Philippines. Available online at https://dilg.gov.ph/PDF_File/reports_resources/DILG-Resources-2012130-2ef223f591.pdf.

Prein, A.F., W. Langhans, G. Fosser, A. Ferrone, N. Ban, K. Goergen, M. Keller, M. Tölle, O. Gutjahr, F. Feser, and E. Brisson. 2015. A review on regional convection-permitting climate modeling: Demonstrations, prospects, and challenges. *Reviews of Geophysics* 53 (2): 323–361.

Rahmat, R., B. Archevarahuprok, C.P. Kang and W. Soe. 2014. *A regional climate modelling experiment for Southeast Asia*, 127. Centre for Climate Research Singapore, Meteorological Service Singapore Report.

Randall, D.A., R.A. Wood, S. Bony, R. Colman, T. Fichefet, J. Fyfe, V. Kattsov, A. Pitman, J. Shukla, J. Srinivasan and R.J. Stouffer. 2007. Climate models and their evaluation. In *Climate change 2007: The physical science basis. Contribution of working group I to the fourth assessment report of the IPCC (FAR)*, 589–662. Cambridge, United Kingdom and New York: Cambridge University Press.

Rogelj, J., D. Shindell, K. Jiang, S. Fifita, P. Forster, V. Ginzburg, C. Handa, H. Kheshgi, S. Kobayashi, E. Kriegler, L. Mundaca, R. Séférian and M.V. Vilariño. 2018. Mitigation pathways compatible with 1.5 °C in the context of sustainable development. In *Global warming of 1.5 °C. An IPCC special report on the impacts of global warming of 1.5 °C above pre-industrial levels and related global greenhouse gas emission pathways, in the context of strengthening the global response to the threat of climate change, sustainable development, and efforts to eradicate poverty*, eds. V. Masson-Delmotte, P. Zhai, H.O. Pörtner, D. Roberts, J. Skea, P.R. Shukla, A. Pirani, W. Moufouma-Okia, C. Péan, R. Pidcock, S. Connors, J.B.R. Matthews, Y. Chen, X. Zhou, M.I. Gomis, E. Lonnoy, T. Maycock, M. Tignor, T. Waterfield (in press).

Seneviratne, S.I., X. Zhang, M. Adnan, W. Badi, C. Dereczynski, A. Di Luca, S. Ghosh, I. Iskandar, J. Kossin, S. Lewis, F. Otto, I. Pinto, M. Satoh, S.M. Vicente-Serrano, M. Wehner and B. Zhou. 2021. Weather and climate extreme events in a changing climate. In *Climate change 2021: The physical science basis. Contribution of working group I to the sixth assessment report of the intergovernmental panel on climate change*, eds. V. Masson Delmotte, P. Zhai, A. Pirani, S.L. Connors, C. Péan, S. Berger, N. Caud, Y. Chen, L. Goldfarb, M.I. Gomis, M. Huang, K. Leitzell, E. Lonnoy, J.B.R. Matthews, T.K. Maycock, T. Waterfield, O. Yelekçi, R. Yu, and B. Zhou. Cambridge University Press (in press).

Stocker, T., ed. 2014. *Climate change 2013: The physical science basis: Working group I contribution to the fifth assessment report of the intergovernmental panel on climate change*. Cambridge, United Kingdom and New York: Cambridge University Press.

Supari, T.F., E. Salimun, E. Aldrian, A. Sopaheluwakan and L. Juneng. 2018. ENSO modulation of seasonal rainfall and extremes in Indonesia. *Climate Dynamics* 51: 2559–2580.

Supharatid, S., T. Aribarg and S. Supratid. 2016. Assessing potential flood vulnerability to climate change by CMIP3 and CMIP5 models: Case study of the 2011 Thailand great. https://doi.org/10.2166/wcc.2015.116.

Supari, F. Tangang, L. Juneng, F. Cruz, J.X. Chung, S.T. Ngai, E. Salimun, M.S.F. Mohd, J. Santisirisomboon, P. Singhruck, T. Phanvan, T. Ngo-Duc, G. Narisma, E. Aldrian, D. Gunawan, and A. Sopaheluwakan. 2020. Multi-model projections of precipitation extremes in Southeast Asia based on CORDEX-Southeast Asia simulations. *Environmental Research* 184: 109350.

Tan, M.L., L. Juneng, F.T. Tangang, N.W. Chan, and S.T. Ngai. 2019. Future hydro-meteorological drought of the Johor River Basin, Malaysia, based on cordex-sea projections. *Hydrological Sciences Journal* 64 (8): 921–933.

Tan, M.L., L. Juneng, F.T. Tangang, N. Samat, N.W. Chan, Z. Yusop and S.T. Ngai. 2020. Southeast Asia hydro-meteorological drought (sea-hot) framework: A case study in the Kelantan River Basin, Malaysia. *Atmospheric Research* 246: 105155.

Tangang, F., S. Supari, J. Chung, F. Cruz, E. Salimun, S. Ngai, L. Juneng, J. Santisirisomboon, J. Santisirisomboon, T. Ngo-Duc, T. Phan-Van, G. Narisma, P. Singhruck, D. Gunawan, A. Aldrian, A. Sopaheluwakan, G. Nikulin, H. Yang, A.R.C. Remedio, D. Sein and D. Hein-Griggs. 2018. Future changes in annual precipitation extremes over Southeast Asia under global warming of 2 °C. *APN Science Bulletin* 8(1). https://doi.org/10.30852/sb.2018.436.

Tangang, F., J. Santisirisomboon, L. Juneng, E. Salimun, J. Chung, Cruz F. Supari, S.T. Ngai, T. Ngo-Duc, P. Singhruck, G. Narisma, J. Santisirisomboon, W. Wongsaree, K. Promjirapawat, Y. Sukamongkol, R. Srisawadwong, D. Setsirichock, T. Phan-Van, E. Aldrian, D. Gunawan, G. Nikulin, and H. Yang. 2019. Projected future changes in mean precipitation over Thailand based on multi-model regional climate simulations of CORDEX Southeast Asia. *International Journal of Climatology* 39: 5413–5436. https://doi.org/10.1002/joc.6163.

Tangang, F., J.X. Chung, L. Juneng, S.E. Supari, S.T. Ngai, A.F. Jamaluddin, M.S.F. Mohd, F. Cruz, G. Narisma, J. Santisirisomboon, T. Ngo-Duc, P. Van Tan, P. Singhruck, D. Gunawan, E. Aldrian, A. Sopaheluwakan, N. Grigory, A.R.C. Remedio, D.V. Sein, D. Hein-Griggs, J.L. Mcgregor, H. Yang, H. Sasaki, P. Kumar. 2020. Projected future changes in rainfall in Southeast Asia based on CORDEX–SEA multi-model simulations. *Climate Dynamics* 55: 1247–1267.

Tibay, J., F. Cruz and F. Tangang, et al. 2021. Climatological characterization of tropical cyclones detected in the regional climate simulations over the CORDEX-SEA domain. *International Journal of Climatology* 2021: 1–17. https://doi.org/10.1002/joc.7070.

Trinh-Tuan, L., J. Matsumoto, F.T. Tangang, L. Juneng, F. Cruz, G. Narisma, J. Santisirisomboon, T. Phan-Van, D. Gunawan, E. Aldrian, and T. Ngo-Duc. 2019. Application of quantile mapping bias correction for mid-future precipitation projections over Vietnam. *SOLA* 15: 1–6.

UNFCCC LDC Expert Group. 2012. A brief overview of the national adaptation plan process. https://unfccc.int/files/adaptation/application/pdf/nap_overview.pdf. Accessed in April 2021.

Valle, D., C.L. Staudhammer, W.P. Cropper Jr., and P.R. Gardingen. 2009. The importance of multimodel projections to assess uncertainty in projections from simulation models. *Ecological Applications* 19: 1680–1692.

Williams, D.N., G. Bell, L. Cinquini, P. Fox, J. Harney and R. Goldstone. 2013. Earth system grid federation: Federated and integrated climate data from multiple sources. In *Earth system modelling*, vol. 6, 61–77. Berlin, Heidelberg: Springer Briefs in Earth System Sciences. Springer. https://doi.org/10.1007/978-3-642-37244-5_7.

Zhang, X., L. Alexander, G.C. Hegerl, P. Jones, A.K. Tank, T.C. Peterson, B. Trewin, and F.W. Zwiers. 2011. Indices for monitoring changes in extremes based on daily temperature and precipitation data. *Wiley Interdisciplinary Reviews: Climate ChAnge* 2: 851–870. https://doi.org/10.1002/wcc.147.

Chapter 2
Technical and Infrastructure Modality for Extreme Climate Early Warning in Indonesia

Edvin Aldrian, Sheila Dewi Ayu Kusumaningtyas, Supari, Danang Eko Nuryanto, and Ardhasena Sopaheluwakan

Abstract The Indonesian archipelago is situated between the Asia and Australia continents and the Pacific and Indian Oceans. It has a typical monsoon climate, with monsoon rainfall generally peaking during boreal winter. The seasonal asymmetries annual cycle is geographically complex and reflects multiscale interactions between lands and seas. Monsoon rainfall exhibits pronounced variability and affecting variability on all timescales from diurnal to interannual and longer in interannual timescale. There are some extreme phenomena in this region. This chapter discusses some phenomena from the daily up to interannual variability for the extreme and how the country, Indonesia, manages the extreme cases. It aims to give educations and introductions to other regions of the world. On intraseasonal and synoptic scales, the region is heavily influenced by the MJO and cold surges especially during the peak of the rainy season, which can interact with each other as well as with in situ synoptic systems such as the Borneo vortex, often leading to torrential rainfall, flash floods, and severe storms, including the possible rare case, a typhoon. The chapter also discusses the type of observation and analyses and the type of instruments for extreme analysis. Further, this chapter introduces major institutions that are involved for early warning for weather and climate in the country.

2.1 Introduction

Indonesia is a tropical archipelago whose climate is mainly monsoonal by definition. It lies between two continents and two oceans. Monsoon is the area's major climate control, simply because it is located in the location above and affected by the solar solstice movement. The situation is similar to land and sea breeze over the coastal area but on a continental scale. Although not all locations are monsoonal, some areas

E. Aldrian (✉)
Agency for Assessment and Application of Technology, BPPT, Jakarta, Indonesia
e-mail: edvin.aldrian@bppt.go.id

S. D. A. Kusumaningtyas · Supari · D. E. Nuryanto · A. Sopaheluwakan
Agency for Meteorology, Climatology, and Geophysics, BMKG, Jakarta, Indonesia

© The Centre for Science & Technol. of the, Non-aligned and Other Devel. Countries 2022
A. S. Unnikrishnan et al. (eds.), *Extreme Natural Events*,
https://doi.org/10.1007/978-981-19-2511-5_2

with two times solar solstice over them will have an equatorial climate, when the same place will experience higher and much extreme weather. The most prominent climate phenomenon for this region is the El Nino–Southern Oscillation (ENSO), modulating the regular monsoonal system and bringing extreme rainfall in high and low variability. On the annual scale, the prominent phenomenon modulated the total rainfall amount on the quasi-biennial scale.

2.2 Annual Cycle

Indonesia experiences a marked seasonal cycle in precipitation characteristic of a monsoon climate. The north–south movement is not exactly straight north–south because Asia and the Australian continent are slightly deflected toward northeast and southeast. Chang et al. (2005) showed that the greater Asian–Australian monsoon region's annual cycle is characterized by two fundamental asymmetries between boreal summer and winter and between boreal spring and fall. The exact monsoonal movement follows the solar solstice location, where the high solar radiation is directly reflected perpendicular to the earth's surface. This precise location of the solstice movement is usually marked with the location of ITCZ or the Inter-Tropical Convergence Zone (Chang et al. 2004, 2005; Wheeler and McBride 2005). According to the solar solstice, ITCZ modulates north and south, but the exact location depends on the wind convergence. Usually, in the equator, the wind flows to the equator as the trade wind. While this wind flows to the west, their convergence zone is where wind flows to the east. The location of ITCZ does not follow the straight line but follows the lowest point where the surface pressure is at the lowest. Since then, the location is highly wet and has much precipitation. The ITCZ does not only draw pressure for extreme but also the main engine in the tropics. One of the large phenomena, such as the Madden Julian Oscillation, is similar to the ITCZ, exerting eastward in the tropics. The MJO propagates to the east in the tropics and is also a significant phenomenon controlling extreme in the tropics. Indonesia's monsoonal processes will eventually bring the wet or rainy season on the half of the year or around months ONDJF, during the dry season during the other second half of the year, or during months MJJAS (Aldrian and Susanto 2003; Giannini et al. 2007). The southern part of Indonesia is a monsoonal type, while the northern part is mostly the equatorial type (see Fig. 2.1). The very different type over a small area exists in the eastern part, mainly related to the Indonesian throughflow. The Indonesian throughflow in the water mass flows from the Pacific Ocean to the Indian Ocean through Indonesia that flows in the eastern part. The incoming water mass over this region determines the type of climate in the area. The monsoonal pattern or type has one clear peak of dry and one the wet season, while the equatorial type has two peaks of dry and wet seasons.

Most climate analysis of Indonesia will eliminate the normal seasonal pattern or the monsoonal pattern first. Then the rest of the analysis is conducted. Over the area, monsoonal patterns occupy more than 70% of the whole pattern, thus need to

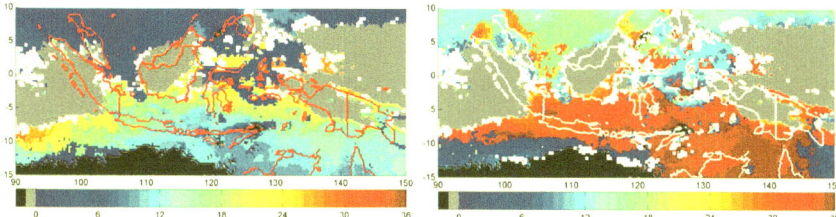

Fig. 2.1 Advancement of monsoon onset in the scale of decade (10 days) in the color bar for dry season (left) onset and wet season (right). Gray color means an area with always > 50 mm per decade (non-monsoonal) and black areas with always < 50 mm per decade (non-monsoonal). One year has 36 decade, and each month has 3 decade. Data from Tropical Rainfall Measuring Mission (TRMM) satellite observation 1998–2010

be eliminated first. In the basic monsoonal criteria, which are the wind differences on the peak of the dry and wet season, i.e., in January and July, most areas will be monsoonal. By modern definition, using the difference of precipitation difference in two seasons, the definition is obscure. The monsoonal region will be remaining in the southern region.

2.3 Definition of Extreme Rainfall of the Area

The Indonesian Agency for Meteorology Climatology and Geophysics BMKG does not clearly define the definition of extreme. By their definition, daily water budget or the precipitation and evaporation account roughly for 5 mm (Wati et al. 2019). This number is from the average evaporation pan number for a day. An excess of water at the surface will occur if the precipitation reaches above 5 mm a day. Although in some areas, the number is less than that. The double and quadruple amount would be excess to rapid surface saturation or flood. This may be one way to define extreme rainfall besides the population definition. Statistically, usually, people will take 5% or even 1% top population as the definition of extreme rainfall. The following definition of extreme rainfall constitutes the return period or ranking of occurrence. If the event often occurs, then they are not part of the local area. The event becomes local normality, and this makes the definition of extreme that could vary locally. The following definition also considers the impact of the environment on the rainfall that occurs. In certain areas, the same rainfall amount could constitute a flood but not to the other area. Although the same amount, one could be considered extreme but not for the other. Occasionally, the duration of extreme even matters, and the duration could increase the vulnerability in certain areas.

There are three types of floods in the areas which are often occurring. First is the local flood or inundation due to long periods of heavy or extreme rainfall pour over the local specific areas. The second type is the flash flood, which occurs when the rainfall falls over an upstream region and causes a flood in the downstream region. Again it is possible that the upstream region has long-term local rainfall and is flooded or spilled over to the downstream region. The third type of flood is due to the inundation of water from the sea or ocean's incoming event. In most cases, it would be the tides or waves of the coast. The inundation is also possible due to local inundation at the land area due to the land subsidence.

In Indonesia where the climate is tropical, the extreme type is dominated by rainfall extreme. The extreme temperature events somehow occurred over specific location, for instance daytime temperature more than 35 °C occurred over Nusa Tenggara which is area with the lowest annual rainfall in the country. Also, the nighttime temperature below freezing point sometimes occurred over Dieng highland, recognized by frost in the morning damaging crops (Pradana et al. 2018; Aini and Faqih 2021). Other types of extremes of this area include extreme wave and air quality pollution. Even though extreme events do not often occur, they often follow the regular monsoonal pattern; they can have the most marked impacts: causing significant devastation to infrastructure, affecting our economy and health, and resulting in the loss of life. It is essential for meteorological communities to improve the understanding and characterization of extreme weather and climate events in time and space with regionally and globally consistent methodologies.

2.4 Extreme Phenomena of the Area

To understand the extreme situations of the area, we also need to understand climate phenomena contributing to extreme situations.

2.4.1 ENSO Variability

El Nino–Southern Oscillation is the most persistent climate phenomenon in the tropics (Hackert and Hastenrath 1986; Aldrian et al. 2007). So much consistency that the phenomenon will come at a certain time of the year and almost has a similar cycle. It starts at the end of boreal spring and reaches a peak at the end of the year and will end in the next year boreal spring. The phenomenon is measured with the anomaly of sea surface temperature over the Pacific Ocean and peaks at the end of the year near Christmas Eve. There are the El Nino and La Nina situations, whereby the first one happens when the anomaly is positive and negative for the second type. The year is declared as the El Nino and La Nina year when the anomaly exceeded a certain threshold over a certain period, usually above three months.

The impact of ENSO will usually be the opposite of Indonesia. When the positive anomaly occurs over the Pacific Ocean, then Indonesia will be a negative anomaly. Consequently, the negative anomaly of sea surface temperature will be less evaporation and means less precipitation. Consequently, there will be dryer situations (see Fig. 2.2). Since the onset and withdrawal of the phenomena are known, there will

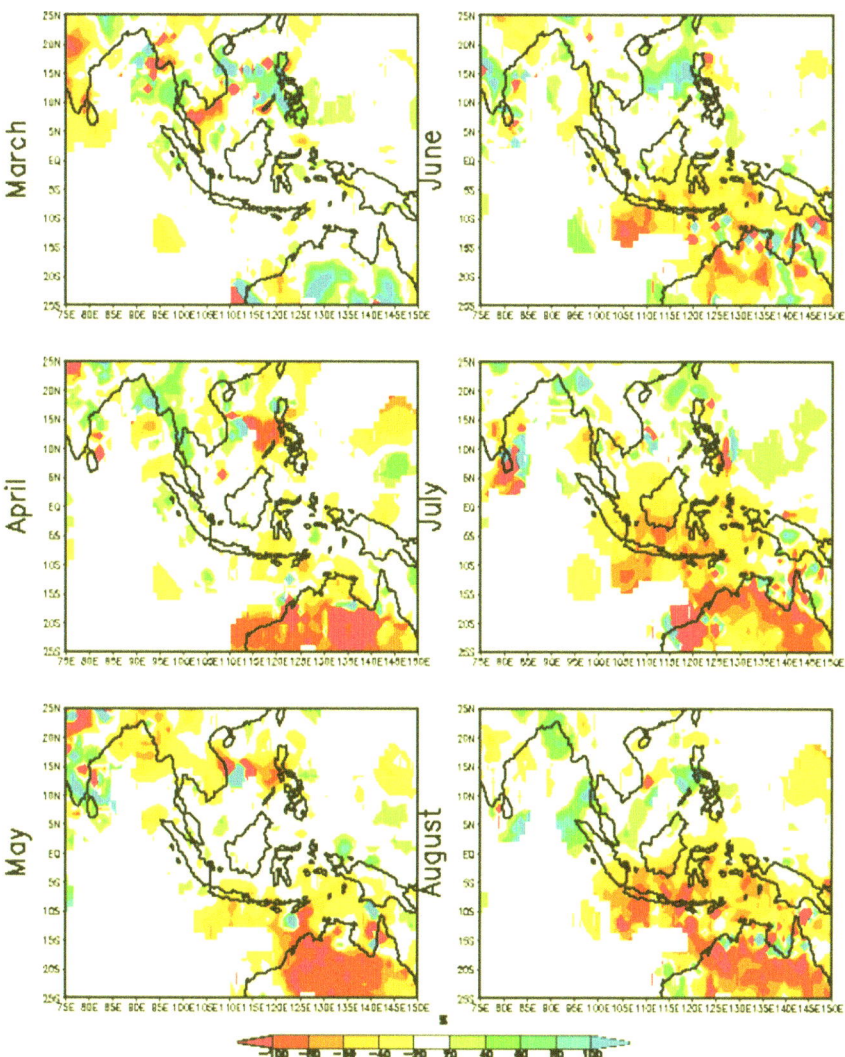

Fig. 2.2 Spatial patterns of rainfall anomaly during specific months of ensemble El Nino year in summer of El Nino year. The ensemble ENSO years are a combination of 1965, 1969, 1972, 1982, 1987, and 1991. The contour scales are given in percentage of rainfall anomaly. For complete spatial analysis of other seasons for El Nino and La Nina cases, please see Aldrian (2002)

be an anomaly during the dry season or MJJAS and the wet or rainy season ONDJF over Indonesia (Aldrian 2002; Supari et al. 2018)). There will be a more evident impact during the dry season than the wet season. This is mainly due to some known fact (mainly due to the ocean surface streamflow overIndonesia) that is not discussed here (Wyrtki 1961).

One of the worst impacts will be El Nino during the dry condition when the actual dryer condition worsens the dry situation. The opposite of the La Nina during the wet or rainy season does not bother much; however, it is a normal perception that the excessive rainfall in the rainy season may be linked to the La Nina season. However, sometimes during the dry season, there is some flooding at some places, as the result of the combined impact of La Nina and wet Madden Julian Oscillation of the area. Usually, the MJO will be longer and wetter during La Nina and drier and shorter during El Nino. Thus in some cases, there will be excessive rainfall during La Nina year, which results in flooding.

On the other end, ENSO, especially El Nino, creates another extreme in the region's climate. The already dry condition is even more dryer in the dry season. The dry spell will increase up to more than 40% during June–July–August and September–October–November season of El Nino years compared to Neutral years (Supari et al. 2018). Hence, there is sometimes extreme dryness. Even worse, the situation is at the end of the dry season, when the dryness reaches its peak. There are many studies on this extreme and all consequences of this kind of extreme. This type of disturbance is pretty much known to scientists in this region. The knowledge of this is pretty much well understood in comparison with the effect on the other side of the extreme. The operational early warning for this case is pretty much better prepared since the timing is almost consistent and still have time for further preparation.

2.4.2 ITCZ Convergence

This is a common phenomenon in the tropics. The convergence occurs due to trade winds from the northern and southern hemispheres converging in the tropics. Sailors know the location as the doldrums or the calms because of its monotonous, windless weather, which is the area where the northeast and southeast trade winds converge. Since the Coriolis force and earth rotation that the trade wind flows to the west (in the northern hemisphere to the southwest and the southern hemisphere to the northwest), then the result of energy balance, the converge is the area where the wind stream flows to east. By definition, this is the definition of the tropical zone. The wind flows eastward, and clouds gather at this place and has lower pressure at the surface than the surrounding.

The Inter-Tropical Convergence Zone or ITCZ is where the extreme occurs, and this area changes position from north to south and back and forth (Pike 1971; Waliser and Sommerville 1994). The back and forth north–south movement is mainly controlled by the solar solstice or roughly corresponding with the thermal equator's location. The exact location is not defined and is mainly controlled by the changing pressure over the tropics. By definition, the solar solstice at a particular area below the solstice's edge around 23.5° latitude will pass twice the solstice. Then the equatorial type of climate will have twice the time where ITCZ passes over them. The other limitation to the twice occurrence will be the border between land and ocean and surface ocean current.

Fortunately, ITCZ is quite active, and is located in Indonesia, during the wet or rainy season. This condition is really in favor of extreme rainfall (Freitas et al. 2017). Combined with the warm sea surface temperature and other disturbances, it will make the ideal condition for extreme situations. Low surface pressure will attract water vapor from surrounding areas. Moreover, trade wind also contributes to the active sea surface temperature. The active or abnormal sea state will also contribute to the higher sea surface temperature. The most common indicator of ITCZ is the presence of eastward wind in the tropics. In most cases, it is not the lowest pressure level area in the tropics or near-equatorial trough, but they are surrounded by some lower pressure areas like vortices and wind saddle points. They are also places where clouds are usually gathered, which eventually create extreme weather conditions.

2.4.3 Madden–Julian Oscillation

The Madden–Julian oscillation is characterized by an eastward progression of large regions of both enhanced and suppressed tropical rainfall, observed mainly over the Indian and Pacific Ocean (Madden and Julian 1971). This phenomenon is one of the most persistent patterns in the tropics besides the ENSO (Zhang and Gottschalck 2002). It travels to the east or propagates eastward, at approximately 4–8 m/s. The anomalous rainfall is usually first evident over the western Indian Ocean. It remains evident as it propagates over the very warm ocean waters of the western and central tropical Pacific (Matthews and Li 2005). The MJO enhances and suppresses rainfall regions at the variability of 30–90 days periodically.

Over the tropics, none of the regions is more pronounced for MJO impact than Indonesia (Hidayat 2016). The enhanced rainfall and reduction are very clear. Furthermore, they are modulated with the presence of ENSO, the El Nino, and La Nina phenomenon. In the El Nino year, the dryer MJO will be extended and less enhanced rainfall during that year. While on the other hand, during La Nina year, the MJO will be much enhanced and sometimes leave with excessive rainfall and flood in some areas (Hidayat 2016). After all, we have for MJO the oscillation between 39 and 90 days oscillation. Although moving eastward, the MJO has different features during the rainy and dry seasons. In the dry season, after leaving the maritime continent, it will go northward, following where the ITCZ is, while in the wet season, MJO will mostly stay in the maritime continent because the ITCZ is in it. Another reason is that the maritime continent is the area with the most southland area, no more island, for example southward Java and Sumatra.

Right now, there is an excellent monitoring mechanism done by the Bureau of Meteorology Australia for MJO (Wheeler and McBride 2005). They measure the average Outgoing Longwave Radiation in the tropics (e.g., see Fig. 2.3). They are undoubtedly the average albedo or the cloud top height in the tropics. The OLR indicator is a nice indicator of how the MJO evolved in the tropics. Since MJO under a consistent movement to the east, we could easily predict by looking presently what is going on in the west of our present location. This is an accurate prediction yet

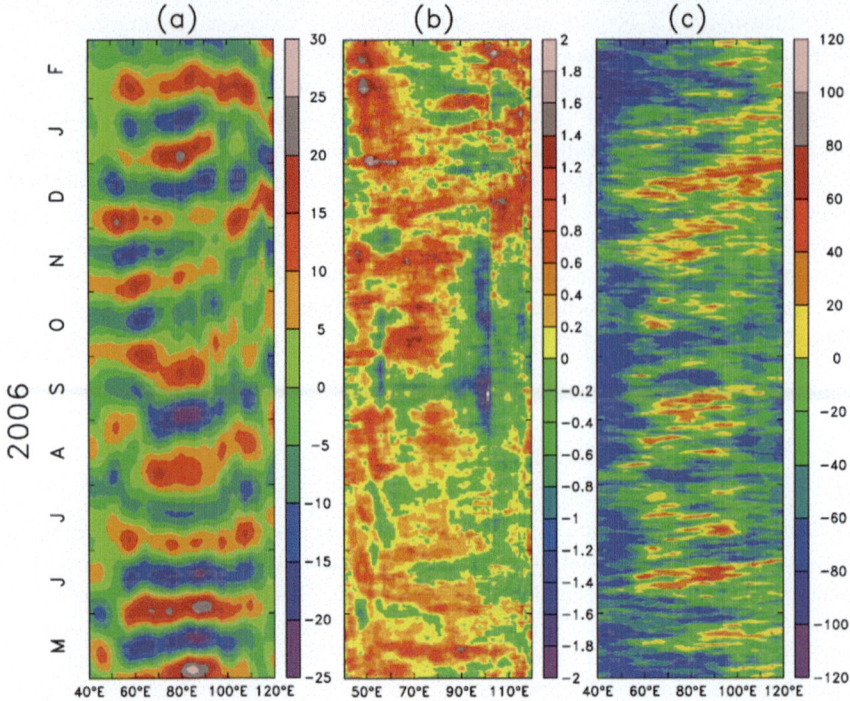

Fig. 2.3 Time–longitude plot of the 5S–5N averaged of **a** 30–60-day bandpass filtered OLR anomalies (Wm^2). The daily anomalies of OLR were filtered using a Lanczos filter with 121 weights. **b** Daily TMI SST anomaly (**c**), **c** The difference of the daily mean interpolated OLR from a value of 200 Wm^2. Red indicates enhanced convective. *Source* Modified from Tangang et al. (2008)

so far. However, right now, there are also zonal or quadrant predictions based on the presence of MJO location so far. Nonetheless, the prediction of upcoming MJO phases, whether dry or wet, will be readily predicted. In the OLR analysis, the colder the number represents, the higher the cloud top height. Then it will represent the wet condition. Then, the higher the values, the OLR average will be the dryer situation with sometimes no cloud at all.

2.4.4 Cold Surge

Cold surge is a unique phenomenon of Southeast Asia. It originally began in the high latitude Siberia in the boreal winter (Chang and Hendon 2006). Sometimes the blocking over this high latitude occurs that stops the Ferrel cell that was originally supposed to move eastward in that latitude. The cold surge can reach equatorial latitudes where it manifests as strengthened northerly or northeasterly winds (Tangang et al. 2008). A corresponding temperature and dry signature may not be present due to the influence warm seawater have on the surge. Depending on the thermodynamic characteristics of the pre-incursion environment, cold surges may also substantially impact convection and rainfall in the tropics (Chen et al. 2002). The cold and dry situations over there will push air moving to the East Asian coastline down to Southeast Asia. Usually, high wind pressure accompanies cold and dry conditions. Sometimes in Hongkong, it occurs during the Chinese New Year Eve. It is a common phenomenon in the boreal winter which mostly occurs in November till March (Hattori et al. 2011). The indicator is the pressure and temperature drop in Hongkong by a few numbers in a short time. Occasionally, there will be snow blizzards in Hongkong. The ocean sea surface temperature that the ocean passes by will be cold and dry. Usually, this surge takes the path Hongkong, Vietnam, the Malaysian peninsula, and the northeast tip of Kalimantan or Borneo Island. Sometimes, the surge could reach the equator and push southward.

The path that the surge passes will be dry and cold. The air southward will be compressed because of that. In some areas in Sumatra, the phenomena leave forest fire conditions if it lasts longer than a week. What happened south of the dry area will be the extreme condition. Although it will be dry and cold in some areas, there will be a possibility that the compressed dry air will cause the wet air south of the area to create extreme rainfall. In some cases, there will be a flood; for example, if the surge could push up to the equatorial line in south Sumatra, then there will be extreme on the west Java, especially the capital city of Jakarta. Some high flood events in Jakarta are attributed to these surge activities. So this phenomenon could result in dry and forest fire and at the same time flood in other areas.

Depending on the strength, sometimes the surge creates more weather phenomena such as vortices along the way. The surge is a very short period event. Sometimes the presence could last several days and create a Mesoscale Convective System, which is even more extreme.

The surge has two stages (Chang et al. 2005):

- The edge, a pressure surge, travels at a speed of ~ 40 m/s, which is much faster than the advective speed and is therefore suggestive of a gravity wave-like propagation;
- The front, a significant drop in dew point that travels at speeds of only ~ 10 m/s.

2.4.5 Tropical Cyclone

Indonesia is not a tropical cyclone path, nor it is located in the cyclogenesis (Chang et al. 2003). The cyclone genesis area is usually an area located more than 10° latitude north and south. The weak earth Coriolis force is the main reason for that. Most locations in Indonesia are far below that number. The problem lies in the cyclone tail. Although not in their paths, the impact due to the presence of tails will create another extreme. In fact, every tropical cyclone induces two tails: the dry and wet tail. The wet tail usually comes with extreme rainfalls (Mulyana et al. 2018). With the impact of climate change, the tail of tropical cyclones is getting stronger, longer, and far reaching, hence creating more severe problems. In some events, the cyclone tails bring a flood situation when it is high extreme enough.

Near Indonesia, there are three locations of tropical cyclone areas: the Philippine archipelago, the Bengal sea off the coast of Bangladesh, and southern Java between the Australian and Indonesia. There was one cyclone developed in the South China sea before at the degree of Singapore and additionally in the Banda sea eastern maritime continent. Both are located within a low latitude area, and the appearances were not common and only lasted a few hours. The appearance follows the cyclone's definition when it is declared by its name when they reach a certain threshold and demise when reaching another threshold.

2.4.6 Mesoscale Convective System

The Mesoscale Convective System (MCS) is a cloud system associated with a convective system complex that generates an adjacent rainfall area of about 100 km in at least one horizontal direction (Houze 2014), which is the life cycle has longer lifetime (greater than 3 h) than the individual convective element (Nuryanto et al. 2019). The majority of heavy rainfall events are triggered by these convective systems (Doswell et al. 1996; Choi et al. 2011; Jeong et al. 2016; Nuryanto et al. 2019). The development of convection over tropical regions is a key component of the global climate system as it vertically transports mass, momentum, and heat with large-scale atmospheric circulation patterns (Chen et al. 1996; Laing and Fritsch 2000; Moncrieff 2010). Interestingly, Indonesia is one of the tropical regions with extensive convective activity. So the attention should be more.

MCSs consist of convective cores and stratiform anvils, whose vertical and horizontal forms develop at different life cycle stages (Houze 1982). In particular, MCSs are embedded in tropical waves (Jakob and Tselioudis 2003), super-clusters of synoptic scale, and the Madden–Julian Oscillation (Nakazawa 1988), affecting the coupling of the atmosphere and atmosphere–ocean over a range of scales (Moncrieff 2010). For 100 years, scientists have been investigating the phenomenon of MCS, and it is still worth studying to this day.

Many MCS studies have been widely carried out in the tropics (e.g., Mohr and Zipser 1996; Zolman and Zipser 2000; Yuan and Houze 2010; Virts and Houze 2015; Putri et al. 2017, 2018; Nuryanto et al. 2017, 2018, 2019). However, there is still very limited research on MCS in Indonesia, i.e., the general life cycle of MCS in IMC (Putri et al. 2017); some cases of MCS events in Sumatra and Java (Putri et al. 2018); and the preliminary propagation study (Nuryanto et al. 2017); as well as MCS kinematic and thermodynamic structures (Nuryanto et al. 2018); also the case of the MCSs (Nuryanto et al. 2019) corresponding to the heavy rainfall event at Greater Jakarta area using satellite data.

The sea breeze merges to the center of the island during the daytime and enhances the inland convection, providing a favorable environment for MCS development over land. During the night, MCS formation is more concentrated over the sea, as the land breeze converges from the surrounding islands to some straits (Qian 2008). Huang et al. (2018) discovered that MCSs occur more frequently over land than over oceans and are more intense. However, except in South America, India, and some parts of tropical Africa, ocean MCSs are usually larger and last longer than land.

Diurnal variation in the size of deep-convective systems over the western Pacific and maritime continent was observed by Chen and Houze (1997). The fractional area with cold TBB (< 208 K) reached a peak over the land in the afternoon, followed by a peak of moderate TBB (235–208 K) and a peak of warm TBB (260–235 K) at midnight. They also observed that the diurnal cycle amplitude was much more significant in the land than in the ocean.

For oceanic MCS, the rain fraction is also greater than for land MCS. In the developing and mature stages, the difference between land and oceanic MCS is especially noticeable, suggesting faster convection changes over land relative to the sea due to a strong diurnal cycle inhibited by the land area (Putri et al. 2017). Although the growth of land and oceanic MCSs varies at the beginning, at the end of their life span, both forms display similar characteristics as the dominant convections of the MCSs disappear.

The MCSs in Indonesia vary in the spatial distribution in different seasons according to different atmospheric conditions (Putri et al. 2017). The frequency of MCS is high across the entire area in December–January–February (DJF), associated with the movement of the wet monsoon from the north. In comparison, the distribution is usually sparser in June–July–August (JJA), with almost no MCS developing over Indonesia's southern part because most of the Indonesia's regions are affected by dry Australian monsoon. The MCS latitudinal distribution description in Fig. 2.4 shows that the hemisphere where the ITCZ resides is located with a high concentration of MCS. However, the maximum occurrence of MCSs persists at around 2° S irrespective of the season, likely due to the richest distribution of land–sea and more complex topography at this latitude over Indonesia (Putri et al. 2017).

The MCS seasonal variation amplitude varies regionally. With a frequency value of 1.5 months^{-1} (10,000 km^2)$^{-1}$, Sulawesi has the most concentrated MCS. Due to its unique surface inhomogeneity, high concentrations of MCSs are likely in this region. Papua (1.4), Borneo (1.3), Sumatra (1.0), and Java (0.9) follow this value. During the dry monsoon season in JJAS, the small value of mean frequency in Java may be attributed to the strong suppression of MCS development (Putri et al. 2017).

Java Island exhibits a peak in seasonal rainfall in boreal winter. During this season, the highest occurrence of MCSs in Java also can be seen (Putri et al. 2017; Nuryanto et al. 2019). MJO's active phase also is well known for enhancing large-scale convective activities throughout the maritime continent of Indonesia (Hidayat and Kizu 2010). MJO, therefore, provided a more favorable atmospheric environment for the growth of the MCSs in this situation.

During the June–July–August (JJA) season, the Sumatra area has minimum MCS occurrence and rainfall peaks (Putri et al. 2017). Combined with a supporting topography and a strong vertical wind shear, the MCSs were initiated in favor of the warm SST around northern Sumatra (Putri et al. 2018). Although during the positive IOD, the convection is restricted to the northern part of Sumatra Island (Fujita et al. 2013),

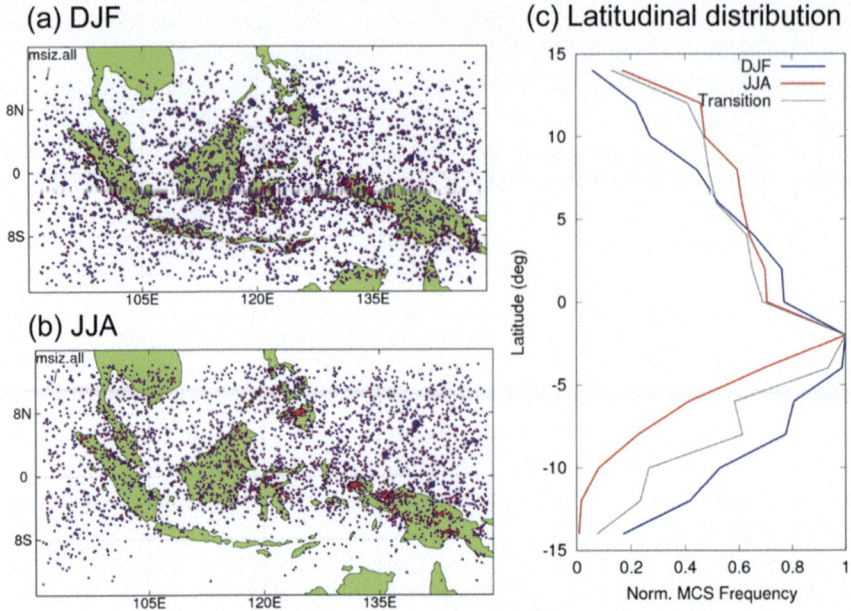

Fig. 2.4 Frequency of MCS in the study area in **a** December–January–February (DJF) and **b** June–July–August (JJA). **c** Latitudinal variation of MCS occurrence in DJF, JJA, and transitional seasons (SON and MAM), normalized by maximum latitudinal frequency in each season. *Source* Modified from Putri et al. (2017)

and the sheer magnitude was between two and three times greater than that around Java (Putri et al. 2018).

2.4.7 Vortices

This is a smaller version of the cyclone. Usually, in Indonesia, it is like a twister version. Local people name it the "puting beliung". They are the size of water sprouts or smaller twisters. On the mesoscale version, the vortices could last several days. The cold surge can inhibit such a version called a Borneo vortex. In some other places, similar conditions may occur. The area is the location of many coasts and mountains, where this combination is perfect for a small twister. This happens when cold air from the hilly mountain meets the warm air from the coast. When there are changes in the local and regional winds, this is usually happening. Although small, the period of its lifetime is less than an hour. It is still dangerous, and the extreme it creates is sometimes devastating. Many of this kind occur in the transition time between two monsoonal periods when the local wind usually in the form of land and sea breeze meets the regional wind usually in the form of mountain wind.

Additionally, the presence of a land–sea breeze will favor the warm uphill breeze from the sea coast. In the blogger scale or mesoscale, the wind changing direction, such as trade wind and surface pressure change, also contributes to the Borneo vortex in the presence of cold surges (Tangang et al. 2008). With the presence of trade wind and ITCZ, sometimes vortices form during the saddle point at the surface pressure level. The combination of this phenomenon with others such as the cold surge or tail of a tropical cyclone will sometimes favor extreme conditions for extreme weather.

2.4.8 Air Quality Extreme

Air quality extreme is the condition whenever the ambient air quality drops to a certain level that causes ambient air to be unable to fulfill its function due to the entry of substances, energy, and/or other components into the air ambient by human activities (President of Indonesia Regulation 41 1999). Air quality extremes are associated with air pollution events. Air pollutant standard index (ISPU), which is a unitless number, is calculated to describe the condition of ambient air quality in a particular location based on the impact on human health, aesthetic value, and other living things (Indonesian Minister of Environment P.14 2020). ISPU established by the Indonesian Ministry of Environment is based on seven major air pollutants as particulate natter (PM10), PM2.5, carbon monoxide (CO), ozone (O_3), sulfur dioxide (SO_2), nitrogen dioxide (NO_2), and total hydrocarbon (HC) (Indonesian Minister of Environment P.14 2020). Each of these pollutants has a national air quality standard set by the Indonesian Ministry of Environment following the World Health Organization (WHO) Air Quality Guidelines (AQGs) to protect public health and is a pivotal component of national risk management and environmental policies. These standards may vary with other countries according to the approach adopted for balancing health risks, technological feasibility, economic considerations, and various other political and social factors, which in turn will depend on, among other things, the level of development and national capability in air quality management (WHO 2006).

Air quality extreme occurs if the concentration of the atmosphere's pollutants is above the national air quality threshold. The increase of pollutants concentration is mainly due to human activities and natural phenomena. Nowadays, Indonesia's population is concentrated in urban areas like Jakarta Greater City and exposed to air pollution (Kusumaningtyas et al. 2021; Permadi and Kim Oanh 2008; Suhadi et al. 2005; Wasi'ah and Driejana 2017). Rapid economic growth coupled with transportation and industrial facilities in urban areas become the crucial sources of pollutants emission such as CO, NO_2, volatile organic compounds (VOCs), and other secondary pollutants (Permadi et al. 2017; Lestari et al. 2020). Meanwhile, recurrent forest and land fires in some parts of Sumatra and Kalimantan also contribute to the worst air pollution due to the release of very high concentrations of greenhouse gasses and aerosols (Eck et al. 2019; Gaveau et al. 2014; Hayasaka et al. 2014). Air quality extreme due to emission from peatland fire in Sumatra and Kalimantan has profoundly perturbed the chemical composition of the atmosphere and proven impacted the regional air quality around neighborhood countries (Kusumaningtyas and Aldrian 2016; Koplitz et al. 2018). The impact of emissions from peatland fires on air quality and livelihood is much higher several orders of magnitude compared to other sources of emissions.

The alteration of atmospheric composition (in terms of physical and chemical properties of gaseous and aerosol) can lead to climate changes. A perturbation to the atmospheric concentration of meaningful greenhouse gas and or aerosols induces a radiative forcing that can affect climate (Brasseur et al. 2003). As commonly known from the Intergovernmental Panel on Climate Change (IPCC), greenhouse gasses have a positive radiative forcing, meaning that they tend to warm the surface due to the absorption of solar radiation. Black carbon emitted from biomass burning is also a strong absorbing aerosol. On the other hand, aerosols in the form of sulfate, nitrate, and carbonaceous aerosols scatter solar radiation, thus cooling the surface. The scattering of solar radiation produces a negative radiative forcing (IPCC 2007; Seinfeld and Pandis 2006; Brasseur et al. 2003). Aerosols from forest/land fires could also affect the climate system by modulating cloud microphysical processes resulting in large perturbations in the regional water cycle and circulation of entire Southeast Asia, including the South China Sea, southern China, and Taiwan (Tsay et al. 2013; Lin et al. 2014). The aerosol burden is strongly influenced by meteorological parameters such as humidity, wind speed and direction, and rainfall. These parameters could be changed in response to El Nino–Southern Oscillation (ENSO) (Yu et al. 2019). During strong El Nino, the drought increases more intensely and extends to more expansive spatial areas. Drought reduces humidity and the absence of precipitation, leaving a drier condition that ease of drained peatland is burned by humans, such as Sumatera and Kalimantan. Therefore, aerosol effects provide positive feedback to ENSO's evolution or the meteorological response to ENSO.

Worst air pollution due to smoke from peatland fires in Sumatera and Kalimantan leaves giant consequences to many sectors. This event impacts the economic, public health, environmental, and political situation among Indonesia and neighboring countries. During the fire in Riau 2013, the air quality index reached 1084, categorized as hazardous. The smoke haze led to the disruption of transportation facilities and airport closure with an estimated loss of ~ 108 thousand USD due to delays and cancelation of domestic and international flights (Kusumaningtyas and Aldrian 2016). An extreme El Nino event coupled with a positive phase Indian Ocean Dipole in 2015 triggered severe drought and eventually inducing peatland burning to spread farther than in typical rainfall years (Eck et al. 2019). Very high aerosol loading as indicated from aerosol optical depth (AOD) value at 550 nm was estimated to reach ~ 11 to ~ 13 during prominent peat burning in Central Kalimantan as presented in Fig. 2.5 (Eck et al. 2019). According to Eck et al. (2019), this AOD value was the highest, and they have no knowledge of such high smoke AOD values having been previously reported in the scientific literature. Recent forest and land fires that coincided with El Nino also reoccurred in Sumatera and Kalimantan in 2019. BMKG recorded the daily average concentration of PM10 in Palembang (South Sumatera) and Pekanbaru site (Riau Province) reached almost 500 $\mu g/m^3$ as presented in Fig. 2.6. This number is

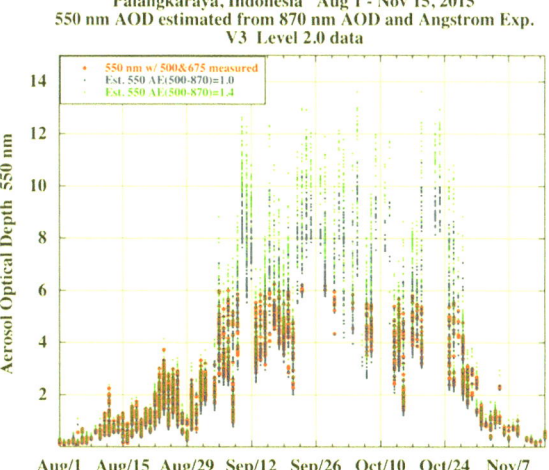

Fig. 2.5 Time series of AOD at 550 nm at the Palangkaraya, Indonesia, site from August 01 through November 15, 2015. Shown in red are AOD at 550 nm computed when measured values at 500 and 675 nm were available. Also shown are estimates of 550 nm AOD, utilizing the measured 870 nm AOD in conjunction with assumed Angstrom exponents (500–870 nm) of 1.0 (blue) and 1.4 (green). *Source* Eck et al. (2019)

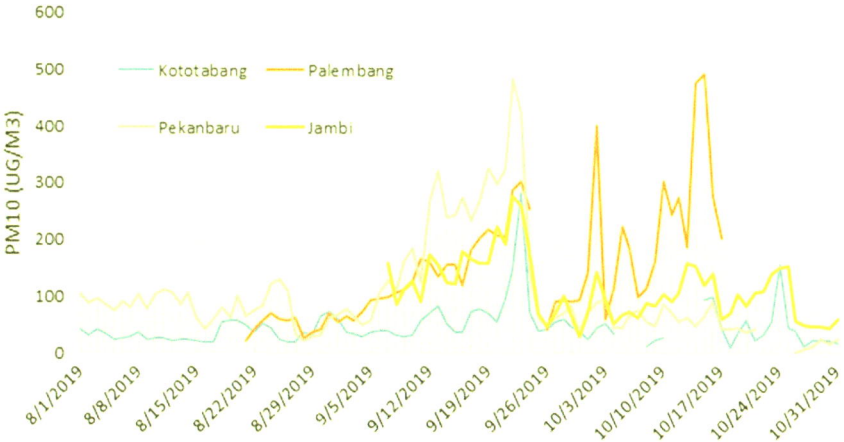

Fig. 2.6 Daily average of PM10 concentration during the burning period in August–October 2019 in several fire-prone areas in Sumatera

three times above the national threshold (150 μg/m^3). Uda et al. (2019) estimated the long-term exposure of PM2.5 from recurrent peat fires and smoke events in Central Kalimantan. They found that there were 648 premature mortality cases per year, which include 55 mortality cases due to chronic respiratory diseases, 266 mortality cases due to cardiovascular diseases, and 95 mortality cases due to lung cancer.

Another event of extreme air quality was the volcanic eruption. The emission plume from Mount Agung eruption in Bali 2017 headed to Denpasar, the capital city, and caused several flights to be delayed and canceled. Economics and tourism sectors were lost. However, due to a lack of air quality monitoring in the field, we could not retrieve information on the concentration of air pollutants.

2.5   Works for Extremes

As for the task and function of BMKG to provide data services for extreme warnings from their observed stations, currently, in 2021, BMKG has 22 climatology stations, more than 120 meteorological stations, and 31 geophysics stations that all supply climate data and more than 5000 meteorological network posts. Additionally, there are automatic weather networks, automatic rain gauge networks, and agricultural automatic weather stations. Currently, there are 697 online instrument networks with more than 120 agricultural automatic weather stations. The BMKG stations usually monitor rainfall, evaporation, wind, relative humidity, and surface pressure.

BMKG also operates more than 40 weather radar networks. Almost all radar networks have been integrated and can be used for the public using their mobile devices. The public may know exactly using their location about the presence of weather conditions, especially rainfall from one hour ago, up to 30 min ahead. The one and half hour animation of the current weather will give the public enough options for their action. The animation also gives a clear situation of the threat of extreme danger for the public.

At the center of the BMKG office, satellite monitoring is also conducted. Satellite images provide information for future warnings and give indicators for the next analysis of extreme. Information on fronts, wet atmosphere condition, and wind direction locally and regionally can be derived from the satellite information. With the help of applications such as windy, the warning can be done easily and faster. BMKG also operates some weather modeling for weather prediction. The model runs from European Center for Medium-Range Weather Forecast or ECMWF, and BMKG obtains only the extreme indicator for analysis from ECMWF. BMKG subscribes to the service of ECMWF. Regularly, BMKG could provide parameter analysis of extremes such as information on probabilistic extreme precipitation.

Several indicators on extremes are derived from services of other meteorological and climatological agencies outside the country. The primary information for El Nino–Southern Oscillation comes from ocean observation in the Pacific Ocean. The Toga COARE (Tropical Ocean Global Atmosphere—Coupled Ocean–Atmosphere Research Experiment) deployed an array of buoys to monitor sea surface and subsurface temperature on the Pacific. From this array of buoys, we derived the ENSO indicators. For ENSO prediction, we produce our BMKG indicators and information from other agencies such as NOAA, JAXA, BoM Australia, and NCEP. Other extreme indicators all also derived from international cooperation, such as the MJO, cold surge, and tropical cyclones. For MJO indicators, usually people follow the BoM Australia MJO monitoring that is now readily available on the Internet. BMKG operates its own Tropical Cyclone Warning Center (TCWC) after being given the authority and operation area for cyclone monitoring.

For forest fire and volcano observation usually, BMKG works in cooperations with other agencies such as the National Agency for Space and Aeronautics and the Ministry for Forestry and Environment. The Volcanic Agency under the Ministry for Energy and Mineral Resources is the last important agency for cooperation.

2.6 Future Extremes

Climate change is characterized by changes in the mean and changes in variability and extremes (Tank 2003). It has been reported that temperature and precipitation extremes will likely be changing due to global warming, which is "extremely likely" influenced by anthropogenic factors (IPCC 2014). Experimental studies using both global and regional models indicated that the temperature increases could increase extreme rainfall events (e.g., Watterson and Dix 2003; Wehner 2004). For the Indonesia case, there are several kinds of literature reporting experiments on the projection of extremes over Indonesia using either statistical (Karlina 2016; Daksiya et al. 2017) or dynamical approach (Jadmiko and Faqih 2014; Chandrasa and Montenegro 2020; Supari et al. 2019).

Karlina (2016) downscaled the global model of HadCM3 using a statistical approach to estimate the future drought over Wonogiri District. She found that the number of drought events at the end of the twenty-first century is projected to be less than that in the historical period. Daksiya et al. (2017) document their study on the possible impact of changing climate on the frequency of daily rainfall extremes in Jakarta. A total of 15 GCMs were involved in their statistical downscaling process. Two statistical methods were applied, i.e., the Long Ashton Research Station-Weather Generator (LARS-WG) and the Statistical Downscaling Model (SDSM). Outputs of those two methods were compared with data from NASA Earth Exchange Global Daily Downscaled Projections (NEX-GDDP), the downscaling product which is freely available. They found that the annual maximum daily rainfall may significantly change in the future, with an average increase as high as 20% in the 100-year return period daily rainfall (see Fig. 2.7). The seasonal daily rainfall maximum increases for the wet season. However, the dry season did not exhibit a consistent increase or decrease across the models suggesting that the annual scale changes are mostly due to changes in the wet season.

Additionally, some dynamical downscaling experiments also contribute to our insight on the future extremes of Indonesia. Jadmiko and Faqih (2014) dynamically downscaled the GCM EH5OM using RegCM3 over Indramayu District. They found that extreme rainfall (rainfall > third quartile) may not change in the future. In a larger experiment, Chandrasa and Montenegro (2020) investigate the potential impact of climate change on rainfall over the whole country. They downscaled the Max Planck's Institute Earth System Model with Medium Resolution (MPI-ESM)

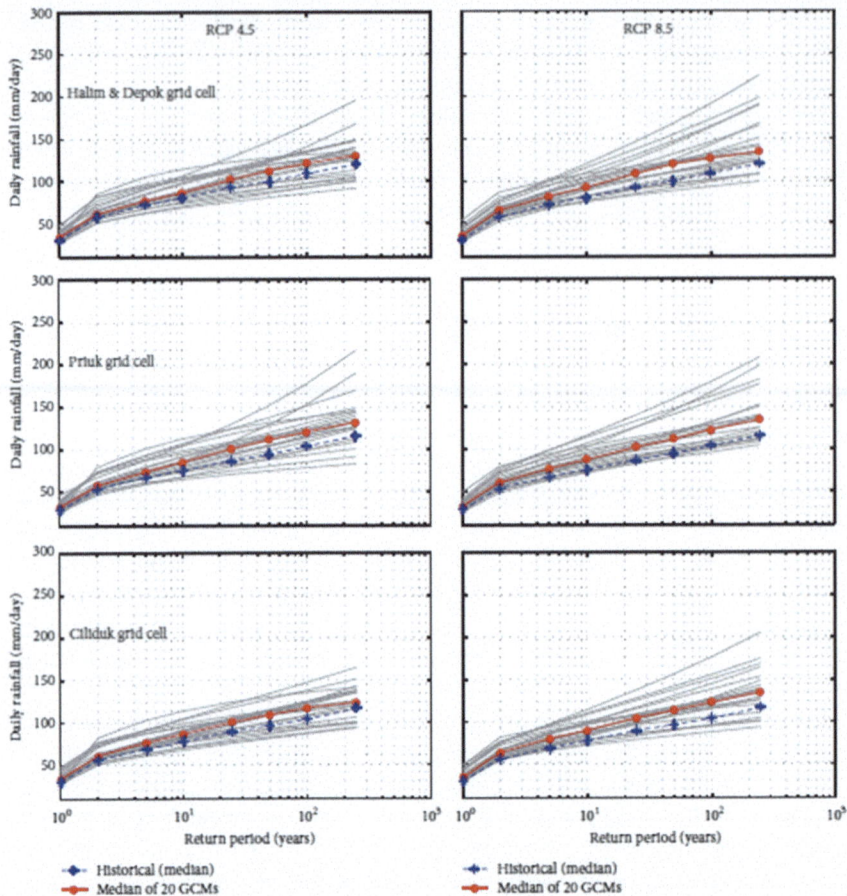

Fig. 2.7 Estimated value of daily rainfall for several return periods under RCP 4.5 (left) and RCP8.5 (right) for locations around Jakarta city. The 20GCM predictions for the period of 2041–2070 are indicated with thick gray lines. *Source* Modified from Daksiya et al. (2017)

using WRF regional model under RCP8.5. Four cumulus schemes were tested to find the best option in simulating the Indonesian climate. Two time intervals were selected to represent the future, the near future (2036–2040), and the far future (2096–2100). They found that during the monsoonal wet season, the number of rainy days and consecutive wet days (CWD) increases in most regions, and contrast, consecutive dry days (CDD) decrease. Additionally, Sumatra may have a higher risk of drought because of the future decreases in soil moisture projected in most of its area for all seasons. Supari et al. (2019) explore the output of downscaling data produced by BMKG to investigate precipitation extremes' possible response under global warming of 2 and 4 °C. They use the Regional Climate Model system RegCM4 to downscale CSIRO Mk3.6. It is reported from their study that under

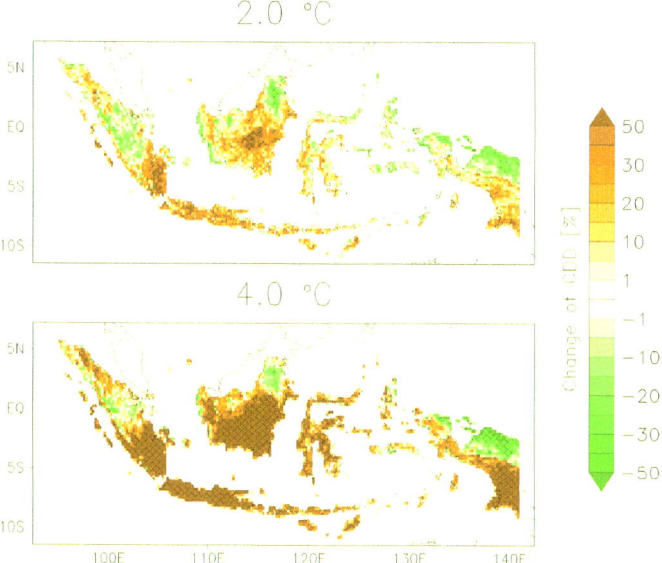

Fig. 2.8 Possible changes in CDD, under two global warming thresholds (2 and 4 °C). *Source* Supari et al. (2019)

these two global warming levels, there will be a decrease in total annual precipitation (PRCPTOT) in most parts of the country. Consistently, the dry spell duration is projected to increase as represented by the CDD index (consecutive dry days, see Fig. 2.8). On the other hand, a mixed increase and decreased tendency is found for the projection of the frequency and precipitation extremes. Seasonally, PRCPTOT tends to decrease during the dry season (June–July–August, JJ) and tends to increase during the wet season (December–January–February, DJF).

A more complex experiment on the future extremes using multi-GCMs, multi-RCMs, and multi-scenarios is documented in Supari et al. (2020) under the CORDEX-SEA Project. The project was run by a collaboration of institutions from countries within the South East Asia region. An ensemble of regional climate simulations consisting of eight members taking from a subset of archived CORDEX-SEA simulations at 25 km spatial resolution was used for their study covering the South East Asia domain, including Indonesia. The assessment of changes was done by comparing precipitation indices during the end of the century (2081–2100) relative to the reference period (1986–2005) under RCP4.5 and RCP8.5 scenarios. They selected four ETCCDI indices, i.e., PRCPTOT, CDD, R50mm, and RX1day, to represent precipitation extremes. They conclude that substantial changes in the characteristics of precipitation extremes may occur over the domain in the future, where changes under RCP8.5 are generally in greater magnitude compared to that under RCP4.5. Those changes include a decrease in PRCPTOT over most countries and consistently an increase of dry spell duration (CDD). In general, the southern part of SEA region will experience a greater magnitude of drying signal in PRCPTOT

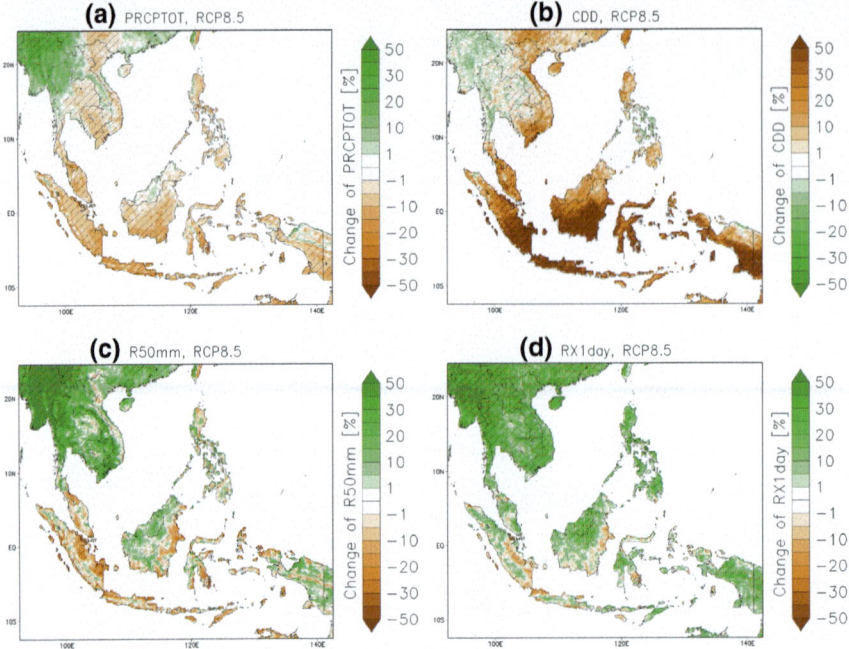

Fig. 2.9 Projected changes in precipitation indices during the end of the century, under RCP8.5. *Source* Supari et al. (2020)

and CDD compared to the northern part (see Fig. 2.9). In contrast, for the index of R50mm and RX1day, they reported a strong significant wetting signal over the northern part, while in the southern part those indices show a mixed weak wetting and drying tendency.

2.7 Institutions for Disaster in Indonesia

After understanding Indonesia's extreme situation and the corresponding climate and air quality phenomena, we now introduce institutions responsible for managing and administering the disaster analysis for climate and air quality.

2.7.1 BMKG or the Agency for MeteorologyClimatology and Geophysics

BMKG has the authority to issue early warnings for future weather and climate. BMKG is like the NOAA version for Indonesia. From this agency, the public will receive extreme warnings. This agency should also issue multiple hazards related to extreme weather and climate and air quality. This agency also provides extreme weather for air navigation and sea state. It provides the meteorological early warning, climate early warning, and air quality. BMKG is part of the World Meteorological Organization or WMO a UN specialized agency for meteorological and climate early warning. BMKG operates many instruments and models to observe and analyze the current weather and climate information. It includes the upper air, surface, and means states of the ocean. For weather prediction, two dynamical numerical models are used by BMKG, i.e., the ECMWF model and the CFS model which is downscaled using WRF.

The chance of extreme weather and climate events is quantified based on the probability forecast of rainfall exceeding certain threshold with the minimum probability to release warning as 70%.

2.7.2 BNPB or the National Disaster Management Authority

BNPB is the specialized agency for disaster management, primarily when a disaster occurs. It also provides disaster management for disaster prevention and precaution if possible. As the BMKG prepares the early warning system, BMNP provides information on the impact and risk level of a disaster. Together they provide the risk-based warning and impact-based forecast, as proposed by the International Strategy for Disaster Risk ISDR by the UN and endorsed by World Meteorological Organization WMO in their Congress.

2.7.3 BPPT or the Agency for Assessment and Application of Technology

BPPT is the clearing technology, audit technology, and technology innovation needed by assessing and applying technologies. One of the products under collaboration with BMKG is the Fire Danger Rating System or FDRS, an early warning system developed for forest fires. BPPT also develops the ASEAN Coordinating Centre for Humanitarian Assistance on Disaster Management (AHA Center and the NeoNet is another version of WMO GEOS (Global Earth Observation) for Indonesia.

2.7.4 LAPAN or the Agency for Space and Aeronautics

LAPAN is an agency like NASA for the USA or JAXA for Japan. LAPAN has the authority for remote sensing services by law in Indonesia. However, some of the remote sensing satellites are also done within BMKG. Precipitation, cloud coverage, and remote sensing analysis for wind direction and water content level are done separately in BMKG. Many remote sensing businesses are similar to the one in BMKG. The information on hotspots for a forest fire is also legally by LAPAN. However, BMKG and the Ministry for Forest and Environment also derived their information using outside satellite observation.

2.7.5 Ministry for Forestry and Environment

This Ministry is vital for the forest fire and the peatland restoration business. Peatland restoration keeps the water level for peatland, which is vital to maintain the capacity not to increase hotspots associated with a forest fire.

2.7.6 Ministry for Energy and Mineral Resources

This Ministry holds the Volcanology Agency, a critical agency for monitoring and declaring the airport operation's air quality index.

2.7.7 Ministry for Public Work and Housing

This Ministry is essential for catchment water management. It is essential for surface water management and control flood and inundation in housing areas.

2.8 Conclusions

This chapter depicts the extremes of climate phenomena in Indonesia. The purpose will be for education and general public information on how the country manages the extreme information. It discusses the climate phenomena that attribute to some extremes and modalities to handle the extreme in Indonesia. The situation might be different from country to country. The definition of extreme and the future extreme type is also discussed here. Then the chapter also introduces institutions for climate extremes. Extreme phenomena described here are not exhaustive, but there are some more. However, major climate extremes phenomena presented here are enough for most causes of extremes. The final section also introduces some works for future extremes based on the model projections. This chapter also discusses one of the notable extremes in air quality, especially during the dry season.

The studies of the extreme in Indonesia monsoon over Indonesia have progressed extensively for the past years. The broad spectrum of spatiotemporal occurences of extreme phenomena and the interrelations between the strong diurnal cycle in the combined land–sea breeze, subseasonal weather types such as the MJO, and ENSO-related interannual variability. Not to mention the timing and frequency of the tropical cyclones, surges, and small scale and mesoscale vortices. The unique feature also includes the influence of the upper ocean streamflow that also controls the climate type (Wyrtki 1961), a feature that may not exist in other regions of the world.

References

Aini, E.N., and A. Faqih. 2021. Frost predictions in dieng using the outputs of sub seasonal to seasonal (S2S) model. *Agromet* 35 (1): 30–38. https://doi.org/10.29244/j.agromet.35.1.30-38.

Aldrian, E. 2002. Spatial patterns of ENSO impact on Indonesian rainfall. *Jurnal Sains & Teknologi Modifikasi Cuaca* 3: 5–15.

Aldrian, E., and R.D. Susanto. 2003. Identification of three dominant rainfall regions within Indonesia and their relationship to sea surface temperature. *International Journal of Climatology* 23: 1435–1452.

Aldrian, E., L.D. Gates, and F.H. Widodo. 2007. Seasonal variability of Indonesian rainfall in ECHAM4 simulations and in the reanalyses: The role of ENSO. *Theoretical and Applied Climatology* 87, no. 1–4: 41–59.

Brasseur, G.P., G.P. Ronald, and A.P.P. Alexander. 2003. Atmospheric chemistry in a changing world, Global Change - The IGBP Series.

Chang, C.-P., C.-H. Liu, and H.C. Kuo. 2003. Typhoon Vamei: An equatorial tropical cyclone formation. *Geophysical Research Letters* 30 (3): 1150.

Chang, C.P., Z. Wang, and H. Hendon. 2006. *The Asian winter monsoon, chapter 3 in the Asian monsoon.* In ed. B. Wang, 89–128. Praxis Publishing.

Chang, C.-P., P.A. Harr, J. McBride, and H.-H. Hsu. (2004). Maritime continent monsoon: Annual cycle and boreal winter variability. In *East Asian monsoon, Chap. 3. World scientific series on meteorology of East Asia*, vol. 2, ed. C.-P. Chang, 107–152. World Scientific.

Chang, C.-P., Z. Wang, J. McBride, and C.H. Liu. 2005. Annual cycle of Southeast Asia—Maritime continent rainfall and the asymmetric monsoon transition. *Journal of Climate* 18: 287–301.

Chandrasa, G.T., and A. Montenegro. 2020. Evaluation of regional climate model simulated rainfall over Indonesia and its application for downscaling future climate projections. *International Journal of Climatology* 40: 2026–2047. https://doi.org/10.1002/joc.6316.

Chen, S.S., R.A.J. Houze, and B.E. Mapes. 1996. Multiscale variability of deep convection in relation to large-scale circulation in TOGA COARE. *Journal of Atmospheric Science* 53: 1380–1409.

Chen, S.S., and R.A.J. Houze. 1997. Diurnal variation and life-cycle of deep convective systems over the tropical Pacific warm pool. *Quarterly Journal of the Royal Meteorological Society* 123: 357–388.

Chen, T.C., M.C. Yen, W.R. Huang, and W.A. Gallus. 2002. An East Asian cold surge: Case study. *Monthly Weather Review* 130 (9): 2271–2290.

Choi, H.Y., J.H. Ha, D.K. Lee, and Y.H. Kuo. 2011. Analysis and simulation of mesoscale convective systems accompanying heavy rainfall: The goyang case. *Asia Pacific Journal of Atmospheric Sciences* 47: 265–279.

Daksiya, V., P. Mandapaka, and E.Y.M. Lo. 2017. A comparative frequency analysis of maximum daily rainfall for a SE Asian region under current and future climate conditions. *Advances in Meteorology.* https://doi.org/10.1155/2017/2620798.

Doswell, C.A., H.E. Brooks, and R.A. Maddox. 1996. Flash flood forecasting: An ingredients-based methodology. *Weather and Forecasting* 11: 560–581.

Eck, T.F., B.N. Holben, D.M. Giles, I. Slutsker, A. Sinyuk, J.S. Schafer, A. Smirnov, M. Sorokin, J.S. Reid, A.M. Sayer, N.C. Hsu, Y.R. Shi, R.C. Levy, A. Lyapustin, M.A. Rahman, S.C. Liew, S.V.S. Cortijo, T. Li, D. Kalbermatter, K.L. Keong, M.E. Yuggotomo, F. Aditya, M. Mohamad, M. Mahmud, T.K. Chong, H.S. Lim, Y.E. Choon, G. Deranadyan, S.D.A. Kusumaningtyas, and E. Aldrian. 2019. AERONET remotely sensed measurements and retrievals of biomass burning aerosol optical properties during the 2015 Indonesian burning season. *Journal of Geophysical Research: Atmospheres* 124 (8): 4722–4740.

Freitas, A.C.V., L. Aímol, T. Ambrizzi, and C.P. de Oliveira. 2017. Extreme intertropical convergence zone shifts over southern maritime continent. *Atmospheric Science Letter* 18: 2–10.

Fujita, M., H.G. Takahashi, and M. Hara. 2013. Diurnal cycle of precipitation over the eastern Indian ocean off Sumatra island during different phases of indian ocean dipole. *Atmospheric Science Letters* 14: 153–159.

Giannini, A., A.W. Robertson, and J.H. Qian. 2007. A role for tropical tropospheric temperature adjustment to ENSO in the seasonality of monsoonal Indonesia precipitation predictability. *Journal of Geophysical Research* 112: D16110.

Gaveau, D.L., et al. 2014. Major atmospheric emissions from peat fires in Southeast Asia during non-drought years: Evidence from the 2013 Sumatran fires Nat. *Science and Reports* 4: 6112.

Hackert, E.C., and S. Hastenrath. 1986. Mechanism of anomalies rainfall in Java. *Monthly Weather Review* 114: 745–757.

Hattori, M., S. Mori, and J. Matsumoto. 2011. The cross-equatorial northerly surge over the maritime continent and its relationship to precipitation patterns. *Journal of the Meteorological Society of Japan* 89A: 27–47.

Hayasaka, H., I. Noguchi, E.I. Putra, N. Yulianti, and K. Vadrevu. 2014. Peat-fire-related air pollution in Central Kalimantan, Indonesia. *Environmental Pollution* 195: 257–266.

Hidayat, R. 2016. Modulation of Indonesian rainfall variability by the Madden–Julian oscillation. *Procedia Environmental Science* 33: 167–177.

Hidayat, R., and S. Kizu. 2010. Influence of the Madden–Julian oscillation on Indonesian rainfall variability in austral summer. *International Journal of Climatology* 30: 1816–1825.

Houze, R.A., Jr., 1982. Cloud clusters and large-scale vertical motions in the tropics. *Journal of the Meteorological Society of Japan* 60: 396–410.

Houze, R.A.J. 2014. *Cloud dynamics*, 2nd ed. Amsterdam: Academic Press.

Huang, X., C. Hu, X. Huang, Y. Chu, Y.-H. Tseng, G.J. Zhang, and Y. Lin. 2018. A long-term tropical mesoscale convective systems dataset based on a novel objective automatic tracking algorithm. *Climate Dynamics* 51 (7–8): 3145–3159.

Indonesian Minister of Environment and Forestry Regulation No P.14. 2020. *Air pollutants standard index*. Jakarta: Ministry of Environment and Forestry of the Republic of Indonesia.

IPCC. 2014. Summary for policymakers. *Climate Change 2014: Impacts, adaptation and vulnerability. Contributions of the Working Group II to the Fifth Assessment Report of the IPCC*, hlm. 1–32. Cambridge, UK: Cambridge University Press. https://doi.org/10.1016/j.renene.2009. 11.012.

Jadmiko, S.D., and A. Faqih. 2014. Dynamical downscaling of Global Climate Model (GCM) output by using Regcm3 model for rainfall projection in Indramayu district. *Journal of Agromet* 28: 9–16.

Jakob, C., and G. Tselioudis. 2003. Objective identification of cloud regimes in the tropical Western Pacific. *Geophysical Research Letters* 30 (21): 2082.

Jeong, J.-H., D.-I. Lee, C.-C. Wang, and I.-S. Han. 2016. Characteristics of mesoscale-convective-system-produced extreme rainfall over southeastern South Korea: 7 July 2009. *Natural Hazards and Earth Systems Sciences* 16: 927–939.

Karlina, K. 2016. Meteorological drought assessment in Wonogiri district. *Journal of the Civil Engineering Forum* 2: 159. https://doi.org/10.22146/jcef.26575.

Koplitz, S.N., L.J. Mickley, D.J. Jacob, M.E. Marlier, R.S. DeFries, D.L. Gaveau, and S.S. Myers. 2018. Role of the Madden–Julian oscillation in the transport of smoke from Sumatra to the Malay Peninsula during severe non-El Nino haze events. *Journal of Geophysical Research: Atmospheres* 123 (11): 6282–6294.

Kusumaningtyas, S.D.A., A.N. Khoir, E. Fibriantika, and E. Heriyanto. 2021. Effect of meteorological parameter to variability of particulate matter (PM) concentration in urban Jakarta city, Indonesia. *IOP: Conference Series: Earth and Environmental Science*.

Kusumaningtyas, S.D.A., and E. Aldrian. 2016. Impact of the June 2013 Riau province Sumatera smoke haze event on regional air pollution. *Environmental Research Letters* 11 (7): 075007.

Laing, A.G., and J.M. Fritsch. 2000. The large-scale environments of the global populations of mesoscale convective complexes. *Monthly Weather Review* 128: 2756–2776.

Lestari, P., S. Damayanti, and M.K. Arrohman. 2020. Emission inventory of pollutants (CO, SO_2, PM2.5, and NOX) in Jakarta Indonesia. *IOP Conference Series: Earth and Environmental Science* 489 (1).

Lin, N., A.M. Sayer, S.H. Wang, A.M. Loftus, T.C. Hsiao, G.R. Sheu, N.C. Hsu, S.C. Tsay, and S. Chantar. 2014. Interactions between biomass-burning aerosols and clouds over Southeast Asia: Current status, challenges, and perspectives. *Environmental Pollution* 195: 292–307.

Madden, R.A., and P.R. Julian. 1971. Detection of a 40–50 day oscillation in the zonal wind in tropical Pacific. *Journal of Atmospheric Science* 28: 702–708.

Matthews, A.J., and H.Y. Li. 2005. Modulation of station rainfall over the western Pacific by the Madden–Julian oscillation. *Geophysics Research Letter* 32: L14827.

Mohr, K.I., and E.J. Zipser. 1996. Mesoscale convective systems defined by their 85-GHz ice scattering signature: Size and intensity comparison over tropical oceans and continents. *Monthly Weather Review* 124: 2417–2437.

Moncrieff, M.W. 2010. The multiscale organization of moist convection and the intersection of weather and climate. Why does climate vary? American geophysical union. *Climate Dynamics* 189: 3–26.

Mulyana, E., M.B.R. Prayoga, A. Yananto, S. Wirahma, E. Aldrian, B. Harsoyo, T.H. Seto, and Y. Sunarya. 2018. Tropical cyclones characteristic in southern Indonesia and the impact on extreme rainfall event. *MATEC Web of Conferences* 229 (5742): 02007.

Nakazawa, T. 1988. Tropical super clusters within intraseasonal variations over the western Pacific. *Journal of the Meteorological Society of Japan* 66: 823–839.

Nuryanto, D.E., H. Pawitan, R. Hidayat, and E. Aldrian. 2017. Propagation of convective complex systems triggering potential flooding rainfall of Greater Jakarta using satellite data. *IOP Conference Series Earth Environmental Science* 54: 012028.

Nuryanto, D.E., H. Pawitan, R. Hidayat, and E. Aldrian. 2018. Kinematic and thermodynamic structures of mesoscale convective systems during heavy rainfall in greater Jakarta. *Makara Journal of Science* 22 (127–136): 1.

Nuryanto, D.E., H. Pawitan, R. Hidayat, and E. Aldrian. 2019. Characteristics of two mesoscale convective systems (MCSs) over Greater Jakarta: A case of heavy rainfall period 15–18 January 2013. *Geoscience Letters* 6: 1.

Permadi, D.A., and N.T. Kim Oanh. 2008. Episodic ozone air quality in Jakarta in relation to meteorological conditions. *Atmospheric Environment* 42 (28): 6806–6815.

Permadi, D.A., A. Sofyan, and N.T. Kim Oanh. 2017. Assessment of emissions of greenhouse gases and air pollutants in Indonesia and impacts of national policy for elimination of kerosene use in cooking. *Atmospheric Environment* 154: 82–94.

Pradana, A., Y.A. Rahmanu, I. Prabaningrum, I. Nurafifa, and D.R. Hizbaron. 2018. Vulnerability assessment to frost disaster in dieng volcanic highland using spatial multi-criteria evaluation. *IOP Conference Series: Earth Environmental Science* 148: 012002.

Pike, A.C. 1971. Intertropicalconvergence zone studied with aninteracting atmosphere and ocean model. *Monthly Weather Review* 99: 469–477.

President of Indonesia Regulation No. 41. 1999. *Air pollution control*. Jakarta.

Putri, N.S., T. Hayasaka, and K.D. Whitehall. 2017. The properties of mesoscale convective systems in Indonesia detected using the Grab 'em Tag 'em Graph 'em (GTG) algorithm. *Journal of the Meteorological Society of Japan* 95: 391–409.

Putri, N.S., H. Iwabuchi, and T. Hayasaka. 2018. Evolution of mesoscale convective system properties as derived from Himawari-8 high-resolution data analyses. *Journal of Meteorological Society of Japan.*

Qian, J.-H. 2008. Why precipitation is mostly concentrated over islands in the maritime continent. *Journal of Atmospheric Science* 65: 1428–1441.

Seinfeld, J.H., and S.N. Pandis. 2006. *Atmospheric chemistry and physics, from air pollution to climate change*, 2nd ed, 1232. New Jersey: Wiley.

Suhadi, D.R., M. Awang, M.N. Hassan, R. Abdullah, and A.H. Muda. 2005. Review of photochemical smog pollution in Jakarta metropolitan Indonesia. *American Journal of Environmental Sciences* 1 (2): 110–118.

Supari, F.T., E. Salimun, E. Aldrian, A. Sopaheluwakan, and L. Juneng. 2018. ENSO modulation of seasonal rainfall and extremes in Indonesia. *Climate Dynamics* 51 (7–8): 2559–2580.https://doi.org/10.1007/s00382-017-4028-8.

Supari, S. A., U.A. Linarka, J. Rizal, R. Satyaningsih, J.X. Chung. 2019. Indonesian climate under 2°C and 4°C global warming: Precipitation extremes. *IOP Conference Series: Earth and Environmental Science* 303. https://doi.org/10.1088/1755-1315/303/1/012048.

Supari, T.F., L. Juneng, F. Cruz, J.X. Chung, S.T. Ngai, E. Salimun, M.S.F. Mohd, J. Santisirisomboon, P. Singhruck, T. PhanVan, T. Ngo-Duc, G. Narisma, E. Aldrian, D. Gunawan, and A. Sopaheluwakan. 2020. Multi-model projections of precipitation extremes in Southeast Asia based on CORDEX-Southeast Asia simulations. *Environmental Research* 184: 109350. https://doi.org/10.1016/j.envres.2020.109350. Epub 2020 Mar 9. PMID: 32179268.

Tangang, F.T., L. Juneng, E. Salimun, P.N. Vinayachandran, Y.K. Seng, C.J.C. Reason, S.K. Behera, and T. Yasunari. 2008. On the roles of the northeast cold surge, the Borneo vortex, the Madden–Julian oscillation, and the indian ocean dipole during the extreme 2006/2007 flood in Southern Peninsular Malaysia. *Geophysical Research Letters* 35: L14S07.

Tank, A.M.G., and G.P. Konnen. 2003. Trends in indices of daily temperature and precipitation extremes in Europe, 1946–99. *Journal of Climate* 16: 3665–3680.

Tsay, S. et al. 2013. From BASE-ASIA toward 7-SEAS: A satellite-surface perspective of boreal spring biomass-burning aerosols and clouds in Southeast Asia. *Atmospheric Environment* 78: 20–34.

Uda, S.K., L. Hein, and D. Atmoko. 2019. Assessing the health impacts of peatland fires: A case study for Central Kalimantan, Indonesia. *Environmental Science and Pollution Research* 26 (30): 31315–31327.

Virts, K.S., and R.A.J. Houze. 2015. Variation of lightning and convective rain fraction in mesoscale convective systems of the MJO. *Journal of Atmospheric Science* 72: 1932–1944.

Waliser, D.A., and R.C.J. Sommerville. 1994. Preferred latitudes of the intertropical convergence zone. *Journal of Atmospheric Science* 15: 1619–1639.

Wasi'ah, N.R., and D. Driejana. 2017. Modeling of tropospheric ozone concentration in urban environment. *IPTEK Journal of Proceedings Series* 3 (6).

Wati, T., S.D.A. Kusumaningtyas, and E. Aldrian. 2019. Study of season onset based on water requirement assessment. *IOP Conference Series: Earth and Environmental Science* 299: 012042.

Watterson, I.G., and M.R. Dix. 2003. Simulated changes due to global warming in daily precipitation means and extremes and their interpretation using the gamma distribution. *Journal of Geophysical Research*: Atmospheres 108 (D13). https://doi.org/10.1029/2002JD002928.

Wehner, M. F. 2004. Predicted twenty-first-century changes in seasonal extreme precipitation events in the parallel climate model. *Journal of Climate* 17 (21): 4281–4290. https://doi.org/10.1175/JCLI3197.1.

Wheeler, M.C., and J. McBride. 2005. Australian–Indonesian monsoon. In: *Intraseasonal variability in the atmosphere-ocean climate system*, Chap. 5, eds. W.K.-M. Lau, and D.E. Waliser, 125–173. Springer-Praxis.

Wyrki, K. 1961. *Physical oceanography of the Southeast Asian waters. Naga report, volume 2, scientific results of marine investigations of the South China Sea and the Gulf of Thailand 1959– 1961.*

World Health Organization. 2006. *Air quality guidelines for particulate matter, ozone, nitrogen dioxide and sulfur dioxide global update 2005: Summary risk assessment.* Germany: World Health Organization.

Yu, X., et al. 2019. Impacts of different types and intensities of El Niño events on winter aerosols over China. *Science of the Total Environment* 655: 766–780.

Yuan, J., and R.A.J. Houze. 2010. Global variability of mesoscale convective system anvil structure from A-train satellite data. *Journal of Climate* 23: 5864–5888.

Zhang, C., and J. Gottschalck. 2002. SST Anomalies of ENSO and the Madden–Julian oscillation in the equatorial Pacific. *Journal of Climate* 15: 2429–2445.

Zolman, J.L., and E.J. Zipser. 2000. A comparison of tropical mesoscale convective systems in El Nino and La Nina. *Journal of Climate* 13: 3314–3326.

Chapter 3
Challenges in Predicting Extreme Weather Events Over the South Asian Region

Someshwar Das

Abstract The South Asian region is prone to extreme weather annually resulting in loss of lives and damage to the properties. While some people have attributed the cause of such events to the global warming and climate change, the fact is that the anomalous weather is a result of the changes in the three-dimensional structures of the atmosphere. While predicting the extreme weather events at precise location, intensity and lead time have been a challenge to the Meteorologists due to the limit of the deterministic forecasts, our observation system should be able to sense the changes in the structure of the atmosphere and the numerical weather prediction models should be able to provide accurate solutions to the future state of the atmosphere with usable skills. Accurate prediction of the extreme weather events requires observations of 3-dimensional structure of atmosphere at good temporal and spatial resolutions, applications of numerical models at cloud-resolving scale, and high-performance computing resources. The observations are required several times a day on routine basis from different sources such as satellites, radars, aircrafts, radiosonde balloons, automatic weather stations, surface meteorological observatories over land, and ocean from ships, and buoys. A review of the challenges faced by the Meteorologists in predicting the extreme weather events over the South Asian region is presented in this article.

Keywords Extreme weather · South Asia · Storm · Prediction · Modeling · NWP

3.1 Introduction

The South Asian region (including Afghanistan, Bangladesh, Bhutan, India, Maldives, Myanmar, Nepal, Pakistan, and Sri Lanka) is frequently affected by different types of extreme weather events, such as severe thunderstorms, dust storms, hailstorms, intense lightnings, cloudbursts, heavy rainfalls (causing floods), Cyclones, strong winds (causing damages to houses, uprooting of trees, snapping of

S. Das (✉)
South Asian Meteorological Association, New Delhi, India
e-mail: somesh03@gmail.com

© The Centre for Science & Technol. of the, Non-aligned and Other Devel. Countries 2022
A. S. Unnikrishnan et al. (eds.), *Extreme Natural Events*,
https://doi.org/10.1007/978-981-19-2511-5_3

electric and telephone lines, etc.), dense fogs (causing poor visibility that affects civil aviation, railways, transport sector and road accidents), scanty rainfall or droughts (affecting agriculture and food productions), heat and cold waves resulting in deaths of several hundred people annually. While some scientists have attributed the cause of such events to the climate change, predicting the extreme weather events at precise location, intensity and lead time have been a challenge to the Meteorologists.

3.1.1   What is Severe?

Severe weather is an anomalous weather event that is rare for the place where it occurs. For example, if the amount of rainfall that occurs in one day at a place "Mawsynram" in Meghalaya occurred in Rajasthan in India, that would be anomalous and an extreme event for Rajasthan. Similarly, if the highest temperature reported at Turbot (53.7 °C in Baluchistan, Pakistan) occurred somewhere in Nepal, that would cause a severe heat wave. The Intergovernmental Panel on Climate Change (IPCC) suggests that "rare" means in the bottom 10% or top 10% of severity for a given event type in a given location. While extreme weather describes unusual weather events that are at the extremes of the historical distribution for a given area, severe weather is generally defined as any aspect of the weather that poses risks to life, property or requires the intervention of authorities.

Also, since severe weather events have always occurred, even before anthropogenic (human-caused) climate change began to be unequivocally present since about 1980, it is impossible to attribute any one extreme event to climate change. Climate change is likely to increase the frequency of extreme weather events, but it will never be possible to point to one such event and say that it was caused by climate change. There is evidence that some weather extremes have already shifted: cold nights have decreased globally, for example, while warm nights have increased (associated with heat waves). Droughts, storm intensity, and heat waves have increased and will continue to do so.

3.1.2   What Are Different Types of Severe Weather Over South Asia?

Different types of severe weather encountered in the South Asian region are thunderstorms (Nor'westers), dust storms, hailstorms, intense lightnings, cloudbursts, heavy rainfalls/flash floods, Cyclones, strong winds, dense fogs, scanty rainfall or droughts, heat, and cold waves. They are briefly described below.

3.1.2.1 Thunderstorms (Nor'westers)

Thunderstorms occur almost everywhere on the earth's surface. It is estimated that at any given time there are about 2000 thunderstorms taking place on the globe (http://www.nssl.noaa.gov/education/svrwx101/thunderstorms/). While they occur both during the summer and the winter seasons, their frequency is highest during the pre-monsoon season over the Indian subcontinent. During the pre-monsoon season, severe thunderstorms generally travel from northwest to southeast direction over the east and northeast parts of India and Bangladesh. Therefore, they are also called the Nor'westers. Three ingredients that must be present for a thunderstorm to occur are moisture, instability, and lifting. Therefore, they also occur during the winter season associated with the Western Disturbances even though the land is cool (Tyagi 2007). Additionally, there is a fourth ingredient (wind shear) for severe thunderstorms. Instability is what allows air in the low levels of the atmosphere to rise into the upper levels of the atmosphere. The instability supports the atmosphere for deep convection and thunderstorms. Instability can be increased through daytime heating. Lifting gives a parcel of air the impetus to rise from the low levels of the atmosphere to the elevation where positive buoyancy is realized. Very often, instability will exist in the middle and upper levels of the troposphere but not in the lower troposphere.

It is lift that allows air in the low levels of the troposphere to overcome low-level convective inhibition. Lift is often referred to as a trigger mechanism. There are many lift mechanisms, some of them are fronts, low-level convergence, low-level warm air advection (WAA), low-level moisture advection, mesoscale convergence boundaries such as outflow and sea breeze boundaries, orographic upslope, frictional convergence, vorticity, and jet streak. All these processes force the air to rise. The region that has the greatest combination of these lift mechanisms is often the location that storms first develop. Moisture and instability must also be considered. A thunderstorm will form first and develop toward the region that has the best combination of: high PBL moisture, low convective inhibition, CAPE, and lifting mechanisms. Thunderstorms often form in clusters with numerous cells in various stages of their life cycle. While each individual cell behaves as a single cell, the prevailing conditions are such that as the first cell matures, it is carried downstream by the upper-level winds and new cell forms upwind of the previous cell. Figure 3.1 illustrates a multi-cell cloud cluster.

3.1.2.2 Dust Storm

The northwest India, Pakistan, and Afghanistan get convective dust storms called locally "Aandhi" during the pre-monsoon season (Joseph et al. 1980). Convective dust storms also occur in the region extending westwards across Pakistan and Arabia to the arid regions of Africa like Sudan, Chad, etc. (Hussain 2005). In this season the lowest atmospheric layers have very high temperatures and relatively low moisture content which makes the thunderstorms to have high bases above the ground of the order of 3–4 km. The ground being dry over long periods, there is loose and fine dust available in plenty. These factors enable severe thunderstorms of northwest India

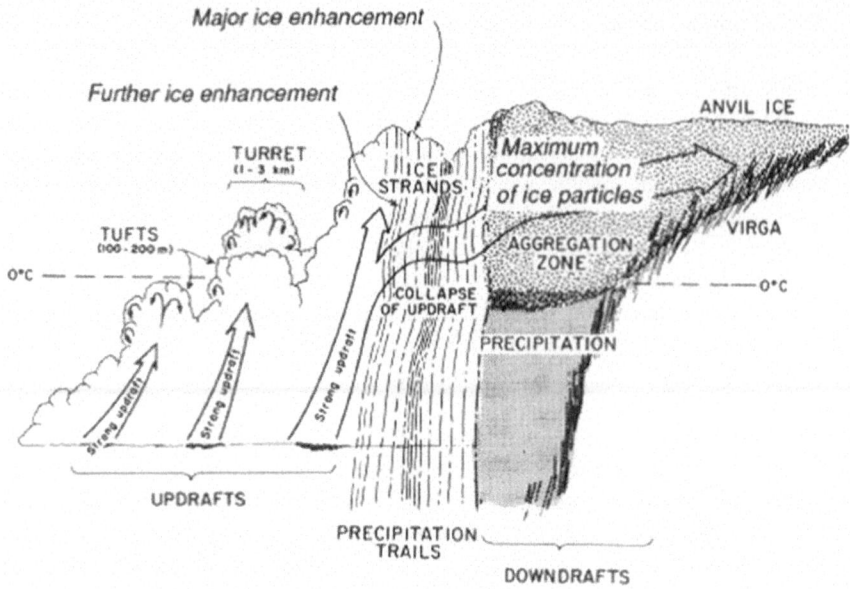

Fig. 3.1 Illustration of a typical multi-cell storm (Houze 1993)

generate dust storms. They are usually brief but can block out the sun, drastically reduce visibility and cause property damage and injuries. Joseph et al. (1980) have done pioneering work on dust storms and the variations in horizontal visibility caused by it, studying 40 cases that occurred at Delhi airport. Studies on the climatology of dust storms and thunderstorms over Pakistan have been carried out by Hussain et al. (2005) and Mir et al. (2006). Their results indicate that extreme eastern and western parts of Northwest Frontier of Pakistan, whole Jammu and Kashmir, and north/north-eastern parts of Punjab share about 65% of the total TS frequency (over Pakistan).

3.1.2.3   Hail Storm

India is among the countries in the world having large frequency of hail. There are about 29 hail days per year of moderate to severe intensity (Nizamuddin 1993). Hail sizes comparable to mangoes, lemons, and tennis balls have been observed. Eliot (1899) found that out of 597 hailstorms in India 153 yielded hailstones of diameter of 3 cm or greater. India and Bangladesh are different from other northern hemisphere tropical stations in that hail is observed in the winter and pre-monsoon seasons with virtually no events after the onset of the southwest monsoon. Chaudhury and Banerjee (1983) show that the percentage of hailstorm days out of thunderstorm days decreases from 5% to less than 2% from March to May for NE India and Bangladesh. Figure 3.2

Fig. 3.2 Hailstorm over Delhi and NCR on 7th February 2019 (1000 UTC-1200UTC). *Source* https://www.livemint.com/news/india/delhi-hailstorm-may-bea-warning-sign-of-climate-change-1549921322894.html

shows an example of severe hailstorm over NOIDA in the National Capital Region (NCR) of Delhi on 7th Feb 2019.

Hail forms only when the height of a thundercloud is very high, water content is large, and updraft velocities large enough to make the size of hail larger. Figure 3.3 illustrates the development of a hail-bearing thunder cloud. Evaporation of shallow clouds results in the formation of cold fronts. The warm moist air rises over the cold fronts thus leading to clouds with great vertical height.

Winter hailstorms are caused due to Western Disturbance (WD). WD is cyclonic circulation associated with subtropical westerly jets in the midlatitude. South westerly jets produce heavy rain in the winter months. WD gets the moisture from Mediterranean Sea and Atlantic Sea. Studies suggest that deep instability in the atmosphere is responsible for formation of hailstones. In winter, the instability is caused due to low-level Available Potential Energy and moisture coming along with upper-level baroclinic instability due to the presence of WD. It lowers the tropopause with increased temperature gradient, thus leading to the formation of winter hailstorms.

3.1.2.4   Intense Lightnings

Each year lightning strikes kill many people and animals. Lightning causes thousands of fires and billions of Rupees in damage to buildings, communication systems, power lines, and electrical systems. Lightning also costs airlines billions of Rupees in flight rerouting and delays. Observations indicate that the number of lightning strikes over the earth per second is about 100 of which 80% are in-cloud flashes and 20% are

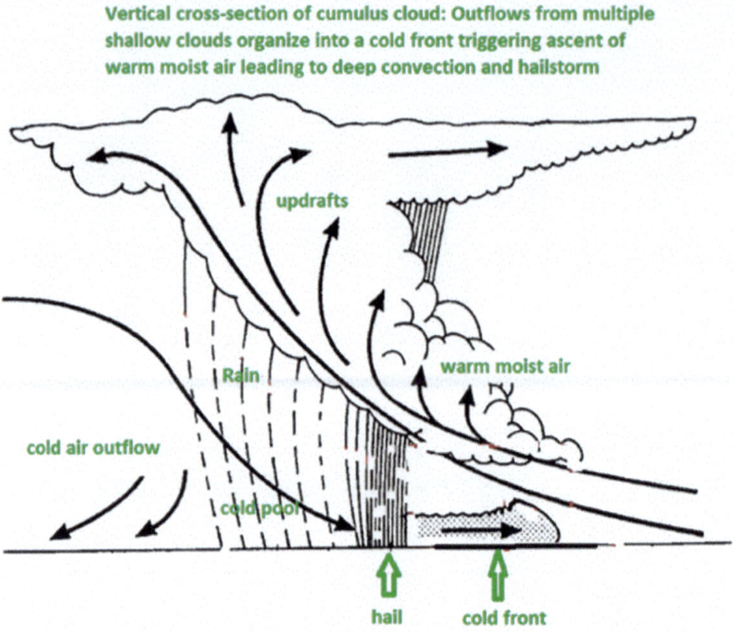

Fig. 3.3 Conceptual description of development of Hailstorm. *Source* Murthy et al. (2017)

cloud-to-ground flashes. This implies that there are about 8,640,000 lightning strikes over the earth per day. Each year, lightning flashes about 1.4 billion times over Earth. The greatest flash density averages only 36 discharges per square kilometer per year. A lightning flash is composed of a series of strokes with an average of about four. The length and duration of each lightning stroke vary but typically average about 30 μs. An average bolt of lightning carries a current of 30 kA, transfers a charge of 5 coulombs, has a potential difference of about 100 MV and dissipates 500 MJ (enough to light a 100-W light bulb for 2 months).

The Lightning Imaging Sensor (LIS) aboard TRMM measures total lightning (intracloud and cloud-to-ground) using an optical staring imager. This sensor identifies lightning activity by detecting changes in the brightness of clouds as they are illuminated by lightning electrical discharges (Christian et al. 1999). Albrecht et al. (2009) constructed climatology maps for the tropical region based on 10 years (1998–2007) of LIS total lightning data. His study showed that more lightning occurs over land than ocean and more lightning occurs near the equator than near the poles. The highest mean flash rate on the earth is 17.43 flash km^{-2} $year^{-1}$ and is located over the Maracaibo Lake in Venezuela (9.625° N, 71.875° W). The maximum flash rate during March–April–May (MAM) is located at Sunamganj, Bangladesh, at the foot of Khasi Hills, Meghalaya, India, before the onset of Indian Monsoon (Albrecht et al. 2009; Das et al. 2010).

3.1.2.5 Cloudbursts

A cloudburst, also known as a rain gush or rain gust is a sudden heavy downpour over a small region, is among the least well known and understood types of mesoscale systems. An unofficial criterion specifies a rate of rainfall equal to or greater than 100 mm per hour featuring high-intensity rainfall over a short period, strong winds, and lightning. A remarkably localized phenomenon affecting an area not exceeding 20–30 km, cloudbursts in India occur when monsoon clouds associated with low-pressure areas travel northward from the Bay of Bengal across the Ganges plains onto the Himalayas and "burst" in heavy downpours (75–100 mm per hour). It represents cumulonimbus convection in conditions of marked moist thermodynamic instability and deep, rapid dynamic lifting by steep orography. Cloudburst events occur at the meso-gamma (2–20 km) scale as defined by Orlanski (1975) and may be difficult to distinguish from thunderstorm. Orographic lifting of moist unstable air releases convective available potential energy (CAPE) necessary for a cloudburst, but the complex interaction between cloud dynamics and orographic dynamics is only beginning to be addressed. Numerical simulation is useful in this regard.

Cloudburst events over remote and unpopulated hilly areas often go unreported. The states of Himachal Pradesh and Uttaranchal in India and Nepal are the most affected due to the steep topography. Most of the damage to property, communication systems, and human causalities result from the flash floods that accompany cloudbursts. Prediction of cloudbursts is challenging and requires high-resolution numerical models and mesoscale observations, high-performance computers, and Doppler weather radar. Societal impact could be markedly reduced if high-resolution measurement (~ 10 km) of atmospheric parameters and vertical profiles are provided through a mesonet observations such as Automatic Weather Station (AWS), Radiosonde/Rowinsonde (RS/RW), and Doppler weather radar. Also, education and training of local administrators to give short-notice warnings would greatly help disaster mitigation.

3.1.2.6 Heavy Rainfall/Flash Flood

Intense rainfall often leads to floods and landslides in the Himalayan region. Heavy rainfall resulting from mesoscale convergence, mesocyclone, cloudburst, and the resulting flash floods are a form of extreme weather event that disrupts human lives during the monsoon season. In recent years many devastating floods occurred due to abnormal rainfall resulting in loss of lives even over the plain regions. Some of them are, for example, the extremely heavy rainfall (993 mm in one day) in Mumbai on 25–26 July 2005 that resulted in cancelation of 700 flights, disrupted rail links, affected an unprecedented 5 million mobile phones, electric poles and ATM machines. Another monsoonal deluge occurred in July 2010 over northern Pakistan that resulted in catastrophic flooding, loss of life and property, and an agricultural crisis. In September 2014 a catastrophic flood occurred in Kashmir Valley that led thousands of people homeless and devastated the agricultural lands. In November–December 2015, a devastating flood affected more than 4 million people, claimed more than 470 lives, and resulted in enormous economic loss in Chennai, India. In August 2018 severe floods affected the south Indian state of Kerala, due to unusually high rainfall during the monsoon season that resulted in the death of over 483 people and evacuation of about a million people. The question is that were these events predictable?

3.1.2.7 Tropical Cyclones

A tropical cyclone is a rotational low-pressure system in tropics when the central pressure falls by 5–6 hPa from the surrounding (Goyal et al. 2016) and maximum sustained wind speed reaches 34 knots (about 62 kmph). It is a vast violent whirl of 150–800 km, spiraling around a center and progressing along the surface of the sea at a rate of 300–500 km a day. These are most devastating phenomenon among all natural disasters. It is accompanied by very strong winds, torrential rains, and storm surges. Out of 80 global annual number—4 form over Bay of Bengal and 2–3 intensify to severe intensity. Nearly 7% of the global TCs form in the North Indian Ocean. About five Tropical Cyclones (TCs) occur in the North Indian Ocean annually including four over the Bay of Bengal and one over the Arabian Sea (Mohapatra and Sharma 2019; Mohapatra et al. 2017; Kotal and Bhattacharya 2013; IMD 2013). Tropical cyclones occur in the North Indian Ocean prominently during the pre-monsoon season (March–April–May) and the post-monsoon season (October–November–December). The Bay of Bengal TCs more often strike Odisha-West Bengal coast in October, Andhra coast in November, and the Tamil Nādu coast in December. Over 60% of the TCs in the Bay of Bengal strike different parts of the east coast of India, 30% strike coasts of Bangladesh and Myanmar and about 10% dissipate over the sea itself (Tyagi et al. 2010). The life period of a TC is about 5–6 days in case of a Very Severe Cyclonic Storm (VSCS) (Kumar et al. 2017). The TC over NIO mostly moves west-northwest wards with wind speed of above 15 kmph. The climatology of heat potential due to TCs has been analyzed by Mohapatra and Kumar (2017). In the past 300 years more than 75% of total cyclones causing death of 5000 or more occurred over North Indian Ocean (Dube et al. 2013). The death toll had exceeded 300,000 and 140,000 in earlier Bangladesh severe cyclones of 1970 and 1991 respectively in which the damages were beyond imagination. The Orissa Super Cyclonic Storm of 29–30 October 1999 had caused over 10,000 human deaths (Mohapatra 2002). Recently there was death toll of 140,000 in Myanmar due to Very Severe Cyclonic Storm (VSCS) Nargis during April–May 2008.

Tropical Cyclones (TCs) are formed when heat is released as moist air rises during evaporation. The resultant condensation of water vapor fuels the formation of cyclonic windstorms. This phenomenon leads to the formation of a warm-core storm system. They are characterized by a low-pressure core that generates thunderstorms.

Observational studies have established that a tropical cyclone is a complex synoptic-scale system (Scale = 1000 km and a few days) of interacting physical and multi-scalar processes in which the genesis of the system is controlled to a large extent by large-scale processes interacting with mesoscale cloud formations, associated deep convection, and air-sea interactions. Organized convection supplies the energy for maintaining and intensification of the system.

The tropical cyclones were devastating a few decades ago causing loss of lives of tens of thousands of people and damage to properties of the order of millions of dollars. In the recent decades, the accuracy of predicting the track, intensity, time, and place of landfall has improved substantially due to the advancement of observation technology, computers, and numerical weather prediction techniques. Uncertainties in the 24–96 h forecast errors have potentially large societal impacts. It is estimated that the cost of evacuation of people is about 1 million dollars per mile in the coastal areas. In view of the high socio-economic importance of the TCs over the Indian region, the Govt. of India (Ministry of Earth Sciences) launched a Forecast demonstration Project (FDP) on landfalling cyclones over the Bay of Bengal in 2008.

3.1.2.8 Windstorms/Strong Winds

Strong winds experienced during severe thunderstorms and cyclones uproot trees, electric poles, telephone lines, and houses, disrupting the lives of the people. A downburst is a strong ground-level wind system that emanates from a point source above and blows radially, that is, in straight lines in all directions from the point of contact at ground level. Microbursts and *macrobursts* are downbursts at very small and larger scales, respectively. Downbursts create vertical wind shear or microbursts, which is dangerous to aviation, especially during landing, due to the wind shear caused by its gust front. Several fatal and historic crashes have occurred over the past several decades due to downbursts. The flight crew training goes to great lengths on how to properly recognize and recover from a microburst/wind shear event. Their lifetime is usually for seconds to minutes.

Wind is stronger when the pressure gradient is high. Strong winds experienced during a thunderstorm are due to the downdrafts beneath the cumulonimbus. Rain from the storm evaporates below the cloud, causing the air to cool beneath it. This cold air is heavy and crashes into the ground below. When it hits the ground, this cold air must turn sideways, and the result is strong winds. These winds are known as microbursts. Wind speeds in microbursts can exceed 160 km h^{-1} and cause significant damage even though they only last for 5–15 min. A typical thunderstorm is made up of a single cumulonimbus (CB) cloud. The CB cloud consists of strong vertical updrafts and downdrafts. Depending upon their intensity (wind speed), they are classified as Gust wind, Squall wind, Light Nor'wester, Moderate Nor'wester, Severe Nor'wester, or a Tornado.

In a typical cyclone, the central pressure can drop to about 900 hPa and the average radius of maximum wind (RMW) is estimated about 47 km (Hsu and Yana 1998). In case of a tornado, the RMW is typically about 46–150 m with an extreme case of about 800 m (USDE 2009; Wurman et al. 2007). The highest peak sustained wind for 1 min is recorded as 345 and 260 km h^{-1} for 10 min (JMA 2017; NHC 2016). Extreme wind speed recorded in a tornado is about 484 ± 32 km/h (F5), the central pressure is estimated about 810 hPa with a pressure drop of 100 hPa (CSWR 2006; Lyons 1997).

3.1.2.9   Dense Fogs

Fog is a major traffic hazard during winter season causing economic loss and casualties. It is estimated that the airlines' loss is above Rs. 18–20 Crores in 7 days period due to the fog. Fog may be defined as a cloud at surface sufficiently thick to reduce visibility to below some threshold. Formation of fog of water particles is a result of a complex interplay of several processes. A description of fog formation must account for the genesis and maintenance of a cloud at the earth's surface. Cloud dynamists often classify fog as a micrometeorological phenomenon because it forms next to the earth in the atmospheric boundary layer, a domain traditionally covered by micrometeorologists. The physical mechanisms responsible for the formation of fog involve three primary processes: (a) Cooling of air to its dew point temperatures (b) Addition of water vapor to the air, and (c) Vertical mixing of moist air parcels having different temperatures.

Fog forecasting using dynamical models is still in a preliminary stage in India. The operational forecasts are generally based on empirical/statistical techniques. Forecasting the location, visibility, time of onset, duration, and dissipation of fog is a challenging issue. As the fog is a small-scale phenomenon, it requires application of very high-resolution model. Recent studies on fog in India have shown significant socio-economic concern due to increase in frequency, persistence, and intensity of fog occurrence over the northern parts of the country. Land use changes and increasing pollution in the region are responsible for growing Fog occurrence.

Considering the importance of Fog in different sectors of human lives, the Govt. of India (Ministry of Earth Sciences) launched a Forecast demonstration Project on Fog in 2008 (FDP-Fog 2008). More recently, the Govt. of India has conducted an intensive field experiment on Fog during the winter months (Dec–Feb) of 2016–2018 called the Winter Fog Experiment (WiFEX). It was a multi-institutional campaign on ground-based measurement at the Indira Gandhi International Airport (IGIA), Delhi, to understand different physical and chemical features of Fog and factors responsible for its genesis, intensity, and duration. The objectives of the WIFEX were to develop better nowcasting (next 6 h) and forecasting of winter fog on various time and spatial scales and help reduce its adverse impact on aviation, transportation and economy, and loss of human life due to accidents.

3.1.2.10   Drought/Flash Drought/Scanty Rainfall

Drought is a phenomenon that depends on the effects of relatively prolonged and abnormal moisture deficiency (Palmer 1965). Drought is controlled by large-scale coupled atmosphere-ocean system directly (or indirectly), which can control precipitation and temperature. For instance, rainfall and temperature anomalies are closely related to the El-Nino Southern Oscillation (ENSO) which is fundamentally driven by changes in the coupled atmosphere-ocean system and leads to various disasters, including droughts and floods (Sigdel et al. 2010; Schubert et al. 2008). The other variables that affect drought include sea surface temperature (SST) and pressure, deforestation, level of CO_2, and levels of other greenhouse gases. Wilhite and Glantz (1985) proposed four categories to measure droughts: (a) meteorological drought due to precipitation shortage (degree of dryness) over a certain period for a specific region, (b) hydrological drought due to the presence of below-average surface and subsurface flow for a longer time duration that accelerates inadequate water supply, (c) agricultural drought due to low soil water availability to support agricultural growth, and (d) socio-economic drought, which defines the imbalances in supply and demand of drought-dependent socio-economic commodities.

The Standardized Precipitation Index (SPI) is mostly adopted in South Asian countries to quantify and monitor droughts. However, the assessment and monitoring of drought using drought indices are more appropriate than the direct use of hydro-meteorological indicators (Chandrasekara et al. 2021). More specifically, indicators are hydro-meteorological variables used to define drought situations such as rainfall and temperature. On the other hand, drought indices are obtained by numerically using hydro-meteorological inputs and the drought indicators. The indices intend to estimate the drought state (i.e., severity, spatio-temporal attributes of drought events) for a certain period. The drought quantification can be conducted using (a) an individual index, (b) multiple indices, and (c) a composite index.

The South Asian countries have experienced frequent drought incidents in recent years. In South Asia, drought occurs frequently in arid and semi-arid regions (Chandrasekara et al. 2021; IWMI 2021). From early 2000 onwards, severe droughts affected vast areas of South Asia, including western India, and southern and central Pakistan. The South Asian regions have been among the perennially drought-prone regions of the world. India, Pakistan, and Sri Lanka have reported droughts at least once in every 3 years in the past five decades, while Bangladesh and Nepal also suffer from frequent droughts. The increased pressure on natural resources, soil degradation, decrease in water resources, and projected future climate change scenarios have become important areas of concern. Proper quantification of drought impacts and monitoring of areas prone to such events are needed for formulating management plans.

Recent studies have identified India among global flash drought hotspots (Chacko Susan; Down To Earth, 10 Nov 2021). Flash drought is the rapid onset or intensification of drought and is set in motion by lower-than-normal rates of precipitation, accompanied by abnormally high temperatures, winds, and radiation. Together, these changes in weather can rapidly alter the local climate. The highest frequency of flash drought occurrence is primarily found within the tropics and subtropics. The risk for flash drought development may continue to increase in certain locations due to increased evaporative demand (PET). Given that flash droughts can develop in only a few weeks, they create impacts on agriculture that are difficult to prepare for and mitigate. Even when environmental conditions seem unfavorable for rapid drought development, a persistent, multi-week lack of rainfall coupled with hot weather can create flash drought development with its associated impacts.

The International Water Management Institute (IWMI) is developing an operational drought monitoring system for South Asia, using remote sensing data from multiple sources. The main monitoring tool is the Integrated Drought Severity Index (IDSI) covering South Asia on a weekly basis, which reflects the effects of droughts as observed through (i) satellite-derived vegetation data, and (ii) the level of dryness expressed by traditional, climate-based drought indices.

3.1.2.11   Heat and Cold Waves

Heat Wave

The WMO has defined a heatwave as five or more consecutive days of prolonged heat in which the daily maximum temperature is higher than the average maximum temperature by 5 °C or more. However, some nations have come up with their own criteria to define a heatwave. Heatwaves form when high pressure aloft (from 3000 to 7600 m) strengthens and remains over a region for several days up to several weeks. The heat waves come in spells of 5–6 days normally and can even go up to 15 days. The Heat Wave Early Warning System and Forecasts are issued based on numerical weather prediction models for different temporal scales in India. Based on these, an outlook is prepared for Mar-May and issued on the last day of February. The central, state, and district agencies review the Heat Wave preparedness coordination based on the seasonal outlook. This outlook is revised at the end of March for April–June. The forecasts for weekly maximum–Minimum temperatures and rainfall are issued once a week out for the next four weeks.

Dehydration, kidney-related diseases, respiratory diseases, heat cramps, and heat stroke are some of the health impacts due to heatwaves. To protect people from heatwave strategies could include keeping the home cooler, staying out of the heat, staying hydrated, and protecting from direct sun. The health impacts of heat are more severe in urban areas, wherein the heat stress-induced deaths in 2100 are estimated to be about 85 per 100,000 globally and above 100 per 100,000 in lower-income groups. The lost productivity from heat stress at work, particularly in developing countries, is expected to be valued at 4.2 trillion USD per year by 2030 (Magotra 2021).

To protect populations from the preventable health impacts of extreme ambient heat a South Asian Heat Health Information Network (SAHHIN) is established, which is a member-driven forum of scientists, practitioners, and policymakers focused on improving capacity to protect populations from the avoidable health risks of extreme heat in a changing climate.

Cold Wave

Cold wave is a localized seasonal phenomenon prevalent in the northern parts of the South Asian region including Afghanistan, Bangladesh, Bhutan, Myanmar, India (except in southern India), and Pakistan. The northern parts of India, especially the hilly regions and the adjoining plain areas are affected by the cold waves (NDMA 2021).

A cold wave occurs in plains when the minimum temperature is 10 °C or below and/or is 4.5° lesser than the season's normal for two consecutive days. Cold wave is also declared when the minimum temperature is less than 4 °C in the plains. Cold waves occur in association with incursion of dry, cold winds from the north into the subcontinent. The northern parts of India, especially the hilly regions and the adjoining plains, are influenced by transient disturbances of the midlatitude westerlies, which often have weak frontal characteristics. A cold wave or frost condition is a rapid fall in temperature within a 24-h period requiring substantially increased protection of agriculture, health, livestock, and other activities.

In India, cold wave has caused 4712 deaths from 2001 to 2019 across various states. IITM data shows an increasing trend of cold waves in the last three decades (1991–2019). With effective planning and interventions, such loss of life could be easily avoided. Especially in the health sector, interventions can be introduced to address cold wave impacts.

3.2 How to Predict the Severe Weather?

Accurate prediction of the extreme weather events requires observations of 3-dimensional structure of atmosphere at good temporal and spatial resolutions, applications of numerical models at cloud-resolving scale, and a high-performance computing system. The observations are required several times a day on routine basis from different sources such as satellites, radars, aircrafts, radiosonde balloons, automatic weather stations, surface meteorological observatories over land, and ocean from ships, and buoys. The observed data are assimilated in a Numerical Weather Prediction (NWP) model at least four times a day in real-time on routine basis. The NWP model consists of complex differential equations representing the physics and dynamics of the atmosphere. The governing equations of the atmosphere are summarized below.

$$\frac{dV}{dt} + fk \times v = -g\nabla_\sigma \cdot Z - \frac{RT}{\sigma P_h + P_t}\sigma\nabla P_h + F \tag{3.1}$$

$$\frac{dT}{dt} = \frac{RT}{c_p(\sigma P_h + P_t)} \cdot \frac{dP}{dt} + \frac{Q}{c_p} \tag{3.2}$$

$$\frac{\partial P_h}{\partial t} = -\int_0^1 \nabla_\sigma (P_h.V)\,d\sigma \tag{3.3}$$

$$gz = gH + R \cdot P_h \int_\sigma^1 \frac{T}{\sigma P_h + P_t} \tag{3.4}$$

$$\dot\sigma = \frac{\sigma}{P_h} \int_0^1 \nabla_\sigma (P_h.V)\,d\sigma - \frac{1}{P_h} \int_0^\sigma \nabla_\sigma (P_h.V)\,d\sigma \tag{3.5}$$

$$\sigma = \frac{(P - P_t)}{P_s - P_t} \tag{3.6}$$

The above equations are written in the terrain following (Sigma) coordinate system defined in (3.6). Equations (3.1, 3.2, 3.3, 3.4, 3.5 and 3.6) are the equations of motion, thermodynamic equation, the continuity equation or pressure tendency equation, and hydrostatic equation. Equation (3.5) is for the vertical velocity derived in Sigma coordinate. The first three equations are used to predict the changes in wind speed, direction, temperature, and surface pressure fields. The geopotential height of an isobaric surface is computed from (3.4). The symbols have their standard meanings.

Some of the severe weather features are diagnosed based on the atmospheric parameters predicted by the above set of equations. For example, the Lightning Potential Index (LPI), is an advanced index for evaluating the potential for lightning activity (Yair et al. 2010). It is strongly correlated with the Convective Available Potential Energy (CAPE) and is calculated based on the dynamics and microphysics of clouds. Figure 3.4 illustrates the correlation between the lightning rate density with the CAPE over a state of India (Maharashtra).

Some of the indices for forecasting the severe thunderstorms and lightning are summarized below.

3.2.1 Lightning Potential Index (J/kg)

Lightning potential index (LPI) is a measure of the potential for charge generation and separation that leads to lightning flashes in convective thunderstorms. The basic formula is as follows (Lynn and Yair 2010; Yair et al. 2010)

$$\text{LPI} = \frac{1}{V} \iiint \epsilon \omega^2 dx\,dy\,dz \tag{3.7}$$

Here, volume of air in the layer between zero and $-20\,°C$ is denoted by V, vertical wind speed (in m s^{-1}) is denoted by ω. ϵ is the function of cloud hydrometeors like mixing ratios of snow, ice, graupel. These are computed by the WRF model (in kg kg^{-1}). It is a dimensionless number whose value lies between 0 and 1 (Lynn and Yair 2010).

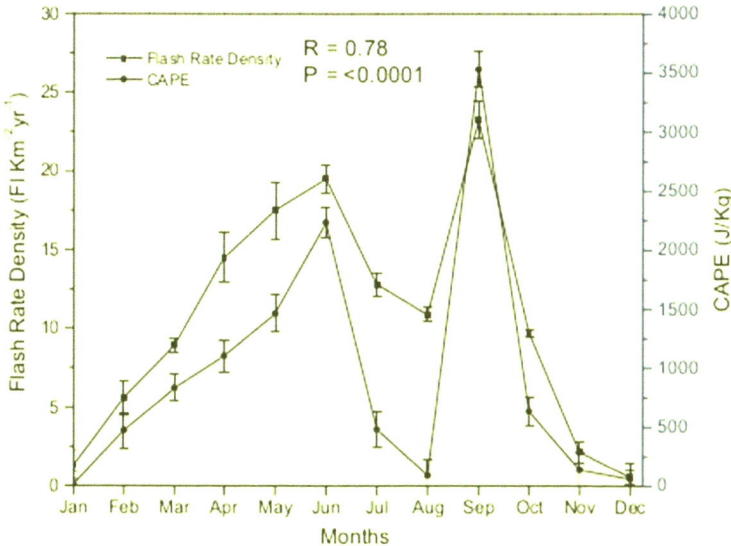

Fig. 3.4 Monthly mean flash rate density and CAPE. Annual variations over the Maharashtra during 10-year period (1998–2007). *Source* Tinmaker et al. (2015)

$$\epsilon = \frac{2(Q_i Q_1)^{0.5}}{Q_i + Q_1} \tag{3.8}$$

Total liquid water mass mixing ratio is Q_1 (kg kg^{-1}). The ice fractional mixing ratio is Q^i (kg kg^{-1}) (Lynn and Yair 2010)

$$Q_i = q_g \left[\frac{(q_s q_g)^{0.5}}{(q_s + q_g)} + \frac{(q_i q_g)^{0.5}}{(q_i + q_g)} \right] \tag{3.9}$$

ϵ is a scaling factor for the cloud updraft. When the mixing ratio of supercooled water and combined ice species are equal then scaling factor reaches to extreme. Calculation of the LPI from the WRF output fields can provide maps of the microphysics-based potential for electrical activity and lightning flashes.

The threat of LPI has been categorized as Low, Moderate, and High.

LOW: LPI < 0.001 and > 0.0005.

Moderate: LPI: < 0.01 and > 0.001.

Threat Level: High: LPI > 0.01.

3.2.2   Lightning Flash Rate Density

In WRF models, lightning predication is depended on electric parameterization scheme. WRF model uses electric scheme defined by McCaul et al. (2009), to generate the lightning flash count. Using this scheme electric field can't be computed. But it explains two different methods for calculating the flash rate, one named graupel flux which is obtained by the multiplication of the upward vertical velocity and graupel mixing ratio at the $- 15°$ C level, even though mass units are absent because the air density is not used in its calculation. The second approach utilizes the relationship between the total storm flash rate and the total volumetric amount of precipitating ice (other hydrometeors). This can be used on a model grid in terms of the vertically integrated ice content in each grid column. Furthermore, a blended version has used that attempts to capitalize on the strength of each approach using weighted averages (0.95 for the first approach; 0.05 for the second approach). Therefore, the total model-estimated lightning flash rate density at one grid point in the model is given as follows (Sandeep et al. 2021).

$$F = k_1 \left(w q_g\right)_m + k_2 \int \rho(qg + qs + qi)dz \qquad (3.10)$$

where $k_1 = 0.042$ m^{-1} is lightning-graupel flux factor (slope of maximum graupel flux and maximum flash density), $k_2 = 0.2$ kg^{-1}m^{2}s^{-1} is lightning-storm ice factor (slope of maximum vertical integrated ice and maximum flash density), w is upward vertical air motion, qg is the graupel mixing ratio, the subscript m attached to the flux implies evaluation at the $- 15$ °C level in the mixed-phase region, q is the local air density, and qg, qs, and qi are the simulated mixing ratios of graupel, snow, and cloud ice, respectively. The integration in Eq. (3.1) is over the full storm depth. Full details about the electric scheme and derivation of constants k_1 and k_2 can be found in McCaul et al. (2009).

3.2.3   SuperCell Composite Parameter (SCP)

It is defined by Carbin et al. (2016) as

$$SCP = (CAPE/1000) * (SRH/50) * (BWD/20),$$

where

CAPE (J/kg)	is the moist unstable convective available potential energy.
SRH (m^2/s^2)	is storm-relative helicity.
BWD	is bulk wind shear between 500 hPa and ground surface.
SCP > 1	implies that the atmosphere is favorable for formation of thunderstorm.

SCP   is characterized in low, moderate, and high threat levels.

LOW: 3 < SCP < 5.
Moderate: 5 < SCP < 7.
High: SCP > 7.

3.2.4   Data Assimilation

All forecast models, whether they represent the state of the weather, the spread of a disease, or levels of economic activity, contain unknown parameters. These parameters may be the model's initial conditions, its boundary conditions, or other tunable parameters which have to be found for a realistic result. Four-dimensional variational data assimilation, or "4d-Var", is a method of estimating this set of parameters by optimizing the fit between the solution of the model and a set of observations that the model is meant to predict (Bannister Ross 2007). In this context, the procedure of adjusting the parameters until the model "best predicts" the observables, is known as optimization. The "four dimensional" nature of 4d-Var reflects the fact that the observation set spans not only three-dimensional space but also a time domain. For weather forecasting, the method of 4d-Var has been adopted by many numerical weather prediction (NWP) agencies as it is flexible enough to allow a range of atmospheric observations of many different types to be digested within a framework of a numerical model of the atmosphere. This has proved to be a valuable tool in estimating the initial conditions of a weather prediction model, which are essential for a good forecast.

There are two types of data assimilations based on the scale of the model: large-scale and convective scale. Although the fundamental data assimilation theory does not depend on the scale of interest, convective-scale data assimilation possesses a number of important differences from the large-scale. First, the main objective of the convective-scale data assimilation for NWP is to improve quantitative precipitation forecasts (QPF) and severe weather systems like thunderstorm, while the major concern for the large-scale is to reduce the error in the prediction of the 500 hPa geopotential height. Second, the major observational data source for the convective scale is from Doppler radar (although other observations are also important) while radiosonde and satellite observations are indispensable for the large-scale. Third, the model constraints are different: for the large-scale, balance constraints, such as geostrophic balance, are good approximations to the full set of atmospheric equations; however, for the convective scale, there are no simple balances and approximations, other than the full set of model equations describing the convective-scale motion. Lastly, synoptic-scale dynamics are quasi-linear (except for subgrid-scale parametrization) for approximately 6 h (Gilmour et al. 2001) while the time scale of a developing thunderstorm can be as short as a few minutes. Because of this number of differences, the implementation of data assimilation methods for the convective scale can be different from the large-scale. The Doppler Weather Radar observations are generally assimilated in rapid update cycle for the convective-scale systems, because of radar's unique capability in sampling the atmospheric convective-scale.

3.3 What Are the Challenges in Observations?

How observations are processed is vital to the provision of monitoring and forecasting services for weather and climate. Assimilation of observational data into NWP models is the established way of exploiting observations for weather forecasting beyond a few hours ahead. Advances in global observation since the 1970s have been accompanied by considerable progress in modeling and data assimilation, resulting in very substantial improvements in the accuracy of forecasts (Simons 2011). Reanalyses of the observations made over past decades with fixed modern assimilation systems provide a widely used record of weather and climate.

The question is how much observations are required to accurately predict a physical phenomenon? The spatial and temporal density of observations required to simulate a phenomenon accurately depends on the spatial and temporal scale of the phenomenon. Figure 3.5 illustrates how there has been a revolutionary improvement of the forecasts of 500 hPa anomaly correlation between observations and the ECMWF forecasts globally due to the increase of observations from satellite during the last three decades.

In the early nineties, forecast skill was considerably lower in the southern hemisphere than in the northern hemisphere. This was because there were fewer weather

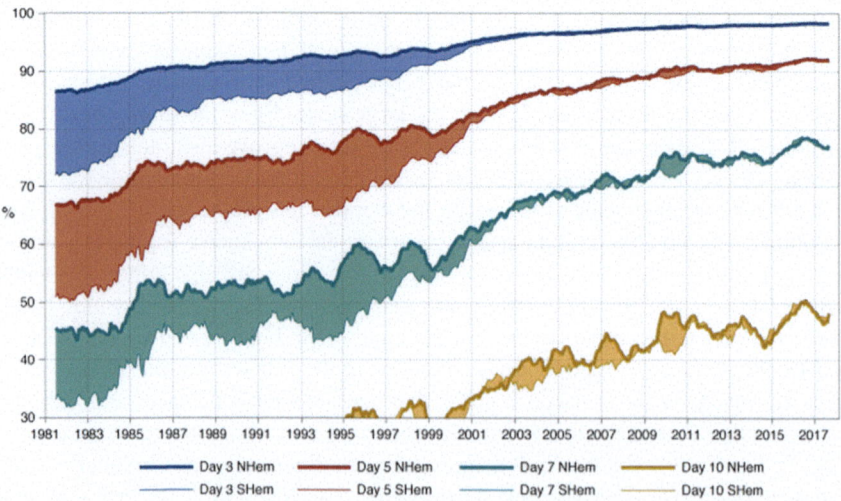

Fig. 3.5 The chart shows the evolution of the 12-month running mean of the anomaly correlation, a measure of skill, for 500 hPa geopotential height forecasts in the northern and the southern hemisphere at various lead times. *Source* ECMWF (2018)

observations in that part of the globe to help initialize forecasts. The gap in skill narrowed when data assimilation at ECMWF moved from 3D-Var to 4D-Var, and again when NOAA-15 satellite was launched. Subsequent satellite launches reduced the gap to virtually zero (ECMWF 2018).

Figures 3.6 and 3.7 illustrates how the simulation of wind fields is improved when more data obtained from ARMEX (Arabian Sea Monsoon Experiment) is assimilated

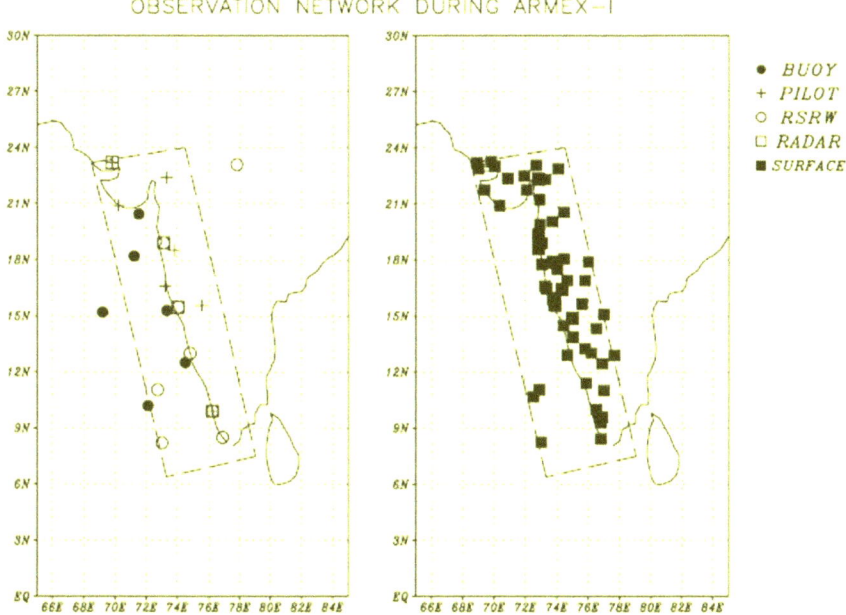

Fig. 3.6 Observations network during the Arabian Sea Monsoon Experiment (ARMEX), June–August 2002. The right panel shows the extra observations from BUOY, PILOT balloons, RSRW, Radar, and Surface observations collected during ARMEX (Das et al. 2007)

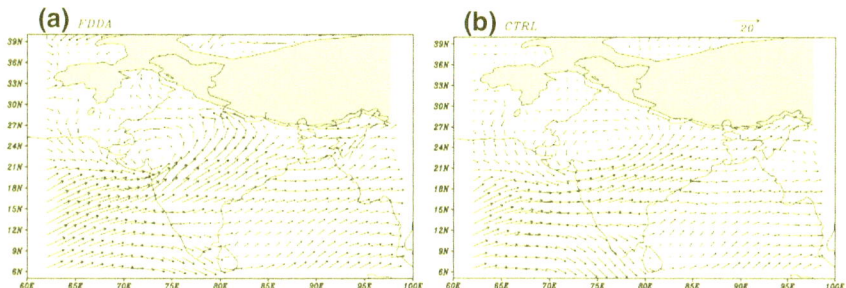

Fig. 3.7 Wind at 850 hPa on 00 UTC 28 June 2002 **a** with FDDA and **b** without FDDA. Terrain heights above 850 hPa are shaded. Reproduced from Das et al. (2007)

into the model during a monsoon depression over the west coast of India (Das et al. 2007).

Extra observations collected from ARMEX (SYNOP, SHIP, BUOY, TEMP, PILOT, AIREP, SATEM, SATOB, ATOVS, SSMI) were assimilated in the MM5 model at 2 km resolution. Figure 3.7 illustrates the 850 hPa wind field analysis obtained by using the FDDA with additional observations collected during the IOP (Intensive Observation Period) and the control analysis without the ARMEX observations. The control analysis contained only the routine surface and upper-air observations including AIREP, SATEM, SATOB, ATOVS, and SSMI. Results clearly show that the location of the vortex is improved in the reanalysis as compared to the control (Fig. 7b). The vortex is also better organized with stronger intensity and wind shear in the southwest sector. The results illustrate that extra observations and higher resolution of the model can improve the skill of the forecasts.

Doppler Weather Radars (DWR) play significant role in forecasting the extreme weather events besides the satellite, upper air, and surface observations. While the satellites cover good parts of the earth (Fig. 3.8), there are significant gaps in the coverage by the DWR and lack of high-density observations over the South Asian region (Figs. 3.10 and 3.11). Owing to this generally, there is a lack of consensus of the monsoon rainfall forecasts by the models (Fig. 3.9).

Figures 3.10 and3.11 shows the present coverage of DWR and surface observatories over the South Asian region. The figures clearly show the lack of observatories, particularly over Afghanistan and the remote areas of the Hindu-Khus Himalayas

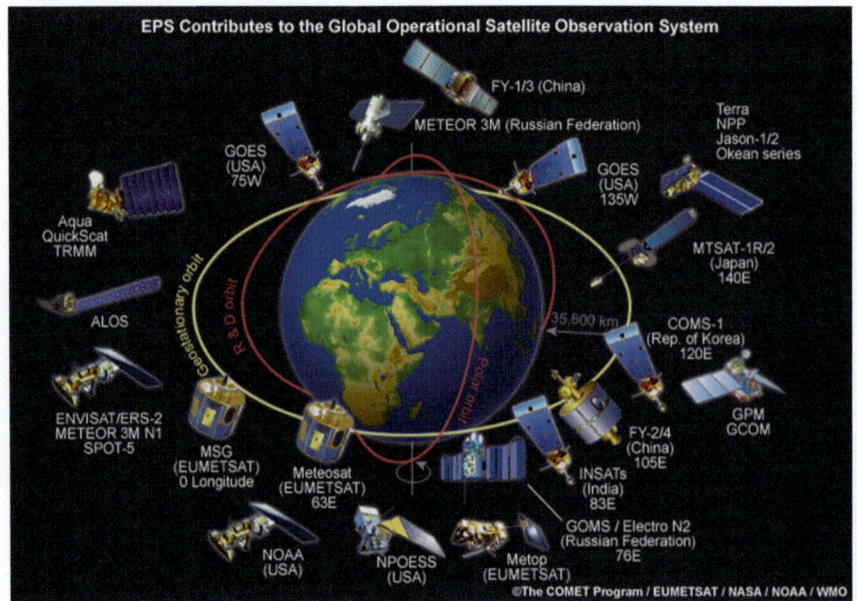

Fig. 3.8 The current global satellite network. *Source* https://www.metlink.org/resource/satellites

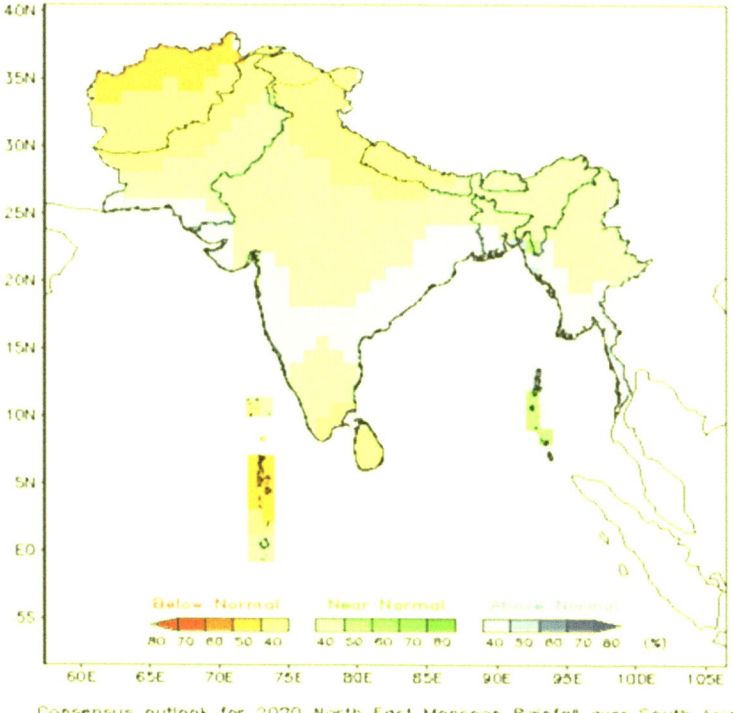

Fig. 3.9 Consensus outlook for 2020 Northeast monsoon rainfall over South Asia

and other mountain regions. It is a big challenge and will take a long time to fill up the data gap regions. Until that happens, we shall keep missing many extreme weather events predicted by the NWP models. While the extreme weather systems are simulated by the present models fairly well, there is still a big challenge in predicting the exact location, time, and intensity of the events at sufficient lead times.

It is another important fact that not all the observations available over the South Asian region are assimilated into the model in real-time. The main reasons for that are not all the observations collected by the stations are available on the GTS (Global Telecommunication System) in real-time and some of those which are available in real-time fail to pass the quality control check by the data assimilation system and thus, get rejected by the model. Since there has been substantial improvement in the availability of global data in real-time during the last one decade, the number of observations assimilated in the model has also increased accordingly. Figure 3.12 illustrates the number of different types of observations assimilated by the model at the National Center for Medium-Range Weather Forecasting (NCMRWF), India in 2013 and 2021.

S. Das

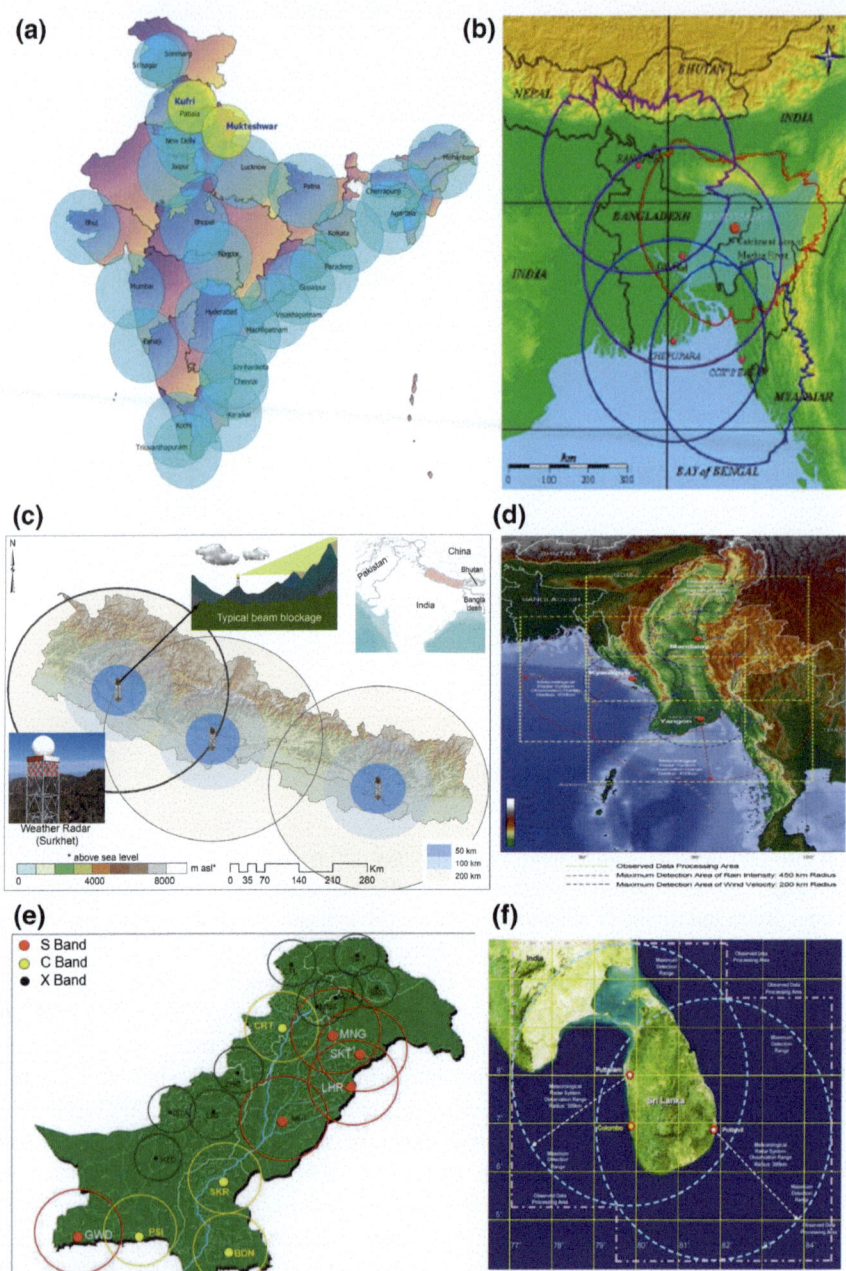

Fig. 3.10 **a**, **b** Doppler weather radar (DWR) networks over India and Bangladesh. **c–i** Doppler weather radar (DWR) networks over Nepal, Myanmar Pakistan, and Sri Lanka (**c–f**). Only the surface network of observatories are shown for Afghanistan, Bhutan, and Maldives as either there is no DWR or they are not functional in those countries

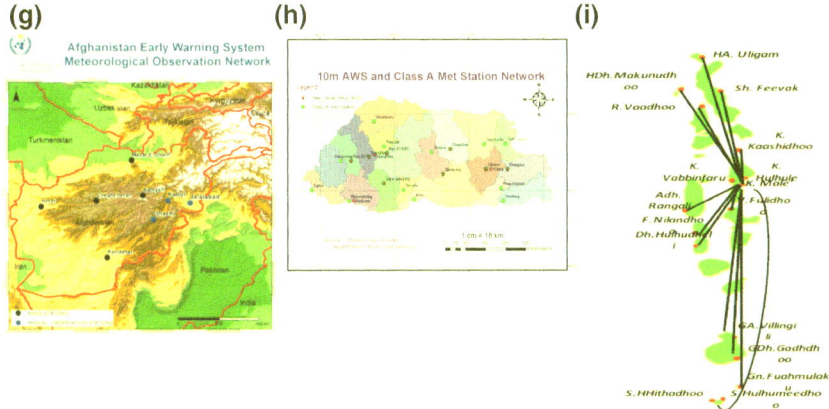

(g)

Afghanistan Early Warning System
Meteorological Observation Network

(h)

10m AWS and Class A Met Station Network

(i)

Fig. 3.10 (continued)

3.4 What Are the Challenges in Modeling?

Understanding, modeling, and predicting weather and climate extremes is identified as a major area necessitating further progress in climate research and has thus been selected as one of the World Climate Research Program (WCRP) Grand Challenges (Zhang et al. 2014; Alexander et al. 2016). In general, the development of an extreme event depends on a favorable initial state, the presence of large-scale drivers, and positive local feedback, as well as stochastic processes (Sillmann et al. 2017a, b). Figure 3.13 illustrates these processes.

The relative importance of these factors varies for different types of extremes. For example, feedbacks for short-lived events (blue) like convective storms are typically associated with unstable atmospheric dynamics, whereas longer duration events (red) like heatwaves or droughts typically involve soil moisture-atmosphere interaction. External factors like global warming can influence extremes through these factors. For example, the increased water vapor in a warmer atmosphere can enhance convective feedback, or increased surface evaporation might amplify heatwaves and droughts.

In a nutshell, weather is a result of the motion of air caused by a balance between various forces. This motion may be described by the equations as discussed in the previous section. Some of these equations are nonlinear partial differential equations. So the question arises if we know the laws and the variables, why can't we make perfect forecasts even with a short lead time? There are many reasons for this. Some of them are (1) Inaccurate Initial Conditions, (2) Multiscale Interactions, (3) Chaos, and Limit on Deterministic Predictability. We discuss these briefly in the following sections, which are mostly based on Goswami (1997).

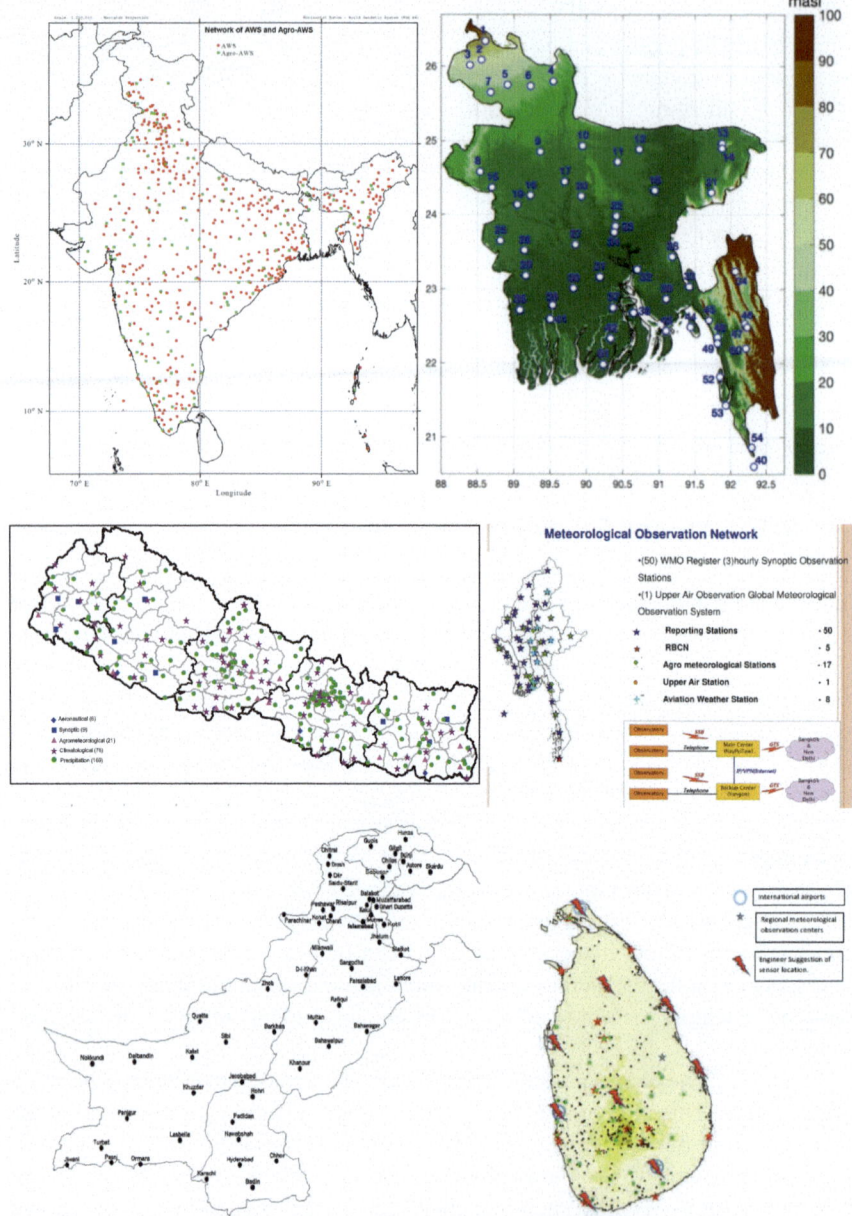

Fig. 3.11 Surface meteorological and automatic weather stations over India, Bangladesh, Nepal, Myanmar, Pakistan, and Sri Lanka

Observation file	Observation details: Global as on 00 UTC, 15 June 2021
Surface	Land SYNOP, Ship SYNOP, Mobile, AWS, BUOY: Tot= 54727
Sonde	TEMP (Land & Ship), PILOT, DROPSONDE, Wind Profilers Total = 3762
Aircraft	AIREP, AMDAR; 97532
Satwind	GOES, Meteosat, MTSat, INSAT-3D, MODIS, MetOp & NOAA =
Scatwind	ASCAT = 606705
GPSRO	(COSMIC, GRAS) = 281617
GOES Radiance	GOES Imager Radiance (Clear)= 352285
ATOVS	MetOp & NOAA satellites (including HRPT data)
IASI	MetOp = 59254
AIRS + HIRS	AQUA: 0 + 316498

Parameter	Data received	Data Assimilated	%
P-surface	29339	25610	87
u, v	408147	292369	71
T	126947	101473	79
q	91958	15367	30
Radiance	2982385	744426	29

ATOVS: Advanced TIROS Operational Vertical Sounder

TIROS: Television Infrared Observation Satellite, 1960

HIRS: High Resolution Infrared Radiation Sounder

AIRS: Atmospheric Infrared Sounder

GRAS: GNSS (Global Navigational Satellite System) Receiver for Atmospheric Sounding.

AMDAR: Aircraft Meteorological Data Relay

Fig. 3.12 Number of different types of observations assimilated by the models at the national center for medium-range weather forecasting (NCMRWF), India in 2013 in the GFS T574L64 model (right) and the NCUM model in June 2021 (left). Only those observations (surface and upper air, i.e., SYNOP, TEMP, SHIP, PILOT, etc.) which are put on the GTS are normally assimilated. Some satellite observations are directly downloaded from the FTP server of the satellite data providing agencies and assimilated

3.4.1 Inaccurate Initial Conditions

To make the forecast for a future time, the initial state of the atmosphere over the whole earth at all heights must be provided as an initial condition. But, as discussed in the previous section, the observations are not sufficient as per the requirements. Satellites are now providing some observations of wind and temperature profiles over the oceans. But there are errors in the retrieval algorithms of wind and temperatures. Therefore, now the Radiance observations are directly assimilated in the model. In addition to instrumental errors in the observations, the large data void regions lead to errors in the specifications of the initial conditions. In principle, the paucity of observations and the inherent instrumental errors in measurement gives rise to errors in the specification of the initial state. As we improve the observing network, this

Processes relevant for simulating and predicting extremes

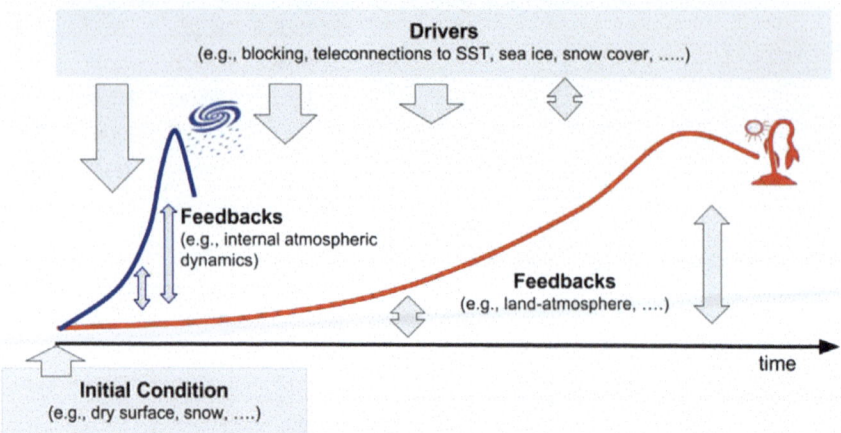

Fig. 3.13 Schematic processes involved in the development of weather extremes (Sillmann et al. 2017a, b)

error may be reduced but we may never be able to totally eliminate errors in the initial conditions!

In addition to errors in the initial conditions, there are some errors in the formulation of the equations themselves. Although the formulation of the adiabatic forces such as the pressure gradient, Coriolis and gravitation are well known, the formulation of the frictional forces and heating requires approximations leading to certain number of inherent errors. The frictional forces involve the turbulent eddies. As these eddies have very small scales, it is formidable to resolve them in a weather prediction model. Therefore, we have to develop a model of how the small-scale eddies influence larger-scale circulation. This process is often known as *parametrisation*. This involves certain approximations leading to errors in the formulation. Similarly, heating by radiation depends in a complex way on moisture distribution, temperature and cloudiness. Approximations are used to represent these effects. Also, the equations are so complex that exact solutions are impossible. So again, approximations are made which produce errors.

Even though many important problems of the atmosphere have been successfully addressed using the governing equations, exact solutions of these equations may still not give us the complete state of the atmosphere as there are other variables not described by these equations. For example, the amount of ozone or aerosols (e.g., dust) is not accounted mostly. These and other variables also affect the state of the atmosphere. While the radiative effects of ozone and aerosols may not be crucial for short-range weather forecasting, they are quite important in determining the mean state of the atmosphere. If we want to understand the behavior of ozone, we must add one equation for ozone concentration and another equation for ozone changes. Ozone can be affected by the concentration of other gases. So we need to introduce more equations. This illustrates that we have to make approximations at some stages and that a perfect model formulation for atmospheric motion is almost impossible. These small but unavoidable imperfections of the equations add another source of errors in the prediction of weather.

The presence of water vapor in the atmosphere makes the weather prediction even more difficult. When water condenses (freezes) it releases 597 (80) calories of heat per gram. This heat transferred to the air represents an important source of energy. Thus thunderstorms, tornadoes, and tropical cyclones all depend on the release of latent heat. Wherever precipitation takes place, condensation of water vapor and release of latent heat occurs. The tropical region receiving tremendous amount of rainfall is a major source of heat for the atmospheric heat engine. Therefore, to be able to predict the weather correctly, we should be able to predict when and where precipitation occurs. This turns out to be the most difficult problem in meteorology. This is partly because precipitation usually occurs from individual clouds which have a typical horizontal size of about 1–10 kms. However, they can form only when the large-scale environment is conducive. For example, clouds cannot form in the regions of large-scale subsidence. Ascending motion in the equatorial region produces subsidence over subtropics inhibiting cloud formation. In other words, again there is interaction between small- and large-scale processes. Moreover, condensation of water vapor into water droplets that fall as rain involves numerous microphysical processes such as condensation nuclei, coagulation of small droplets into bigger droplets, and others. In a large-scale model, explicit calculation of these processes is impossible. Again, we make certain approximations leading to errors in our model formulations. In fact, the interaction between the small-scale cumulus cloud and the large-scale environment is not yet fully understood.

3.4.2 Multiscale Interactions

The atmosphere is a giant laboratory in which different phenomena with a wide range of time and space scales coexist. There are phenomena ranging from turbulent eddies with a horizontal scale of a few meters and time scale of few seconds to large weather disturbances (e.g., depressions and tropical cyclones) with a horizontal scale of about 1000 km and time scale of about a week. In addition, there are larger-scale phenomena such as the meridionally narrow cloud bands extending thousands of kilometers in the east-west direction often seen in cloud pictures, known as the *intertropical convergence zone* (ITCZ), with horizontal scale of about 10,000 km and time scale of weeks to months.

However, if we examine the kinetic energy in different scales of motion in the atmosphere, we notice that except for the annual cycle due to external solar forcing, the large-scale weather disturbances with time scales of a few days are the most energetic. The primary aim of weather prediction is therefore to predict the large-scale weather disturbances (called synoptic disturbances) correctly. However, the biggest hurdle is that these weather disturbances owe their very existence to smaller-scale processes. For example, the tropical cyclone owes its existence and strength to the condensation of a large amount of water vapor. The water vapor is produced by evaporation and sucked into the cyclone by frictional convergence due to small-scale turbulent eddies. As the moisture goes up, it forms a large number of cumulonimbus clouds, individually having a horizontal scale of about 10 km. These individual clouds organize themselves into spiral bands having a horizontal scale of several hundred kilometers. Although the small-scale processes are not energetic themselves, without them the large-scale systems cannot be sustained. Therefore, there is continuous interaction going on between the small and large scales. Existence of multiple scales from turbulent eddies ($L \sim 1$ Mt) to tropical cyclones ($L \sim 1000$ km) in the atmosphere and the fact that one scale of motion depends on the other makes it hard to models.

However, it is almost impossible to model all the scales of motion together. There are two major problems. First, the physical laws governing the evolution of some of these small-scale turbulent processes are not well known. Therefore, even if we wanted to model them in detail, we will have to make certain approximations. The second problem is technical in nature. If we want to resolve all the scales of motion, up to let us say 2 m, we have to solve the same equations over the entire globe with a grid spacing of at least 1 m. This means there will be about 5×10^{15} points over the entire globe. Then we have to consider at least 20 vertical levels. Thus, the governing equations will have to be solved in about 10^{17} grid points in every time step! This is a formidable task even for the fastest supercomputer in the foreseeable future!

Thus, the large-scale models of the atmosphere cannot resolve the small-scale eddies. But their effect on the larger scales must be taken into account in some way. As mentioned earlier this is the problem *of parametrisation* of the sub-grid scale processes. To be successful, we must understand clearly how the small scales influence the large ones. There are some major lacunae in our understanding of this field. Over the last three decades, great strides have been made in parametrisation of rain formation and its effect on the large-scale environment and formulations of evaporation and frictional forces. However, there is a lot more to be done in this area.

3.4.3 Chaos and Limit on Deterministic Predictability

Suppose that the uncertainties in the formulation of the governing equations were not there and that the model of the atmosphere represented by the governing equations was perfect, could we then predict the atmosphere indefinitely in advance? The fact is that even if the equations governing the atmospheric motions were known exactly, due to the intrinsic nonlinearity of the system, weather prediction would be limited to about two weeks in advance. E. N. Lorenz of Massachusetts Institute of Technology (Lorenz 1969) showed that even if the equations are perfect, infinitesimal unavoidable errors in the initial conditions can make a forecast differ significantly from the observations within a period of about two weeks. Such a divergence of a forecast from observation is characteristic of all *nonlinear* systems and is known as *deterministic chaos*. Lorenz's original work has opened up a whole new field of research on deterministic chaos. Thus, even if the model was perfect, intrinsic nonlinearity of the atmosphere would restrict our ability to predict the weather to about two weeks. This limit of deterministic predictability is different in different nonlinear systems. It depends on the instabilities present in the system and the nature of the nonlinearity of the system.

While these intrinsic problems may never allow the Meteorologists to make perfect forecasts, as we shall see in the next sections that, tremendous progress has been made in weather forecasting over the past four decades.

3.4.4   Ensemble Forecasting System

Since the atmosphere behaves like a chaotic system, a little change in the initial conditions of the model can lead to large differences in the forecasts. Therefore, the challenge is to get accurate initial conditions that are as close to the real atmosphere as possible. The initial conditions are obtained by assimilating observations in the model in real-time. However, how good are those initial conditions depend on the number of observations assimilated/rejected by the model, the type of assimilation system (3D-var, 4D-var, EnKF, etc.), model resolution, update cycle (6, 3, 1 hourly), etc. Since the accuracy of the initial conditions remains uncertain, the problem is mitigated through ensemble forecasting. An ensemble prediction system usually includes a control forecast and a good number of perturbed forecasts. Initial conditions for other ensemble members are generated by adding perturbations (or errors) to the analysis. During the early stage of the forecast, error grows more or less linearly with time and the deterministic forecast shows good skill. Beyond this range of linear error growth, deterministic forecast loses its skill, but ensemble mean (or average) can be treated as a single forecast representing the best available estimate of the future atmosphere. Spread in the forecast is a measure of disagreement between the ensemble members. A good agreement among the members results in less spread and a good reason to become confident about the forecast. The third important aspect of ensemble prediction is that it provides a quantitative basis for probabilistic forecasting.

Ensemble forecasting methods vary in different operational centers around the world mostly by the way in which initial condition perturbations are generated. The simplest way to generate perturbations is to add random (Monte Carlo) noise to the original analysis. By construction, perturbations generated by Monte Carlo method do not include the "growing errors of the day". A second class of methods that take care of growing errors in the initial perturbations were developed, tested, and implemented at operational centers around the world. "Breeding" and "singular vector" methods of perturbation generation lie in this class. Breeding vectors (BVs) are used to generate perturbations to the initial condition at NCEP and the singular vector (SV) approach is used at ECMWF. In Met Office, UK, Ensemble Transform Kalman Filter (ETKF) is used in its Global and Regional Ensemble Prediction System

(MOGREPS) to generate initial perturbations. This method is similar to the error breeding method.

The NCMRWF global ensemble prediction system (NEPS-G) has a horizontal resolution of 12 km and the model has 70 vertical levels reaching up to the height of 80 km. A total of 23 ensemble members (22 +1 members lagged ensemble; 11 members from current day 00 UTC and control +11 members from previous day 12 UTC) constitute this ensemble system. The perturbations for all the ensemble members are generated by ETKF system four times a day (at 00, 06, 12 and 18 UTC) from the previous 6 h short forecast of the evolved perturbations for the variables u, v, θ, q and exner pressure on all levels. These analysis perturbations are added to the reconfigured analysis from the four-dimensional variational data assimilation system (4D-VAR) of Unified Model. A 10-days forecast of NEPS-G is routinely generated based on 00 UTC initial conditions which include a control forecast with the 4D-VAR analysis and 22 ensemble member forecasts with 22 perturbed initial conditions.

3.4.5 Horizontal and Vertical Resolution

The microphysical and dynamical processes involved in most of the extreme weather events operate at very small horizontal and time scales. The scale of eddies (dust devils) is about a few meters, while the clouds in thunderstorms, hailstorms, and cloudbursts range from one to tens of km in clustered form. Their vertical extent may also go from middle to high atmospheric levels 10–15 km. To resolve them properly the horizontal resolutions of the model must be about 1 or 2 km while vertically there must be 30–50 levels. This demands millions of calculations per second while solving the nonlinear partial differential equations, and thus we must have a Supercomputer to be able to forecast the extreme events accurately on time.

3.4.6 High-Performance Computing

As discussed in the previous sections, we need to have a High-Performance Computing (HPC) system to be able to forecast the extreme weather events accurately on time. The NCMRWF, IMD, and IITM under the Ministry of Earth Sciences, Govt. of India today have together up to about 8 Peta Flops computing systems. They are used to run complex NWP models on routine basis operationally. Yet, we know that there are errors in the forecasts of the location, intensity, and time of occurrences of the extreme weather events. It implies that we must have dense network of observations, high-resolution models (~ 1 km horizontally), rapid update cycle (1 hourly) for data assimilation and a good ensemble prediction system (more than 50 members) to be able to get better forecasts of the events. The models must be run many times a day with latest observations. Problems are still complicated over the mountains due to steep topography. Thus, accurate forecasting of the extreme weather events requires very High-Performance Computing resources (Supercomputers).

3.4.7 AI and ML in Extreme Weather Forecasting

Some efforts are made recently to apply AI (Artificial Intelligence) and ML (Machine Learning) techniques in Extreme Weather Forecasting. Numerical modeling of weather and climate has dominated meteorological forecasting, to the extent that only 10 years ago prediction using statistical empirical models seemed rather antiquated. With the recent advances in deep neural networks and other related machine learning techniques, statistical empirical prediction is firmly back in vogue (Chantry et al. 2021). This raises a number of questions. Will the methods of artificial intelligence replace numerical models? Or can they be used to supplement or enhance numerical models?

We know that the weather forecasting is done at different time scales (Nowcasting, Short-range, Medium-range, Sub-seasonal, Seasonal, and Climate). Nowcasting applications appear to be a good starting point for Hard AI (i.e., replacing all predictions by AI). On these timescales, physical constraints, such as conservation laws, can be ignored as errors will not accumulate significantly over a couple of hours. Furthermore, the wealth of Internet data, for example, in the form of mobile phone data, is increasingly available and usable for weather predictions. Assimilating this data using conventional methods will be very difficult due to the high number of measurements and large errors. In promising recent work, machine learning methods are becoming competitive for short timescales. These approaches combine the data assimilation and forecasting problems, directly using observations (e.g., satellites) as inputs to the prediction system.

For short and medium-range predictions, an approach using Hard AI becomes more challenging. Skillful forecasts on these timescales would implicitly require the representation of the equations of motion and the interactions between features of the system (such as clusters of deep convection, gravity waves, and midlatitude jets). It is in principle possible to learn the equations of motion from data using machine learning methods, however, challenges with this approach include satisfying conservation principles and stability of the simulations. Forecast skill is found to decrease after a few days, such that it is currently difficult to envisage Hard AI outperforming and hence potentially replacing numerical models on medium-range timescales. However, for short-range, this approach can produce skillful forecasts indicating potential for Hard AI to compete with numerical models.

One of the biggest challenges for Hard AI on short and medium-range timescales is the lack of high-quality data that is available for training. For seasonal predictions, machine learning methods are more likely to beat conventional prediction systems. Here, the systematic errors of models are sufficiently large that AI may soon be competitive. However, the same challenges of training datasets size still exist for the seasonal prediction problem. Adopting the Hard AI approach on the climate change timescale is particularly difficult. The problem is inherently one of extrapolation, predicting climates not yet observed. It is a major challenge for machine learning techniques. One key question for climate change is whether cloud feedbacks are positive or negative. There is no indication from existing data that global cloud cover is increasing or decreasing. Hence a single black-box AI scheme would struggle in giving us meaningful projections of the future.

Several infrastructure requirements to support machine learning applications within weather and climate modeling have been identified (Chantry et al. 2021). These include,

- To design standard methods that enable researchers to easily link Python and Fortran programs as most machine learning methods are trained and used within Python code (based on machine learning libraries such as TensorFlow or Keras) while weather and climate models are typically based on Fortran code.
- To develop benchmark training datasets to allow for a qualitative comparison between machine learning approaches
- To make better use of heterogeneous hardware when running weather and climate models. Most current weather forecasting models run on CPU hardware only, whereas machine learning solutions are typically faster on GPU hardware. This may generate tension for weather and climate computing centers with their own supercomputing facilities.

It is still unclear how the workflow of weather and climate modeling will change in the coming 5–10 years. However, it is very likely that machine learning will be used in many components within the workflow. The new research work will mostly focus on Medium AI (using tools that have learned from observations or via the emulation of existing kernels of weather forecasting models), or Soft AI (to allow for improvements in computing efficiency), looking to improve parameterization kernels of weather and climate models or optimize existing kernels. There are several works concerning the problem of data assimilation, which is an important and expensive part of weather prediction. Contributions also take about probabilistic forecasting, a key approach to forecasting on all timescales, as well as the exciting topic of machine learning for post-processing.

3.5   What is the Current Status of Severe Weather Forecasting?

Despite many intrinsic problems and limitations of predictability, Meteorologists have continued to improve the skills of the weather forecasts and save millions of lives annually. In this section, we shall illustrate the forecasting skills of some of the extreme weather events that occurred over the South Asian region in recent years, i.e., some cases of severe thunderstorms, lightning, cloudburst, tornado, cyclone, and heavy rainfall causing catastrophic floods.

3.5.1 Numerical Prediction of Severe Thunderstorms/Nor'westers

Thunderstorms are mesoscale phenomena. They are also referred to as Nor'westers over the Eastern, North-eastern parts of India and Bangladesh. Mesoscale weather systems refer to those which are smaller than synoptic-scale (about 1000 km) and larger than the cumulus scale (~ 1 km). Mesoscale models are designed for the simulation and prediction of mesoscale weather systems. Such models remain important for an operational numerical weather prediction center, because they can be run at very high resolution on a nested grid with a wide variety of options for the parameterization of physical processes. The global models do not have such privileges and, they are very expensive to run at high resolutions. Moreover, at finer resolution, the mesoscale models are also capable of assimilating large number of high-resolution observations available from present-day satellites and Doppler radars. The mesoscale models such as the Weather Research and Forecasting (WRF) model can be configured to run from global to cloud-resolving scale for simulation of thunderstorms and cloud cluster properties. Since the early 1990s, several important changes took place in mesoscale modeling. Presently, mesoscale models have been developed with a wide variety of flexibilities in terms of changing horizontal and vertical resolutions, nesting domains, and choosing options for different physical parameterization schemes, i.e., MM5, WRF, RAMS, ETA, ARPS, HIRLAM, etc. (Anthes 1990; Dudhia 1993; Cotton et al. 1994; Mesinger 1996; Toth 2001; Case et al. 2002). These models require initial and boundary conditions from a large-scale/global model and may be used for forecasting up to 72 h. Such models can be run at cloud-resolving scale for simulation of thunderstorms and cloud cluster properties.

Das et al. (2015a) studied several cases of Nor'westers that formed over northeast India and adjoining Bangladesh region during the pre-monsoon season employing observations from ground-based radar, Tropical Rainfall Measuring Mission (TRMM) satellite, and synoptic stations. Subsequently, they tried to simulate the storms using the WRF model. Figure 3.14 depicts the precipitation rates retrieved from the Dhaka radar of Bangladesh Meteorological Department (BMD) for different thunderstorm days. Following the classification of Parker and Johnson (2000), it is found that almost all the Nor'westers as well as the squall lines of East/Northeast India and Bangladesh belong to the trailing and parallel stratiform (TS/PS) categories. TS type squall-line is more common in this region (Dalal et al. 2012). Precipitations were simulated by the model (Fig. 3.15) for all the observed cases presented in Fig. 3.14. The model shifted the areas of precipitations both in time and location. But the intensities of the precipitation rates are simulated very well. The model simulated the storms generally 3–4 h ahead of the observations. Several sensitivity experiments were conducted by Das et al. (2015a) with different combinations of cumulus parameterizations, cloud microphysics, and planetary boundary layer schemes to examine the root mean square errors (RMSE) of forecasts. The NOAH scheme was used for land surface processes in all the experiments. They also analyzed the skill scores of forecasts based on the Taylor and Box-Whisker diagrams.

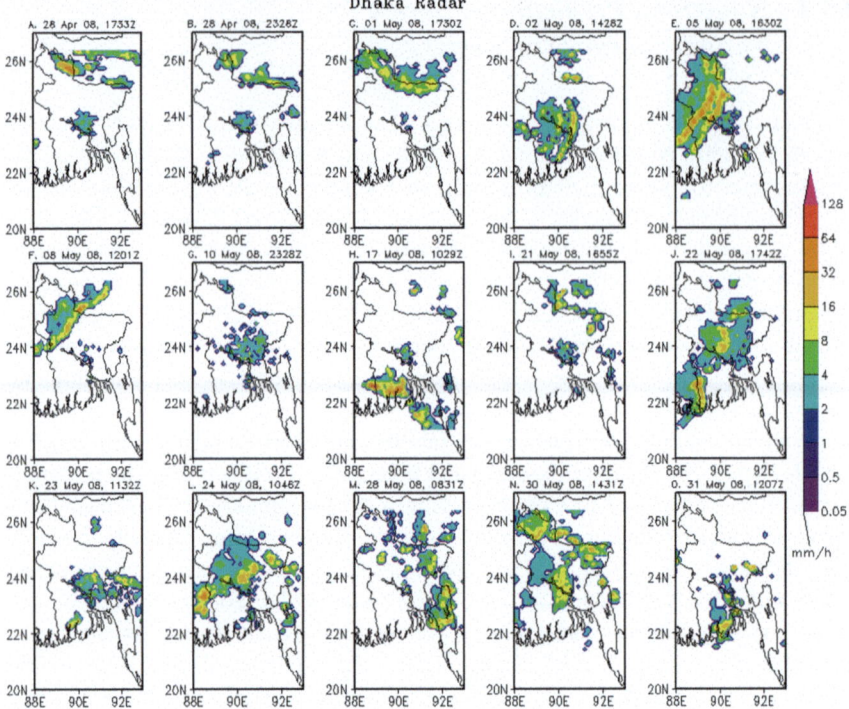

Fig. 3.14 Precipitation rate (mm h^{-1}) retrieved from the Dhaka radar on the days of Nor'westers in 2008 (Reproduced from Das et al. 2015a)

Their results showed highest correlation coefficient between model and observations, lowest RMSE, and reasonable standard deviation using the combinations of no-cumulus (explicitly resolved convection), Milbrandt (Cloud Microphysics), and YSU (Planetary Boundary Layer) schemes.

3.5.2 Lightning Threat Forecasts

Lightning is a characteristic of severe weather and is often associated with convective processes that lead to hail and heavy rainfall. Hence, the spatial pattern of lightning occurrence is of substantial interest for emergency services, insurance wildfire management, and the energy sector. Lightning mainly occurs in the cumulonimbus clouds. Lightning is reckonable as posing a significant threat to human life, as per the annual lightning report of IMD 2020–2021. Lightning strikes over India have increased by 34, and 168% rise in Bihar. The total death toll reported in Bihar is 401 during 2020–2021, which is the maximum among all the states. Lightning is generally observed in the second half of the day. However, during extreme cloud

WRF Model Rainfall

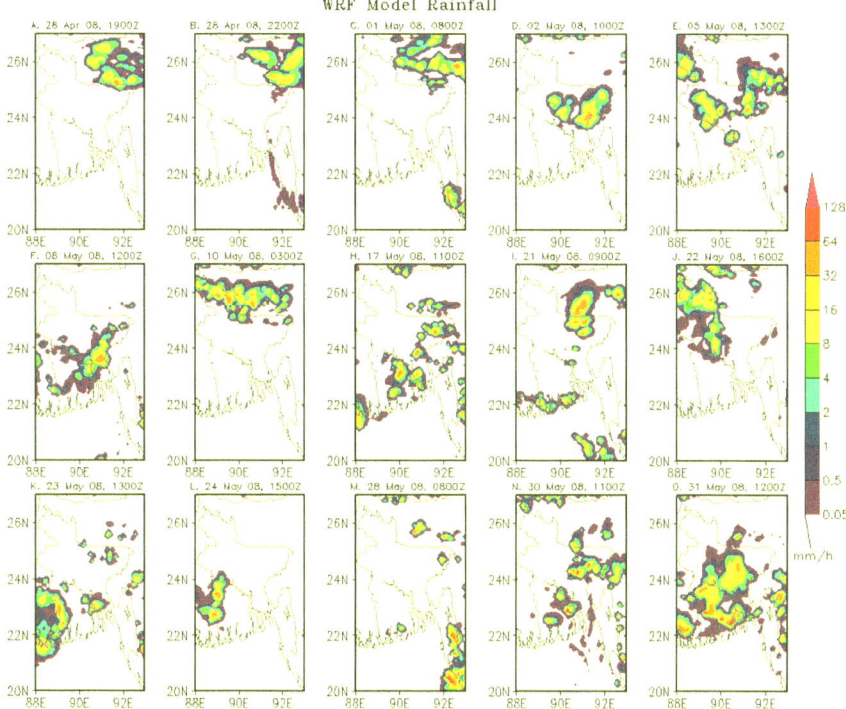

Fig. 3.15 Precipitation rate (mm h^{-1}) simulated by the WRF model on the days of the Nor'westers in 2008 (Reproduced from Das et al. 2015a)

movement with high wind speed, particularly during pre-monsoon (March-May) and post-monsoon (October–November) lightning may occur at any time. It also depends on local geographical factors such as topography, river basins, vegetation cover, and coastal regions. Upper-air thermodynamic instability and cloud microphysical parameters are good indicators for predicting lightning (Mccaul et al. 2009).

Lightning observations are available from the Lightning Imaging Sensors (LIS) on board some of the satellites like TRMM (Tropical Rainfall Measuring Mission), GOES (Geostationary Observation Environmental Satellite), ISS (International Space Station), and ground-based lightning networks. Not many countries of South Asia have good lightning detection networks over the ground. In India, ground-based lightning network is maintained by IITM and NCESS under the Ministry of Earth Sciences. IMD provides map of real-time lightning strikes over the South Asian region up to last 30 min superimposed on the INSAT3D brightness cloud IR imagery (see Fig. 3.16).

Intense lightnings occurred at many places in Rajasthan and UP-Bihar region on 12 July 2021, which killed 11 people and injured many on the Amer fort, Jaipur. IITM, NCMRWF, and IMD provides 3 hourly accumulated lightning flash counts on real-time experimental basis predicted by their NWP models (Fig. 3.17). IITM also

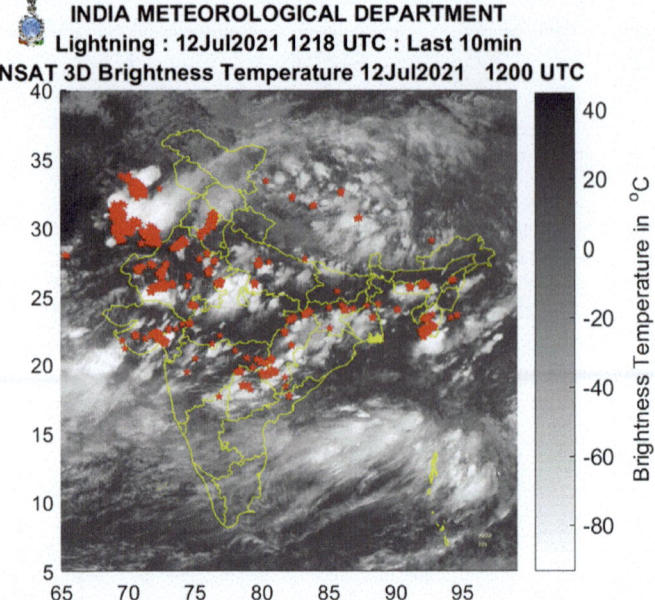

Fig. 3.16 Lightning strikes over the South Asian region superimposed on the INSAT3D brightness temperature at 12 UTC on 12 July 2021. *Source* IMD

provides 3 hourly lightning threats based on Lightning potential index predicted by the models. These products can be very useful for providing lightning warnings to the public in real-time.

3.5.3 *Cloudburst Forecasts*

Cloudbursts occur in India when monsoon clouds associated with low-pressure areas travel northward from the Bay of Bengal across the Ganges plains onto the Himalayas and "burst" in heavy downpours (75–100 mm per hour). It represents cumulonimbus convection in conditions of marked moist thermodynamic instability and deep, rapid dynamic lifting by steep orography. The associated convective cloud can extend up to a height of 15 km above the ground. Cloudburst events over remote and unpopulated hilly areas often go unreported. The states of Himachal Pradesh and Uttaranchal are the most affected due to the steep topography. Most of the damage to property, communication systems, and human causalities result from the flash floods that accompany cloudbursts. Prediction of cloudbursts is challenging and requires high-resolution numerical models and mesoscale observations, high-performance computers, and Doppler weather radar. Cloudbursts over the Himalayan region have

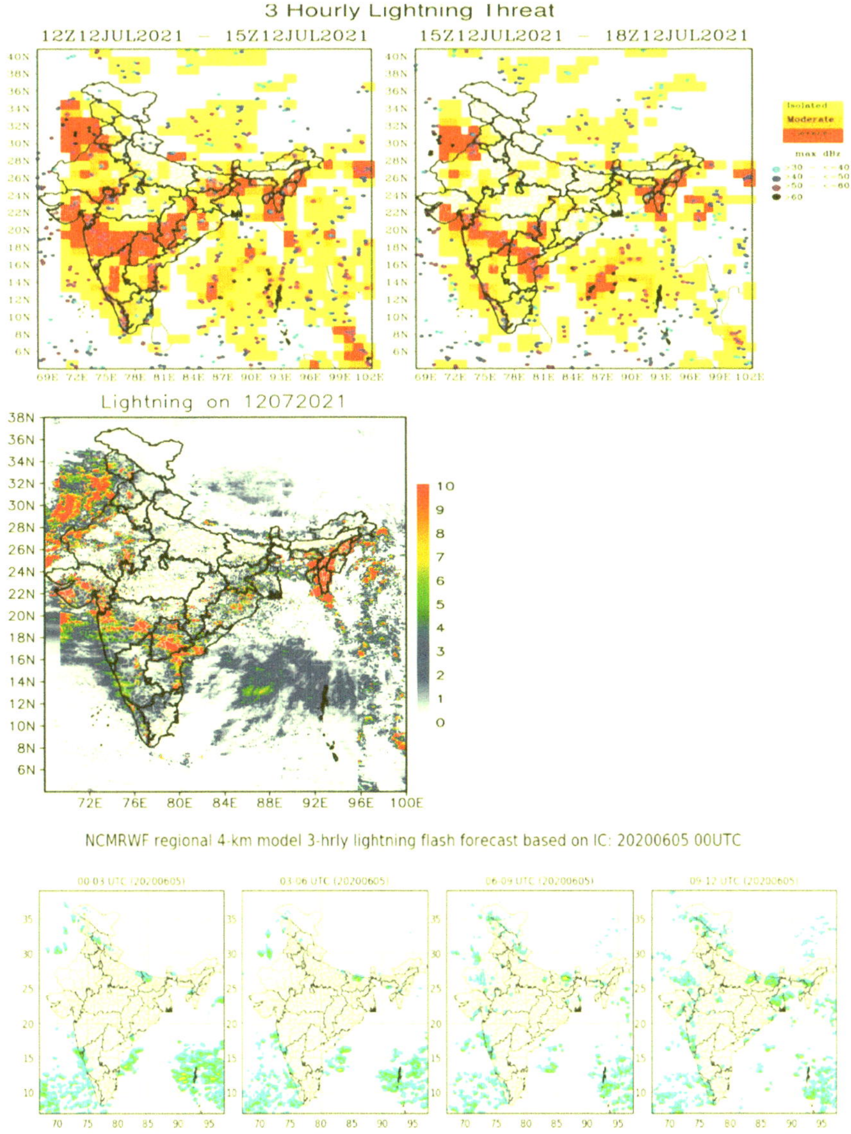

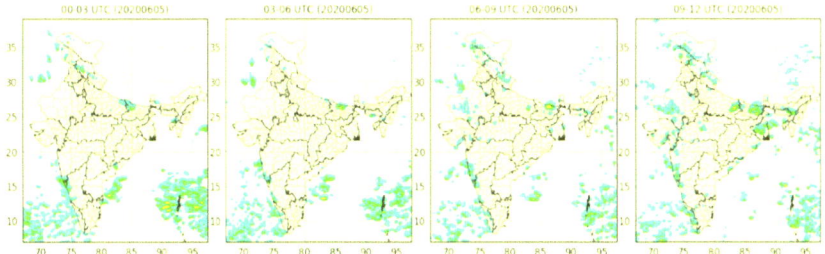

Fig. 3.17   Three hourly lightning threats based on lightning potential index and flash counts on 12 July 2021 (upper and middle panel). The lower panel shows three hourly lightning flash forecasts based on NCMRWF 4 km resolution regional model for another case of 5th June 2020

been studied by many researchers (Das et al. 2006; Ashrit 2010; Chevuturi et al. 2015; Dimri et al. 2017; Sarkar et al. 2017). One of the events is demonstrated below.

A cloudburst leading to heavy rainfall occurred over Uttarkashi (30.73° N, 78.45° E) in the Western Himalayas on 3rd August 2012. Sarkar et al. (2017) investigated the event using the WRF model at nested 9 and 3 km horizontal resolutions. They also studied the impact of data assimilation (DA) on the simulation of the heavy rainfall episode. Figure 3.18 illustrates the simulated rainfall by the model with and without using the DA at 3 km inner domain.

The diagram also shows the observed rainfall (merged IMD-NCMRWF-TRMM). The WRF model run with DA could predict a heavy rainfall greater than 24 cm (Fig. 3.18a) at a location close to and southeast of Uttarkashi. The location of rainfall is better predicted in the WRF simulation with DA.

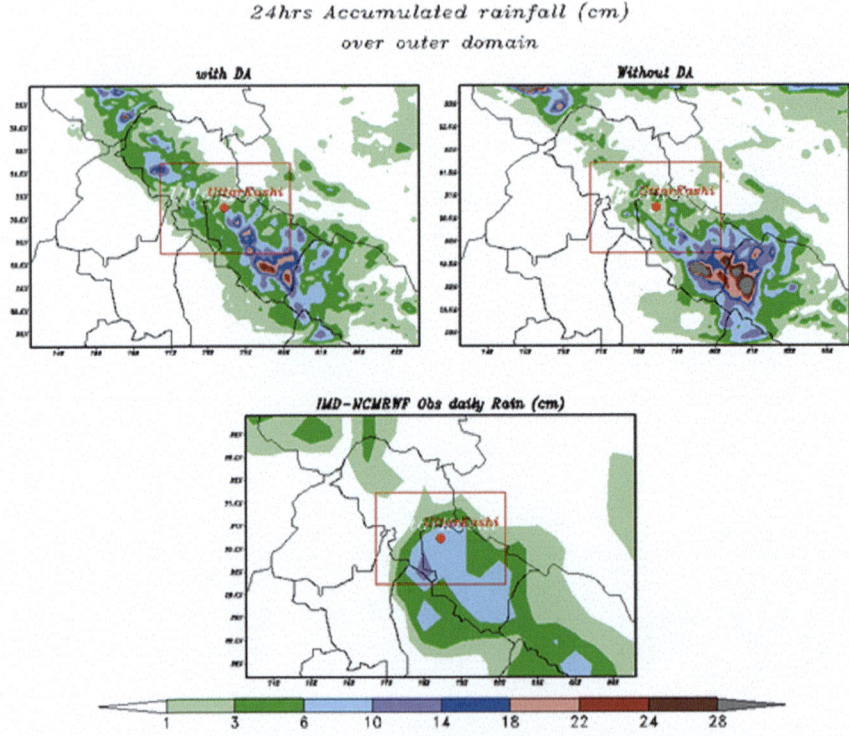

Fig. 3.18 WRF simulated 24 h accumulated rainfall (cm) **a** with GTS data assimilation, **b** without data assimilation and **c** IMD-NCMRWF merged rainfall over inner domain (Sarkar et al. 2017)

3.5.4 A Historic Tornado of Nepal

A deadly storm struck the Bara and Parsa districts of Nepal on 31st March 2019 in the evening around 13–14 UTC, which was the first officially recorded tornado in the nation. A supercell thunderstorm spawned the tornado in the Chitwan National Park. Traveling along a 90 km path, which was visible from space (Sentinel-2satellite) and reaching a maximum width of 200 m, the tornado caused considerable damage in the Bara and Parsa districts of Nepal. It caused 28 deaths, 1176 suffered injuries, and damaged 2600 homes according to press reports. Tornadoes are common during the pre- and post-monsoon months in the Ganges Basin to the south of Nepal. During this time, the convective available potential energy and wind shear are high and conducive to the development of rotating thunderstorms.

Figure 3.19 illustrates the cloud imageries from Himawari satellite and INSAT 3DR Brightness temperature. The diagrams show a big cloud cluster moved over the Bara and Parsa districts of southern Nepal from northern hills. The INSAT 3DR also shows that simultaneously there was another system over northern Bangladesh causing severe thunderstorms over there.

An attempt was made to simulate the severe storm using the WRF model with nested domains at 18 and 6 km resolutions. The Convective Available Potential Energy (CAPE), the Storm Relative Environmental Helicity (SREH), and the Bulk Richardson Number Shear (BRNSHR) are considered to be good indicators for the formation of supercell tornadoes (Das et al. 2015b, 2016). Figure 3.20 depicts

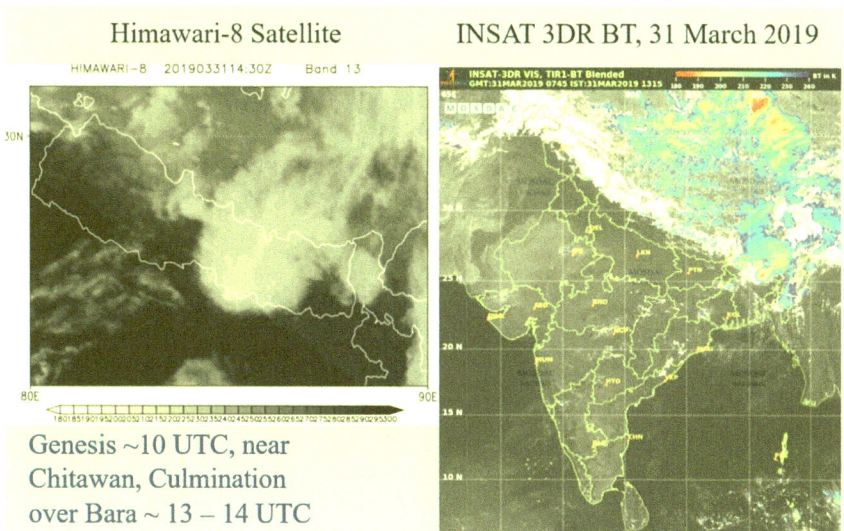

Fig. 3.19 Cloud imageries from Himawari satellite and INSAT 3DR brightness temperature on 31st March 2019, 14: 30 UTC

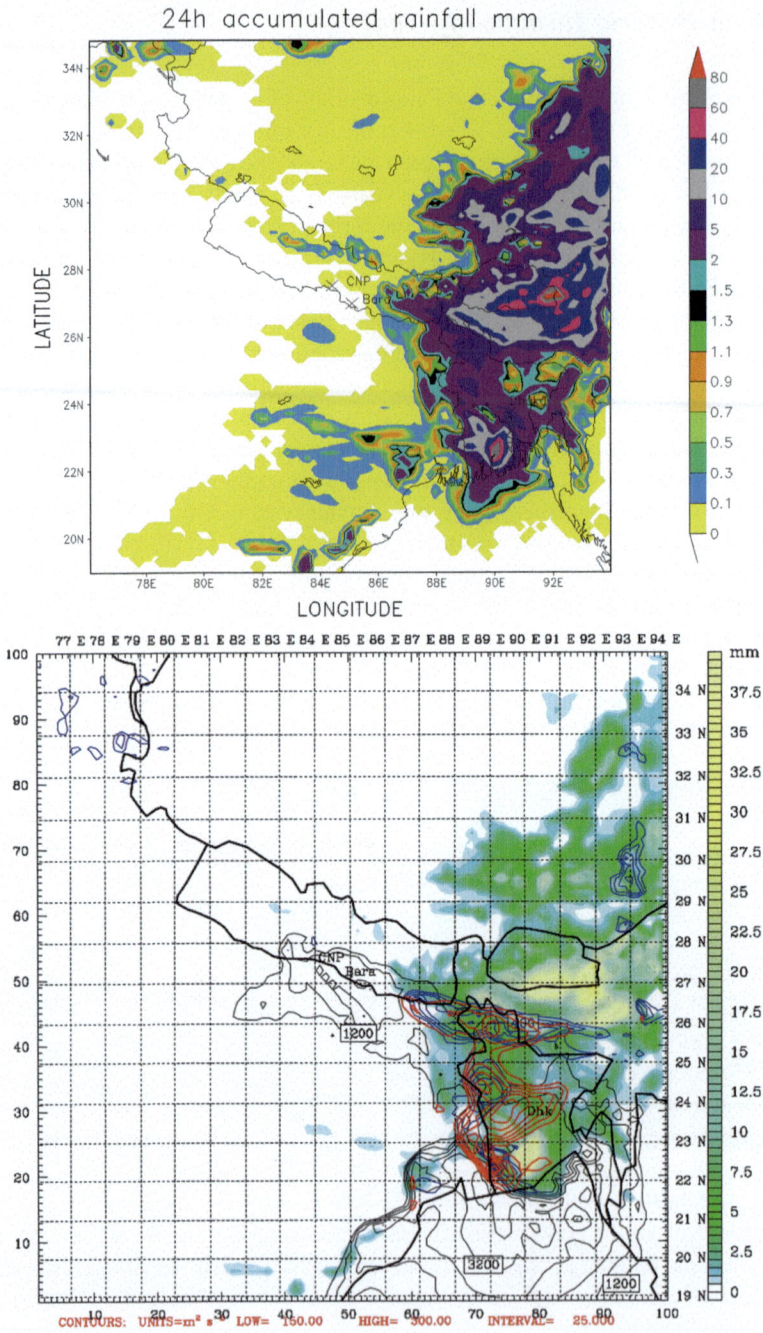

Fig. 3.20 The 24 h accumulated rainfall (upper panel) and CAPE, SREH, and BRNSH (lower panel) as simulated by the model. CAPE ($> 1200\ \text{J kg}^{-1}$ is in blue contours), SREH ($> 300\ \text{m}^2\ \text{s}^{-2}$ in green contours) and BRNSHR ($> 150\ \text{m}^2\ \text{s}^{-2}$ in red contours) as simulated by the model for 12 UTC, 31 March 2019. Rainfall is shaded

the accumulated rainfall and the three parameters (CAPE, SREH, and BRNSH) simulated by the model.

Note that the model simulated some rainfall near the location of the event, but its intensity was not as expected. Moreover, the values of CAPE, SREH, and BRNSH simulated by the model are very weak and they do not show environment conducive for the formation of a tornado. The supercell storms are expected over the region where CAPE is greater than 1200 J kg^{-1}, SREH is greater than 300 m^2 s^{-2}, and BRNSHR is greater than 150 m^2 s^{-2}. Tornadoes are likely where the three lines (blue, green, and red) coincide with each other in Fig. 3.20. While the model could not simulate the tornado well, it simulated the severe thunderstorms over northern Bangladesh very well as seen from Fig. 3.20.

Possible Causes of the failure of the model to produce the tornado could be the lack of good initial conditions for the model. The quality of initial conditions was not conducive due to lack of surface and upper-air data at high resolution over the mountains. As discussed earlier, NWP is an initial value problem. It must get good initial conditions to be able to forecast an event successfully. There is no Doppler Weather Radar (DWR) in the vicinity of the event. The nearest DWR is located in Patna (India), but unfortunately, the Radar was down for maintenance on that day. These results emphasize the need for convective scale Rapid Cycle Data Assimilation with good quality dense observations over the mountainous region.

3.5.5 Tropical Cyclone Forecasting

There is a substantial improvement in the skill of forecasting the cyclones in the last two decades mainly due to the improvement in the observation technology, understanding, and modeling capability. The improvements have also occurred due to increase in the computing resources (HPF). Here we illustrate the example of Extremely Severe Cyclonic Storm (ESCS) "Fani", which formed over the Bay of Bengal and had a long track history crossing the Odisha coast in India, recurving thereafter and reaching up to Bangladesh during 26 April–4 May 2019 (IMD 2019). Figure 3.21 depicts the satellite imagery of ESCS FANI over Bay of Bengal and the Radar reflectivity from the DWR paradeep for near landfall time.

Observed and forecast track with cone of uncertainty and wind distribution based on 0530 IST of 30th April (72 h prior to landfall) of ESCS FANI indicating accurate landfall prediction near Puri is presented in Fig. 3.22. The "cone of uncertainty (COU)"—also known colloquially as the "cone of death", "cone of probability" and "cone of error"—represents the forecast track of the center of a cyclone and the likely error in the forecast track based on predictive skill of past years. The COU in the forecast of IMD was introduced with effect from the TC, "WARD" during December 2008 (Mohapatra et al. 2012).

There is a continuous reduction in the errors of the landfall point forecasts, landfall time, intensity, and track errors forecast by the model as illustrated in Figs. 3.23 and 3.24 (Mohapatra et al 2012, 2013a, 2015, 2019).

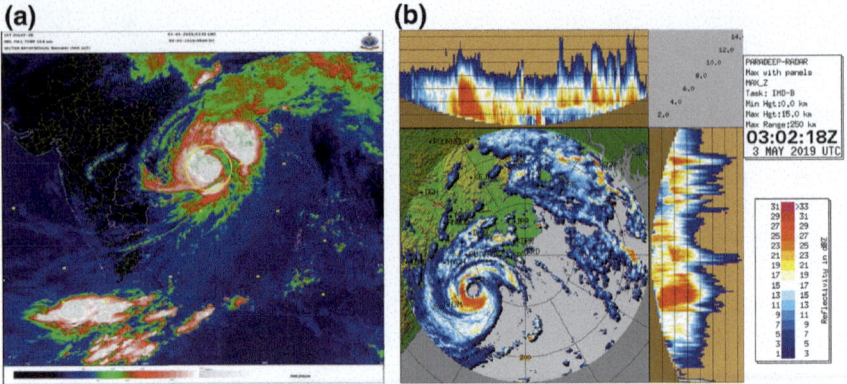

Fig. 3.21 Typical **a** satellite imagery for ESCS FANI over Bay of Bengal and **b** radar imagery from DWR paradeep for ESCS FANI over Bay of Bengal near landfall time

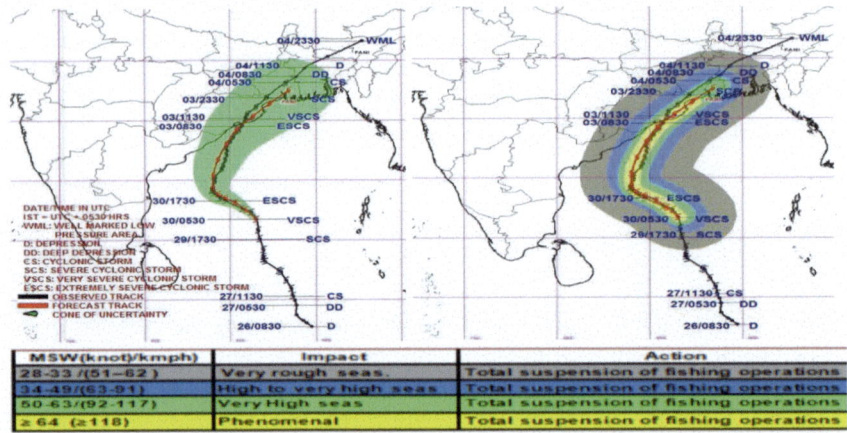

Fig. 3.22 Observed and forecast track based on 0530 h IST of 30th April (72 h prior to landfall) of ESCS FANI indicating accuracy in landfall predictions near Puri (Odisha)

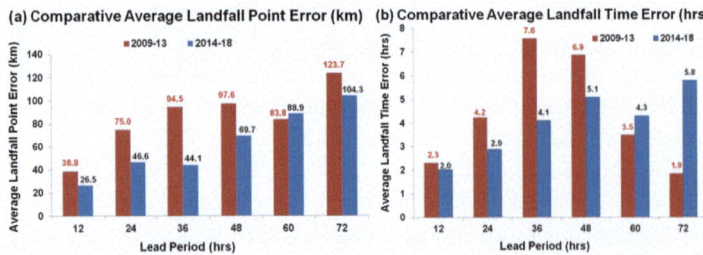

Fig. 3.23 Average landfall **a** point forecast error (km) and **b** time forecast error (h) during 2014–18 as compared to that during 2009–13. *Source* Mohapatra and Sharma (2019)

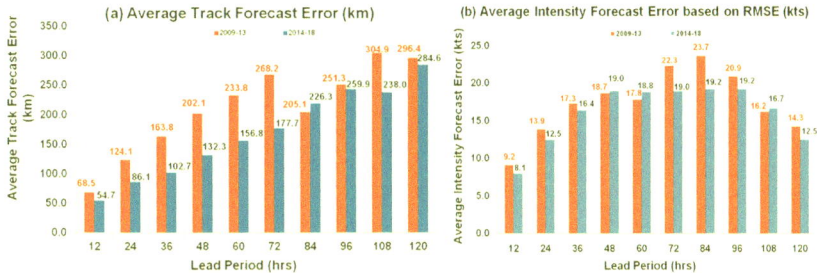

Fig. 3.24   Average track forecast **a** error (km) and **b** average intensity forecast error RMSE (knots) during 2014–18. *Source* Mohapatra and Sharma (2019)

IMD also issues Extended Range Outlook giving 15 days probabilistic cyclogenesis forecasts based on a Multi-Model Ensemble Prediction System (MMEPS). The product is available at http://www.rsmcnewdelhi.imd.gov.in/images/bulletin/eroc.pdf.

3.5.6   Heavy Rainfall Causing Catastrophic Floods

In Sect. 1.2.6 we discussed about some of the recent episodes of heavy rainfall that caused catastrophic floods over the South Asian region. We shall discuss one of them here to illustrate the state-of-art in heavy rainfall/flood forecasting. We illustrate the case of the monsoonal deluge that occurred in July 2010 over northern Pakistan and resulted in catastrophic flooding, loss of life and properties. Almost 20 million people needed shelter, food, and emergency care due to the flood that affected the Indus Valley of Pakistan in July–August 2010. The July–August 2010 floods were the worst ever known to have occurred in that region. The event was investigated by many researchers (Houze et al. 2011; Webster et al. 2011, 2013).

Webster et al. (2011) used the NOAA CPC Morphing Technique (CMORPH) Precipitation Product to analyze the results and the European Center for Medium-Range Weather Forecasts (ECMWF) 15-day Ensemble Prediction System (EPS) to assess whether the rainfall over the flood-affected region was predictable (Figs. 3.25 and 3.26). The ECMWF EPS forecasts consisted of 51 ensemble members initialized twice per day (00 and 12 UTC), each ensemble member having a 15-day forecast horizon. The horizontal resolution of the model was 50×50 km from 0 to 10 days and then 80 km $\times$ 80 km from day 10–15 (Buizza et al. 2007). A multi-year analysis showed that Pakistan rainfall was highly predictable out to 6–8 days.

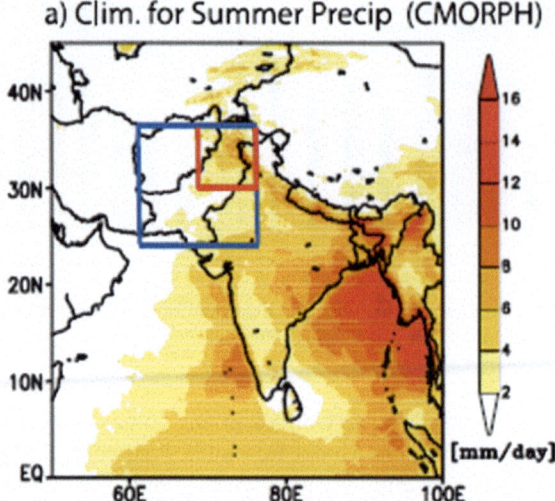

Fig. 3.25 May-August CMORPH precipitation [mm/day] climatology for 2003–2010. The blue rectangle covers entire Pakistan (62° N–74° N; 24° E–36° E) and the red rectangle covers the northern Pakistan (70° E–74° E, 30° N–36° N) where the rainfall exceeded 20 mm/day (Webster et al. 2011)

3.5.7   How Reliable Are Weather Forecasts?

Accuracy of Weather forecasts depends on the scale (temporal and spatial) of a phenomenon and the lead time of forecasts. For example, a short-lived small-scale phenomena like turbulence, dust devils, tornadoes, thunderstorms, and cloud-bursts are difficult to forecast at longer lead time, while the long-lived large-scale phenomena like the cyclones, wide-spread rainfall, monsoon circulations, western disturbances, etc., are easy to forecast even at long lead time of 7–10 days in advance. In general, a five-day forecast can accurately predict the weather about 60% of the time and a one-day forecast can accurately predict the weather approximately 80% of the time. However, a 10-day or longer forecast is only right about half the time. Recent developments in the observation and modeling techniques have improved the skills of the weather forecasts substantially.

Analysis of skill scores of forecasting thunderstorms (Fig. 3.27) shows that there is a continuous improvement in the Ratio Score, Probability of Detection (POD), False Alarm Rate (FAR), Critical Success Index (CSI), Equitable Threat Score (ETS) of convective storms in 24 h over the Indian region during 2016–2020 (IMD 2020).

Results show that there is a continuous improvement in the forecast skills of the annual average track errors, landfall point errors, time of landfall errors, and intensity errors of the cyclones over the Indian region (Fig. 3.28) since 2003 (IMD 2020). While there is more accuracy in cyclone prediction as it is a larger system, there has also been a lot of improvement in the accuracy of heavy rainfall warning

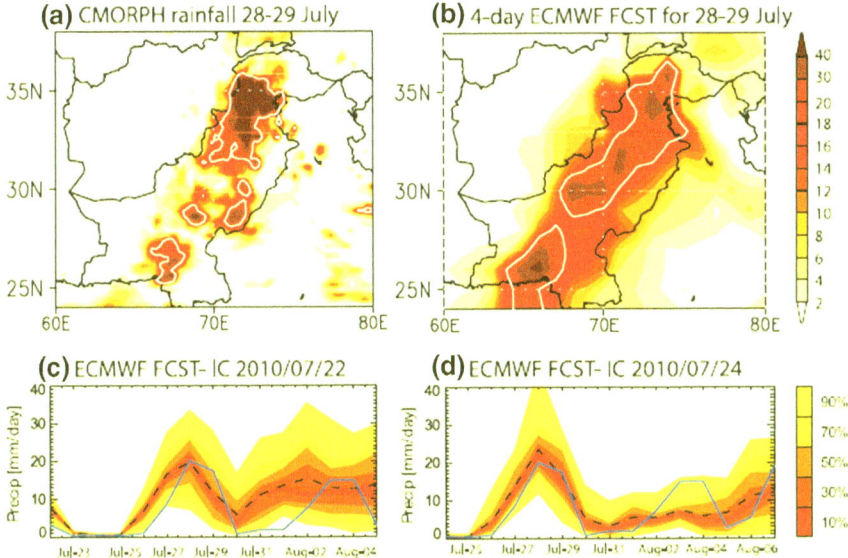

Fig. 3.26 Total precipitation [mm/day] for **a** CMORPH over 28–29 July 2010 and **b** ECMWF ensemble mean of the forecast initialized four days previously (July 24, 2010) for the same time period. White contour shows 20 mm/day. ECMWF 15-day forecast of the precipitation [mm/day] in the red rectangle (Fig. 3.25) initialized on July **c** 22nd, and **d** 24th, 2010. Black dashed line shows the ensemble mean. Colored shading depicts the probability of precipitation rate based on the 51 ensemble members. Dark blue line represents the observed CMORPH precipitation averaged for the same region (Webster et al. 2011)

Fig. 3.27 Comparative 24 h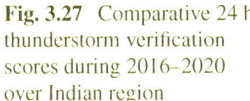
thunderstorm verification
scores during 2016–2020
over Indian region

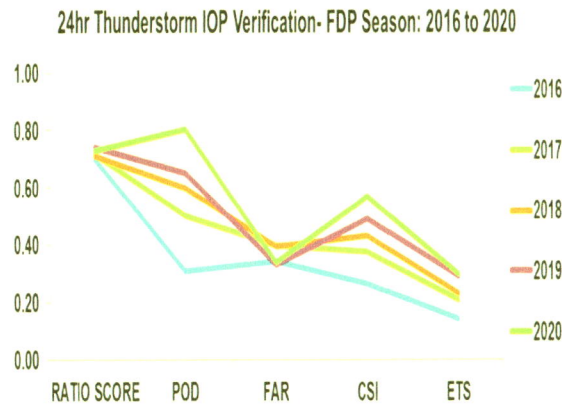

from around 50% in 2014 to 77% in 2020. IMD started issuing lightning forecasts since 2019. The accuracy of lightning alerts is about 88% in 3 h. The accuracies of forecasts also vary from season to season. Generally, the winter weather systems are more predictable than that of the summer season.

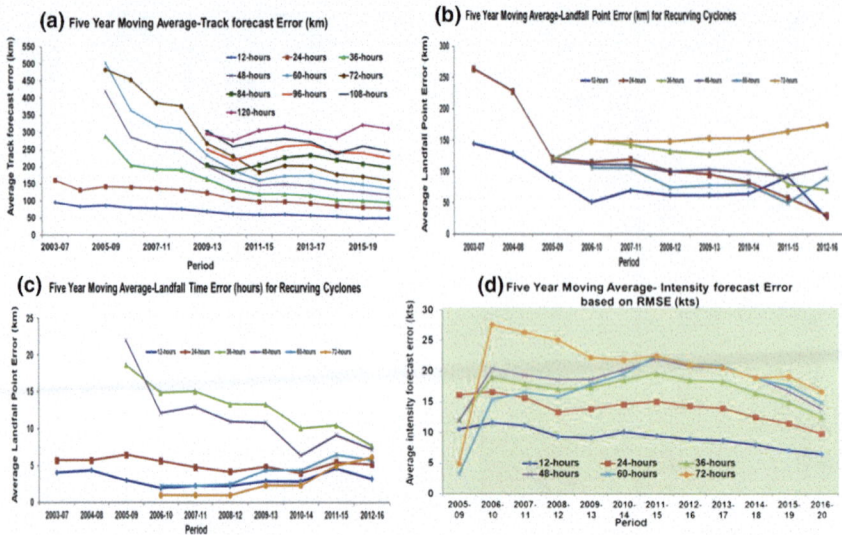

Fig. 3.28 Five year moving average of RMSE of forecasts at different lead times from 12 to 120 h, **a** track error (km), **b** landfall point error (km), **c** time of landfall error (hour), and **d** intensity at the landfall (knots). *Source* IMD (2020)

3.5.8 What is the Projection of Extreme Weather Events in a Changing Climate?

According to IPCC AR6 assessments, the frequency and intensity of hot extremes have increased and those of cold extremes have decreased on the global scale since 1950. The frequency and intensity of hot extremes will continue to increase and those of cold extremes will continue to decrease, at both global and continental scales and in nearly all inhabited regions with increasing global warming levels. Heavy precipitation will generally become more frequent and more intense with additional global warming. This will result in an increase in the frequency and magnitude of pluvial floods—surface water and flash floods—(*high confidence*). Climate change has contributed to decreases in water availability during the dry season over a predominant fraction of the land area due to evapotranspiration increases (*medium confidence*). Precipitation deficits and changes in evapotranspiration (ET) govern net water availability. A lack of sufficient soil moisture, sometimes amplified by increased atmospheric evaporative demand (AED), results in agricultural and ecological drought. Lack of runoff and surface water results in hydrological drought. Increases in evapotranspiration have been driven by AED increases induced by increased temperature, decreased relative humidity, and increased net radiation (*high confidence*). Several regions will be affected by more severe agricultural and ecological droughts even if global warming is stabilized in a range of 1.5°–2 °C of global warming (*high confidence*). The proportion of intense Tropical Cyclones

(TCs), and peak wind speeds of the most intense TCs will increase on the global scale with increasing global warming (*high confidence*) according to IPCC. The total global frequency of TC formation will decrease or remain unchanged with increasing global warming (*medium confidence*). Future wind speed changes are expected to be small, although poleward shifts in the storm tracks could lead to substantial changes in extreme wind speeds in some regions (*medium confidence*). There is *low confidence* in past trends in characteristics of severe convective storms, such as hail and severe winds, beyond an increase in precipitation rates. Fire weather conditions (compound hot, dry, and windy events) have become more probable in some regions (*medium confidence*) and there is *high confidence* that they will become more frequent in some regions at higher levels of global warming.

3.6 What is the Severe Weather Forecast Demonstration Project?

As the observation technology is improving rapidly, field campaigns are frequently organized to collect intensive observations on extreme weather events. Field observations of severe weather phenomena are collected to improve our understanding of their structures and life cycles and develop better models for forecasting the events. Over the years several field experiments have been conducted both over the Indian region and outside. IMD had conducted three field experiments during 1929–1941 to study the outbreak of severe convective storms (IMD 1944; Tyagi et al. 2012). A number of field experiments have been conducted since then in USA and elsewhere to understand and predict the convective storms.

The Severe Weather Forecasting Demonstration Project (SWFDP) of the World Meteorological Organization (WMO) is an initiative to strengthening capacity of National Meteorological and Hydrological Services (NMHSs) in developing and least developed countries including Small Island Developing States to deliver improved forecasts and warnings of severe weather to save lives, livelihoods, and property (https://public.wmo.int/en/programmes/severe-weather-forecasting-progra mme-swfp). SWFDP was initiated in 2006 to make the NWP products, including Ensemble Prediction System (EPS) products, of the most advanced Global Data-Processing and Forecasting System Centers available to all WMO Members. The Project made global-scale products available to Regional Specialized Meteorological Centers (RSMC) that integrate and synthesize them in order to provide daily guidance for short-range and medium-range forecasts of hazardous weather conditions and weather-related hazards to NMHSs in their geographical region. Thus, NMHSs are enabled to issue effective severe weather warnings to disaster management and civil protection authorities in their respective countries. Currently, SWFP covers over 80 developing countries.

Several field experiments have been conducted in India under the Indian Climate Research Program (ICRP). We shall briefly discuss here three field experiments on

(1) Thunderstorm (SAARC STORM), (2) Tropical Cyclone (FDP-Cyclone), and (3) Winter Fog Experiment (WiFEX).

3.6.1 SAARC STORM

Realizing the importance of the pre-monsoon thunderstorms and their socio-economic impact, the India Department of Science and Technology started the nationally coordinated Severe Thunderstorm Observation and Regional Modeling (STORM) program in 2005. It is a comprehensive observational and modeling effort to improve understanding and prediction of severe thunderstorms (STORM 2005). The STORM program was a multi-year exercise and was quite complex in the formulation of its strategy for implementation. Two pilot experimental campaigns were conducted during the pre-monsoon seasons (April–May) of 2006 and 2007 (Mohanty et al. 2006, 2007). However, the weather knows no political boundaries. Since the neighboring South Asian countries are also affected by the Nor'westers, the STORM program was expanded to cover the South Asian countries under the South Asian Association for Regional Cooperation (SAARC) in 2009 (Das et al. 2014). The STORM program covered all the SAARC countries in three phases (Fig. 3.29).

In the 1st phase, the focus was on Nor'westers that form over the eastern and north-eastern parts of India, Bangladesh, Nepal, and Bhutan. In the 2nd phase, the dry convective storms/dust storms and deep convection that occur in the western parts of India, Pakistan, and Afghanistan were investigated. Similarly, in the 3rd phase, the maritime and continental thunderstorms over southern parts of India, Sri Lanka, and Maldives were investigated. Thus, overall, the SAARC STORM program covered investigations about formation, modeling, and forecasting, including nowcasting of severe convective weather in the pre-monsoon season over South Asia. Pilot field experiments were conducted during 1–31 May of 2009–14 jointly with the SAARC countries. The program was put on hold after the closure of the SAARC Meteorological Research Center, Dhaka in 2015. The SAARC STORM program is now continued as annual exercise within India during the pre-monsoon season.

3.6.2 Forecast Demonstration Project (FDP)—Cyclone

During the past few years, huge technological advancements have been achieved in the world to observe the inner core of the cyclone. Accordingly, a program was evolved for improvement in prediction of track and intensity of tropical cyclones over the Bay of Bengal resulting in the Forecast Demonstration Project (FDP-BOBTEX 2008). FDP program was aimed to demonstrate the ability of various NWP models to assess the genesis, intensification, and movement of cyclones over the north Indian ocean with enhanced observations over the data-sparse region and to incorporate modifications into the models which could be specific to the Bay of Bengal based

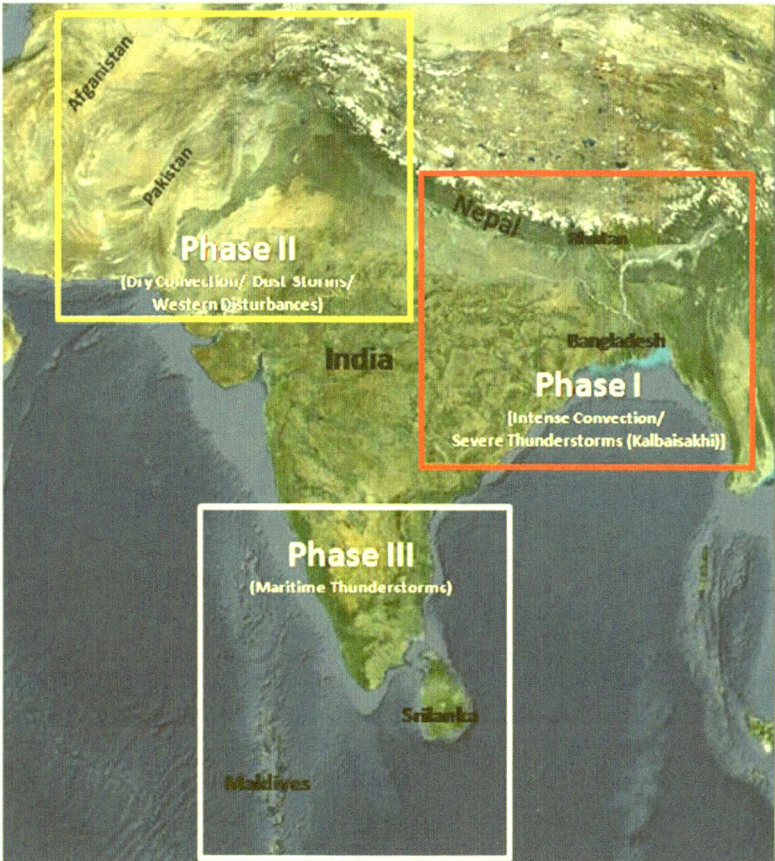

Fig. 3.29 The South Asian countries: Afghanistan, Bangladesh, Bhutan, India, Maldives, Nepal, Pakistan, and Sri Lanka, which participated in the SAARC STORM program (Das et al. 2014)

on the in-situ measurements (Mohapatra et al. 2013b). FDP Program was scheduled in three phases, viz., (i) Pre-pilot phase (15 Oct–30 Nov 2008, 2009), (ii) Pilot phase (15 Oct–30 Nov 2010–2012) and (iii) Final phase (15 Oct–30 Nov 2013–14). India also has plan of probing the cyclones with hired aircraft and dropsonde experiments. To accomplish the above objectives, initiative was made with priorities on (i) observational upgradation, (ii) modernization of cyclone analysis and prediction system, (iii) cyclone analysis and forecasting procedure, (iv) warning products generation, presentation, and dissemination, (v) confidence-building measures and capacity building.

Various strategies were adopted for improvement of observation, analysis, and prediction of cyclone. Several national institutions participated for joint observational, communicational, and NWP activities during the pre-pilot and pilot phases of FDP campaign during 2008–11. The comparison of observational systems before and after FDP indicates a significant improvement in terms of Radar, Automatic Weather Station (AWS), high wind speed recorders over the region. It has resulted in reduction in monitoring and forecasting errors. The performance of NWP models has improved along with the introduction of NWP platforms like IMD GFS, WRF, HWRF, and ensemble prediction system (EPS). Salient features of achievements along with the problems and prospects of this project are discussed in Mohapatra et al. (2013b).

3.6.3   Winter Fog Experiment (WiFEX)

Fog is one of the major weather hazards, impacting aviation, road transportation, economy, and public life in the northern parts of India. During the winter of 2013–2014 a total of 140 flights were canceled, 143 were diverted and 363 were delayed from the Indira Gandhi International Airport (IGIA), New Delhi causing a loss of Rs. 120 million to the aviation sector (Kulkarni 2016). Radiation and advection fog is common in the winter season. Radiation fog forms generally in the rear sector of a western disturbance (WD), whereas advection fog develops in the forward sector of the WD. The WD causes light to moderate rain during winter, which advects large amount of moisture into the lower troposphere. In addition, irrigation for winter crops adds significant amount of moisture into the lower atmosphere. As soon as the lower-level ridge line forms over the northern region, it enforces persisting stable (calm winds) and clear atmosphere conditions and lower surface temperature leading to the formation of strong surface-based inversions, which facilitate fog formation and its sustenance. The land surface processes and emission sources in the Indo-Gangetic Plains (IGP) contribute to moisture supply and high concentrations of pollutants, which favor hazy/foggy conditions for extended periods. Sometimes the fog continues for weeks over a vast area with only partial lifting in the late afternoon. Combinations of these factors introduce formidable challenges for fog forecasting over the region.

The Winter Fog Experiment (WiFEX) was conducted at IGIA and Indian Agriculture Research Institute (IARI), New Delhi between 15 December 2015 and 15 February 2016. Details of the experiment are given in Ghude et al. (2017). The two sites are about 12 km apart. The airport site has a vast open area with frequent formation of fog during the winter season. The IARI site has more vegetation cover. It is characterized by irrigated agricultural fields about 1 km in radius and surrounded by densely populated residential areas. A set of in-situ and remote sensing instruments were deployed at IGIA and IARI campus from four Indian research/educational institutions: Indian Institute of Tropical Meteorology, Pune; IMD, Delhi; Indian Institute of Science Education and Research (IISER), Mohali, and IARI, Delhi. The instruments measured ground and surface properties, surface layer meteorology, atmospheric profiles, microphysical parameters, radiation, aerosol chemical, and optical properties of fog. The aim of such a combination was to measure simultaneously all key processes in fog genesis and lifecycle. The data collected from WiFEX are being used for research, development, and improvement of the models.

3.7   Summary and Discussion

South Asia (including Afghanistan, Bangladesh, Bhutan, India, Maldives, Myanmar, Nepal, Pakistan, and Sri Lanka) is home to an annual cycle of powerful, destructive weather, including severe thunderstorms, hailstorms, lightnings, cloudbursts, cyclones, heavy rainfall, flash flood, drought, heatwave and cold wave, etc., causing losses of lives and damages to properties. Extreme weather events triggered by climate change cost South Asia, billions of dollars annually according to a report by the State of the Climate in Asia 2020. The extreme weather caused over 5000 human deaths in 2020. While the frequent anomalous weather may be blamed on the effects of global warming and climate change, any such event is routed to the changes in the structure of the atmosphere. Our observation systems are supposed to sense any such change in the structure of the atmosphere and the weather and climate models are supposed to predict such changes with sufficient lead time, so that warnings may be issued to the public. Hence, it is a challenge to the Atmospheric Scientists.

Failure to observe the changes in the structure of the atmosphere implies that our observation network/technology is insufficient. We have discussed that the observation network is insufficient in many South Asian countries and in the remote areas and over the mountains. The discrete observation points have coarse resolution and hence, often they are unable to represent the sub-grid scale weather phenomena accurately. There is also a problem in accurately representing a continuous weather by the discrete observation points in space and time. These are somehow managed by the data assimilation. Many countries in South Asia do not even have a proper data assimilation system.

We discussed that weather is a result of the motion of air caused by a balance between various forces. This motion is described by the governing equations of the atmosphere based on the principles of hydrodynamics and thermodynamics. Some of the equations are nonlinear partial differential equations whose solutions are to be obtained numerically by making billions of calculations on the fastest supercomputers. So, regardless of the causes of extreme weather events, if we know the laws and the variables, why can't we make perfect forecasts even with a short lead time? The main reasons for this are (1) Inaccurate Initial Conditions, (2) Multiscale Interactions, (3) Chaos and Limit on Deterministic Predictability. We discussed that the numerical weather prediction is an initial value problem. Little change in the accuracy of the initial condition can cause large errors in the solutions of the model. There are uncertainties in the formulation of the governing equations. But even if the model was perfect, intrinsic nonlinearity of the atmosphere would restrict our ability to predict the weather to about two weeks, because there is a limit to the deterministic predictability due to chaos.

Despite the limitations in the observations and the modeling, the techniques of weather forecasting have continuously evolved in the past 3–4 decades. We have seen that the number of deaths that used to occur during tropical cyclones has reduced substantially. People have gradually started trusting the accuracy of weather forecasts as their skills have improved due to gradual improvements in the observation technology such as Satellites, Doppler Weather Radars, Automatic Weather Stations, etc., and improvements in our understanding of the atmosphere, data assimilation system and modeling techniques. The numerical weather prediction centers are providing a wide range of products and advance warnings of severe weather systems like the thunderstorms, lightnings, cyclones, heavy rainfall, strong winds, etc. The quality of forecasts will keep improving with time due to further improvements in the technology, modeling techniques, and computing resources.

Acknowledgments I would like to thank Dr. Amitava Bandopadhyay, Director, NAM S&T Centre, New Delhi for inviting me to write this article. Some of the results presented in this paper are taken from the M.Sc. thesis of my students Mr. P. P. Musaid and Mr. Anish Kumar who carried out their projects with me.

References

Albrecht, R., S. Goodman, D.E. Buechler, and T. Chronis. 2009. Tropical frequency and distribution of lightning based on 10 years of observations from space by the lightning imaging sensor (LIS). In *4th conference on meteorological applications of lightning data, 89th AMS annual meeting*, 10–15 Jan 2009, Phoenix Arizona, USA.

Alexander, L.V., X. Zhang, G. Hegerl, and S.I. Seneviratne. 2016. Implementation plan for WCRP grand challenge on understanding and predicting weather and climate extremes—The "extremes grand challenge". Version, June 2016 available from https://www.wcrp-climate.org/images/documents/grand_challenges/WCRP_Grand_Challenge_Extremes_Implementation_Plan_v201 60708.pdf.

Anthes, Richard A. 1990. Recent applications of the Penn State/NCAR mesoscale model to synoptic, mesoscale, and climate studies. *Bulletin of the American Meteorological Society* 71 (11): 1610–1629.

Ashrit, R. 2010. Investigating the Leh 'cloudburst'. NCMRWF research report, NMRF/RR/10/2010.

Bannister Ross, N. 2007. Elementary 4D-VAR, DARC technical report no. 2. Data Assimilation Research Centre, University of Reading, UK.

Buizza, R., J.-R. Bidlot, N. Wedi, M. Fuentes, M. Hamrud, G. Holt, and F. Vitart. 2007. The new ECMWF VAREPS (variable resolution ensemble prediction system). *Quarterly Journal Royal Meteorological Society* 133: 681–695. https://doi.org/10.1002/qj.75.

Carbin, G.W., Tippett M.K., Lillo S.P., and Brooks H.E. 2016. Visualizing long-range severe thunderstorm environment guidance from CFSv2. *Bulletin American Meteorology Social* 97:1021–1032.

Case, Jonathan L., John Manobianco, Timothy D. Oram, Tim Garner, Peter F. Blottman, and Scott M. Spratt. 2002. Local data integration over East-Central Florida using the ARPS data analysis system. *Weather and Forecasting* 17 (1): 3–26.

Centre for Severe Weather Research (CSWR). 2006. Doppler on wheels. Retrieved 2006-12-29.

Chandrasekara, S.S.K., H.-H. Kwon, M. Vithanage, J. Obeysekera, and T.-W. Kim. 2021. Drought in South Asia: A review of drought assessment and prediction in South Asian countries. *Atmosphere* 12: 369. https://doi.org/10.3390/atmos12030369.

Chantry, M., H. Christensen, P. Dueben, and T. Palmer. 2021. Opportunities and challenges for machine learning in weather and climate modelling: Hard, medium and soft AI. *Philosophical Transactions of the Royal Society A* 379: 20200083. https://doi.org/10.1098/rsta.2020.0083.

Chaudhury, A., and A.K. Banerjee. 1983. A study of hailstorms over NE India. *Vayu Mandal* 13: 91–95.

Chevuturi, A., A.P. Dimri, A.K. Someshwar Das, and D. Niyogi. 2015. Numerical simulation of intense precipitation event over Rudraprayag in Central Himalayas during 13–14 Sep 2012. *Journal of Earth System Sciences* 124 (7): 1545–1561.

Christian, H.J., Blakeslee, R.J., Goodman, S.J., Mach, D.A., Stewart, M.F., Buechler, D.E. et al. 1999. The lightning imaging sensor. In: Proceeding of 11th International Conference on Atmospheric Electricity, National Aeronautics and Space Administration, Guntersville, vol AI:746–749.

Cotton, William R., Gregory Thompson, and Paul W. Mieike. 1994. Real-time mesoscale prediction on workstations. *Bulletin of the American Meteorological Society* 75 (3): 349–362.

Dalal, S., D. Lohar, S. Sarkar, I. Sadhukhan, and G.C. Debnath. 2012. Organizational modes of squall-type convective systems during pre-monsoon season over eastern India. *Atmospheric Research* 106: 120–138.

Das, Someshwar, R. Ashrit, and M.W. Moncrieff. 2006. Simulation of a Himalayan cloudburst event. *Journal of Earth System Science* 115 (3): 299–313.

Das, Someshwar, R. Ashrit, Mitchell W. Moncrieff, M. Dasgupta, J. Dudhia, C. Liu, and S.R. Kalsi. 2007. Simulation of intense organized convective precipitation observed during the Arabian sea monsoon experiment (ARMEX). *Journal of Geophysical Research* 112: D20117. https://doi.org/10.1029/2006JD007627.

Das, Someshwar. 2010. *Climatology of thunderstorms over the SAARC region*. SMRC No. 35. Available from SAARC Meteorological Research Centre E-4/C, Agargaon, Dhaka-1207, Bangladesh.

Das, Someshwar, U.C. Mohanty, Ajit Tyagi, D.R. Sikka, P.V. Joseph, L.S. Rathore, A. Habib, S. Baidya, K. Sonam, and A. Sarkar. 2014. The SAARC STORM—A coordinated field experiment on severe thunderstorm observations and regional modeling over the South Asian region. *Bulletin of the American Meteorological Society*. https://doi.org/10.1175/BAMS-D-12-00237.1.

Das, M.K., Someshwar Das, M.A.M. Chowdhury, and S. Karmakar. 2015a. Simulation of Tornado over Brahmanbaria on 22 March 2013 using Doppler weather radar and WRF model. *Geomatics, Natural Hazards and Risk*. https://doi.org/10.1080/19475705.2015.1115432.

Das Someshwar, A.S., Mohan K. Das, M.M. Rahman, and M.N. Islam. 2015b. Composite characteristics of Nor'westers based on observations and simulations. *Atmospheric Research*. https://doi.org/10.1016/j.atmosres.2015.02.009.

Das Someshwar, M.N.I., and Mohan K. Das. 2016. Simulation of severe storms of tornadic intensity over Indo-Bangla region. *MAUSAM (Quarterly Journal of Meteorology, Hydrology & Geophysics)* 67 (2): 479–492.

Dimri, A.P., A. Chevuturi, D. Niyogi, R.J. Thayyen, K. Ray, S.N. Tripathi, A.K. Pandey, and U.C. Mohanty. 2017. Cloudbursts in Indian Himalayas: A review. *Earth-Science Reviews* 168 (2017): 1–23.

Dube, S.K., Poulose Jisme, and A.D. Rao. 2013. Numerical simulation of storm surge associated with severe cyclonic storm in the Bay of Bengal during 2008–2011. *Monsoon* 64: 193–202.

Dudhia, J. 1993. A non-hydrostatic version of the Penn State/NCAR mesoscale model: Validation tests and simulation of an Atlantic cyclone and cold front. *Monthly Weather Review* 121: 1493–1513.

ECMWF. 2018. https://www.ecmwf.int/en/about/media-centre/news/2018/noaa-satellite-launch-20-years-ago-marked-start-new-era.

Eliot, J. 1899. Hailstorm in India during the period 1883–1897 with a discussion on their distribution. *Indian Meteor Memory* 6: 237–315.

FDP-BOBTEX. 2008. *Forecast demonstration project (FDP)-Bay of Bengal tropical cyclone experiment. Science Plan*, 78. Ministry of Earth Sciences, Govt. of India.

FDP-Fog. 2008. *Forecast demonstration project (FDP) on fog. Science Plan*, 47. Ministry of Earth Sciences, Govt. of India.

Ghude, S.D., G.S. Bhat, T. Prabha, R.K. Jenamani, D.M. Chate, P.D. Safai, A.K. Karipot, M. Konwar, P. Pithani, V. Sinha, P.S.P. Rao, S.A. Dixit, S. Tiwari, K. Todekar, S. Varpe, A.K. Srivastava, D.S. Bisht, P. Murugavel, K. Ali, U. Mina, M. Dharua, Y. Jaya Rao, B. Padmakumari, A. Hazra, N. Nigam, U. Shende, D.M. Lal, P. Acharja, R. Kulkarni, C. Subharthi, B. Balaji, M. Varghese, S. Bera, and M. Rajeevan. 2017. Winter fog experiment over the Indo-Gangetic plains of India. *Current Science* 112: 767–784.

Gilmour, I., L.A. Smith, and R. Buizza. 2001. Linear regime duration: Is 24 hours a long time in synoptic weather forecasting? *Journal of Atmospheric Science* 58: 3525–3539.

Goswami, B.N. 1997. The challenge of weather prediction. *Resonance—Journal of Science Education* 2 (1). https://www.ias.ac.in/article/fulltext/reso/002/01/0008-0015.

Goyal, S., M. Mohapatra, A. Kumar, S.K. Dube, K. Rajendra, and P. Goswami. 2016. Validation of a satellite-based cyclogenesis technique over the North Indian ocean. *Journal of Earth System Science* 125 (7): 1353–1363.

Houze, Jr., R.A., K.L. Rasmussen, S. Medina, S.R. Brodzik, and U. Romatschke. 2011. Anomalous atmospheric events leading to the summer 2010 floods in Pakistan. *Bulletin of the American Meteorological Society* 92 (3), 291–298.https://doi.org/10.1175/2010BAMS3173.1.

Houze, Robert, A., Jr. 1993. *Cloud dynamics*, 573. Academic Press.

Hsu, S.A., and Zhongde Yana. 1998. A note on the radius of maximum winds for hurricanes. *Journal of Coastal Research* 12 (2): 667–668.

Hussain, A., H. Mir, and M. Afzal. 2005. Analysis of dust storms frequency over Pakistan during 1961–2000. *Pakistan Journal of Meteorological* 2: 49–68.

IMD. 2013. *Cyclone warning services: Standard operation procedure*. New Delhi: Cyclone Warning Division IMD.

IMD. 2019. RSMC report on extremely severe cyclonic storm "FANI" over east-central equatorial Indian ocean and adjoining southeast Bay of Bengal (26 April–04 May, 2019): Summary.

IMD. 2020. India Meteorological Department, Ministry of Earth Sciences, annual report 2020.

IMD. 1944. Nor'wester of Bengal. India Meteorological Department Tech. Note 10. 17 pp.

IWMI. 2021. International Water Management Institute, contact: Giriraj Amarnath, www.iwmi.org.

JMA. 2017. Western North Pacific typhoon best track file 1951–2017. Japan Meteorological Agency. 2010–01–13. Retrieved 2010–01–13.

Joseph, P.V., D.K. Raipal, and S.N. Deka. 1980. ANDHI, the convective dust storm of northwest India. *Mausam* 31: 431–442.

Kotal, S.D., and S.K. Bhattacharya. 2013. Tropical cyclone genesis potential parameter (GPP) and its application over the north Indian Sea. *Mausam* 64: 149–170.

Kulkarni, R.G. 2016. Wintertime fog in Delhi and its effect on aviation economy. M. Sc. Project Report, Savitribai Phule University, Pune.

Kumar, S.V.J., S.S. Ashthikar, and M. Mohapatra. 2017. Life period of cyclonic disturbances over the north Indian ocean during recent years. In *Tropical cyclone activity over the North Indian ocean*, ed. M. Mohapatra, B.K. Bandyopadhyay, and L.S. Rathore, 181–198. New Delhi Germany: Capital Publishers, Springer.

Lorenz, E. 1969. The predictability of a flow which possesses many scales of motion. *Tellus* 21: 289–307. https://doi.org/10.1111/j.2153-3490.1969.tb00444.x.

Lynn, B., and Y. Yair. 2010. Prediction of lightning flash density with the WRF model. *Advances in Geosciences* 23: 11–16. https://doi.org/10.5194/adgeo-23-11-2010.

Lyons, Walter A. 1997. *The handy weather answer book*. Detroit: Visible Ink Press.

Magotra, R. 2021. *Proceedings of the webinar on "heat waves in South Asia"*. South Asian Meteorological Association (SAMA), 6 June 2021.

McCaul, E.W., Jr., S.J. Goodman, K.M. LaCasse, and D.J. Cecil. 2009. Forecasting lightning threat using cloud-resolving model simulations. *Weather and Forecasting* 24 (3): 709–729.

Mesinger, Fedor. 1996. Improvements in quantitative precipitation forecasts with the eta regional model at the national centers for environmental prediction: The 48-km upgrade. *Bulletin of the American Meteorological Society* 77 (11): 2637–2650.

Mir, H., A. Hussain, and Z.A. Babar. 2006. Analysis of thunderstorms activity over Pakistan during (1961–2000). *Pakistan Journal of Meteorological* 3: 13–32.

Mohanty, U.C., and Coauthors. 2006. *Weather summary during pilot experiment of severe thunderstorms observations and regional modeling (STORM) programme*, 177. India Department of Science and Technology Report.

Mohanty, U.C., and Coauthors. 2007. *Weather summary during pilot experiment of severe thunderstorms observations and regional modeling (STORM) programme*, 179. India Department of Science and Technology Report.

Mohapatra, M., and V.V. Kumar. 2017. Interannual variation of tropical cyclone energy metrics over North Indian ocean. *Climate Dynamics* 48: 1431–1445.

Mohapatra, M., D.C. Gupta, N.K. Chanchalani, and S.K. Dastidar. 2002. Orissa super cyclone, 1999—A case study. *Journal of Indian Geophysical Union* 6: 93–106.

Mohapatra, M., D.P. Nayak, and B.K. Bandyopadhyay. 2012. Evaluation of cone of uncertainty in tropical cyclone track forecast over North Indian ocean issued by India Meteorological Department. *Tropical Cyclone Research and Review* 1: 331–339.

Mohapatra, M., D.P. Nayak, R.P. Sharma, and B.K. Bandyopadhyay. 2013a. Evaluation of official tropical cyclone track forecast over north Indian ocean issued by India Meteorological Department. *Journal of Earth System Science* 122 (3): 589–601. https://doi.org/10.1007/s12040-013-0291-1.

Mohapatra, M., D.R. Sikka, B.K. Bandyopadhyay, and Ajit Tyagi. 2013b. Outcomes and challenges of forecast demonstration project (FDP) on landfalling cyclones over Bay of Bengal. *Mausam* 61 (1): 1–12.

Mohapatra, M., D.P. Nayak, Monica Sharma, R.P. Sharma, and B.K. Bandyopadhyay. 2015. Evaluation of official tropical cyclone landfall forecast issued by India Meteorological Department. *Journal of Earth System Science* 124 (4): 861–874. https://doi.org/10.1007/s12040-015-0581-x.

Mohapatra, M., B. Geetha, and M. Sharma. 2017. Reduction in uncertainty in tropical cyclone track forecasts over the North Indian ocean. *Current Science* 112 (9): 1826–1830.

Mohapatra, M., and Monica Sharma. 2019. Cyclone warning services in India during recent years: A review. *MAUSAM* 70 (4): 635–666.

Murthy, B.S., R. Latha, and H. Madhuparna. 2017. WRF simulation of a severe hailstorm over Baramati: A study of space–time evolution. *Meteorological Atmosphere Physics*. https://doi.org/10.1007/s00703-0170516-y.

NDMA. 2021. *National guidelines for preparation of action plan—Prevention and management of cold wave and frost*, 74. National Disaster Management Authority, Govt. of India.

NHC. 2016. The Northeast and North Central Pacific hurricane database 1949–2016. United States National Oceanic and Atmospheric Administration's National Weather Service.

Nizamuddin, S. 1993. Hail occurrences in India. *Weather* 48: 90–92.

Orlanski, I. 1975. A rational subdivision of scales of atmospheric processes. *Bulletin American Meteorology Social* 56:527–530.

Palmer, W.C. 1965. Research Paper No. 45, Meteorological Drought; Superintendent of Documents, U.S. Government Printing Office: Washington, DC, USA.

Parker, M.D., and R.H. Johnson. 2000. Organizational modes of midlatitude mesoscale convective systems. *Monthly Weather Review* 128: 3413–3436.

Sandeep, Araveti, A. Jayakumar, M. Sateesh, Saji Mohandas, V.S. Prasad, and E. Rajagopal. 2021. Assessment of the efficacy of lightning forecast over India: A diagnostic study. *Pure and Applied Geophysics*. https://doi.org/10.1007/s00024-020-02627-5.

Sarkar, Abhijit, Devajyoti Dutta, Paromita Chakraborty, and Someshwar Das. 2017. Numerical diagnosis of situations causing heavy rainfall over the Western Himalayas. *Modeling Earth Systems and Environment* 3 (2): 515–531. https://doi.org/10.1007/s40808-017-0310-3.

Schubert, S.D., M.J. Suarez, P.J. Pegion, R. Koster, and J.T. Bacmeister. 2008. Potential predictability of long-term drought and pluvial conditions in the U.S. Great plains. *Journal of Climate* 21: 802–816.

Sigdel, M., and M. Ikeda. 2010. Spatial and temporal analysis of drought in Nepal using standardized precipitation index and its relationship with climate indices. *Journal of Hydrology Meteorological* 7: 59–74.

Sillmann, Jana, Thordis Thorarinsdottir, Noel Keenlyside, Nathalie Schaller, and Lisa Alexander et al. 2017a. *Understanding, modelling and predicting weather and climate extremes: Challenges and opportunities. Weather and climate extremes,* 65–74. Elsevier. https://doi.org/10.1016/j.wace.2017a.10.003.

Sillmann, J., Thordis Thorarinsdottir, Noel Keenlyside, Nathalie Schaller, and Lisa Alexander et al. 2017b. *Understanding, modeling and predicting weather and climate extremes: Challenges and opportunities. Weather and climate extremes,* 65–74. Elsevier. https://doi.org/10.1016/j.wace.2017b.10.003.

Simons, A. 2011. From observations to service delivery: Challenges and opportunities. *WMO Bulletin* 60 (2): 2011.

STORM. 2005. STORM science plan. India Department of Science and Technology Report, 118. Available online at www.imd.gov.in/SciencePlanofFDPs/STORM%20Science%20Plan.pdf.

Tinmaker, M.I.R., M.Y. Aslam, and D.M. Chate. 2015. *Lightning activity and its association with rainfall and convective available potential energy over Maharashtra.* India: Nat Hazards. https://doi.org/10.1007/s11069-015-1589-x.

Toth, Zoltan. 2001. Meeting summary: Ensemble forecasting in WRF. *Bulletin of the American Meteorological Society* 82 (4): 695–698.

Tyagi, A. 2007. Thunderstorm climatology over Indian region. *Mausam* 58: 189–212.

Tyagi, A., D.R. Sikka, Suman Goyal, and Mansi Bhowmick. 2012. A satellite based study of pre-monsoon thunderstorms (Nor'westers) over eastern India and their organization into mesoscale convective complexes. *Mausam* 63 (1): 29–54.

Tyagi, Ajit, M. Mohapatra, B.K. Bandyopadhyay, and Naresh Kumar. 2010. Inter-annual variation of frequency of cyclonic disturbances landfalling over WMO/ESCAP panel member countries", WMO Technical Document, WMO/TD-No. 1541 WWRP-210-2, 1-7, WMO, Geneva.

United States Department of Energy (USDE). 2009. *Natural Phenomena hazards design and evaluation criteria for Department of Energy: E.2.2 additional adverse effects of tornadoes,* E7. Retrieved 2009–11–20.

Webster, P.J., V.E. Toma, and H.-M. Kim. 2011. Were the 2010 Pakistan floods predictable? *Geophysical Research Letters* 38: L04806. https://doi.org/10.1029/2010GL046346.

Webster, P.J., and Kristofer Y. Shrestha. 2013. *An extended-range water management and flood prediction system for the Indus river basin system: Application to the 2010–2012 floods,* 68. Project report.

Wilhite, D.A., and M.H. Glantz. 1985. 1985: Understanding the drought phenomenon: The role of definitions. *Water International* 10: 111–120.

Wurman, Joshua, C. Alexander, P. Robinson, and Y. Richardson. 2007. Low-level winds in tornadoes and potential catastrophic tornado impacts in urban areas. *Bulletin of the American Meteorological Society. American Meteorological Society* 88 (1): 31–46.

Yair, Y., B. Lynn, C. Price, V. Kotroni, K. Lagouvardos, E. Morin, A. Mugnai, and M.D.C. Llasat. 2010. Predicting the potential for lightning activity in Mediterranean storms based on the weather research and forecasting (WRF) model dynamic and microphysical fields. *Journal of Geophysical Research* 115: D04205. https://doi.org/10.1029/2008JD010868.

Zhang, X., G. Hegerl, S. Seneviratne, R. Stewart, F. Zwiers, and L. Alexander. 2014. *WCRP grand challenge: Science underpinning the prediction and attribution of extreme events.* Available at https://www.wcrp-climate.org/images/documents/grand_challenges/GC_Extremes_v2.pdf.

Part II
Extreme Rainfall Events and Thunderstorms

Chapter 4
Statistical Characteristics of Extreme Rainfall Events Over the Indian Subcontinent

P. C. Anandh, Naresh Krishna Vissa, and Bhishma Tyagi

Abstract The understanding of various characteristics of rainfall is essential for water resources management. However, the highly varying nature of rainfall is constrained in accurate estimation of rainfall over a particular region. Such variability in rainfall leads to either floods or droughts, and both are potentially catastrophic. Moreover, anthropogenic climate change further complicates understanding the various characteristics of rainfall. Thus, understanding this chaotic nature of rainfall is not only interesting but also challenging. Over the Indian subcontinent, considerable attention is given to understanding the various statistical characteristics of seasonal, daily and extreme rainfall. However, the daily rainfall distribution inequality is not fully understood across the seasons over the Indian subcontinent. In this regard, the present study aims to employ the GINI inequality index to understand the daily rainfall concentration. The investigation of spatial and temporal variation in daily rainfall concentration helps in understanding atmospheric processes that influence rainfall variability. In addition, the peak over threshold method and two-parameter gamma distributions are employed to understand the statistical character and its relationship with rainfall distribution over the Indian subcontinent for all the seasons. The results would have significant implications in the flood and drought forecasting and water resources planners for sustainable water resources management.

Keywords Extreme rainfall events · GINI index · Gamma distribution · Indian summer monsoon · Indian subcontinent

P. C. Anandh · N. K. Vissa (✉) · B. Tyagi
Department of Earth and Atmospheric Sciences, National Institute of Technology Rourkela, Sundargarh, Odisha 769008, India
e-mail: vissanaresh@gmail.com; vissan@nitrkl.ac.in

© The Centre for Science & Technol. of the, Non-aligned and Other Devel. Countries 2022
A. S. Unnikrishnan et al. (eds.), *Extreme Natural Events*,
https://doi.org/10.1007/978-981-19-2511-5_4

4.1 Introduction

4.1.1 Extreme Weather Events and Climate Change

The frequent occurrences of extreme weather events (EWE) pose a significant challenge to sustainable living and development across the globe. For instance, according to the global climate risk report 2021, more than 475,000 people lost their lives between 2000 and 2019 due to eleven thousand EWE, which occurred globally, with US$2.56 trillion direct losses in the economy (Eckstein et al. 2021). Even though many factors are responsible for the EWE, the role of anthropogenic climate change is invariably evident in most of the EWE that occurred in recent decades. In addition, there is a high probability of increasing frequency, intensity and duration of EWE due to the rising carbon footprint in the climate systems. The lack of long-term data in environmental, climatic indicators and socio-economic fields makes it difficult to assess the sector wise risk associated with EWE, either in advance or after. Therefore, for efficient mitigation and adaption, the scientific understanding of the various temperament of the EWE process, drivers and their relationship with human-induced climate change is essential and inevitable.

4.1.2 Precipitation Extremes and Climate Change

Among the various EWE, the precipitation extremes are most disruptive and cause widespread devastation. The intensification of the global water cycle due to the increased moisture availability in the atmosphere, according to the Clausius–Clapeyron relation (Trenberth 2012). Such as per degree rise in temperature increases seven per cent of moisture in the atmosphere. Thus, increasing moisture content, other climate variables and changes in large-scale circulation patterns result in precipitation extremes. The future climatic simulations as coupled model intercomparison project phase 5 (CMIP5) and phase 6 (CMIP6) highlight increasing in the precipitation extremes in not only daily time scale but also in sub-daily time scales (Prein et al. 2017; Li et al. 2020). The results are in accordance with the observational evidence with low bias. In addition, global warming intensifies and increases the frequency of ocean-atmospheric coupled phenomena such as Madden Julian Oscillation (MJO), El Niño–Southern Oscillation (ENSO) and Indian Ocean Dipole (IOD) than earlier with associated atmospheric circulation patterns (Subramanian et al. 2014; Cai et al. 2014a, b, 2015). It is also a significant concern. Thus, the precipitation extremes occurring with the large-scale climatic drivers are stronger and destructive.

4.1.3   Precipitation Extremes-Indian Scenario

There have been several attempts to understand extreme rainfall events and intensity from observations and modelling efforts at the national level. Based on daily rainfall data, trend analysis was conducted by previous studies. Rakhecha and Soman (1994) reported the significant increase of extreme rainfall events over the west coast of India (north of 12° N) and the central parts of the peninsula. Under a global warming environment, increases in extreme rain events and decreased trend of frequency of moderate events are reported over central India during the summer monsoon season from 1951 to 2000 (Goswami et al. 2006). The weakening of the Indian summer monsoon (ISM) circulation is evident from analysing rain events over six homogenous rainfall zones (Dash et al. 2009). Analysis of long-term historical rainfall data trends reveals that monsoon rainfall has decreased; whereas, the rest of the season's rainfall increased at the national level (Kumar et al. 2010). Pattanaik and Rajeevan (2010) examined the variability of extreme rainfall events during ISM. Over the Himalayas, rainfall events with one-day duration have been reported during both excess and drought rainfall years; moreover, these extreme rainfall events were reported in the ENSO years (Nandargi and Dhar 2011). Extreme rainfall and flash floods are increasing at an alarming rate except in some parts of central India (Guhathakurta et al. 2011). Jain and Kumar (2012) identified the increasing trend of rainfall over the east coast of India. Spatial trends of summer monsoon rainfall climatology over the northeast and west coast of the Indian region suggest increasing trends. The western Himalayas and north-central Indian regions, respectively, show decreasing trends (Kishore et al. 2016). Ghosh et al. (2016) reported the contrasting rainfall trends in the spatial variability of means and extremes during the ISM. Their findings contradict the conventional notion of 'dry areas becoming drier and wet areas becoming wetter under global warming in India. Over the monsoon core regions, threefold increases of extreme rainfall are reported during the ISM, and few studies advocate that increases of wet spells during ISM (Singh et al. 2014; Roxy et al. 2017). Recently, the southward shift in the ERE during ISM, and primarily concentrating over the southern India regions due to the eastward shift in the moisture flux over the Indian Ocean regions (Suman and Maity 2020) Similarly, during the other seasons, the occurrences of precipitation extremes are escalating. The frequency of the number of tropical cyclones during the pre-monsoon and post-monsoon increases and gets stronger (Deshpande et al. 2021). The influence of climate change on the thunderstorms frequencies and other climate variables was also reported over eastern and northeast India during the pre-monsoon seasons (Sahu et al. 2020). Similarly, during the winter monsoon over the Himalayan and northern Indian regions, the probability of extremes is more due to changes in western disturbances characteristics (Madhura et al. 2015).Based on the above studies thus the occurrences of precipitation extremes are prevalent across the seasons and regions. Thus, the qualitative and quantitative analysis of extreme rainfall using various statistical methods would enhance the understanding of precipitation extremes. Thus, the present study investigates the statistical characteristics using percentile threshold methods, Gamma probability and GINI index over the Indian subcontinent.

4.2 Data and Methodology

4.2.1 The Tropical Rainfall Measuring Mission (TRMM)-Multi-satellite Precipitation Analysis (TMPA)

The tropical measuring mission (TRMM)-based multi-satellite precipitation analysis (TMPA) 3B42 version 7 (Huffman et al. 2007) daily rainfall (mm day^{-1}) data has been used to study the various statistical characteristics of rainfall over the Indian region. The TRMM TMPA data has a spatial resolution of $0.25° \times 0.25°$ with a temporal resolution at three hours. The present study was conducted from 1998 to 2018. It is one of the high-quality data available over the tropics and sub-tropics for an extended period. The algorithm combines the various satellites high-quality microwave and infrared readings. It includes new techniques such as surface reference, modified attenuation characteristics for solid and melting layers and non-uniform beam filling corrections in version 7 of TRMM TMPA. The data product has been validated with the rain gauges and found in agreement with the IMD rainfall data. In addition, the representation of rainfall has been improved in the latest version 7 compared to version 6 over the various sub-regions over the Indian regions and neighbouring oceans (Prakash et al. 2015, 2016). However, overestimation of rainfall was reported over the Himalayan orography at higher latitudes (Bharti et al. 2016). The data product enhanced the knowledge of various rainfall characteristics over the Indian subcontinent since its inception. For instance, the primary component of the hydrological cycle, such as the long-term diurnal cycle of rainfall, has been investigated over the Indian subcontinent. The rainfall "hot spot regions" during the different octets has been well discussed across the seasons with the influence of intraseasonal and interannual phenomena (Varikoden et al. 2012; Anandh et al. 2018). The studies have also proved the ability of TRMM precipitation data to investigate the weather extremes over the Indian subcontinent, such as extreme rainfall events and droughts (Yaduvanshi et al. 2015; Bharti et al. 2016; Anandh and Vissa 2020).

4.3 Methodology

4.3.1 Daily Extreme Rainfall Indices

The percentile-based threshold method or the peak over threshold method used to identifies the daily extreme rainfall events. Such as aggregating the rainfall events, which are greater than particular threshold values over every grid point. Notably, the daily threshold values will be different with each grid point. In this study, extreme daily rainfall is defined as exceeding 95th percentile threshold (hereafter, R95) values over every grid point. The frequency and intensity of the R95 values would vary with every grid point. The R95 values were calculated for each season as defined by the Indian Meteorological Department (IMD), such as winter (January–February), pre-monsoon (March–May), ISM (June–September) and post-monsoon (October–December). All the analysis was carried out only for the wet day, i.e. rainfall ≥ 1 mm day^{-1}. The sum of total extreme rainfall to the total seasonal rainfall at each gird gives the extreme rainfall contribution (ERC) (Mishra and Singh 2010; Anandh and Vissa 2020). The number of rainy days that exceeds the R95 threshold over the particular grid points gives the frequency of extreme events over each grid point (Anandh and Vissa 2020).

4.3.2 GINI Index

The disparity in rainfall distribution is a common phenomenon; particularly, over the Indian subcontinent. The GINI index will accurately quantify such disparity in rainfall concentration. In general, the GINI index is widely used in economics to study the inequality in income or wealth distributions. However, due to its robustness, insensitive to the scaling or probability distribution provides easy interpretation between the different climatic zones. So, the index has been widely used in hydrology and climatic sciences. The values of GINI coefficient vary between 0 and 1, "0" represents the even distribution of rainfall; whereas, "1" signifies the uneven rainfall distribution. Such as the lowers GINI values on a rainy day are equally distributed within the seasons; whereas, the higher values signify most of the rainfall occurs in fewer days. The GINI index is also computed for the seasonal wise as described in Sect. 3.1 (Martin-Vide 2004; Rajah et al. 2014; Singh et al. 2017). The GINI index was calculated using the formula given in Eq. (4.1), where n signifies the number of observations and y_i represents the rainfall amount belongs any observations i.

$$G = \frac{1}{n}\left(n + 1 - 2\left(\frac{\sum_{i=1}^{n}(n + 1 - i)y_i}{\sum_{i=1}^{n} y_i}\right)\right) \tag{4.1}$$

4.3.3 Gamma Probability Distribution Function (Gamma PDF)

The daily rainfall consists of light, moderate and heavy rainfall events and such rainfall distributions are highly asymmetric and positively skewed. The Gamma probability distribution elucidates the rainfall distribution in two parameters: the scale and the shape. Thus, these two parameters are easy to interpret and describe rainfall characteristics between different climate zones. Among all the many probability distributions, the rationale to choose the Gamma PDF is as follows. (A) It has been widely used to describe rainfall distribution in both observed and modelled data. (B) The lower limit of the Gamma distribution on the left side is bounded to zero so that the possibility of negative rainfall values is absent. (C) The values of the Gamma PDF positively skewed, i.e. more values on the right side of the distribution. (D) Flexibility in the distribution of the shape parameter, such as the values may vary from one to higher values. Such variations have many rainfall regimes. (E) The different time scales of rainfall, such as daily to seasonal, could be easily modelled using the two-parameter Gamma PDF with relatively high accuracy.

The Gamma PDF is calculated by Eq. (4.2), and α denotes the shape parameter; whereas, the β denotes the scale parameter. τ denotes the Gamma PDF (Wilks 1990; May 2004; Husak et al. 2007).

$$f(x) = \frac{1}{\beta^{\alpha}.\Gamma(a)}.x^{\alpha-1}.e^{-x/\beta} \quad x, \alpha, \beta > 0 \tag{4.2}$$

4.4 Results and Discussion

4.4.1 Seasonal Spatial Distribution of R95 Threshold Values, Number of Extreme Rainfall and Extreme Rainfall Contribution

4.4.1.1 Winter Season

The spatial distributions of R95 percentile threshold values are shown in Fig. 4.1 for the winter season. A large spatial difference is found in the distributions of R95 values. Defining the seasonal extreme rainfall value at every grid point is essential instead of the fixed threshold value for larger regions. *Oceanic regions*: Over the south-western Bay of Bengal (BoB) and eastern equatorial Indian Ocean (EIO) regions, the R95 values are higher than 80 mm day^{-1}; whereas, over the other oceanic regions, the R95 values are less than 40 mm day^{-1}. *Land regions*: The higher values of R95 are observed over the south of 18° N and north of 28° N, such as over the southern east

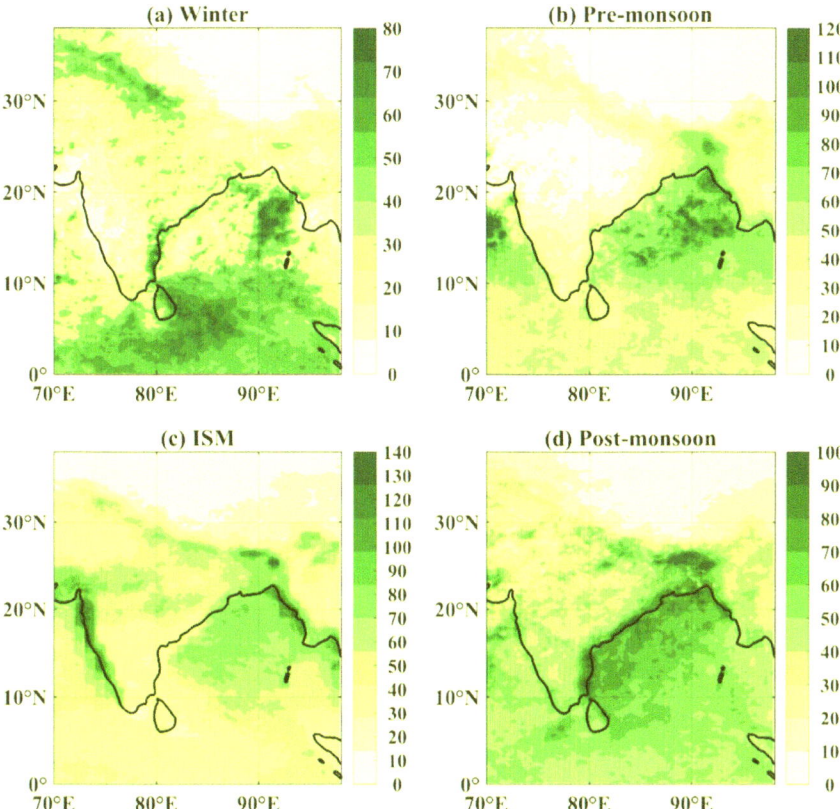

Fig. 4.1 Extreme rainfall (R95 percentile) threshold values for **a** winter, **b** pre-monsoon, **c** Indian summer monsoon (ISM) and **d** post-monsoon

coastal regions of BoB and north-western Himalayan orography (60 mm day^{-1}). Over the other regions, the R95 threshold values prevail less than 30 mm day^{-1}.

The total number of extreme rainfall events during the winter season is shown in Fig. 4.2a. Over the oceanic regions, the south-western BoB regions and eastern equatorial Indian Ocean regions, the events are comparatively higher during the winter season (~ 40–50 events). Over the western equatorial Indian Ocean, regions ~ 25 ERE are evident; whereas, over the northern regions of BoB and Arabian Sea (AS), the events are less during the winter season. Over the land regions, the ERE were prevalent over the Himalayan orography; whereas, the Indian subcontinent ERE were comparatively lesser during the winter season. The extreme rainfall contribution for the winter is shown in Fig. 4.3a. Over the Himalayan orography and southern east coastal regions, the ERC are higher; whereas, over the oceanic regions, equatorial Indian Ocean regions ~ 30% of the total rainfall is associated with the ERE. The higher ERC may be associated due to less number of events. The major rain-bearing systems over the Himalayan mountains are western disturbances during the winter

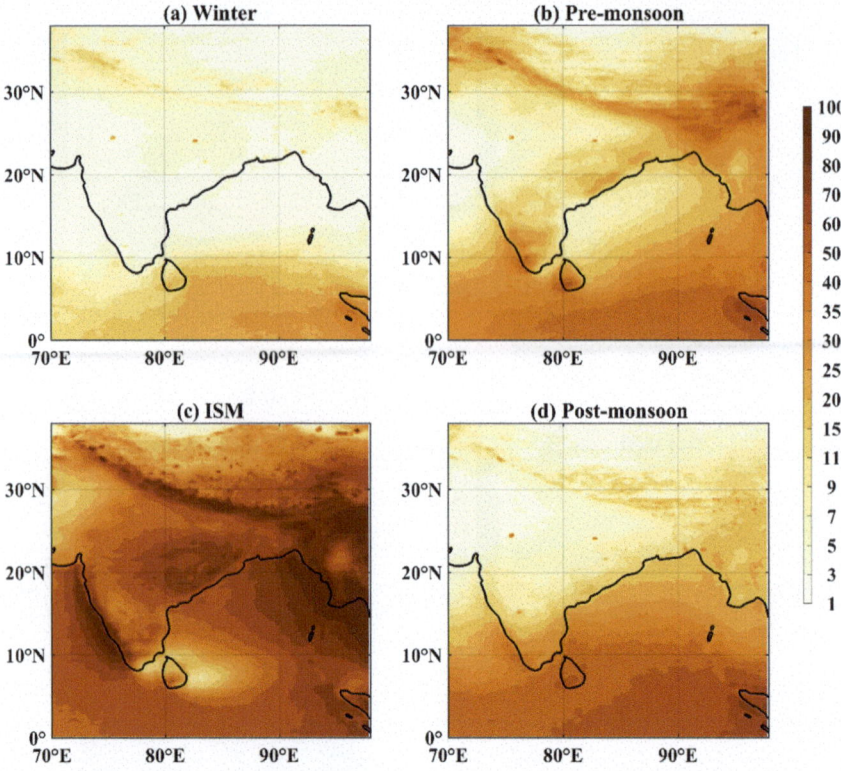

Fig. 4.2 Total number of extreme rainfall events during **a** winter, **b** pre-monsoon, **c** Indian summer monsoon (ISM) and **d** post-monsoon

monsoon and the late withdrawal of post-monsoon season (Raj 2003) and intraseasonal convective activity (Vialard et al. 2011) might be responsible for the ERE over the southern BoB regions.

4.4.1.2 Pre-monsoon Season

The spatial distribution of R95 for the pre-monsoon is shown in Fig. 4.1b, respectively. During this season, over the oceanic regions, the northern BoB and AS regions, the R95 values are higher (~ 100 mm day^{-1}); whereas, over the southern of BoB (AS) and EIO, the R95 values are ~ 40 mm day^{-1}. Over the land regions, peak values are evident over northeast India (NEI), the Arkan Mountains of Myanmar and the Ganges Delta regions (~ 80–100 mm day^{-1}). The secondary peak is evident over the Himalayan mountains, western (eastern) Ghats orography and coastal regions (30–60 mm day^{-1}). The frequency of extreme rainfall events during the pre-monsoon is shown in Fig. 4.2b, respectively. Oceanic regions, the maximum ERE over the

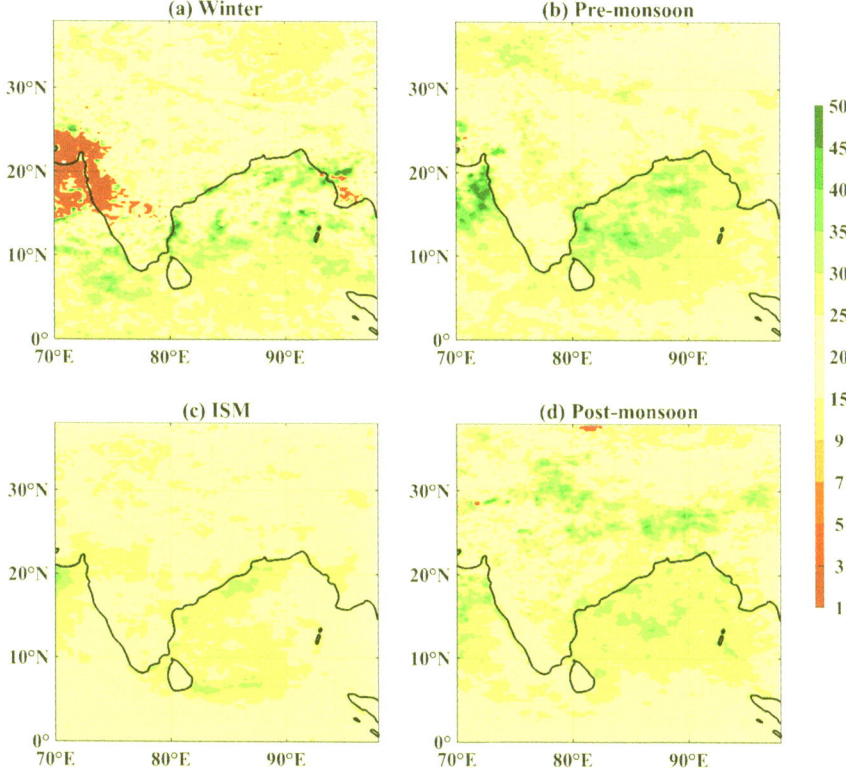

Fig. 4.3 Extreme rainfall contribution for **a** winter, **b** pre-monsoon, **c** Indian summer monsoon (ISM) and **d** post-monsoon

EIO regions (~ 70) and decreases towards the poles. The presence of active winter monsoons over the equatorial regions might influence the frequency of ERE. Over the land, the frequency of events is higher over the NEI, Himalayan mountains, western Ghats (WG) and eastern Ghats (EG) mountains.

Similarly, over the Chota Nagpur plateau and the Ganges Delta regions, events frequency were higher. The formation of the strong localized mesoscale convective system and tropical cyclones during pre-monsoon might be the plausible reasons for the ERE over these regions (Tyagi et al. 2011; Vissa et al. 2013). In addition, the western disturbances might influence the ERE over the high altitude regions of the Himalayan mountains and occasionally over the northern Indian plains. The seasonal ERC is shown in Fig. 4.3b. 20–30% total rainfall occurred as extremes during the pre-monsoon over the major rainfall zones of land and oceanic regions. Over the off-shore regimes of coastal regions, Chota Nagpur regions and the north-western Himalayan region, the ERC are notably higher, are as shown in Fig. 4.3b.

4.4.1.3   Indian Summer Monsoon

The spatial distribution of R95 is shown in Fig. 4.1c. Significant spatial heterogeneity of R95 is evident during this ISM. Over the oceanic regions, the poleward increase of R95 values is evident over the BoB. Such as the highest R95 values found over the northern BoB regions, followed by the central BoB regions (~ 80–100 mm day^{-1}). Over the other oceanic regions, values are less than 20–50 mm day^{-1}. Over the land regions, the orography regions of Himalayan, NEI and WG show the R95 prominent peak (~ 100 mm day^{-1}), followed by the monsoon core regions, where the values are R95 peak lies around 40–70 mm day^{-1}.

The distribution of the ERE events during the ISM is shown in Fig. 4.2c. Over the oceanic regions, except the south-western BoB region, the ERE frequency is comparatively higher during ISM. For instance, off-shore regions of west and east regions of BoB and AS, more than 100 ERE are evident. Over the land regions, the maximum number of events found over the orographic regions of Himalayan arc, NEI and WG regions more than 100 events were evident. The secondary peak was evident over the monsoon core regions and the EG orographic regions. The ERC during the ISM is given in Fig. 4.3c. ERC over major rainfall zone of oceans and land regions contributes about 20–30% to the seasonal totals during ISM.

Over the higher altitude regions (WG, Arakan mountains and Himalayas mountains), the interaction between moisture-laden low-level jet and orography leads to high precipitation occurrences over the windward side (Shige et al. 2017). Over the WG, the mechanism for the ERE and associated persistent convection is mainly attributed to stratiform heating, low-level moisture convergence, higher latent heat fluxes and strong wind shear between the low and middle troposphere (Maheskumar et al. 2014). During the active monsoon conditions, rain-fed systems such as off-shore vortices, on-shore vortices, monsoon depressions, tropical convergence zone, low-pressure systems, cloud bursts and mid-tropospheric cyclones are primarily responsible for the ERE over the Indian monsoon core region (Francis and Gadgil 2006; Pattanaik and Rajeevan 2010; Maheshkumar et al. 2014; Flynn et al. 2017; Fletcher et al. 2020). Over central India, the ERE is associated with the monsoon trough, monsoon depressions and low-pressure systems (Ajayamohan et al. 2010). The presence of the monsoon trough acts as a zone of active convection can lead to severe thunderstorms, lightning and flash floods (Sato 2013).

4.4.1.4   Post-monsoon Season

The spatial variations of R95 values and ERE contribution to the seasonal total are shown in Figs. 4.1d and 4.3d, respectively. Unlike other seasons, higher R95 (50–90 mm day^{-1}) is found along the east coast of India, south-western BoB and Bangladesh. Results are in concurrence with the findings of Hamada et al. (2014). Over the eastern equatorial Indian Ocean and the Arabian Sea, the R95 range 40–60 mm day^{-1}; however, low R95 values are evident over the other regions. The poleward decreases of ERE are evident over the oceanic regions, such that more ERE was found over the EIO and southern BoB regions, as shown in Fig. 4.2d. Over the land regions, the frequency of events is higher over the southern Indian peninsula and Himalayan orography (< 60). In contrast, the number of events over the other regions is less during post-monsoon seasons, as shown in Fig. 4.2d. The ERE contributes ~ 20–40% of the seasonal rainfall during the post-monsoon season (Fig. 4.3d).During the post-monsoon season, major rain-producing weather systems over the Indian subcontinent are tropical cyclones over the BoB, thunderstorms and active northeast monsoon season associated convective systems. Recent studies indicate the influence of interannual phenomenon and intraseasonal over the post-monsoon ERE (Vissa et al. 2013; Boyaj et al. 2018; Anandh and Vissa 2020).

4.4.2 Gamma Probability Density Function

The Gamma PDF was employed to delineate the extreme rainfall zones during all the seasons. Theoretically, the shape (α)-dominated areas signify the symmetric distribution of rainfall and fewer extremes; whereas, in the scale (β) dominated rainfall regimes, the extreme rainfall events are more prevalent. *Oceanic regions*: Over the BoB regions, the scale-dominated regions are more during the pre-monsoon, ISM and post-monsoon seasons; whereas, during winter, the scale values are high over the south-western BoB and central EIO regions shown in Fig. 4.4. Similarly, over the AS regions, scale values are dominant across the seasons except during winter. Land and Coastal regions: During ISM, scale values are high over the northern and central BoB regions (AS), and shape values are low. The monsoon rain-bearing systems such as monsoon depressions, off-shore trough and mid-tropospheric cyclones prevalent

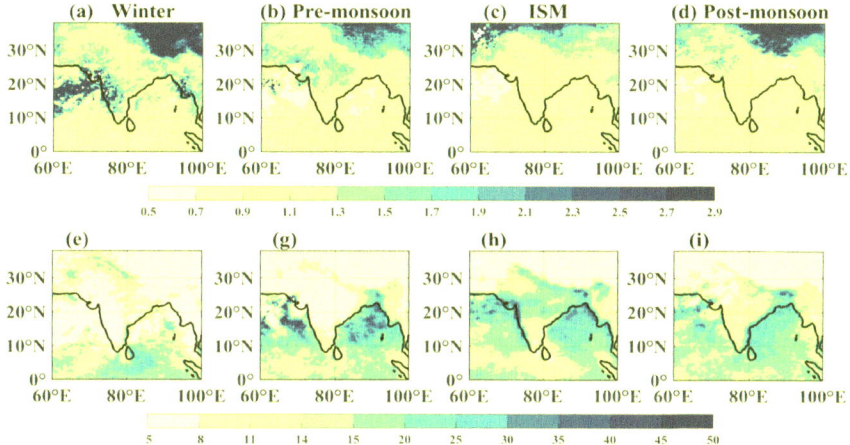

Fig. 4.4 The shape and scale parameter from a gamma probability distribution for **a** winter, **b** pre-monsoon, **c** Indian summer monsoon (ISM) and **d** post-monsoon

over the regions would result in more extreme rainfall, resulting in dominant scale values. Whereas, over the south-western BoB and south-eastern AS during post-monsoon, the scale values are dominant, where the frequency of the ERE and the tropical low-pressure system is prevalent. *Land and Coastal regions*: During ISM, the large-scale and low-shape values are prominent over the geographically distinct region of the Indian subcontinent. Over the WG and Himalayan mountains, NEI, monsoon core regions and west (northern east) coastal regions, the scale values are dominant than shape values and signify more rainfall extremes. Similarly, during the post-monsoon, east coastal regions of BoB, the scale values significantly. Overall, the "rainfall hot spot" regions are consistent with the higher R95 extreme event distributions over the oceanic and land regions. A detailed description of the significant rainfall zones and their plausible rain-bearing systems is given in Table 4.1 for all the seasons.

4.4.3   GINI Index

GINI index was employed to get insights regarding the inequality in rainfall distribution across the seasons (Fig. 4.5). The perusal of results provides valuable insights over the relative rainfall contribution during all the seasons. Oceanic regions: The high GINI values are evident during the winter, and pre-monsoon over the northern BoB regions indicates fewer rainy days. In contrast, during the ISM and post-monsoon season, values between around 0.5–0.75 signify moderate to heavy rainfall occurrences. Notably, the extreme rainfall zones of different seasons could be easily identified using GINI index. For instance, the GINI values are higher during winter and ISM (> 0.8) over the southern western BoB and adjoining coastal regions denotes the rainfall that occurred in the fewer rainy days and might be associated with extreme rainfall. Over the EIO regions, the values are comparatively higher during winter and pre-monsoon (> 0.65) than ISM and post-monsoon seasons (< 0.65), which denotes the moderate to heavy rainfall events during the former and moderate to light rainfall events during the latter. Over the land regions: The Himalayan orography shows more than 0.6 during all seasons denotes the absence of light rainfall over these regions, and most events occur as moderate to heavy rainfall events. The GINI values lie between (0.5 and 0.8) during ISM, which signifies moderate to heavy rainfall events over the monsoon core regions. The results are consistent with Rajah et al. (2014), who reported increasing extremes rainfall and decreasing the light rainfall globally using the GINI index.

Table 4.1   Major rain-bearing systems over distinct geographic parts of India across the seasons (adopted from Anandh and Vissa 2020)

Regions	Pre-monsoon	ISM	Post-monsoon	Winter
Himalayas	Western disturbances and deep-convective mesoscale convective systems (MCS) associated with thunderstorms (Dimri et al. 2015)	The interaction synoptic low-pressure systems with Himalayan orography, extratropical Rossby breaking (Romatschke et al. 2010; Houze et al. 2017)	Western disturbances, thunderstorms and remnants of tropical cyclones convection (Dimri et al. 2015)	Western disturbances (Dimri et al. 2015)
Central India and Indo-Gangetic plains	Nor'westers with deep-convective cores and tropical cyclones (Tyagi et al. 2011; Vissa et al. 2013)	Tropical convergence zone (TCZ) and synoptic-scale monsoon low-pressure systems (Krishnamurthy and Ajayamohan 2010; Romatschke et al. 2010)	Tropical cyclones formed over BoB and AS regions (Vissa et al. 2013; Anandh and Vissa 2020)	Western disturbances (Dimri et al. 2015)
Southern India and Deccan plateau	Thunderstorms, MCS and TCZ (Virts and Houze 2016)	TCZ and low-pressure systems (Krishnamurthy and Ajayamohan 2010)	Tropical cyclones and convective systems associated with the northeast monsoon (Vissa et al. 2013; Anandh and Vissa 2020)	Low-level troughs related to the easterlies and the late withdrawal of northeast monsoon (Vialard et al. 2011)
Bay of Bengal and east coast of India	A tropical cyclone, isolated and widespread thunderstorms (Vissa et al. 2013)	Low-pressure systems, MCS and TCZ (Romatschke et al. 2010; Virts and Houze 2016)	Tropical cyclones and the convective systems associated with the northeast monsoon (Vissa et al. 2013; Anandh and Vissa 2020)	The late withdrawal of northeast monsoon associated convection and easterly trough (Vialard et al. 2011)

(continued)

Table 4.1 (continued)

Regions	Pre-monsoon	ISM	Post-monsoon	Winter
The Arabian Sea, west coast of India and WG	Thunderstorms and tropical cyclones (Vissa et al. 2013)	Northward propagation of TCZ, off-shore trough and vortex and mid troposphere cyclones (Francis and Gadgil 2006)	Thunderstorms, tropical cyclones and convective systems associated with the northeast monsoon (Vissa et al. 2013; Anandh and Vissa 2020)	The troughs are related to the easterly winds

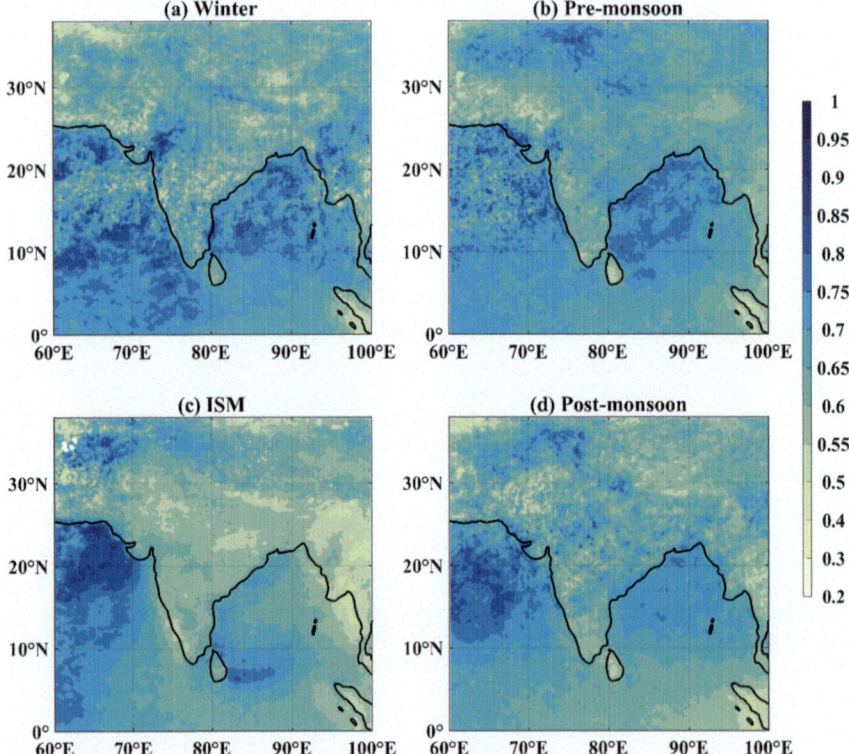

Fig. 4.5 The GINI index values for **a** winter, **b** pre-monsoon, **c** Indian summer monsoon (ISM) and **d** post-monsoon

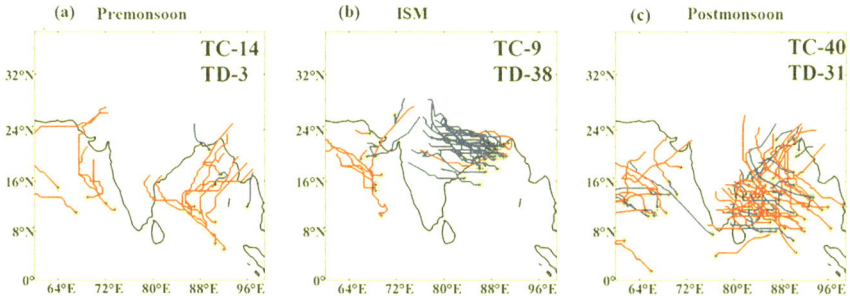

Fig. 4.6 India Meteorological Department (IMD) best tracks for tropical low-pressure systems for **a** winter, **b** pre-monsoon, **c** Indian summer monsoon (ISM) and **d** post-monsoon (adopted from Anandh and Vissa 2020)

4.4.4   Tracks of the Low-Pressure Systems

The tracks of the low-pressure systems such as tropical cyclones and tropical depressions for the pre-monsoon post-monsoon seasons are given in Fig. 4.6. During pre-monsoon, the frequencies of the tropical cyclones (14) are higher than the tropical depression (3), as shown in Fig. 4.6a. During ISM, there were 38 tropical depressions, and nine tropical cyclones occurred. The high vertical wind shear during ISM does not favour the formation of tropical cyclones. However, over the post-monsoon season, the activity of tropical cyclones and tropical depressions is more pounced, such as 41 and 31. Overall, 71 tropical low-pressure systems occurred post-monsoon season, followed by ISM (47) and pre-monsoon seasons (17). The tropical low-pressure systems are significant rain-bearing systems over the Indian subcontinent and eventually would cause ERE (Vissa et al. 2021). The genesis and tracks of these low-pressure systems agree with the extreme rainfall zones as elucidated in the previous sections. For instance, during ISM, the tracks of the low-pressure systems (Fig. 4.6b) are matched with the regions of high ERE frequency and scale values. Notably, over the regions of northern BoB and central India. Similarly, during the post-monsoon season (Fig. s4.6c), in BoB regions and the eastern coastal regions, the genesis and tracks of the tropical low-pressure systems are higher. They agree with the extreme rainfall statistics as discussed in the previous sections.

4.5   Conclusion

The sudden catastrophe would affect the people's socio-economic well-being even after cessation. Therefore, to mitigate the vast impact, the understanding of various characteristics of weather and climate extremes is essential. In this regard, the present study investigated the various extreme rainfall characteristics over the Indian subcontinent. The frequency of extreme rainfall events is higher during the ISM, followed by the post-monsoon and pre-monsoon season. The finding from the present study highlights the prevailing of extremes across the seasons. The identified "rainfall hot spot" regions are consistent gamma probability distributions. In addition, the GINI index employed to understand the inequality rainfall distributions shows that the occurrences of extremes are prevalent and the decrease in the little to moderate rainfall events in the Indian subcontinent. The findings from the present study will help for the water management and mitigation measures. The further scope of the work is to employ the high-resolution global precipitation measurement data to understand the hourly extreme rainfall characteristics over the Indian subcontinent. In addition, the coupled model intercomparison project phase 6 models simulations and projections will be used to understand the rainfall extremes in future scenario.

Acknowledgements P. C. Anandh acknowledges the National Institute of Technology for the technical and financial support for the research work. The authors would like to acknowledge Goddard Earth Sciences Data and Information Services Center (GES DISC) for providing Tropical Rainfall Measuring Mission (TRMM) data and the Indian Meteorological Department for providing tropical cyclones best tracks. NKV would like to acknowledge the Science and Engineering Research Board (SERB), Government of India (Grant Ref: ECR/2016/001896). Authors acknowledging anonymous reviewers and editors for the constructive and meticulous comments to improve the manuscript.

References

Ajayamohan, R.S., W.J. Merryfield, and V.V. Kharin. 2010. Increasing trend of synoptic activity and its relationship with extreme rain events over central India. *Journal of Climate* 23 (4): 1004–1013.

Anandh, P.C., and N.K. Vissa. 2020. On the linkage between extreme rainfall and the Madden–Julian oscillation over the Indian region. *Meteorological Applications* 27 (2): e1901.

Anandh, P.C., N.K. Vissa, and C. Broderick. 2018. Role of MJO in modulating rainfall characteristics observed over India in all seasons utilizing TRMM. *International Journal of Climatology* 38 (5): 2352–2373.

Bharti, V., C. Singh, J. Ettema, and T.A.R. Turkington. 2016. Spatiotemporal characteristics of extreme rainfall events over the Northwest Himalaya using satellite data. *International Journal of Climatology* 36 (12): 3949–3962.

Boyaj, A., K. Ashok, S. Ghosh, A. Devanand, and G. Dandu. 2018. The Chennai extreme rainfall event in 2015: The Bay of Bengal connection. *Climate Dynamics* 50 (7): 2867–2879.

Cai, W., S. Borlace, M. Lengaigne, P. Van Rensch, M. Collins, G. Vecchi, A. Timmermann, A. Santoso, M.J. McPhaden, L. Wu, and M.H. England. 2014a. Increasing frequency of extreme El Niño events due to greenhouse warming. *Nature Climate Change* 4 (2): 111–116.

Cai, W., A. Santoso, G. Wang, E. Weller, L. Wu, K. Ashok, Y. Masumoto, and T. Yamagata. 2014b. Increased frequency of extreme Indian Ocean Dipole events due to greenhouse warming. *Nature* 510 (7504): 254–258.

Cai, W., G. Wang, A. Santoso, M.J. McPhaden, L. Wu, F.F. Jin, A. Timmermann, M. Collins, G. Vecchi, M. Lengaigne, and M.H. England. 2015. Increased frequency of extreme La Niña events under greenhouse warming. *Nature Climate Change* 5 (2): 132–137.

Dash, S.K., M.A. Kulkarni, U.C. Mohanty, and K. Prasad. 2009. Changes in the characteristics of rain events in India. *Journal of Geophysical Research: Atmospheres* 114 (D10).

Deshpande, M., V.K. Singh, M.K. Ganadhi, M.K. Roxy, R. Emmanuel, and U. Kumar. 2021. Changing status of tropical cyclones over the north Indian Ocean. *Climate Dynamics*: 1–23.

Dimri, A.P., D. Niyogi, A.P. Barros, J. Ridley, U.C. Mohanty, T. Yasunari, and D.R. Sikka. 2015. Western disturbances: A review. *Reviews of Geophysics* 53: 225–246.

Eckstein, D., V. Künzel, L. Schäfer. 2021. *Global climate risk index 2021. Who suffers most from extreme weather events, 2000–2019*, 1–52. Germnay: Germanwatch e.V.

Fletcher, J.K., D.J. Parker, A.G. Turner, A. Menon, G.M. Martin, C.E. Birch, A.K. Mitra, G. Mrudula, K.M. Hunt, C.M. Taylor, and R.A. Houze. 2020. The dynamic and thermodynamic structure of the monsoon over southern India: New observations from the INCOMPASS IOP. *Quarterly Journal of the Royal Meteorological Society* 146 (731): 2867–2890.

Flynn, W.J., S.W. Nesbitt, A.M. Anders, and P. Garg. 2017. Mesoscale precipitation characteristics near the Western Ghats during the Indian Summer Monsoon as simulated by a high-resolution regional model. *Quarterly Journal of the Royal Meteorological Society* 143 (709): 3070–3084.

Francis, P.A., and S. Gadgil. 2006. Intense rainfall events over the west coast of India. *Meteorology and Atmospheric Physics* 94 (1): 27–42.

Ghosh, S., H. Vittal, T. Sharma, S. Karmakar, K.S. Kasiviswanathan, Y. Dhanesh, K.P. Sudheer, and S.S. Gunthe. 2016. Indian summer monsoon rainfall: Implications of contrasting trends in the spatial variability of means and extremes. *PLoS ONE* 11 (7): e0158670.

Goswami, B.N., V. Venugopal, D. Sengupta, M.S. Madhusoodanan, and P.K. Xavier. 2006. Increasing trend of extreme rain events over India in a warming environment. *Science* 314 (5804): 1442–1445.

Guhathakurta, P., O.P. Sreejith, and P.A. Menon. 2011. Impact of climate change on extreme rainfall events and flood risk in India. *Journal of Earth System Science* 120 (3): 359–373.

Hamada, A., Y. Murayama, and Y.N. Takayabu. 2014. Regional characteristics of extreme rainfall extracted from TRMM PR measurements. *Journal of Climate* 27 (21): 8151–8169.

Houze, R.A., L.A. McMurdie, K.L. Rasmussen, A. Kumar, and M.M. Chaplin. 2017. Multiscale aspects of the storm producing the June 2013 flooding in Uttarakhand, India. *Monthly Weather Review* 145 (11): 4447–4466.

Huffman, G.J., D.T. Bolvin, E.J. Nelkin, D.B. Wolff, R.F. Adler, G. Gu, Y. Hong, K.P. Bowman, and E.F. Stocker. 2007. The TRMM multi-satellite precipitation analysis (TMPA): Quasi-global, multiyear, combined-sensor precipitation estimates at fine scales. *Journal of Hydrometeorology* 8 (1): 38–55.

Husak, G.J., J. Michaelsen, and C. Funk. 2007. Use of the gamma distribution to represent monthly rainfall in Africa for drought monitoring applications. *International Journal of Climatology* 27 (7): 935–944.

Jain, S.K., and V. Kumar. 2012. Trend analysis of rainfall and temperature data for India. *Current Science*: 37–49.

Kishore, P., S. Jyothi, G. Basha, S.V.B. Rao, M. Rajeevan, I. Velicogna, and T.C. Sutterley. 2016. Precipitation climatology over India: Validation with observations and reanalysis datasets and spatial trends. *Climate Dynamics* 46 (1–2): 541–556.

Krishnamurthy, V., & Ajayamohan, R.S. 2010. Composite structure of monsoon low pressure systems and its relation to Indian rainfall. *Journal of Climate* 23(16): 4285–4305.

Kumar, V., S.K. Jain, and Y. Singh. 2010. Analysis of long-term rainfall trends in India. *Hydrological Sciences Journal* 55 (4): 484–496.

Li, C., F. Zwiers, X. Zhang, G. Li, Y. Sun, and M. Wehner. 2020. Changes in annual extremes of daily temperature and precipitation in CMIP6 models. *Journal of Climate*: 1–61.

Madhura, R.K., R. Krishnan, J.V. Revadekar, M. Mujumdar, and B.N. Goswami. 2015. Changes in western disturbances over the Western Himalayas in a warming environment. *Climate Dynamics* 44 (3–4): 1157–1168.

Maheskumar, R.S., S.G. Narkhedkar, S.B. Morwal, B. Padmakumari, D.R. Kothawale, R.R. Joshi, C.G. Deshpande, R.V. Bhalwankar, and J.R. Kulkarni. 2014. Mechanism of high rainfall over the Indian west coast region during the monsoon season. *Climate Dynamics* 43 (5–6): 1513–1529.

Martin-Vide, J. 2004. Spatial distribution of a daily precipitation concentration index in peninsular Spain. *International Journal of Climatology* 24 (8): 959–971.

May, W. 2004. Variability and extremes of daily rainfall during the Indian summer monsoon in the period 1901–1989. *Global and Planetary Change* 44 (1–4): 83–105.

Mishra, A.K., and V.P. Singh. 2010. Changes in extreme precipitation in Texas. *Journal of Geophysical Research: Atmospheres* 115 (D14).

Nandargi, S., and O.N. Dhar. 2011. Extreme rainfall events over the Himalayas between 1871 and 2007. *Hydrological Sciences Journal* 56 (6): 930–945.

Pattanaik, D.R., and M. Rajeevan. 2010. Variability of extreme rainfall events over India during southwest monsoon season. *Meteorological Applications* 17 (1): 88–104.

Prakash, S., A.K. Mitra, I.M. Momin, D.S. Pai, E.N. Rajagopal, and S. Basu. 2015. Comparison of TMPA-3B42 versions 6 and 7 precipitation products with gauge-based data over India for the southwest monsoon period. *Journal of Hydrometeorology* 16 (1): 346–362.

Prakash, S., A.K. Mitra, E.N. Rajagopal, and D.S. Pai. 2016. Assessment of TRMM-based TMPA-3B42 and GSMaP precipitation products over India for the peak southwest monsoon season. *International Journal of Climatology* 36 (4): 1614–1631.

Prein, A.F., R.M. Rasmussen, K. Ikeda, C. Liu, M.P. Clark, and G.J. Holland. 2017. The future intensification of hourly precipitation extremes. *Nature Climate Change* 7 (1): 48–52.

Raj, Y.E.A. 2003. Onset, withdrawal and intra-seasonal variation of northeast monsoon over coastal Tamil Nadu, 1901–2000. *Mausam* 54 (3): 605–614.

Rajah, K., T. O'Leary, A. Turner, G. Petrakis, M. Leonard, and S. Westra. 2014. Changes to the temporal distribution of daily precipitation. *Geophysical Research Letters* 41 (24): 8887–8894.

Rakhecha, P.R., and M.K. Soman. 1994. Trends in the annual extreme rainfall events of 1 to 3 days duration over India. *Theoretical and Applied Climatology* 48 (4): 227–237.

Romatschke, U., S. Medina, and R.A. Houze. 2010. Regional, seasonal, and diurnal variations of extreme convection in the South Asian region. *Journal of Climate* 23 (2): 419–439.

Roxy, M.K., S. Ghosh, A. Pathak, R. Athulya, M. Mujumdar, R. Murtugudde, P. Terray, and M. Rajeevan. 2017. A threefold rise in widespread extreme rain events over central India. *Nature Communications* 8 (1): 1–11.

Sahu, R.K., J. Dadich, B. Tyagi, and N.K. Vissa. 2020. Trends of thermodynamic indices thresholds over two tropical stations of north-east India during pre-monsoon thunderstorms. *Journal of Atmospheric and Solar-Terrestrial Physics* 211: 105472.

Sato, T. 2013. Mechanism of orographic precipitation around the Meghalaya Plateau associated with intraseasonal oscillation and the diurnal cycle. *Monthly Weather Review* 141 (7): 2451–2466.

Singh, D., M. Tsiang, B. Rajaratnam, and N.S. Diffenbaugh. 2014. Observed changes in extreme wet and dry spells during the South Asian summer monsoon season. *Nature Climate Change* 4 (6): 456–461.

Singh, J., S. Sekharan, S. Karmakar, S. Ghosh, P.E. Zope, and T.I. Eldho. 2017. Spatio-temporal analysis of sub-hourly rainfall over Mumbai, India: Is statistical forecasting futile? *Journal of Earth System Science* 126 (3): 38.

Subramanian, A., M. Jochum, A.J. Miller, R. Neale, H. Seo, D. Waliser, and R. Murtugudde. 2014. The MJO and global warming: A study in CCSM4. *Climate Dynamics* 42 (7–8): 2019–2031.

Suman, M., and R. Maity. 2020. Southward shift of precipitation extremes over south Asia: Evidences from CORDEX data. *Scientific Reports* 10 (1): 1–11.

Trenberth, K.E. 2012. Framing the way to relate climate extremes to climate change. *Climatic Change* 115 (2): 283–290.

Tyagi, B., V.N. Krishna, and A.N.V. Satyanarayana. 2011. Study of thermodynamic indices in forecasting pre-monsoon thunderstorms over Kolkata during STORM pilot phase 2006–2008. *Natural Hazards* 56 (3): 681–698.

Varikoden, H., B. Preethi, and J.V. Revadekar. 2012. Diurnal and spatial variation of Indian summer monsoon rainfall using tropical rainfall measuring mission rain rate. *Journal of Hydrology* 475: 248–258.

Vialard, J., P. Terray, J.P. Duvel, R.S. Nanjundiah, S.S.C. Shenoi, and D. Shankar. 2011. Factors controlling January–April rainfall over southern India and Sri Lanka. *Climate Dynamics* 37 (3): 493–507.

Virts, K.S., and R.A. Houze. 2016. Seasonal and intraseasonal variability of mesoscale convective systems over the South Asian monsoon region. *Journal of the Atmospheric Sciences* 73 (12): 4753–4774.

Vissa, N.K., P.C. Anandh, V.S. Gulakaram, and G. Konda. 2021. Role and response of ocean–atmosphere interactions during Amphan (2020) super cyclone. *Acta Geophysica*: 1–14.

Vissa, N.K., A.N.V. Satyanarayana, and B.P. Kumar. 2013. Intensity of tropical cyclones during pre-and post-monsoon seasons in relation to accumulated tropical cyclone heat potential over Bay of Bengal. *Natural Hazards* 68 (2): 351–371.

Wilks, D.S. 1990. Maximum likelihood estimation for the gamma distribution using data containing zeros. *Journal of Climate* 3 (12): 1495–1501.

Yaduvanshi, A., P.K. Srivastava, and A.C. Pandey. 2015. Integrating TRMM and MODIS satellite with socio-economic vulnerability for monitoring drought risk over a tropical region of India. *Physics and Chemistry of the Earth, Parts a/b/c* 83: 14–27.

Chapter 5
Complexities of Extreme Rainfall in the Philippines

Lyndon Mark P. Olaguera, Faye Abigail T. Cruz, Julie Mae B. Dado, and Jose Ramon T. Villarin

Abstract Extreme rainfall events are among the prevalent hazards in the Philippines. Better understanding, monitoring, as well as accurate forecasting of such events can help communities to prepare for and mitigate their disastrous impacts. In this chapter, some examples of past extreme rainfall events in the Philippines are presented, which highlight the complexity in the various weather systems contributing to extreme rainfall. Further research is still needed, particularly to improve understanding of the environment that leads to extreme rainfall events. Advances in science, data, and tools are important in these efforts, including improvements in forecasting extreme rainfall events in the Philippines, which can help reduce disaster risk and increase resilience in the country, especially in the face of projected increases in extreme rainfall events in a globally warmer future.

Keywords Extreme rainfall · Philippines · Monsoon · Tropical cyclone · Thunderstorm · Cold surge

5.1 Introduction

The Philippines is highly exposed to extreme rainfall hazards, including tropical cyclones. The archipelago is situated within the Asian monsoon region and along the path of tropical cyclones (TC), predominantly originating from the Western North Pacific. In addition, the high vulnerability of communities and sectors that are exposed to these hazards increases the country's risk to disasters. Based on data from the International Disaster Database, EM-DAT, the Philippines has experienced at least 432 natural disaster events from 1990 to January 2020, about 80% due to storms and floods (Brucal et al. 2020). Damage costs associated with TCs over the

L. M. P. Olaguera · F. A. T. Cruz (✉) · J. M. B. Dado · J. R. T. Villarin
Manila Observatory, Quezon City, Philippines
e-mail: fcruz@observatory.ph

L. M. P. Olaguera · J. R. T. Villarin
Department of Physics, Ateneo de Manila University, Quezon City, Philippines

© The Centre for Science & Technol. of the, Non-aligned and Other
Devel. Countries 2022
A. S. Unnikrishnan et al. (eds.), *Extreme Natural Events*,
https://doi.org/10.1007/978-981-19-2511-5_5

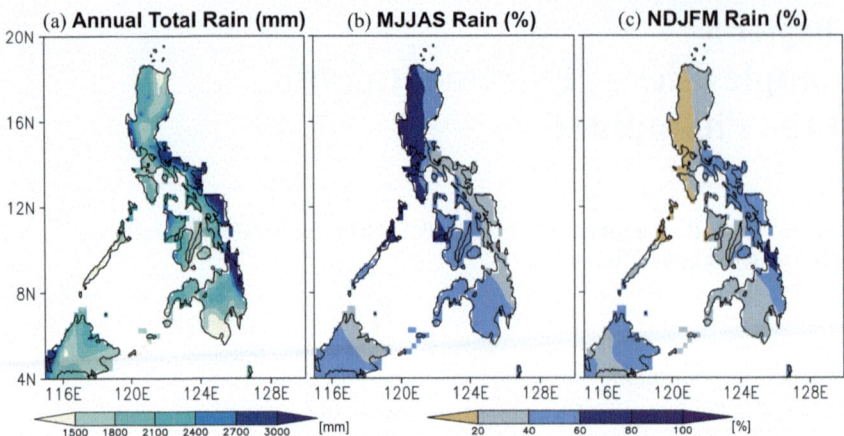

Fig. 5.1 **a** Annual total rainfall (mm) from 1981–2010 and percentage (%) of the annual total rain during the months of **b** May to September (MJJAS) and **c** November to March (NDJFM). *Source* APHRODITE (Yatagai et al. 2012)

Philippines have also been increasing in recent decades (Cinco et al. 2016). Hence, better understanding, monitoring, as well as accurate, timely, and reliable forecasts, are needed in order to prepare for extreme rainfall events that can potentially cause flooding, landslides, and other hydrometeorological disasters.

Figure 5.1 shows the climatological (1981–2010) annual total rainfall in the Philippines, where rainfall is higher (above 3000 mm) along the eastern coast of the country. The spatial pattern of seasonal rainfall in the country is influenced by the interaction of the prevalent winds with local topography (Matsumoto et al. 2020). In Luzon Island, the Cordillera mountain range and Sierra Madre mountain range that run over the western and eastern sides of the island, respectively, induce monsoon blocking effects, such that the western coast receives about 40–80% of its annual rainfall during the southwest monsoon (SWM, locally known as *Habagat*) season from as early as May until September, while areas along the eastern coast receive 20–40% of its annual rainfall (Fig. 5.1b). In contrast, areas along the eastern coast receive 40–80% of its annual rainfall during the northeast monsoon (NEM, locally known as *Amihan*) season from November to March, with less rainfall received in the western coast (Fig. 5.1c). Apart from the monsoons, TCs, both landfalling and non-landfalling, also contribute to rainfall in the country. On average, about 19–20 TCs pass through the Philippine Area of Responsibility (PAR) every year, nine of which make landfall in the Philippines (Cinco et al. 2016). Analysis of TC tracks indicates that TCs are more frequent during July to September and usually traverse the north and central Philippines (see Fig. 9 of Cinco et al. 2016). Higher percentage of TC-induced rainfall can also be found in these areas, where it accounts for more than 40% of the total annual rainfall (Cinco et al. 2016; Bagtasa 2017). Furthermore, there is also rainfall contribution from local convective systems (e.g., Bañares et al. 2021).

The interaction between a non-landfalling TC and the prevailing winds can lead to high rainfall amounts, often higher than the monthly climatological normal, which can be considered as extreme. Systems can interact with each other such that TCs can further enhance the SWM (Cayanan et al. 2011), a cold surge vortex enhancing the NEM (Olaguera et al. 2021a), or the active convection associated with the Madden–Julian Oscillation (MJO) coinciding with an enhanced NEM in the Philippines (Pullen et al. 2015). Furthermore, the interaction of these systems with local topography affects rainfall distribution and intensity over the Philippines (Chang et al. 2005). Other systems also induce extreme heavy rainfall events over the country such as cold surges (Lim et al. 2017; Abdillah et al. 2021), cold surge shearline (formerly locally termed as "tail-end of a cold front") (Olaguera et al. 2021a, b), cold surge vortices, i.e., westward propagating cyclonic circulations that are weaker than tropical cyclones (Chen et al. 2013, 2015), and localized thunderstorms embedded within the inter-tropical convergence zone (ITCZ; Yumul et al. 2011).

In this chapter, some examples of past extreme rainfall events in the country are discussed to highlight the complexity of these events, as well as identify possible ways forward in improving understanding. Section 5.2 presents several notable extreme weather events in the country and the potential mechanisms that induced them. Current challenges and directions for future research are discussed in Sect. 5.3, and the concluding remarks in Sect. 5.4.

5.2 Historical Extreme Rainfall Events

Extreme rainfall can be characterized by indices based on actual values or thresholds exceeded, in terms of magnitude, intensity, duration, and/or frequency. Several indices have been defined by the Expert Team on Climate Change Detection and Indices (ETCCDI) (Karl et al. 1999; Peterson 2005), which have been used to examine trends and changes in rainfall extremes globally and for particular areas. From the 1950s to 2000s, there has been an increasing trend in the annual total rainfall and annual number of days with rainfall above 50 mm over many parts of the Philippines (Endo et al. 2009). Villafuerte et al. (2014) also noted increasing trends in the maximum 5-day rainfall and decreasing trends in the length of dry spells during the July to September season from 1951 to 2010, particularly in stations located in the center and northwest of the country. Bagtasa (2017) found increasing trends in the TC-induced rainfall by more than 1 mm day^{-1} per decade since 2000 in most parts of the country.

In this section, some examples of past extreme rainfall events in the Philippines have been selected based on their notable impacts such as flooding and to show the different weather systems that can result in extreme rainfall. A working definition of extreme is used based on exceedances of certain thresholds. For example, these events have exceeded the climatological monthly total rainfall, or the monthly maximum 5-day precipitation. It should also be noted that only cases of "wet" extremes are discussed here.

5.2.1 Tropical Cyclones

In the Philippines, tropical cyclones (TC) are classified by the Philippine Atmospheric, Geophysical and Astronomical Services Administration (PAGASA), the country's official weather bureau, according to the maximum 10-min sustained wind speed (Table 5.1; Cinco et al. 2016). Tropical cyclones, both landfalling and non-landfalling, remain one of the most destructive natural phenomena that induce heavy to extreme rainfall, resulting in flooding and landslides in many regions around the globe including the Philippines. Both rainfall and strong winds associated with landfalling TCs have devastating impacts, as well as induce storm surges as these approach the coastline (Cruz and Narisma 2016; Takagi and Esteban 2016). Over Luzon Island, TCs before landfall that are slow-moving or have high intensity (i.e., typhoons) tend to induce more extreme precipitation than TCs that are fast-moving or have low intensity, especially during the June to September season (Racoma et al. 2022).

The Philippines has been ranked second in terms of TC landfalls following China (see Table 5.1 of Fudeyasu et al. 2014). However, no significant long-term trends in both the annual number of TCs that enter PAR and those that make landfall over the whole Philippines have been found (Cinco et al. 2016). Recent analysis of Basconcillo and Moon (2021) indicated about 210% increase in the TCs over the Philippines during the December to February season from 2012 to 2020 and as much as 480% increase in Mindanao Island, where the TC number is climatologically lower than those in the central and northern Philippines (Cinco et al. 2016).

Modes of variability also add complexity to TC variability in the country. Kubota and Chan (2009) compiled historical records of TCs over the Philippines and analyzed the interdecadal variability of landfalling TCs from 1902 to 2005. They found a close relationship between the annual landfalling TCs and the different phases of El Niño-Southern Oscillation (ENSO) and the Pacific Decadal Oscillation (PDO). That is, the annual number of landfalling TCs during the low PDO phase decreases (increases) in El Niño (La Niña) years. This relationship becomes unclear during the high phase of PDO. Their monthly analysis shows a larger difference in TC landfalls during

Table 5.1 PAGASA tropical cyclone intensity scale

Category	Intensity (maximum sustained wind speed at 10-min average)
Super Typhoon (STY)	>220 kph (>120 kts)
Typhoon (TY)	118–220 kph (64–120 kts)
Severe Tropical Storm (STS)	89–117 kph (48–63 kts)
Tropical Storm (TS)	62–88 kph (34–47 kts)
Tropical Depression (TD)	≤61 kph (≤33 kts)

Note: This table is based on previous PAGASA tropical cyclone classification before March 2022
https://bagong.pagasa.dost.gov.ph/information/about-tropical-cyclone#classification-of-tc

the September to November (SON) season between the El Niño and La Niña years compared to the June to August (JJA) season. They also found that the annual TC landfalls over the Philippines had a periodicity of about 32 years before 1939 and became shorter to 10−22 years after 1945. Similar to Cinco et al. (2016), they also found no significant trend in the annual landfalling TCs. Wu et al. (2004) examined the impacts of ENSO on the landfalling TCs over the Western North Pacific (WNP). They found that the TC landfalls are greatly reduced (increased) during SON in El Niño (La Niña) years. They suggested that the decreased TC landfalls during El Niño years are related to the eastward shift in the mean genesis location of TCs and that a break in the 500-hPa subtropical high appears around 130°E, which favors more recurving TCs toward Korea and Japan. On the other hand, the increase in landfalling TCs during La Niña years is related to the westward shift of the mean genesis position and westward expansion of the WNP subtropical high (Wang and Chan 2002). Takagi and Esteban (2016) examined the trends of landfalling TCs over the Philippines from 1945−2013, while focusing on the unusual characteristics of Super Typhoon Haiyan, the most devastating TC on record. They divided the Philippine domain by 2° latitude and found a significant increasing trend in the annual landfalling TCs along 10°N − 12°N and suggested that this might be related to the southward shift of the inter-tropical convergence zone (ITCZ) between 1949 and 2005, which is conducive to the increase in TCs in the lower latitudes over the WNP. Fudeyasu et al. (2014) and Cinco et al. (2016) found two peaks in TC landfall in July and October and a remarkable decrease in TC landfall around August, highlighting the seasonality in the TC landfalls over the country. Kubota and Wang (2009) and Lyon and Camargo (2009) also found a difference in ENSO phase-TC activity relationship around the Philippines before and after September.

One example of an extreme rainfall event due to a TC is Typhoon (TY) Ketsana, which made landfall in the Philippines in late September 2009 and brought exceptionally high rainfall leading to widespread flooding from central to southern Luzon including Metropolitan Manila (Abon et al. 2011). The estimated damage cost to agriculture and infrastructure was about PhP 11 billion (approx. US $244 million) (Cruz and Narisma 2016; https://reliefweb.int/report/philippines/philippines-ndcc-update-situation-report-no-48-tropical-storm-ondoy-ketsana-and; last accessed January 5, 2022). Figure 5.2a shows the track of TY Ketsana based on the JMA archive, while Fig. 5.2b shows the spatial distribution of accumulated rainfall from September 25 to 27, 2009, based on the TRMM (TMPA) 3B42 v7 rainfall estimates. It developed into a Tropical Depression (TD) on September 25, 2009, and made landfall over the Philippine landmass on September 26, 2009, as a Tropical Storm (TS) (Cruz and Narisma 2016). The lifetime maximum 10-min wind speed attained by this TC is 130 kph, after it exited the Philippine landmass. The spatial distribution of rainfall shows that the high amounts of rainfall were mostly concentrated along the path of the TC between 12°N and 15°N. The highest 3-day accumulated rainfall (555 mm) was recorded at Science Garden, Quezon City, synoptic station of PAGASA from September 25 to 27, 2009, which exceeded the climatological mean monthly rainfall in this station (451 mm) from 1981 to 2010.

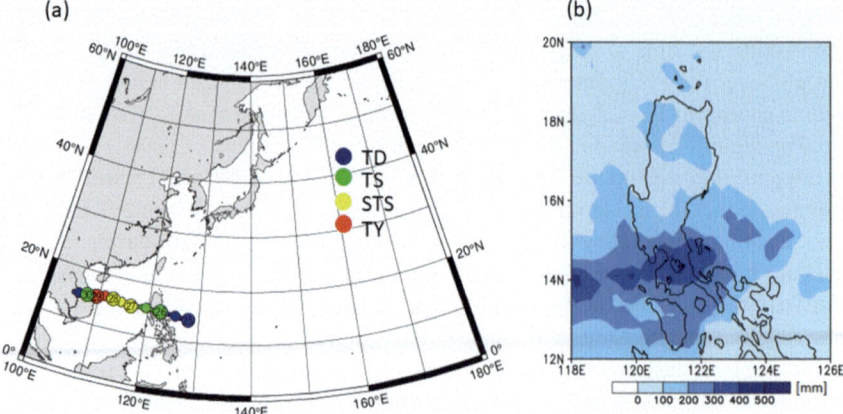

Fig. 5.2 **a** Track of Typhoon Ketsana from September 26 to 30, 2009, provided by the Japan Meteorological Agency (JMA) and archived by the National Institute of Informatics "Digital Typhoon" (http://agora.ex.nii.ac.jp) and **b** accumulated rainfall from September 25 to 27, 2009, based on TRMM (TMPA) 3B42 v7

The interaction of TCs with topography can further enhance rainfall, especially on the windward side of mountains, induced by orographic lifting. In the Philippines, Racoma et al. (2016) and Minamide and Yoshimura (2014) examined the impacts of topography on the TC-induced rainfall and track over Luzon and Mindanao Islands, respectively, using numerical model simulations. In Luzon Island, the Sierra Madre mountain range can affect the distribution of TC rainfall, as well as slow down TC movement, resulting in higher and prolonged rainfall over land (Racoma et al. 2016). Minamide and Yoshimura (2014) found that the presence of the mountains was essential for the heavy rainfall induced by Severe Tropical Storm (STS) Washi in 2011 over Mindanao Island. Espinueva et al. (2012) also examined the impacts of STS Washi in 2011 and noted that the disastrous flooding brought by this TC is due to a combination of climatic, environmental, and societal factors. The short-duration rainfall events to the north of the river basin, encroachment in the sandbars and river banks due to urbanization, and coincidence with high tide were some of the factors noted in their study.

5.2.2 Tropical Cyclone-Enhanced Southwest Monsoon

During the SWM season, non-landfalling TCs located east and north of the Philippines may enhance the prevailing southwesterly winds and cause extensive flooding over the western coast of the country (e.g., Cayanan et al. 2011; Lagmay et al. 2015). Over this area, high precipitation events (HPEs) associated with landfalling TCs accounted for no more than 15% of the total identified HPEs from 1958 to 2017, indicating a higher percentage contribution from non-landfalling TCs (Bagtasa 2019).

For daily rainfall above the 85th percentile, there are also significant increasing trends in the mean annual number of HPE days by about 6% per decade and in the annual total precipitation of the HPEs by about 12.7% per decade (Bagtasa 2019).

Recent examples of enhanced SWM events occurred in August 2012 and 2013. In August 2013, extensive flooding occurred over Metro Manila and other cities along the west coast of Luzon Island due to continuous rains brought by the interaction of the SWM and Tropical Storm (TS) Trami located northeast of Luzon Island. Figure 5.3a shows the spatial distribution of the accumulated rainfall based on TRMM (TMPA) 3B42v7 rainfall estimates from August 18–22, 2013. Higher rainfall amounts can be seen on the western coast of Luzon Island, with maximum rainfall observed west of Metro Manila. A similar case occurred in August 2012 (Fig. 5.3b) when the SWM was enhanced by Typhoon Haikui. In both cases, the total rainfall for these events has exceeded the average monthly total rainfall in August over Metro

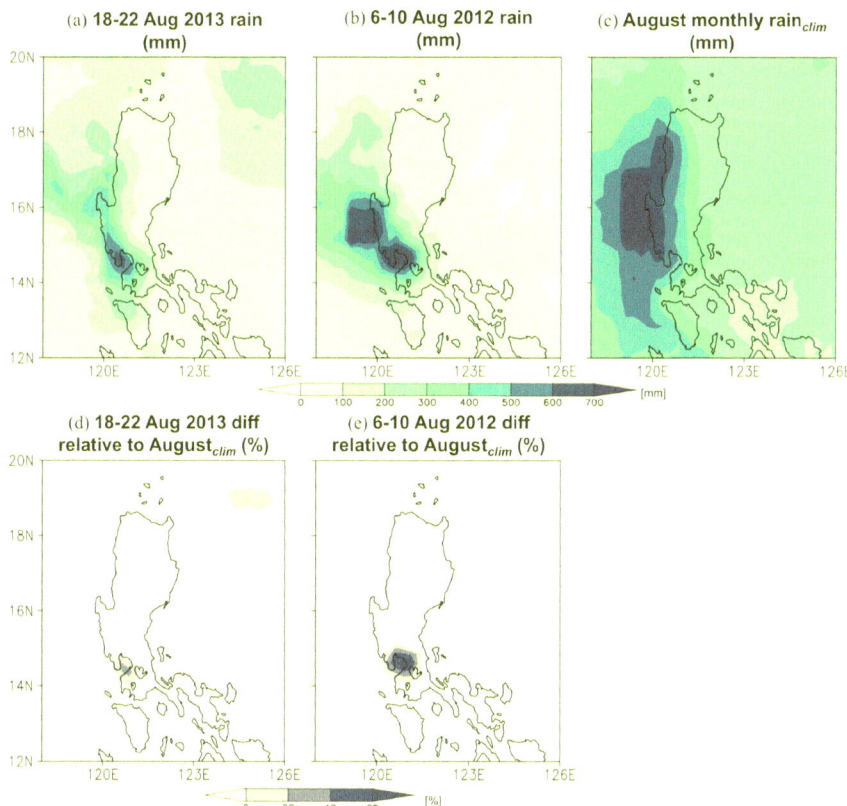

Fig. 5.3 **a** Five-day accumulated rainfall (mm) from August 18–22, 2013; **b** five-day accumulated rainfall (mm) from August 6–10, 2012; **c** monthly total rainfall in August averaged from 1998 to 2015; percent difference of **d** August 18–22, 2013 and **e** August 6–10, 2012 rainfall relative to the average August monthly total. *Source* TMPA (3B42_Daily; Huffman et al. 2016)

Manila (Fig. 5.3d, e). Observations from synoptic stations of PAGASA recorded an accumulated rainfall of 1067.4 mm from August 18–22, 2013, in Sangley Point, Cavite, and 1007.4 mm from August 6–10, 2012, in Science Garden, Quezon City. These rainfall amounts exceeded the climatological monthly rain values of both stations (Lagmay et al. 2015). Lagmay et al. (2015) proposed that the interaction of SWM with the stratovolcanoes west of the city resulted in a dispersive tail of clouds that were pulled by the tropical cyclones northeast of the Philippines toward Metro Manila and brought high rainfall over the city.

Using spectral decomposition analysis, Cayanan et al. (2011) isolated the TC contribution to the enhancement of the SWM westerlies. They found that when TCs are located northeast of Luzon Island, the SWM westerlies are enhanced over the western coast resulting in enhanced vertical ascent in that area. When TCs are located to the north or northwest of Luzon Island, strong northwesterly winds are generated that converge with the prevailing SWM westerlies. Both cases result in enhanced rainfall over the western coast of Luzon Island. However, this enhanced SWM rainfall induced by a TC can also occur in other parts of the western seaboard of the country. On June 20, 2008, as Typhoon Fengshen moved westward across central Philippines, anomalously high rainfall was observed over Panay Island, which was attributed to the interaction of SWM with the typhoon for the western and southern part of the island and to rainfall from cloud bands associated with the typhoon for the northern part of the island (Yumul et al. 2012). In particular, Iloilo City received 354 mm of rainfall that day, exceeding both the climatological monthly average, and the previous high recorded at 319.8 mm in July 1994 (Yumul et al. 2012).

5.2.3   Tropical Cyclone-Enhanced Northeast Monsoon

Unlike the SWM-TC interaction, the enhancement of the NEM by TCs has received less attention. There are cases when TCs located in the central and southern Philippines enhance the interaction between the northeasterly winds and the topography on the eastern coast of the country. One such example is Typhoon Kammuri in December 2019 that enhanced the NEM and brought heavy rainfall and flooding over northern Luzon. Figure 5.4a shows the track of Typhoon Kammuri from November 25, 2019, to December 6, 2019, based on the Japan Meteorological Agency (JMA) best track archive. This tropical cyclone entered PAR as a Typhoon Category (max 10-min. sustained winds at ~167 kph), made landfall over Gubat, Sorsogon on December 2, 2019, and exited the Philippine landmass on December 4, 2019. Even though the TC was already west of the Philippines, it still enhanced the NEM and caused flooding over northern Luzon until December 6, 2019 (http://floodlist.com/asia/philippines-northeast-monsoon-floods-cagayan-december-2019; last accessed January 5, 2022).

Figure 5.4b, c shows the interaction of the TC with a cold front to the north of Luzon Island at 06Z December 2, 2019, before making landfall. The location of the cold front can be determined by the negative meridional gradient of equivalent potential temperature as shown in Fig. 5.4c. In this case, the cold front is located

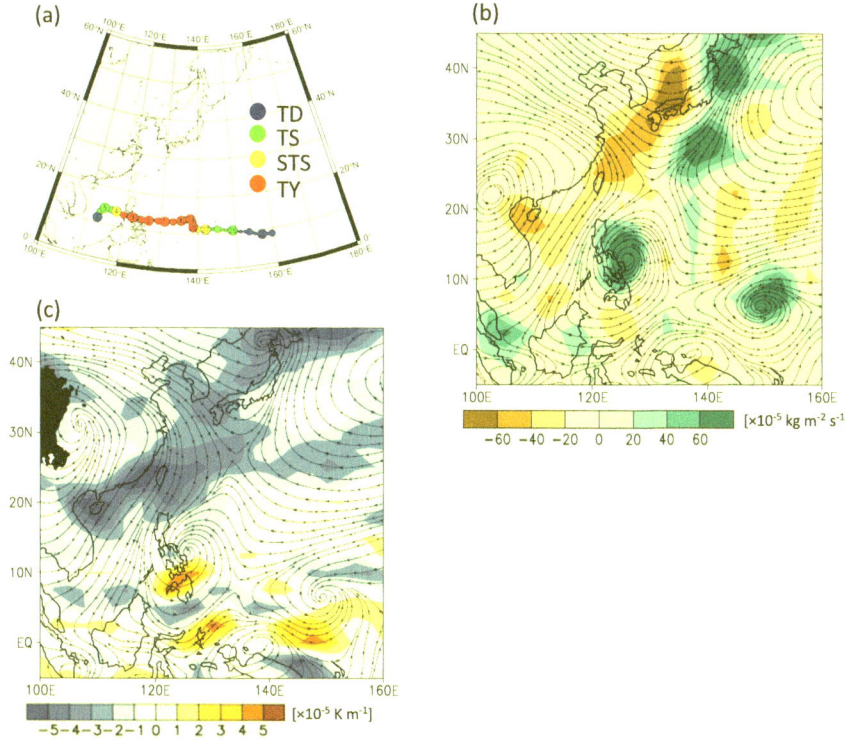

Fig. 5.4 a Track of TY Kammuri from November 25 to December 6, 2019, provided by the Japan Meteorological Agency (JMA) and archived by the National Institute of Informatics "Digital Typhoon" (http://agora.ex.nii.ac.jp). **b** Vertically integrated moisture flux (streams; $\times 10^{-5}$ kg m^{-1} s^{-1}) and convergence (VIMFC; shades; $\times 10^{-5}$ kg m^{-2} s^{-1}) at 06Z December 2, 2019, and **c** meridional gradient of equivalent potential temperature (shades; $\times 10^{-5}$ K m^{-1}) and winds at 925 hPa (streams; m s^{-1}) at 06Z December 2, 2019. The dataset used here is based on NCEP-DOE AMIP-II Reanalysis (https://psl.noaa.gov/data/gridded/data.ncep.reanalysis2.pressure.html)

to the north of the Philippines. Enhanced moisture flux convergence can be seen to the east of the Philippines around the TC and to its north. Enhanced ascent usually occurs along and ahead of the cold front as the warm air is forced to rise there. In the case of Typhoon Kammuri, the southerlies along its eastern flank, which brings the warm air, interacts with the cold air from the mid-latitudes leading to enhanced ascent and subsequently, rainfall. This will further add moisture to the rainfall brought by the TC. Lee et al. (2017) analyzed satellite infrared images of Super Typhoons Haiyan (in 2013) and Hagupit (in 2014) and suggested that the cold fronts in both cases might have contributed to their intensification to Super Typhoon categories. However, no physical mechanisms were provided in their paper as this is still an ongoing research. The NEM-TC interaction is also experienced in other regions with complex topography. In October 1998, the interaction of the NEM and

Typhoon Babs with the terrain induced heavy rainfall over the north and northeastern Taiwan (Wu et al. 2009).

5.2.4 Cold Surges, Vortices, and Shearlines

From November to March, the areas over the eastern coast of the Philippines experience HRF events, i.e. heavy rainfall events that often lead to flooding. The HRF events are associated with cold surges (e.g., Chang et al. 2005; Pullen et al. 2015; Lim et al. 2017) due to the southward expansion of the Siberian high and their interaction with the topography. Abdillah et al. (2021) recently identified four dominant pathways of cold surges over the tropics that include South China Sea (SCS) type (northerly surge over the SCS alone), Philippine Sea (PHS) type (northerly surge over the PHS alone), SCS-PHS type (northerly surge over the SCS and PHS), and blocked type (northerly surges other than the two regions) based on the 160 cold surges that they classified for 40 winters from 1979/80 to 2018/19. They identified 19 cases (11.9%) for the PHS type, which is comparable to those over the SCS with 18 cases (11.3%). Lim et al. (2017), on the other hand, examined the interaction between the MJO and cold surges over Southeast Asia using satellite data from 1998 to 2012. They found that the presence of the MJO over the Maritime Continent provides more favorable conditions for extreme rainfall in the region. In addition, they showed that the rainfall extremes associated with cold surges are more likely to occur over the southeastern coast of the Philippines when both MJO and cold surge are present. This is also corroborated by an earlier study of Pullen et al. (2015), where they showed that the presence of MJO, cold surge, and La Niña condition then contributed to the enhanced rainfall over the eastern coast of the Philippines in February 2008.

Chen et al. (2015) found that westward propagating cold surge vortices (CSVs; low-level cyclonic circulation) induce heavy rainfall events over Malaysia, Vietnam, and the Philippines. These vortices are formed by the interaction of the East Asian cold surges during the NEM season and the easterly waves or by the interaction between the East Asian cold surges and Borneo's topography. The CSVs only form in the vicinity of the Philippines and over Borneo (i.e., the so-called Borneo Vortex), but the formation mechanism of the CSVs in the two regions is different. Chen et al. (2013) examined the winter rainfall of Malaysia from 1979 to 2009 and found that the CSVs alone contribute about 33% (34%) to the mean total November–December (December–February) rainfall over Peninsular Malaysia (West Borneo).

The contribution of CSVs to the NEM rainfall of the Philippines has not been quantified previously. Olaguera et al. (2021a) examined a heavy rainfall event on January 16, 2017 (JAN2017) when a westward propagating cyclonic circulation interacted with the trailing end of a cold frontal system, forming a cold surge shearline (formerly locally known as "tail-end of a cold front") that induced urban flooding over Cagayan de Oro City in northern Mindanao Island. The spatial distribution of the meridional gradient of equivalent potential temperature can be used to determine the location of the cold front as shown in Fig. 5.5a for the JAN2017 HRF event. In this

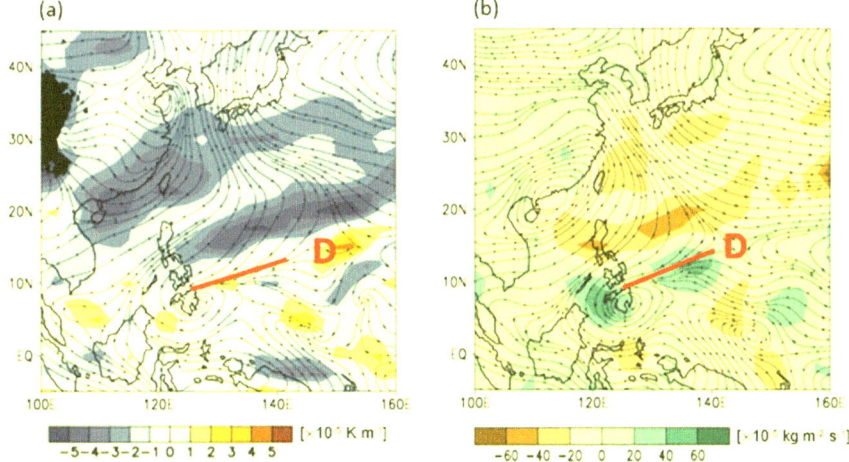

Fig. 5.5 Spatial distribution of **a** meridional gradient of equivalent potential temperature (shades; $\times 10^{-5}$ K m^{-1}) and winds at 925 hPa (streams; m s^{-1}) and **b** vertically integrated moisture flux (streams; $\times 10^{-5}$ kg m^{-1} s^{-1}) and convergence (shades; $\times 10^{-5}$ kg m^{-2} s^{-1}) at 06Z January 16, 2017. The "D" indicates the location of the col of the deformation zone. The axis of dilatation is indicated by the thick red line. The dataset used here is based on NCEP-DOE AMIP-II Reanalysis (https://psl.noaa.gov/data/gridded/data.ncep.reanalysis2.pressure.html)

event, the location of the cold front is just to the north of Mindanao Island (Fig. 5.5a). A deformation zone formed by two anticyclonic circulations oriented in a northwest-southeast direction (East Asia and Central Eastern Pacific) and two cyclonic circulations oriented in a northeast-southwest direction (North Pacific and Mindanao Island) can be depicted in this event. The axis of dilatation of the deformation zone, where the southerlies from the eastern flank of the cyclonic circulation converge with the northeasterlies, extends from 145 °E to the northern coast of Mindanao Island. This is also where the enhanced moisture convergence (Fig. 5.5b) and, hence, high rainfall amounts were found by Olaguera et al. (2021a). The warm air from the tropics is forced to ascend along the boundary of the cold front because it is less dense than the cold air. It is also worth mentioning that the deformation zone evolves at a fast pace in this case. At 12Z January 16, 2017, the axis of dilatation is already located to the south of Mindanao Island and dissipates afterward (not shown). Hence, datasets with high temporal resolution are required to investigate this kind of event. Continuous rainfall that is also associated with a cold surge shearline induced multiple landslides and swelling of drainage systems in the same city between January 1 and 18, 2009 (JAN2009; Faustino-Eslava et al. 2011; Olaguera et al. 2021b). During this event, the 99th percentile of daily rainfall records of January from 1979 to 2017 was exceeded for several days in Cagayan de Oro city and Hinatuan on the eastern coast of Mindanao Island. Olaguera et al. (2021b) attributed this event to the southward expansion of the anticyclone to the north of Luzon Island, enhanced meridional and zonal advection of cold and warm temperatures to the northeast of Mindanao Island,

and enhanced moisture convergence along the cold surge shearline. They identified 15 similar cases of this extreme weather event that occurred during January from 1979 to 2017. The results of their study demonstrate that extreme weather events in this region during the northeast monsoon season may occur even without the presence of such westward propagating cyclonic circulations to the south of Mindanao Island.

To understand the ocean–atmosphere interaction of a cold surge event, Ogino et al. (2018) conducted radiosonde observations from the Hakuho-maru research vessel over the Philippine Sea in December 2012 and compared their results with a previous campaign over the East China Sea in February 1974. They found that the strong horizontal advection of cold and dry air is more dominant below 850 hPa level. They further showed that although such strong horizontal advection of cold air suppressed the local convective instability, it was balanced by the eddy heat and moisture transport from the sea surface. Hence, the convective instability was maintained and still led to the observed enhanced rainfall over the eastern coast of the Philippines during their observation period. Nonetheless, their findings as well as the multi-scale interaction (Pullen et al. 2015) of the systems involved add to the complexity of extreme rainfall events in the country.

5.2.5   Local Thunderstorms

Localized thunderstorms occur throughout the year in the Philippines (Flores and Balagot 1969). According to Yumul et al. (2011), some areas in the country may experience up to 125 thunderstorms per year. However, there are limited studies of these events. In Metro Manila, high rainfall due to short-lived thunderstorms can result in flooding (Lagmay et al. 2017). Bañares et al. (2021) found that a number of these thunderstorms are localized rain events that occurred in the afternoon in Metro Manila during the months of April to May in the 2013–2014 period. Changes in meteorological variables recorded by stations in the city suggest that these thunderstorms appear to move in the direction of the prevailing wind. During the afternoon thunderstorm event on May 8, 2013, stations in Metro Manila recorded a large decrease in temperature (up to 4.2 °C) some minutes before high rainfall was detected (Bañares et al. 2021).

5.3   Challenges and Directions for Future Research

Advances in observations and numerical modeling have aided our understanding of the weather systems that contribute to extreme rainfall in the Philippines, but some challenges remain. Available synoptic stations of PAGASA are mostly located in low elevation regions, with less coverage for mountainous areas (Villafuerte et al. 2021). In recent years, there have been efforts to install automatic weather stations

(AWS) by both public and private institutions for real-time monitoring of rainfall and other weather variables (Hilario et al. 2021). Gridded datasets can also be used to augment gaps in either spatial or temporal coverage, but these datasets also have limitations. In a comparison study of gridded datasets, Peralta et al. (2020) suggest TRMM 3B42v7 can be used to study extreme rainfall in the Philippines, but also note some limitations. A denser network of stations providing data at sub-hourly resolutions that is maintained over a long time period is also essential in monitoring and further understanding thunderstorms, especially in cities. With a comprehensive dataset, extreme rainfall events can be identified and be analyzed to determine their characteristics and contributing factors.

Further research is also needed to improve understanding of the environment that leads to extreme rainfall events. Villafuerte et al. (2015) showed an increase in the intensity of extreme rainfall related to ENSO. Moreover, multi-scale interactions among the different drivers of rainfall extremes such as monsoons, ITCZ, MJO, PDO, and ENSO remain to be a challenge in tropical meteorology. For example, Cayanan et al. (2011) only examined a limited number of enhanced SWM cases. How TCs enhance the SWM in a climatological sense as well as the underlying atmospheric conditions that enhance rainfall need further investigation. This question may be addressed by high-resolution numerical simulations, as demonstrated by Olaguera and Narisma (2015) for the case of the enhanced SWM in 2013.

Modeling and forecasting extreme rainfall are challenging given the complex multi-scale interactions that involve synoptic-scale to mesoscale systems, and local-scale features, such as topography and land use (Dado and Narisma 2019). Numerical models are useful tools but still tend to have model biases, for example, in terms of amount and spatial distribution of extreme rainfall or tracks and intensities of tropical cyclones (e.g., Espinueva et al. 2012; Cruz and Narisma 2016; Racoma et al. 2016). Cruz and Narisma (2016) demonstrated that TC Ketsana and its associated rainfall can be simulated using the Weather Research and Forecasting (WRF) model but also noted some shortcomings in simulating the amount, timing, and spatial distribution of rainfall, as well as TC track and intensity. Calibration of the model, e.g., adjusting coefficients of parameters in the cumulus parameterization scheme (Tolentino and Bagtasa 2021) and assimilation of observed data into the model (Hou et al. 2013) are some ways to improve forecasting accuracy, but this is dependent on the availability and quality of observation datasets. In addition to classical methods of numerical weather predictions, methods such as analog forecasting (Bagtasa 2021) are explored for best estimates of intense rainfall events associated with tropical cyclones.

Extreme rainfall events translate to other hazards such as flooding and landslides. Early warning systems are therefore crucial especially for communities situated near river basins, mountainous regions, and even cities. Data from AWS and radar are used for real-time monitoring of these extreme events, but these data can also be used for streamflow and flood simulations (Abon et al. 2016). Doppler radar and AWS data can also be used to train and develop a nowcasting model, which can be a viable option for monitoring probable extreme rainfall events (Panaligan et al. 2016).

5.4 Conclusion

Extreme rainfall events in the Philippines are caused by different weather systems at multiple scales, ranging from local thunderstorms to monsoons and tropical cyclones and their interactions. There has been some progress toward understanding these events; however, more research is still needed, including more in-depth climatological studies at seasonal timescales to quantify the contribution of these weather systems to extreme rainfall, which require better temporal and spatial coverage of observation data and improved numerical models.

With continued global warming, heavy precipitation events are projected to intensify by about 7% per degree of warming at the global scale (IPCC 2021). In the Philippines, Villafuerte et al. (2015) already estimated an increase of about 4.3% in extreme rainfall intensity per 1 °C near-surface global warming over the period 1911–2010, which implies a potential increase in extreme rainfall as global temperatures increase. Recent projections for the Philippines indicate this projected increase in extreme rainfall in some areas, even though drier conditions overall are projected in the country in both moderate and high emissions scenarios (DOST-PAGASA et al. 2021). While this chapter has focused on "wet" extremes, analysis on "dry" extremes such as droughts, including its projected changes, will also need to be examined. Hence, it is important to continue efforts in advancing the science, data, and tools to improve understanding and forecasting of extreme rainfall events in the Philippines to help reduce disaster risk and increase resilience in the country.

Acknowledgements This research is supported by the High-Definition Clean Energy, Climate, and Weather Forecasts for the Philippines project. The authors wish to thank Francia Avila, Marcelino Villafuerte II, and Erica Bañares for the discussions on this book chapter, and to Prof. Asanobu Kitamoto of the National Institute of Informatics for his permission to use images obtained from the Digital Typhoon website (http://agora.ex.nii.ac.jp/digital-typhoon/index.html.en). The authors also acknowledge the reviewers whose comments have helped improve this chapter.

References

Abdillah, M.R., Y. Kanno, T. Iwasaki, and J. Matsumoto. 2021. Cold surge pathways in East Asia and their tropical impacts. *Journal of Climate* 34 (1): 157–170. https://doi.org/10.1175/JCLI-D-20-0552.1.

Abon, C.C., C.P.C. David, and N.E.B. Pellejera. 2011. Reconstructing the Tropical Storm Ketsana flood event in Marikina River, Philippines. *Hydrology and Earth System Sciences* 15: 1283–1289. https://doi.org/10.5194/hess-15-1283-2011.

Abon, C.C., D. Kneis, I. Crisologo, A. Bronstert, C.P.C. David, and M. Heistermann. 2016. Evaluating the potential of radar-based rainfall estimates for streamflow and flood simulations in the Philippines. *Geomatics, Natural Hazards and Risk* 7 (4): 1390–1405. https://doi.org/10.1080/19475705.2015.1058862.

Bagtasa, G. 2017. Contribution of tropical cyclones to rainfall in the Philippines. *Journal of Climate* 30 (10): 3621–3633. https://doi.org/10.1175/JCLI-D-16-0150.1.

Bagtasa, G. 2019. Enhancement of summer monsoon rainfall by tropical cyclones in northwestern Philippines. *Journal of the Meteorological Society of Japan* 97(5): 967–976. https://doi.org/10. 2151/jmsj.2019-052.

Bagtasa, G. 2021. Analog forecasting of tropical cyclone rainfall in the Philippines. *Weather and Climate Extremes* 32, 100323. https://doi.org/10.1016/j.wace.2021.100323.

Bañares, E.N., G.T.T. Narisma, J.B.B. Simpas, F.T. Cruz, G.R.H. Lorenzo, M.O.L. Cambaliza, and R.C. Coronel. 2021. Seasonal and diurnal variations of observed convective rain events in Metro Manila, Philippines. *Atmospheric Research* 258: 105646. https://doi.org/10.1016/j.atm osres.2021.105646.

Basconcillo, J., and I.J. Moon. 2021. Recent increase in the occurrences of Christmas typhoons in the Western North Pacific. *Scientific Reports* 11: 7416. https://doi.org/10.1038/s41598-021-868 14-x.

Brucal, A., V. Roezer, D.S. Dookie, R. Byrnes, M.L.V. Ravago, F. Cruz, G. Narisma. 2020. *Disaster impacts and financing: local insights from the Philippines.* London: Grantham Research Institute on Climate Change and the Environment and Centre for Climate Change Economics and Policy, London School of Economics and Political Science.

Cayanan, E.O., T.-C. Chen, J.C. Argete, M.-C. Yen, and P.D. Nilo. 2011. The effect of tropical cyclones on southwest monsoon rainfall in the Philippines. *Journal of the Meteorological Society of Japan* 89A: 123–139. https://doi.org/10.2151/jmsj.2011-A08.

Chang, C.P., Z. Wang, J. McBride, and C.H. Liu. 2005. Annual cycle of Southeast Asia—Maritime Continent rainfall and the asymmetric monsoon transition. *Journal of Climate* 18 (2): 287–301. https://doi.org/10.1175/JCLI-3257.1.

Chen, T.C., J.D. Tsay, M.C. Yen, and J. Matsumoto. 2013. The winter rainfall of Malaysia. *Journal of Climate* 26 (3): 936–958. https://doi.org/10.1175/JCLI-D-12-00174.1.

Chen, T.C., J.D. Tsay, J. Matsumoto, J. Alpert. 2015. Development and formation mechanism of the Southeast Asian winter heavy rainfall events around the South China Sea. Part I: Formation and propagation of cold surge vortex. *Journal of Climate* 28(4): 1417–1443. https://doi.org/10. 1175/JCLI-D-14-00170.1.

Cinco, T.A., R.G. de Guzman, A.M.D. Ortiz, R.J.P. Delfino, R.D. Lasco, F.D. Hilario, E.L. Juanillo, R. Barba, and E.D. Ares. 2016. Observed trends and impacts of tropical cyclones in the Philippines. *International Journal of Climatology* 36 (14): 4638–4650. https://doi.org/10.1002/joc. 4659.

Cruz, F.T., and G.T. Narisma. 2016. WRF simulation of the heavy rainfall over Metropolitan Manila, Philippines during tropical cyclone Ketsana: A sensitivity study. *Meteorology and Atmospheric Physics* 128: 415–428. https://doi.org/10.1007/s00703-015-0425-x.

Dado, J.M., and G.T. Narisma. 2019. The effect of urban expansion in Metro Manila on the southwest monsoon rainfall. *Asia-Pacific Journal of Atmospheric Sciences.* https://doi.org/10.1007/s13143-019-00140-x.

DOST-PAGASA, Manila Observatory and Ateneo de Manila University. 2021. *Philippine Climate Extremes Report 2020: Observed and Projected Climate Extremes in the Philippines to Support Informed Decisions on Climate Change Adaptation and Risk Management,* 145. Philippine Atmospheric, Geophysical and Astronomical Services Administration, Quezon City, Philippines.

Endo, N., J. Matsumoto, and T. Lwin. 2009. Trends in precipitation extremes over Southeast Asia. *SOLA* 5: 168–171. https://doi.org/10.2151/sola.2009-043.

Espinueva, S.R., E.O. Cayanan, and N.C. Nievares. 2012. A retrospective on the devastating impacts of Tropical Storm Washi. *Tropical Cyclone Research and Review* 1 (2): 163–176.

Faustino-Eslava, D.V., G.P. Yumul Jr., N.T. Servando, C.B. Dimalanta. 2011. The January 2009 anomalous precipitation associated with the "Tail-end of the Cold Front" weather system in Northern and Eastern Mindanao (Philippines): Natural hazards, impacts and risk reductions. *Global and Planetary Change* 76(1–2): 85–94. https://doi.org/10.1016/j.gloplacha.2010.12.009.

Flores, J.F., and V.F. Balagot. 1969. *Climate of the Philippines, World Survey of Climatology,* vol. 8, 159–213. Amsterdam: Elsevier Scientific Publishing Company.

Fudeyasu, H., S. Hirose, H. Yoshioka, R. Kumazawa, and S. Yamasaki. 2014. A global view of the landfall characteristics of tropical cyclones. *Tropical Cyclone Research and Review* 3 (3): 178–192. https://doi.org/10.6057/2014TCRR03.04.

Hilario, M.R.A., L.M. Olaguera, G.T. Narisma, and J. Matsumoto. 2021. Diurnal characteristics of summer precipitation over Luzon Island, Philippines. *Asia-Pacific Journal of Atmospheric Sciences* 57: 573–585. https://doi.org/10.1007/s13143-020-00214-1.

Hou, T., F. Kong, X. Chen, H. Lei. 2013. Impact of 3DVAR data assimilation on the prediction of heavy rainfall over Southern China. *Advances in Meteorology*, 1–17. https://doi.org/10.1155/2013/129642.

Huffman, G.J., D.T. Bolvin, E.J. Nelkin, R.F. Adler. 2016. *TRMM (TMPA) Precipitation L3 1 day 0.25 degree x 0.25 degree V7*, ed. Andrey Savtchenko, Goddard Earth Sciences Data and Information Services Center (GES DISC). Accessed 17 November 2020. https://doi.org/10.5067/TRMM/TMPA/DAY/7.

IPCC. 2021. Summary for policymakers. In: *Climate Change 2021: The Physical Science Basis. Contribution of Working Group I to the Sixth Assessment Report of the Intergovernmental Panel on Climate Change ed.* Masson-Delmotte, V., P. Zhai, A. Pirani, S.L. Connors, C. Péan, S. Berger, N. Caud, Y. Chen, L. Goldfarb, M.I. Gomis, M. Huang, K. Leitzell, E. Lonnoy, J.B.R. Matthews, T.K. Maycock, T. Waterfield, O. Yelekçi, R. Yu, and B. Zhou. Cambridge University Press, Cambridge, United Kingdom and New York, NY, USA, 3–32 https://www.ipcc.ch/report/ar6/wg1/about/how-to-cite-this-report.

Karl, T.R., N. Nicholls, and A. Ghazi. 1999. Clivar/GCOS/WMO workshop on indices and indicators for climate extremes workshop summary. *Climatic Change* 42: 3–7. https://doi.org/10.1023/A:1005491526870.

Kubota, H., and J.C.L. Chan. 2009. Interdecadal variability of tropical cyclone landfall in the Philippines from 1902 to 2005. *Geophysical Research Letters* 36 (12): L12802. https://doi.org/10.1029/2009GL038108.

Kubota, H., and B. Wang. 2009. How much do tropical cyclones affect seasonal and inter annual rainfall variability over the western North Pacific? *Journal of Climate* 22 (20): 5495–5510. https://doi.org/10.1175/2009JCLI2646.1.

Lagmay, A.M.F., G. Bagtasa, I.A. Crisologo, B.A.B. Racoma, and C.P.C. David. 2015. Volcanoes magnify Metro Manila's southwest monsoon rains and lethal floods. *Frontiers in Earth Science*. https://doi.org/10.3389/feart.2014.00036.

Lagmay, A.M., J. Mendoza, F. Cipriano, P.A. Delmendo, M.N. Lacsamana, M.A. Moises, N. Pellejera III., K.N. Punay, G. Sabio, L. Santos, J. Serrano, H.J. Taniza, and N.E. Tingin. 2017. Street floods in Metro Manila and possible solutions. *Journal of Environmental Sciences* 59: 39–47. https://doi.org/10.1016/j.jes.2017.03.004.

Lee, Y.S., Y.A. Liou, J.C. Liu, C.T. Chiang, and K.D. Yeh. 2017. Formation of winter superty-phoons Haiyan (2013) and Hagupit (2014) through interactions with cold fronts as observed by multifunctional transport satellite. *IEEE Transactions on Geoscience and Remote Sensing* 55 (7): 3800–3809. https://doi.org/10.1109/TGRS.2017.2680418.

Lim, S.Y., C. Marzin, P. Xavier, C.P. Chang, and B. Timbal. 2017. Impacts of boreal winter monsoon cold surges and the interaction with MJO on Southeast Asia rainfall. *Journal of Climate* 30 (11): 4267–4281. https://doi.org/10.1175/JCLI-D-16-0546.1.

Lyon, B., and S.J. Camargo. 2009. The seasonally-varying influence of ENSO on rainfall and tropical cyclone activity in the Philippines. *Climate Dynamics* 32: 125–141. https://doi.org/10.1007/s00382-008-0380-z.

Matsumoto, J., L.M.P Olaguera, D. Nguyen-Le, H. Kubota, and M.Q. Villafuerte. 2020. Climatological seasonal changes of wind and rainfall in the Philippines. International *Journal of Climatology* 40(11): 4843–4857. https://doi.org/10.1002/joc.6492.

Minamide, M., and K. Yoshimura. 2014. Orographic effect on the precipitation with Typhoon Washi in the Mindanao Island of the Philippines. *SOLA* 10 (1): 67–71. https://doi.org/10.2151/sola.2014-014.

Ogino, S.Y., P. Wu, M. Hattori, N. Endo, H. Kubota, T. Inoue, and J. Matsumoto. 2018. Cold surge event observed by radiosonde observation from the research vessel "Hakuho-maru" over the Philippine Sea in December 2012. *Progress in Earth and Planetary Science* 5: 9. https://doi.org/10.1186/s40645-017-0163-4.

Olaguera, L.M., J. Matsumoto, J.M.B. Dado, and G.T.T. Narisma. 2021a. Non-tropical cyclone related winter heavy rainfall events over the Philippines: Climatology and mechanisms. *Asia-Pacific Journal of Atmospheric Sciences* 57: 17–33. https://doi.org/10.1007/s13143-019-00165-2.

Olaguera, L.M., G.T. Narisma. 2015. *Improving extreme weather forecast through observation data assimilation*, 1–89. Graduate Thesis, Ateneo de Manila University.

Olaguera, L.M.P., M.E. Caballar, J.C. De Mata, L.A.T. Dagami, J. Matsumoto, H. Kubota. 2021b. Synoptic conditions and potential causes of the extreme heavy rainfall event of January 2009 over Mindanao Island, Philippines. *Natural Hazards*, https://doi.org/10.1007/s11069-021-04934-z.

Panaligan, D., J.A. Razon, J. Caro, C.P. David. 2016. Using machine learning to provide rapid rainfall forecasts based on radar-derived data. *Theory and Practice of Computation*, 117–131. https://doi.org/10.1142/9789814730464_0010.

Peralta, J.C.A.C., G.T.T. Narisma, and F.A.T. Cruz. 2020. Validation of high-resolution gridded rainfall datasets for climate applications in the Philippines. *Journal of Hydrometeorology* 21 (7): 1571–1587. https://doi.org/10.1175/JHM-D-19-0276.1.

Peterson, T.C. 2005. Climate change indices. *WMO Bulletin* 54 (2): 83–86.

Pullen, J., A.L. Gordon, M. Flatau, J.D. Doyle, C. Villanoy, and O. Cabrera. 2015. Multiscale influences on extreme winter rainfall in the Philippines. *Journal of Geophysical Research: Atmospheres* 120 (8): 3292–3309. https://doi.org/10.1002/2014JD022645.

Racoma, B.A.B., C.P.C. David, I.A. Crisologo, and G. Bagtasa. 2016. The change in rainfall from tropical cyclones due to orographic effect of the Sierra Madre Mountain Range in Luzon, Philippines. *Philippine Journal of Science* 145 (4): 313–326.

Racoma, B.A.B., N.P. Klingaman, C.E. Holloway, R.K. Schiemann, and G. Bagtasa. 2022. Tropical cyclone characteristics associated with extreme precipitation in the Northern Philippines. *International Journal of Climatology*, 42 (6): 3290–3307. https://rmets.onlinelibrary.wiley.com/action/showCitFormats?doi=10.1002%2Fjoc.7416.

Takagi, H., and M. Esteban. 2016. Statistics of tropical cyclone landfalls in the Philippines: Unusual characteristics of 2013 Typhoon Haiyan. *Natural Hazards* 80: 211–222. https://doi.org/10.1007/s11069-015-1965-6.

Tolentino, J.T., and G. Bagtasa. 2021. Calibration of Kain-Fritsch cumulus scheme in weather research and forecasting (WRF) model over Western Luzon. *Philippines. Meteorology and Atmospheric Physics* 133 (3): 771–780. https://doi.org/10.1007/s00703-021-00779-0.

Villafuerte, M.Q., II., J. Matsumoto, I. Akasaka, H.G. Takahashi, H. Kubota, and T.A. Cinco. 2014. Long-term trends and variability of rainfall extremes in the Philippines. *Atmospheric Research* 137: 1–13. https://doi.org/10.1016/j.atmosres.2013.09.021.

Villafuerte, M.Q., II., J. Matsumoto, and H. Kubota. 2015. Changes in extreme rainfall in the Philippines (1911–2010) linked to global mean temperature and ENSO. *International Journal of Climatology* 35 (8): 2033–2044. https://doi.org/10.1002/joc.4105.

Villafuerte, M.Q., II., J.C.R. Lambrento, C.M.S. Ison, A.A.S. Vicente, R.G. de Guzman, and E.L. Juanillo. 2021. ClimDatPh: An online platform for Philippine climate data acquisition. *Philippine Journal of Science* 150 (1): 53–66.

Wang, B., and J.C.L. Chan. 2002. How strong ENSO events affect tropical storm activity over the western North Pacific. *Journal of Climate* 15 (13): 1643–1658. https://doi.org/10.1175/1520-0442(2002)015%3c1643:HSEEAT%3e2.0.CO;2.

Wu, M.C., W.L. Chang, and W.M. Leung. 2004. Impacts of El Niño-Southern Oscillation events on tropical cyclone landfalling activity in the western North Pacific. *Journal of Climate* 17 (6): 1419–1428. https://doi.org/10.1175/1520-0442(2004)017%3c1419:IOENOE%3e2.0.CO;2.

Wu, C.C., K.K.W. Cheung, and Y.Y. Lo. 2009. Numerical study of the rainfall event due to the interaction of Typhoon Babs (1998) and the northeasterly monsoon. *Monthly Weather Review* 137 (7): 2049–2064. https://doi.org/10.1175/2009MWR2757.1.

Yatagai, A., K. Kamiguchi, O. Arakawa, A. Hamada, N. Yasutomi, and A. Kitoh. 2012. APHRODITE: Constructing a long-term daily gridded precipitation dataset for Asia based on a dense network of rain gauges. *Bulletin of the American Meteorological Society* 93: 1401–1415. https://doi.org/10.1175/BAMS-D-11-00122.1.

Yumul, G.P., Jr., N.A. Cruz, N.T. Servando, and C.B. Dimalanta. 2011. Extreme weather events and related disasters in the Philippines, 2004–08: A sign of what climate change will mean? *Disasters* 35 (2): 362–382. https://doi.org/10.1111/j.1467-7717.2010.01216.x.

Yumul, G.P., Jr., N.T. Servando, L.O. Suerte, M.Y. Magarzo, L.V.V. Juguan, and C.B. Dimalanta. 2012. Tropical cyclone-southwest monsoon interaction and the 2008 floods and landslides in Panay island, central Philippines: Meteorological and geological factors. *Natural Hazards* 62: 827–840. https://doi.org/10.1007/s11069-012-0109-5.

Chapter 6
A Case Study of an Unexpected Extreme Rainfall Event on September 1, 2020, in Sri Lanka

A. R. P. Warnasooriya, M. M. P. Mendis, and Malth Fernando

Abstract Sri Lanka is in the tropical belt and has monsoonal climate. Southwest monsoon is the longest monsoon season (May to September) which brings about one-third of the total annual rainfall of the country. Usually, southwest monsoon starts to weaken gradually in September. There are some monsoon-related phenomena which are very much prominent in deciding the strength of the monsoon. Low-level Jet (LLJ), Tropical Easterly Jet (TEJ) and monsoon surge are among them. On September 1, 2020, unexpected very heavy rainfall occurred, more than 250 mm within 24 h, in southwestern part of Sri Lanka causing a local flood situation and affecting more than 7000 people. This study is focused on analyzing the synoptic situation on September 1, 2020. Synoptic observation, ERA and JRA 55 Reanalysis data are used to generate synoptic charts. Behavior of wind patterns at different levels of the atmosphere and change of atmospheric pressure were analyzed. Satellite data were also analyzed, and analysis clearly showed that this extreme rainfall was associated with a north–south-oriented trough in the southwestern monsoon flow from the Arabian Sea.

Keywords Monsoon · Synoptic · Extreme rainfall · Monsoon surge · Vorticity · Wind shear

6.1 Introduction

Many parts of Sri Lanka are vulnerable to extreme events leading to flood and drought (Arunasalam et al. 2019). The yearly monsoons, associated floods and landslides cause the most damages in Sri Lanka in terms of economic impact and human casualties (UNDRR 2019). Sri Lanka comprehensive disaster management program 2014 noted that marginal increase of figures on loss of life during the past five years is partly due to the high-intensity precipitation leading to flash floods (MoDM 2014). On average over long term, Sri Lanka's sector-specific losses per year from natural

A. R. P. Warnasooriya (✉) · M. M. P. Mendis · M. Fernando
Department of Meteorology, Colombo, Sri Lanka
e-mail: rashanthie@yahoo.com

© The Centre for Science & Technol. of the, Non-aligned and Other Devel. Countries 2022
A. S. Unnikrishnan et al. (eds.), *Extreme Natural Events*,
https://doi.org/10.1007/978-981-19-2511-5_6

disasters (flood, landslide, cyclone and drought) are estimated at US$380 million (GFDRR 2016). It is further mentioned that annual expected losses (AEL) are the highest for flood peril, with an AEL of US$240 million (GFDRR 2016).

Sri Lanka is an island located close to the equator (Fig. 6.1). It is influenced by two monsoon and weather systems that originate within the inter-tropical convergence zone (ITCZ). The two monsoon seasons are the southwest monsoon (SWM) from May to September and the northeast monsoon (NEM) from December to February. These two monsoons account for around 57% of annual total rainfall and are different in the sense that in southwest monsoon season, warm equatorial maritime air predominates over the country. In northeast monsoon season, the air masses are replaced by the cool tropical continental air which originates from Siberian high-pressure cell.

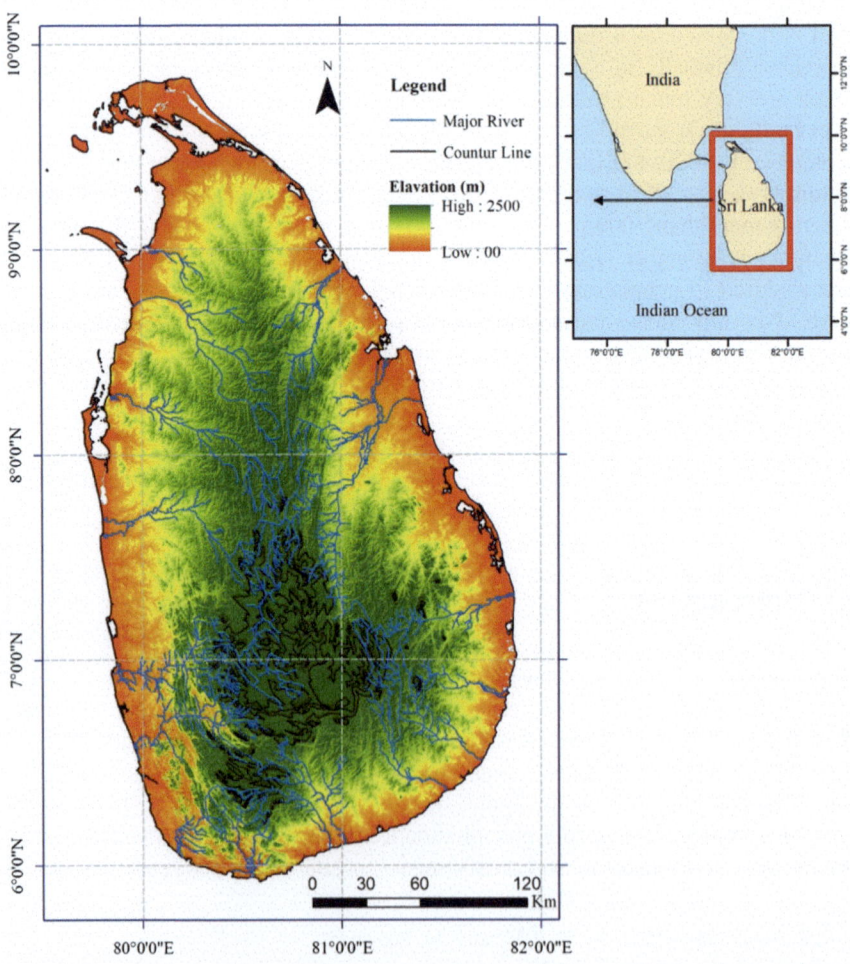

Fig. 6.1 Topography and the location of Sri Lanka

The normal onset of the SW monsoon is around the last week of May and with-drawal starts in the middle of September, and the NEM establishes around last week of November and recedes from last week of February. The NEM brings rainfall to the northern and eastern parts of the country. Within these monsoons, there are two inter-monsoon seasons from March to April and from October to November. Weather systems such as convergence, convection and depressions bring intense rain during the inter-monsoon seasons.

During the SWM season, heavy rainfall is received by low land areas in the southwestern part of Sri Lanka and the western slopes of the hill country. Southwest monsoon rains can be experienced at any time of the day and night, sometimes inter-mittently, mainly in the southwestern part of the country. Amount of rainfall during this season varies from about 100 mm to over 3000 mm. The highest rainfall received in the mid-elevations of the western slopes (Ginigathhena—3267 mm, Watawala—3252 mm, Norton—3121 mm). Rainfall decreases rapidly from these maximum regions toward the higher elevation and in Nuwara-eliya drops to 853 mm. The variation toward the southwestern coastal area is less rapid, with the southwestern coastal belt experiencing between 1000 and 1600 mm of rain during this 5-month long period (source: Department of Meteorology).

There are some monsoon-related phenomena which are very much prominent in deciding the strength of the monsoon. Low-level Jet (LLJ), Tropical Easterly Jet (TEJ) and monsoon surge are among them. These two synoptic-scale jets are associated with the southwest monsoon and located at 850 hPa and 150 hPa, respec-tively, over Southern India (Joseph and Raman 1966; Find later 1966, 1967; Mokashi 1974; Desai et al. 1976). Usually, a Low-level Atmospheric Jet is supposed to be a nocturnal phenomenon. Joseph and Raman (1966) first examined the LLJ features over the peninsular India followed by Desai et al. (1976), which confirms the exis-tence of a westerly LLJ stream with strong vertical and horizontal wind shear. Weather phenomena change substantially when there is a significant increase in wind speed (Fang et al. 2007).

Monsoon surge is a meteorological phenomenon where a weather event happened when there is a substantial increase in wind speed or precipitation (Xiang et al. 2007). These surges can be classified as northeast or southwest monsoon. Southwest monsoon surge while depressions or tropical storms in Bay of Bengal brings heavy localized rainfalls to Sri Lanka particularly over southwestern part of the country. According to joined study conducted by the researchers Dr. J Williams and Dr. G Jung, in 1993, monsoon surges have been categorized weak, moderate and strong or deep by considering maximum wind speed, extent and characteristics of weather. Southwest monsoon surge is characterized according to the following categories by Williams and Jung (1993) (Table 6.1; Fig. 6.2).

On September 1, 2020, showery condition was enhanced over Sri Lanka with very heavy falls exceeding 200 mm reported within 24 h at several places in southwest quarter of the country. Highest rainfall received during 24 h was 286 mm at Ketendola (Galle district). According to Disaster Management Center (DMC), Sri Lanka, total 7166 families were affected and 31 houses were partially damaged by strong winds

Table 6.1 SWM surge is characterized according to the following categories by Williams and Jung (1993)

Monsoon surge categories	Southwesterly winds	Extent	Characteristics of weather
Weak monsoon surge	Maximum 15 knots/7.7 mps	1524 m	Isolated cumulonimbus and dense cirrostratus near the thunderstorms
Moderate monsoon surge	Maximum 25 knots/12.9 mps	4572 m, strong easterlies above 6096 m	Nimbostratus clouds, imbedded cumulonimbus and moderate patchy rainfall
Strong or deep monsoon surge	Maximum 30–50 knots/15–25 mps	10,668 m	Nimbostratus clouds, imbedded cumulonimbus and abundant rainfall

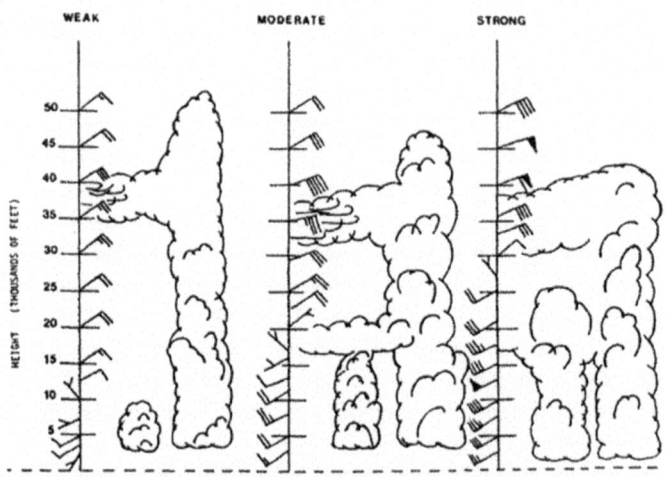

Fig. 6.2 Vertical wind profile associated with weak, moderate and strong monsoon surges (Williams and Jung 1993)

and heavy rain in Galle district. Some property damages were also reported at some places in northwestern, southern and western parts of the country due to strong winds.

Understanding of dynamics of various hazardous weather systems is crucial part of skillful weather forecasting to provide an accurate indication of developing such a weather event from hours to days ahead. They are, therefore, one of the most relevant components of routine and severe weather forecasting and warnings at National Meteorological and Hydrological Services (NMHSs) to provide more reliable information to Disaster Management Authorities in order to reduce loss of lives and properties. Emergency managers as well as general public need to know where and when damaging weather conditions will impact their region to perform an earlier response. If extreme weather events are more common in certain months or regions of the service area, the utility can better prepare in terms of staffing and infrastructure development. This is where the importance of weather forecasting in disaster management comes in.

National Meteorological and Hydrological services (NMHS) improvements to reduce disaster losses in developing countries—Benefit Coast Ratios (BCRs)—range from 4 to 1 to 36 to 1 (WMO-No. 1153).

The objective of this study is to identify and document the possible synoptic and environmental characteristics of the heavy rainfall events on the September 1, 2020, over western part of Sri Lanka.

Figure 6.3 shows the spatial distribution of rainfall in Sri Lanka on September 1, 2020. It can be seen that rainfall is concentrated on the western and southern parts of the country. The actual rainfall amount was observed to reach more than 250 mm/day at some places in western and southern provinces. It is also observed that there is no significant rainfall reported from southwestern slopes of the mountainous region located in the middle of the country. It is suggested that the orographic effect did not play a remarkable role in this heavy rainfall event.

Figure 6.4 shows hourly rainfall distribution reported at automated rain gauge located in Salawa area (marked by a black star in Fig. 6.3) from 7.30 p.m. on September 1, 2020, to 11.30 a.m. on September 2, 2020. Accordingly, the rainfall started from around 08.30 p.m. on Sep. 1, 2020, and continued until 10.30 a.m. on September 2, 2020. Very intense rainfall was reported within the period of 9.00 p.m.–10.30 p.m. on September 1, 2020.

6.2   Data and Methodology

The large-scale atmospheric pattern associated with the very heavy rainfall was reconstructed using ERA, JRA 55 and NCEP Reanalysis data and spatial resolution of the data of JRA 55 and NCEP was ($2.5° \times 2.5°$) and ($1° \times 1°$), respectively. Interactive Tool Analysis of Climate Systems (ITACS) software, which is developed by Japan Meteorological Agency (JMA), was used to study the data. Rainfall analysis during the heavy rainfall event was done using more than 250 daily observation data, and ArcGIS was used to develop the spatial rainfall distribution map.

Grid Analysis and Display System (GrADS) was used to analyze the horizontal and vertical profiles of the atmosphere. Diagrams displaying amounts of precipitation were plotted by using ArcGIS software.

6.3   Results and Discussion

The data have been used to explore the possible causes and mechanisms behind the occurrence of the heavy rainfall event over Sri Lanka on September 1, 2020. Figures 6.5, 6.6, 6.7, 6.8, 6.9, 6.10, 6.11, 6.12, 6.13, 6.14, 6.15, 6.16 and 6.17 contain

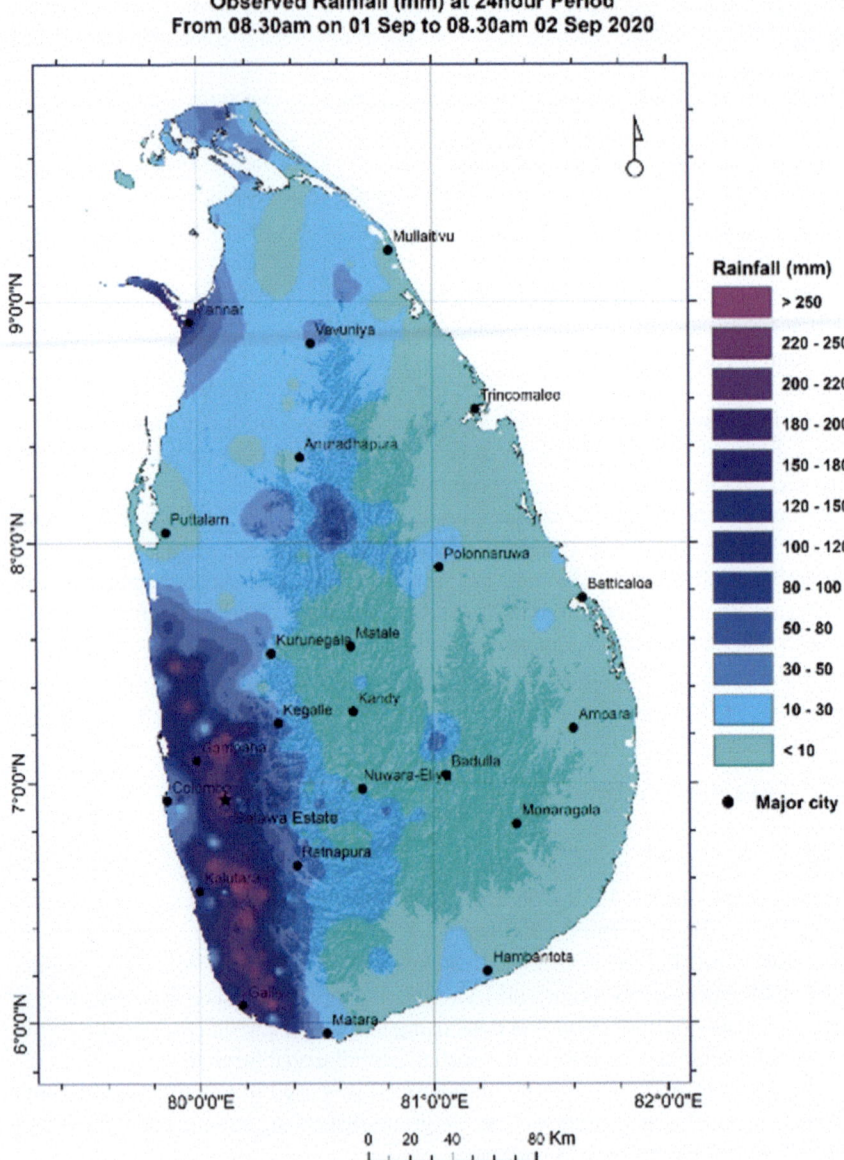

Observed Rainfall (mm) at 24hour Period
From 08.30am on 01 Sep to 08.30am 02 Sep 2020

Fig. 6.3 Spatial daily rainfall distribution over Sri Lanka on September 1, 2020. *Data*—Department of Meteorology

Fig. 6.4 Hourly rainfall distribution reported at automated rain gauge located in Salawa area (marked by a black star in Fig. 6.3) from 7.30 p.m. on September 1, 2020, to 11.30 a.m. on September 2, 2020

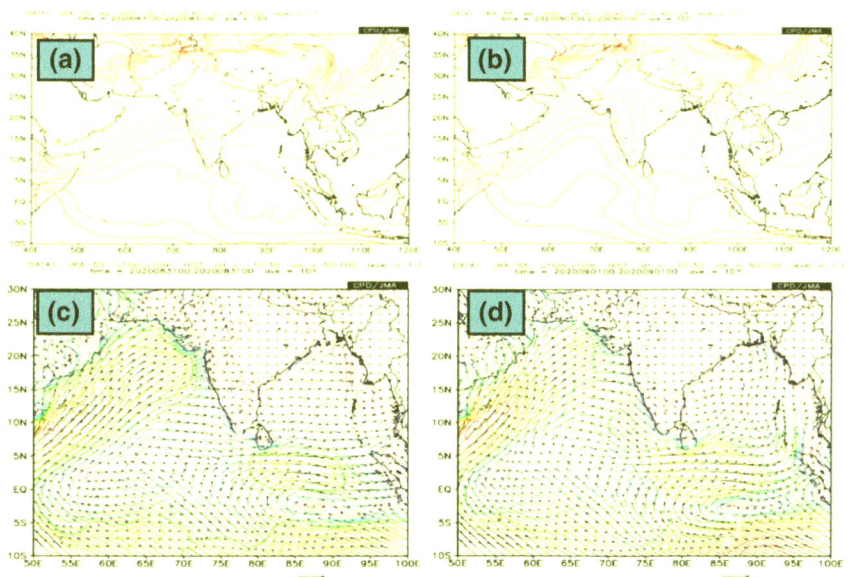

Fig. 6.5 a, b Daily average mean sea-level pressure (hPa) distribution on August 31, 2020, and on September 1, 2020, and **c, d** surface wind (m/s) distribution on August 31, 2020, and on September 1, 2020

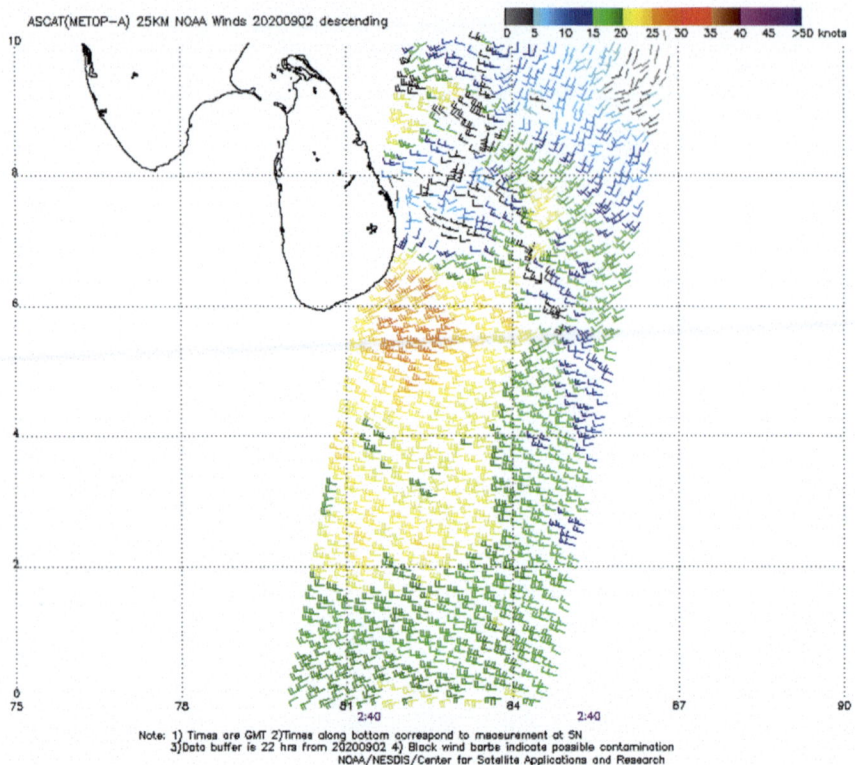

Fig. 6.6 ASCAT sea surface winds at 02.40 UTC on September 2, 2020

the analysis of various synoptic maps of the flow variables on September 1, 2020. The results are described in the following section in details.

6.3.1 Synoptic Situation

(a)   Mean Sea-level Pressure and Wind

The mean sea level (Fig. 6.5a, b), the pressure distribution, the isobars which are indicated by colors, suggest an increase in pressure gradient, while surface wind pattern (Fig. 6.5c, d) shows an enhancing wind speed over the vicinity of Sri Lanka on September 1, 2020.

Figure 6.7 is also evident that the strong winds around 25-30kts over just south of Sri Lanka during the morning on September 2, 2020.

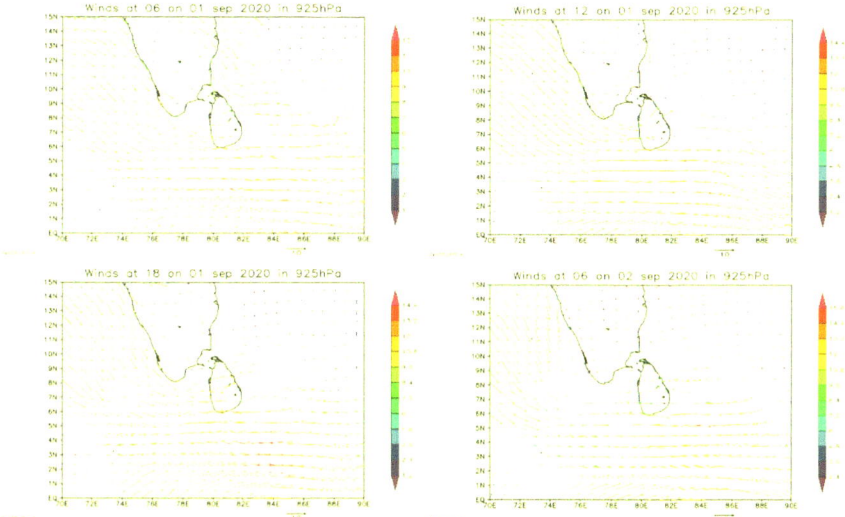

Fig. 6.7 Wind direction and speed (m/s) at 925 hPa level at 0600, 1200, 1800 UTC on September 1, 2020, and 0600Z on September 2, 2020

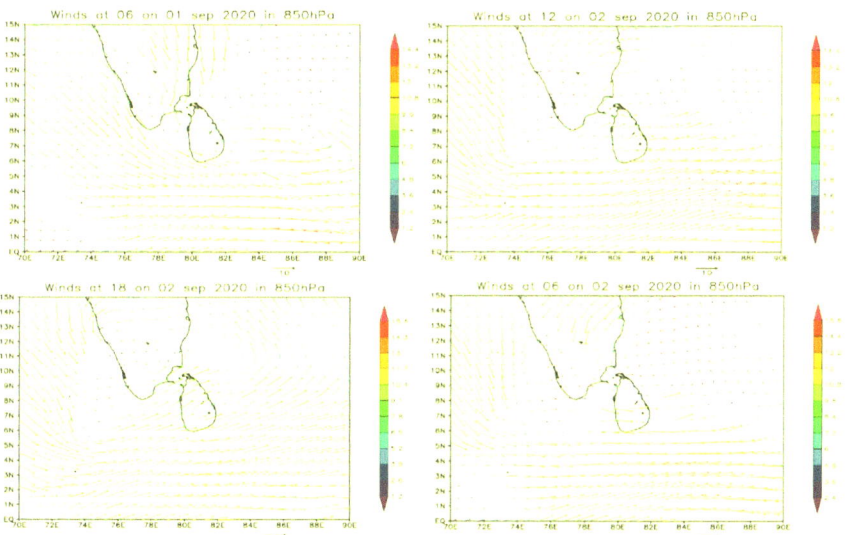

Fig. 6.8 Wind direction and speed (m/s) at 850 hPa level at 0600, 1200, 1800 UTC on September 1, 2020, and 0600Z on September 2, 2020

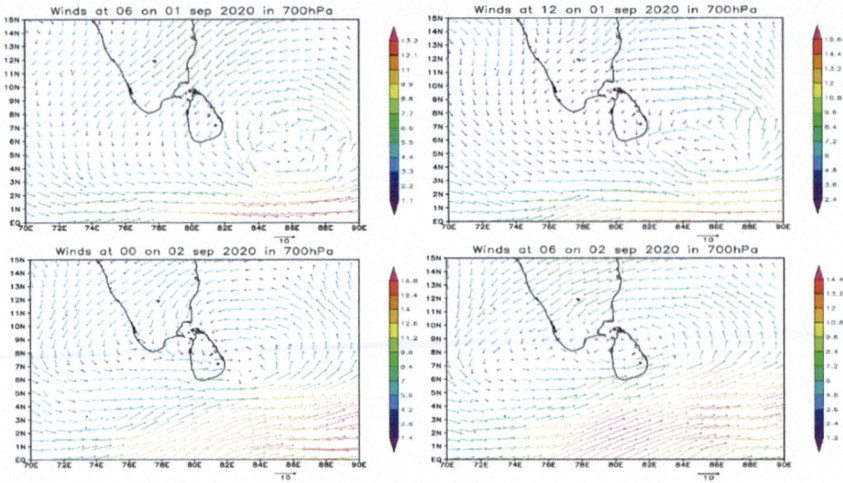

Fig. 6.9 Wind direction and speed (m/s) at 700 hPa level at 0600, 1200, 1800 UTC on September 1, 2020, and 0600Z on September 2, 2020

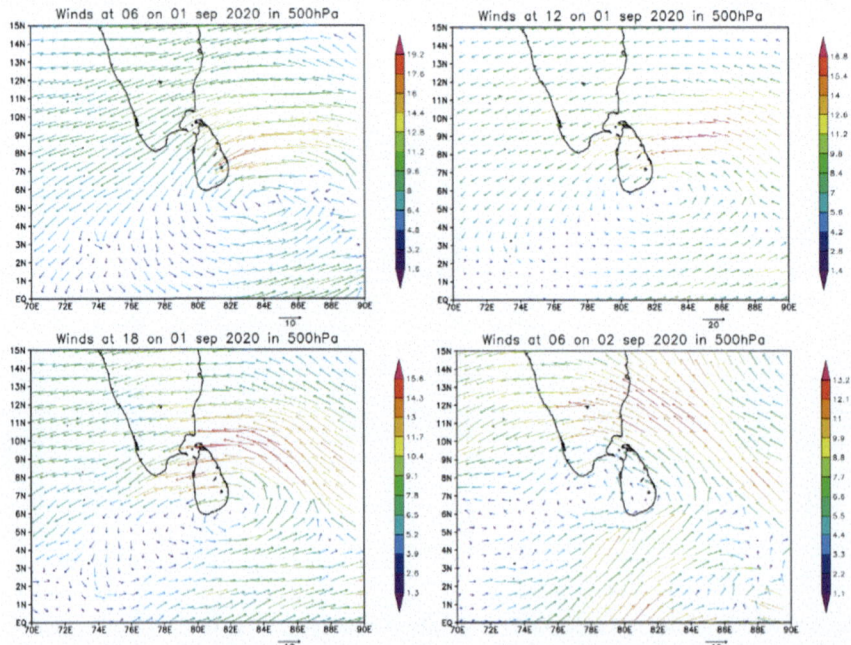

Fig. 6.10 Wind direction and speed (m/s) at 500 hPa level at 0600, 1200, 1800 UTC on September 1, 2020, and 0600Z on September 2, 2020

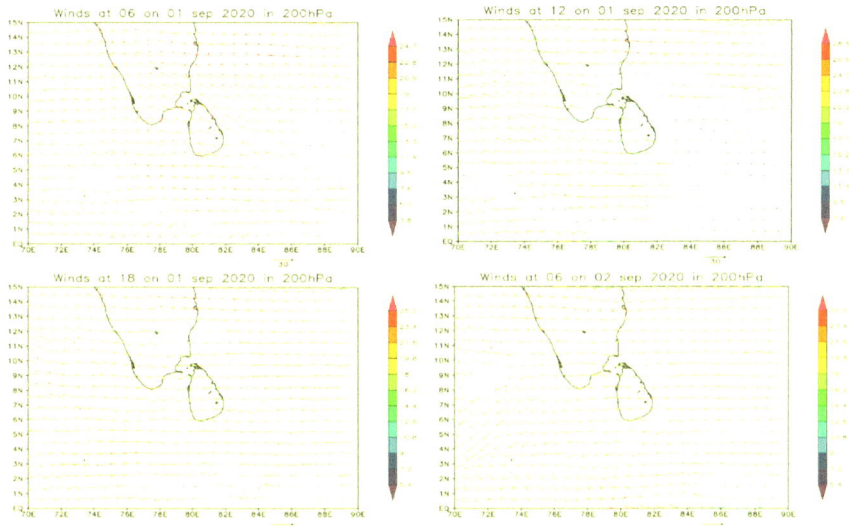

Fig. 6.11 Wind direction and speed (m/s) at 200 hPa level at 0600, 1200, 1800 UTC on September 1, 2020, and 0600Z on September 2, 2020

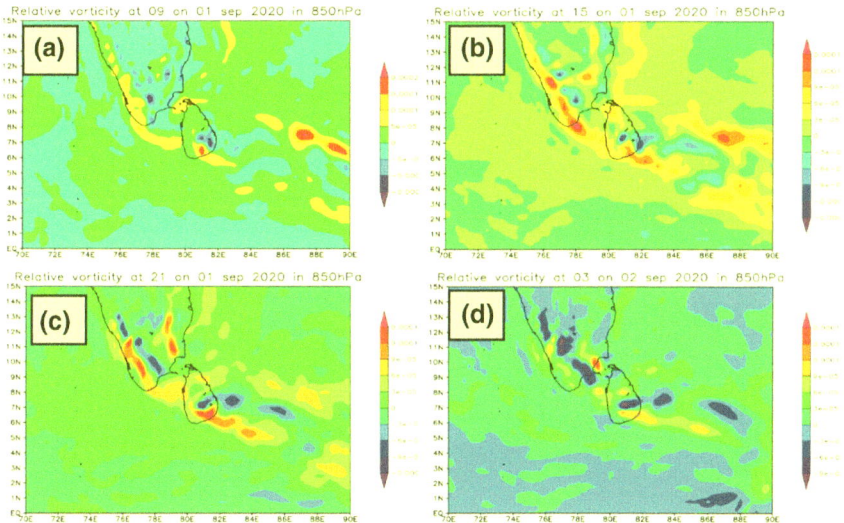

Fig. 6.12 Relative vorticity at 850 hPa level at 0900, 1500, 2100 UTC on September 1, 2020, and 0300Z on September 2, 2020

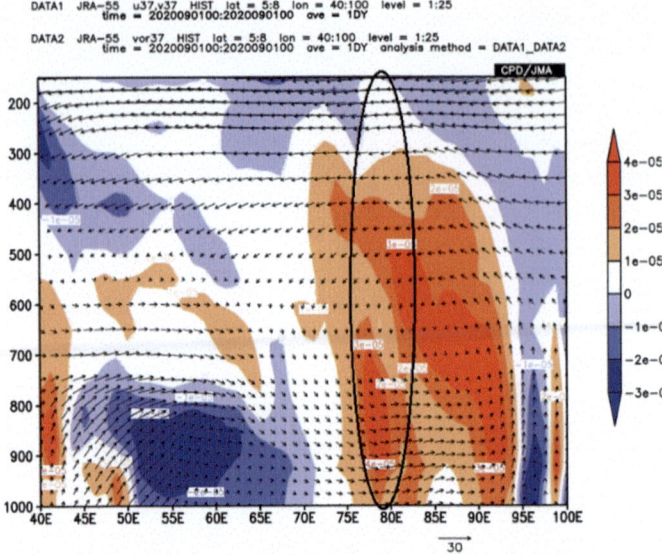

Fig. 6.13 Vertical cross section of winds and relative vorticity on September 1, 2020, over the region of 5 N–8 N

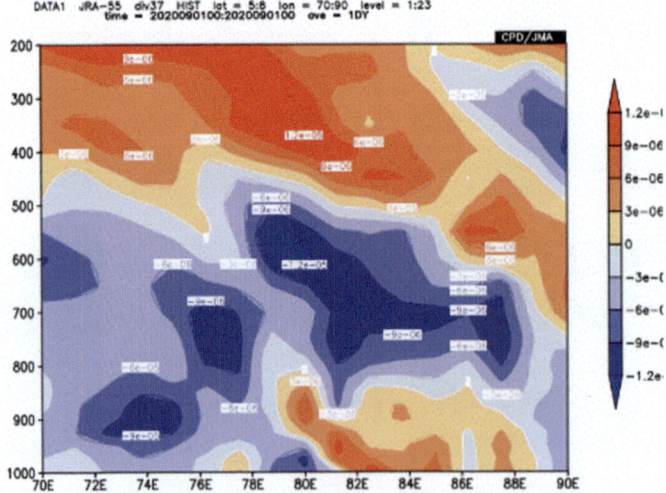

Fig. 6.14 Daily average vertical profile of divergence on September 1, 2020

(b)   Lower, Middle and Upper-level Wind

Figures 6.7 and 6.8 show the temporal change in 925 and 850 hPa level wind speed and directions during the heavy rainfall event over the country. Horizontal wind shear over the western part of the country is also clearly evident during the event over these

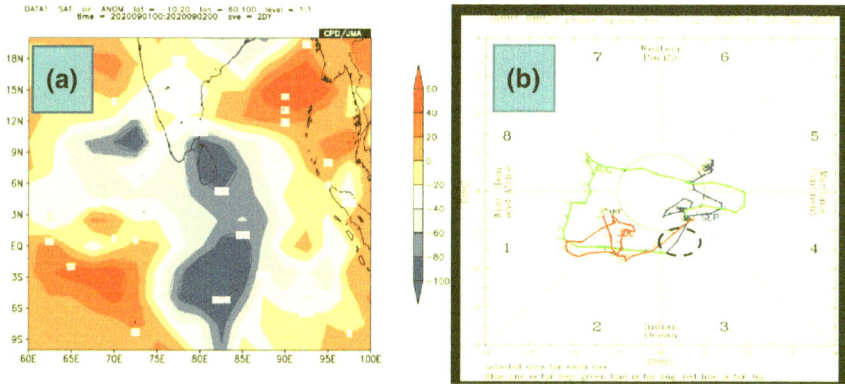

Fig. 6.15 **a** Outgoing longwave radiation (OLR) W/m^2 during September 1–2, 2020 and **b** track of Madden Julian Oscillation (MJO) during July to September 2020. *Source* BOM Australia

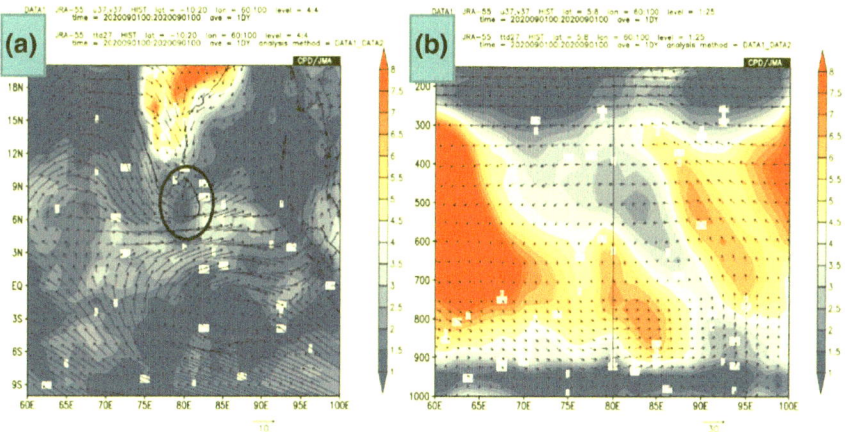

Fig. 6.16 **a** Dew point depression (*T-Td*) (K) analysis at 925 hPa level and **b** vertical profile of dew point depression within latitudes (5–8)°N on September 1, 2020

both layers. It can be seen that a low-level (westerly winds and northwesterly) wind convergence in 925 hPa level over southwestern part of the country during the event, and it was gradually decreasing from 0600 UTC on September 2, 2020.

In contrast with the lower levels, light variable winds over 700 hPa level (Fig. 6.9) were seen during the event over the country. A mild cyclonic circulation with horizontal wind shear was observed over the vicinity of Sri Lanka during the event in 700 hPa level providing favorable condition for the cloud formation.

Figure 6.10 shows the 500 hPa wind pattern during the event. It is evident that the strong easterlies with significant horizontal wind shear over 500 hPa level presence over Sri Lanka during the event. A movement of mild cyclonic circulation from east to west during the event was observed.

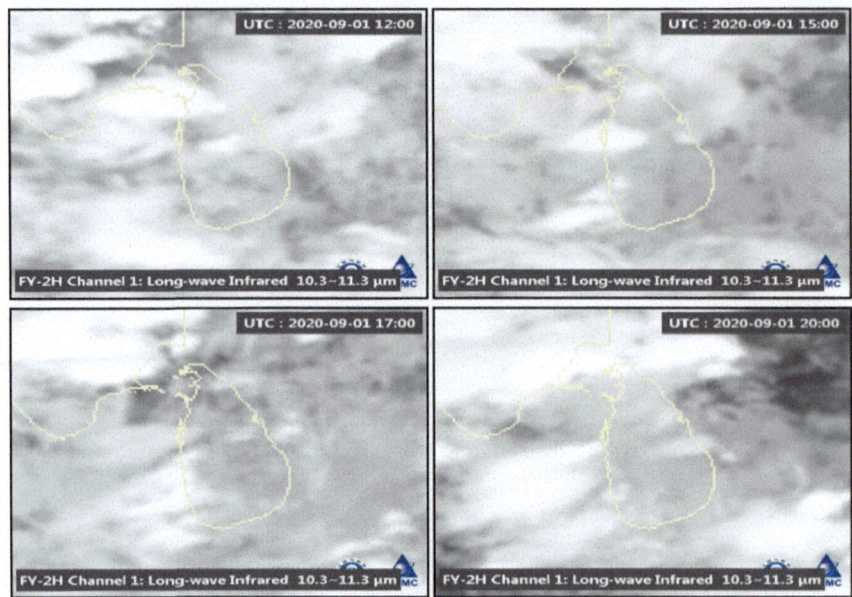

Fig. 6.17 Chinese FY 2H satellite imageries during 1200 UTC-2000 UTC on September 1, 2020

Figure 6.11 depicts the upper level, 200 hPa level, wind pattern and the speed during the event. There were strong winds (around 40–50kts) and diverge pattern over western part of the country before and after the rainy condition. It was seen that the wind speed was reduced to some extent and diverge patter was not seen during the heavy rainfall.

(c)   Relative Vorticity

Relative vorticity consists of two parts, due to curved flow through the base of a trough and due to wind shear. Horizontal wind shear is a difference in wind speeds; when the wind speeds are greater to the south of the parcel, it denotes the cyclonical spin of the parcel and vice versa. Figure 6.12a–d shows the temporal development and spatial movement of relative vorticity field along the southwestern coastal areas with maximum at around 1500 UTC, when intense rainfall occurred over the southwestern parts of the country. Figure 6.13 illustrates the vertical profile of relative vorticity over the region of 5 N–8 N. This plot readily reveals the favorable vertical updraft for the cloud development. As such it is suggested that very favorable conditions prevail in terms of vorticity over Sri Lanka for cloud formation during the event.

(d)   Low-level Convergence and Upper-level Divergence

The daily average vertical profile of divergence on September 1, 2020, is given in Fig. 6.14. It is found that a distinct area of low-level convergence (negative divergence) extends from Arabian Sea to western part of the country only up to 900hpa level and a positive divergence within 900–800hpa level. A well-defined area of the mid-tropospheric convergence with positive upper-level divergence, which is favorable for the cloud development, is evident in the divergence field. If the lifting strength is strong enough, the convective activity of convective boundary layer (CBL) can only shift the LLJ structure to higher heights by modifying from beneath. Previous study done by Mr. Sun et al. (2021) revealed that the regions of extreme precipitation are accompanied by the low-level convergence, positive vorticity disturbance and upper-level divergence disturbance.

(e)   Outgoing Longwave Radiation (OLR) and MJO

The outgoing longwave radiation (OLR/W/m^2) during September 1–2, 2020, is shown in Fig. 6.15a. It was found that the region of Sri Lanka is covered by low OLR of the order around 80–100 W/m^2. It is evident that the dense clouds are characterized over Sri Lanka.

Figure 6.15b shows the track and amplitude of Madden Julian Oscillation (MJO) index, which is an eastward propagating intra-seasonal oscillation with a low frequency on 30–60 day scale, during July to September 2020. The influence of MJO on precipitation could lead to extreme events such as record-breaking precipitation. On a global scale, extreme rainfall events during active MJO periods (situations in strong convective activity anomalous related to MJO) are about 40% higher than its quiescent periods (Jonas et al. 2004). Figure 6.16b revealed that the MJO exists in phase 3 (start date of blue line) with amplitude of more than 1 on September 1, 2020. Accordingly, MJO also plays a substantial role in the enhancement of precipitation associated with the monsoon surge over the Sri Lanka on September 1, 2020.

(f)   Horizontal and Vertical Profile of Dew Point Depression (T-Td) and Wind

Dew point depression (T-Td) is a useful indicator of how moist the air is. Figure 6.16a contains the analyzed dew point depression field at 925 hPa level, and it indicates that heavy incursion of moisture into the western part of the country. A leading tongue of high moisture, about 90–100%, extending from the Arabian Sea into western part of Sri Lanka is an exceptional feature of this map (Fig. 6.16b). A narrow zone of high moisture content, like a moisture river, is seen over and in the vicinity of Sri Lanka within 1000–900 hPa levels. In contrast, comparatively low amount of moisture in the middle level (800–650 hPa) of the troposphere is evident. A more detailed vertical wind profiles analysis could have been done radiosonde observations been available at that time.

The FY 2H satellite imageries are presented in Fig. 6.17. Deep convective clouds are observed over western and northwestern coastal areas of Sri Lanka and the vicinity, which are supported by the outgoing longwave radiation fields (Fig. 6.15a).

This study finds that potential atmospheric factors and relevant mechanisms are associated with monsoon surges, which induce very heavy rainfalls and strong winds over Sri Lanka. These extreme rainfall events and strong windy conditions can bring serious economic losses with great social damages. The analysis provides a useful reference in the predicting of such type of extreme events, which offers decisive information for disaster preparation and rescue planning.

6.4   Conclusions

Based on this study, the following conclusions can be drawn.

The actual rainfall amount was observed to reach more than 250 mm/day at some places in western and southern provinces. It is also observed that there is no significant rainfall reported from southwestern slopes of the mountainous region 2(b) located in the middle of the country. It is suggested that the orographic effect did not play a remarkable role in this heavy rainfall event.

Lower and mid-tropospheric analysis and observations suggest that the high intense rainfall was brought about by combination of factors. The results indicate that the short-duration very heavy rainfall was set-off by an anomalously active and as an influence of large-scale southwest monsoon activity with the following characteristics:

- Lower-level and mid-tropospheric-level analysis suggested that vertical and horizontal wind shear was present over Sri Lanka during the event.
- The strong positive relative vorticity was present up to middle tropospheric level during the event.
- Lower and middle-level convergence and upper-level divergence were favorable for this precipitation event.
- Lower and middle-level cyclonic circulations provided favorable environment for cloud development in this event.
- A narrow zone of high moisture content, like a moisture river, is seen over and in the vicinity of Sri Lanka within 1000–900hpa levels. In contrast, comparatively low amount of moisture in the middle level (800–650 hPa) of the troposphere is evident. During the monsoon season, these low-level updrafts aid in pumping more moisture into the free atmosphere and help in the formation of convective clouds

It is suggested that the high relative vorticity, horizontal and vertical wind shear, low-level convergence, upper-level divergence, favorable low-level moisture over Sri Lanka preceded the development of heavy rainfall. Presence of MJO in phase 3 with higher amplitude was favorable for enhancement of activity in the Bay of Bengal. As defined by Williams and Jung (1993), the monsoon event can be classified between weak and moderate monsoon surges (Table 6.1).

According to knowledge, there has not been much work done on the extreme rainfall associated with monsoon surge over Sri Lanka. However, there are some notable studies performed related to extreme rainfall events over Sri Lanka (Premalal et al. 2014; Warnasooriya et al. 2018).

Further study is required to understand more about such cases. It will be interesting to study more such cases of rainfall events to provide better forecast support for predicting such anomalous events. High-resolution data are needed to identify localized effects.

Acknowledgements The JRA 55 Reanalysis dataset was provided by the JMA and is available at http://jra.kishou.go.jp/JRA-55/index_en.html. Preliminary analysis in the present study used the ITACS system developed by the JMA. ERA Reanalysis data were provided by ECMWF. Satellite images were obtained from the Chinese FY 2H satellite.

References

Arunasalam, T., R. Suppiah, and U. Sonnadara. 2019. Trends in extreme rainfall events in Sri Lanka, 1961–2010. *Journal of the National Science Foundation of Sri Lanka* 47(3).

BOM (Bureau of Meteorology-Australia). 2021. *Madden-Julian Oscillation (MJO).* Source link: http://www.bom.gov.au/climate/mjo.

Desai, B.N., N. Rangachari, and S.K. Subramanian. 1976. Structure of low-level jet stream over the Arabian Sea and the Peninsula as revealed by observations in June and July during the monsoon experiment (MONEX) 1973 and its probable origin. *Indian Journal of Meteorology, Hydrology & Geophysics* 27: 263–274.

Disaster Management Center, Ministry of Disaster Management, Sri Lanka, Asia Pacific Alliance for Disaster Management Sri Lanka (A-PAD SL). 2016. People of Japan, Report on "'Impacts of Disasters in Sri Lanka 2016", Published by: The Consortium of Humanitarian Agencies (CHA) 01, Gower Street, Colombo 05, Sri Lanka. ISBN: 978-995-1041-54-0.

Fang, X., X. Zheng, F. Zhu, Y. Li, and Z. Li. 2007. Southwest monsoon surge and its weather impacts. *Journal of Tropical Meteorology* 13(1), 41–44. 1006-8775(2006)02-0041-04.

Findlater, J. 1966. Cross-equatorial jet streams at low level over Kenya. *Meteorological Magazine* 95: 353–364.

Findlater, J. 1967. Some further evidence of cross-equatorial jet streams at low level over Kenya. *Meteorological Magazine* 96: 216–219.

GFDRR. 2016. *Fiscal disaster risk assessment and risk financing options, Sri Lanka.* Global Facility for Disaster Reduction and Recovery (GFDRR), the international bank for reconstruction and development, Washington: the World Bank group.

Jonas, C., D.E. Waliser, K.M. Lau, and W. Stern. 2004. Global occurrence of extreme precipitation and The Maddan—Julian Oscillation: Observation and pradictability. *Journal of Climate* 17: 4575–4589. https://doi.org/10.1175/3238.1.

Joseph, P.V., and P.L. Raman. 1966. Existence of low-level westerly jet stream over peninsular India during July. *Indian Journal of Meteorology, Hydrology & Geophysics* 17: 437–471.

MoDM. 2014. *Sri Lanka Comprehensive Disaster Management Programme (SLCDMP) 2014–2018.* Colombo: Ministry of disaster management.

Mokashi, R.Y. 1974. The axis of the tropical easterly jet stream over India and Cyclone. *Indian Journal of Meteorology & Geophysics* 25 (1): 55–68.

Premala, K.H.M.S., A.R. Warnasooriya, and A.C.M. Rodrigo. 2015. Synoptic analysis of catastrophe heavy rain and strong winds over Sri Lanka on 01st June 2014, Sri Lanka. *Journal of Meteorology*, 50–63. www.meteo.gov.lk.

Rogers, D., G. Love, and B. Stewart. 2017. *Meteorological and hydrological services in Sri Lanka: A review, World Bank: A World Bank report.* Eckstein, D., M. L. Hutfils, and M. winges. 2018. *Global Climate Risk Index 2019: Briefing paper*, 5–12. Bonn, Germanwatch eV Office.

Sun, J., X. Yao, G. Deng, and Y. Liu. 2021. Characteristics and synoptic patterns of regional extreme rainfall over the central and eastern Tibetan plateau in boreal summer. *Atmosphere* 12, 379.

UNDRR. 2019. *Disaster risk reduction in Sri Lanka, status report 2019*, Bangkok, Thailand: United Nations Office for Disaster Risk Reduction (UNDRR), Regional Office for Asia and the Pacific.

Warnasooriya, A.R., A.C.M. Rodrigo, and K.H.M.S. Premalal. 2018. *Case study of flash flood assessment of 14th November 2014 in Colombo due to short period high intense rainfall: Climate events and disaster mitigation from policy to practice*, 81–91. New Delhi, India: NAM S & T Center.

Williams, J., and G. Jung. 1993. *Forecaster's Handbook for the Philippine Islands and Surrounding Waters*, 85–114. Naval Research Lab. Marine Met. Div.

World Meteorological Organization. 2015. WMO No.1153, *Valuing weather and Climate; economic assessment of meteorological and hydrological services*.

Xiang, F., X.J. Zheng, F. K. Zhu, Y. Li, and Z.Z. Li. 2007. Southwest monsoon surge and its weather impact. *Journal of Tropical Meteorology* 13(1, Article ID: 1006-8775 02–0041-04), 41–44.

Chapter 7
Thermodynamic Changes in the Atmosphere Associated with Pre-monsoon Thunderstorms Over Eastern and North-Eastern India

Bhishma Tyagi, Rajesh Kumar Sahu, Manoj Hari, and Naresh Krishna Vissa

Abstract During the pre-monsoon season (March–May), eastern and north-eastern India receives frequent thunderstorms. These thunderstorm events are responsible for sudden wind gusts, high rainfall and associated lightning, which destroy property, the environment and human habitats. As thunderstorms are short span events with high vertical cloud development, they make noticeable changes in the thermodynamics of the atmosphere. Identifying these thermodynamical changes may enhance the understanding genesis of these convective events and their propagation and would improve forecasting of these events more accurately. The present work reviews the atmospheric thermodynamical changes due to pre-monsoon thunderstorms over eastern and north-eastern India. The review found that the study region experiences frequent multi-cell clusters or squall line thunderstorms. The existence of small hills and plateaus in the area acts as triggering agents for thunderstorm initiation (e.g. Chota Nagpur Plateau and Hazaribagh Plateau in eastern India), and the Bay of Bengal is the source of moisture supply. The convective instability indices work well over eastern India but fail to differentiate the thunderstorms over north-eastern India. Various studies reveal that the frequency and intensity of these thunderstorms are changing over the region due to global warming. The pre-monsoon thunderstorms over eastern and north-eastern India inspired researchers to formulate new thermodynamic indices listed in the present review. The study also discusses the use of re-analysis data to calculate thermodynamic indices and their closeness to calculations based on in-situ data over the region.

B. Tyagi (✉) · R. K. Sahu · M. Hari · N. K. Vissa
Department of Earth and Atmospheric Sciences, National Institute of Technology Rourkela, Rourkela, Odisha 769008, India
e-mail: tyagib@nitrkl.ac.in

R. K. Sahu
e-mail: rajeshkumar_sahu@nitrkl.ac.in

M. Hari
e-mail: manoj_h@nitrkl.ac.in

N. K. Vissa
e-mail: vissan@nitrkl.ac.in

© The Centre for Science & Technol. of the, Non-aligned and Other
Devel. Countries 2022
A. S. Unnikrishnan et al. (eds.), *Extreme Natural Events*,
https://doi.org/10.1007/978-981-19-2511-5_7

Keywords Thunderstorms · Thermodynamic indices · Radiosonde · Re-analysis data · Climate change · Nor'westers · Kalbaishakhi · Eastern and North-Eastern India

Abbreviations

ANN	Artificial Neural Network
AWS	Automatic Weather Station
BOYD	Boyden Index
BPI	Bordoichila Prediction Index
BRN	Bulk Richardson Number
CAPE	Convective Available Potential Energy
CBHT	Cloud Base Height
CIN	Convective Inhibition
CMIP5	Coupled Model Intercomparison Project Phase 5
C-NON	Correct Non-occurrence
CSI	Critical Success Index
CTI	Cross Totals Index
DCI	Deep Convective Index
DPT	Dew Point Temperature
DPT850	Dew Point Temperature at 850 hPa
DWR	Doppler Weather Radar
FAR	False Alarm Ratio
GPS	Global Positioning System
HI	Humidity Index
hPa	Hectopascal
HSS	Heidke Skill Scores
IMD	India Meteorological Department
KI	K Index
LCL	Lifting Condensation Level
LFC	Level of Free Convection
LI	Lifted Index
LIDAR	Light Detection and Ranging
MCT	Modified CTI
MEI	Modified Energy Index
MKI	Modified KI
MODIS	Moderate Resolution Spectral Radiometer
MPI	MODIS Profile Index
MR	Miss Rate
MSWI	Modified SWEAT
MTT	Modified TTI
MVT	Modified VTI
NCAPE	Normalised Convective Available Potential Energy

NSS	Normalised Skill Score
NTD	Non-Thunderstorm Day
POD	Probability of Detection
RADAR	Radio Detection and Ranging
RAMS	Regional Atmospheric Modelling System
RH700	Relative Humidity at 700 hPa
RINO	Bulk Richardson Number
RSS	Rank Sum Score
STORM	Severe Thunderstorm Observations and Regional Modelling
SAARC-STORM	South Asian Association for Regional Cooperation- STORM
SHEAR	Vertical Wind Shear
SHOW	Showalter Index
SODAR	Sound Detection and Ranging
SWEAT	Severe Weather Threat Index
SWISS	Combined Stability and Wind Shear Index for Thunderstorms in Switzerland
TD	Thunderstorm Day
TPI	Composite Stability Index
TSS	True Skill Statistic
TTI	Total Totals Index
VTI	Vertical Total Index
WRF-ARW	Advanced Research Weather Research and Forecasting model
WRF-NMM	Weather Research and Forecasting-Non-hydrostatic Mesoscale Model

7.1 Introduction

Thunderstorms are characterised by violent thunder, lightning, wind gusts and associated rainfall resulting in loss of life and destruction (Brooks 2013; Harel and Price 2020). The global frequency of thunderstorms is very high, with around 1000 events every hour over the tropics (Williams 2005), resulting in associated lightning damages to infrastructure, property, lives and the cause of forest fires (Christian et al. 2003; Brooks 2013). Research on climate change impact on tropical thunderstorms has not gained prime focus, despite their association with ~80% of total global lightning (Finney et al. 2018). Maximum thunderstorms are observed over the African region (Harel and Price 2020), followed by the Indian subcontinent (Mezuman et al. 2014). As thunderstorms are associated with vertically developed clouds, assessing atmospheric thermodynamics from surface to the upper level is one common approach to understand their development and progress (Haklander and Delden 2003). The vertical profiling of atmospheric thermodynamic parameters is done by various instruments (i.e. radiosonde, light detection and ranging: LIDAR, radio detection and ranging: RADAR, sound detection and ranging: SODAR) over a place, which in

turn will give the meteorological parameters from surface to upper level for calculating the thermodynamic indices (Kunz 2007; Tyagi et al. 2013). Even the surface
observations using an automatic weather station (AWS) help to analyse the lifting
condensation level (LCL), which can substantially give a fair idea about the thunderstorm occurrence (Samanta et al. 2020). The Doppler weather radar proved to
be a valuable instrument for real-time monitoring of updrafts/downdrafts associated
with thunderstorm cell development and movement over a region (Pradhan et al.
2012). The variability of these thermodynamic parameters with different stages of
thunderclouds and associated stability of the atmosphere made them one choice for
successful now casting of thunderstorms (Sahu et al. 2020a).

The presence of instability is one of the main ingredients for the development
of a thunderstorm event (da Silva et al. 2019). The atmospheric instability can be
conditional, latent, potential or a combination of these instabilities (Kunz 2007).
Depending on the instability type, different thermodynamic parameters or instability indices have been developed by researchers to account for the thermodynamic
state of the atmosphere. As each of the indices are representing instability in the
atmosphere, each of them has their own strengths and mostly combines the buoyancy, moisture, temperature and wind properties in the low and medium levels of
the atmosphere (Tajbakhsh et al. 2012). Over time, the development of thermodynamic indices involves processing directly observed parameters with cloud characteristics (i.e. Cloud base height and Cloud top temperature) to parameters involving
multi-layer observations of temperature and humidity from vertical profiles of atmospheric observations (McCann 1994; Romps et al. 2014). Thermodynamic indices
aim to understand the associated atmospheric instability and assess thundercloud
development/potential to produce thunderstorms (Peppier 1988). In general, a set of
thermodynamic indices will be employed for the case study of thunderstorms or any
convective activity (Sánchez et al. 2009). Various thermodynamic indices developed
over time, which proved their utility to predict thunderstorm occurrence in different
regions. An important point to note is that these thermodynamic indices have space
and time variations and their threshold may change from place to place or season
to season (Sahu et al. 2020a). Depending on the region and local environmental
factors, the accuracy of any index for differentiating the convective activity may
change (Vujović et al. 2015; Bikos et al. 2016). Despite these factors, the thermodynamic indices are popular for quick assessment of thunderstorm occurrence over any
region and are successful for global prediction also (Robinson et al. 2013; Schefczyk
and Heinemann 2017).

Thunderstorm occurrence is significantly high in the Indian region, and the
convective system occurs in almost all seasons in different parts of the country. There
has been proper documentation of thunderstorm climatology over India from time
to time. Manohar and Kesarkar (2003) analysed thunderstorm climatological variations with considering the contrasting features of the eastern and western Indian
regions. Their study reveals that eastern India has more favourable conditions for
thunderstorm generation, and the frequency is higher than in the western region. It
was found that thunderstorm frequency is highest during the monsoon season over

the Indian area; however, the rainfall due to pre-monsoon thunderstorms is characterised by vertically developed cumulonimbus clouds by continental convection (Manohar and Kesarkar 2004). Eastern and north-eastern India received the highest number of thunderstorms during the pre-monsoon season than other parts (Manohar and Kesarkar 2005). Tyagi (2007) prepared the thunderstorm climatology over India, and the study reveals higher thunderstorm occurrence over eastern and north-eastern India during the pre-monsoon season. Singh et al. (2011) reported the thunderstorm climatology for April and May 1981–2008 over eastern and north-eastern India. The study reported that a higher frequency of thunderstorms is experienced in May than in April, and the region experiences both increasing and decreasing trends in thunderstorm frequency in different sub-regions. The thunderstorm-related causalities are also highest with pre-monsoon thunderstorms over eastern and north-eastern India (Bhardwaj et al. 2017).

In eastern and north-eastern India, pre-monsoon thunderstorms are locally known as Kalbaishakhi or Bordoichila (STORM Science Plan 2015). Due to wind from northwest to south-east directions, these thunderstorms are also known as Nor'westers (Das et al. 2014a, b). It was noticed that wind speeds could be 140–150 km/h associated with these Nor'westers, which uprooted trees, damaged buildings and agriculture and resulted in the loss of life (Sahu et al. 2020a). The types of these Nor'westers are multi-cell clusters over eastern and north-eastern India (Mohanty et al. 2007). For more than a century, these Nor'westers were studied over the Indian subcontinent for deciphering the mechanism and improvement of their prediction (The Meteorological Office 1920; Sohoni 1928; Roy and Chatterji 1929; Brooks 1934; Pramanik and Alipore 1939; Desai 1950; Gupta 1952; Koteswaram and Srinivasan 1958; Rao and Raman 1961). However, there were not many observations available at different locations, and with the availability of data sets, most of the studies were focused on Kolkata (Ghosh et al. 2008). In 2006, the Department of Science and Technology, Government of India launched a field experiment, supplemented by regional modelling abbreviated as Severe Thunderstorm and Regional Modelling (STORM), focused on the east and north-east regions of India for studying the pre-monsoon thunderstorms. In the initial phase of the programme, other regions of India and neighbouring countries were not covered, which were added to STORM in 2009, and modified as SAARC-STORM (Das et al. 2014a, b). SAARC-STORM collected a lot of meteorological data helpful in understanding the movement of these Nor'westers over the region and resulted in a better prediction of these events (Das et al. 2014a, b; Madala et al. 2016).

Thermodynamic indices adoption for understanding Nor'westers has proved their usefulness over the region in recent decades (Tyagi et al. 2011, 2013; Guha et al. 2013; Sahu et al. 2020a, b). With the advanced knowledge of Nor'westers and inputs from STORM observations, the India Meteorological Department (IMD) showed a significant improvement in the now casting of thunderstorms all over India by incorporating thermodynamic indices threshold analysis (Ray et al. 2015). Although these Nor'westers are short-lived phenomena, they significantly change the soil temperature and soil heat flux (Tyagi and Satyanarayana 2010), surface energy balance (Tyagi 2012; Tyagi et al. 2012, 2014; Tyagi and Satyanarayana 2015), turbulent kinetic energy and surface layer properties (Tyagi and Satyanarayana 2013a, 2013b, 2019), temporal and spatial variability of coherent structures (Tyagi and Satyanarayana, 2014a, b) for the daily scales. With climate change, meteorological parameters over any place are changing, impacting the frequency and intensity of natural hazards globally (Diffenbaugh et al. 2013). The changing climate also changes the existing threshold values of thermodynamic indices globally (Kunz et al. 2009). The eastern and north-eastern Indian region also reveals these changes over the last few decades (Sahu et al. 2020a, b). Though the studies accounting for the climate change impact on thermodynamic indices over the Indian subcontinent are less, the Indian region exhibited pronounced changes for high rainfall events and flood risks (Guhathakurta et al. 2011; Sarkar and Maity 2020).

The present chapter discusses usefulness of the thermodynamic indices for thunderstorm prediction over eastern and north-eastern India, along with a discussion of study sites and various indices used for prediction. The chapter will further progress with summarising newly developed indices for Nor'westers and the impact of climate change studies on various indices thresholds. Section 7.2 of this chapter will discuss the study region, topographical variations, moisture and heat availability over the region, availability of various observational data sets over the region and the classical Nor'westers nomenclature over the region. Section 7.3 will elaborate on the use and utility of various thermodynamic indices over the region, focusing on new indices development for Nor'westers. Section 7.4 will discuss our understanding of useful thermodynamic indices over the region, proposed threshold values, changing threshold values with time and frequency and intensity changes of Nor'westers. Finally, Sect. 7.5 will conclude the understanding of thermodynamic indices associated with Nor'westers over the eastern and north-eastern Indian region.

7.2 Study Region and Availability of In-Situ Observations

The regions for analysing pre-monsoon thunderstorms for the present study are eastern and north-eastern India. In the present study, eastern India comprises Odisha, West Bengal and Jharkhand states. The states considered are Arunachal Pradesh, Assam, Manipur, Meghalaya, Mizoram, Nagaland, Sikkim and Tripura for north-eastern India. These states are home to many hills and plateaus, of which the main are Chota Nagpur Plateau, Hazaribagh Plateau, Garhjat Hills, Malayagiri Hills in eastern India. In north-eastern India, the prominent hills are Garo Hills, Shillong Plateau, Naga Hills, Khasi Hills, Jaintia Hills, Lushai Hills and Mizo Hills. The north-eastern states and parts of eastern India also encounter foothills of the Himalayan region, with bordering countries of Nepal and Bhutan, respectively. The primary source of moisture transport to the region is the Bay of Bengal, apart from local rivers and tributaries (STORM Science Plan 2005). During the pre-monsoon season, the region experiences hot, scorching summers, where the air temperatures may go well above 45 °C on individual days (Sahu et al. 2020a). Various small hills and plateaus in the region act as triggering agents for initiating thunderstorms (STORM Science Plan 2005).

According to Ghosh et al. (2008) and Tinmaker and Ali (2012), the Nor'westers have a definite pattern of movement, based on earlier observations of the India Meteorological Department (IMD) (IMD TN 10 1944). These thunderstorms are denoted by Types A, B, C and D, respectively, based on their origin location, with Type A originated over Chota Nagpur Plateau, Type B originated over Nepal region, Type C originated over Shillong Plateau, Lushai and Jaintia hills and Type D originated over Garo and Khasi hills. The direction of movement for these A, B, C, D types of thunderstorms also differs. The study region with prominent hills and plateau is listed along with different colour arrows for Types A, B, C, D thunderstorms and their travel directions as per IMD TN (1944) in Fig. 7.1. Eastern India experiences Type A thunderstorms, mostly with fewer occurrences of Type B. North-eastern India, on the other hand, experiences Types B, C and D thunderstorms with the rare encounter of Type A thunderstorms during pre-monsoon season. As the India Meteorological Department started research on Nor'westers more than a century ago, the

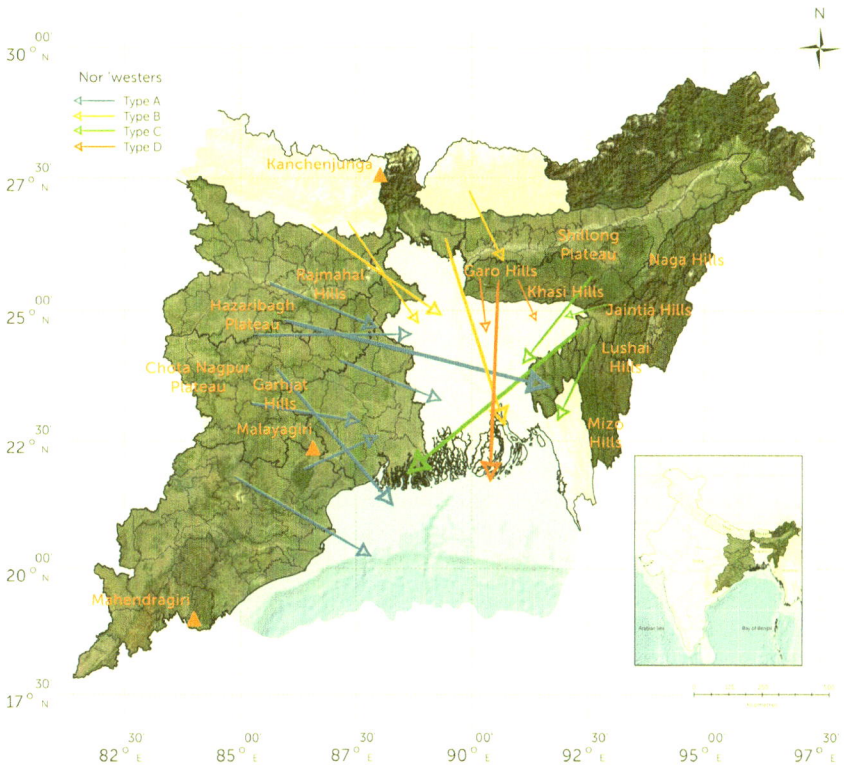

Fig. 7.1 Eastern and North-eastern region with prominent hills/plateau names and location /movement-based Types A, B, C, D thunderstorms depicted by arrows (after IMD TN 1944)

Table 7.1 Seasonal thunderstorm statistics over eastern and north-eastern India as per literature

Sl. no.	References	Eastern/North-Eastern India				Comments
		Pre-monsoon	Monsoon	Post-monsoon	Winter	
1	Manohar and Kesarkar (2004)	2.62/5.79	6.22/6.25	1.0/1.7	0.62/0.59	Area averaged values
2	Manohar and Kesarkar (2005)	456.30/469	1444/675	113.3/79.2	107.5/48.1	Arial total number of thunderstorms
3	Tyagi (2007)	20–30/30–40	40–70/50–70	3–7/2–14	2–4/ > 5	Seasonal thunderstorms frequency
4	Singh and Bhardwaj (2019)	6–12/ > 15	16- > 20/ > 20	1–5/3–5	0.8–1.6/ 2.4–4.0	Seasonal thunderstorms frequency
5	Mahanta and Yamane (2020)	11.33	3.63	2.25	2.0	Average seasonal values only over Assam

availability of in-situ surface observations in eastern and north-eastern India is significant. Though the region has various rivers and tributaries present, they are primarily dry/having less water availability during the summer season. They may not serve for enough moisture transportation to produce thunderstorms over the region. Table 7.1 discusses the seasonal variations of thunderstorms over eastern and north-eastern India. There are differences about the total number of thunderstorms over the region by different researchers, which arises mainly due to different time periods considered in their study and the area averaged for formulating the eastern and north-eastern regions.

Figure 7.2 shows the available automatic weather stations (AWSs) at various eastern and north-eastern India locations from IMD (http://www.imdpune.gov.in), where the information is available for more than fifty years. It is to be noted that we have not listed surface observations of state of Bihar in Fig. 7.2. As shown in Fig. 7.2, 88 stations have long-term availability of surface observations, which are very useful for understanding the climatological variations of meteorological variables over the region and assessing the genesis of thunderstorm activity variation. The names of these stations are provided in Table 7.3 in Appendix 1.

Other than surface observations, the upper air observations help to understand the convective updrafts, downdrafts associated with thunderstorms. The data sets are also helpful for understanding the thermodynamical variations of the atmosphere during different epochs of thunderstorm events (Madala et al. 2013, 2016). IMD set up a radar network over various coastal and inland stations to improve the prediction capability from natural hazards (e.g. thunderstorms, tropical cyclones, heavy rainfall).

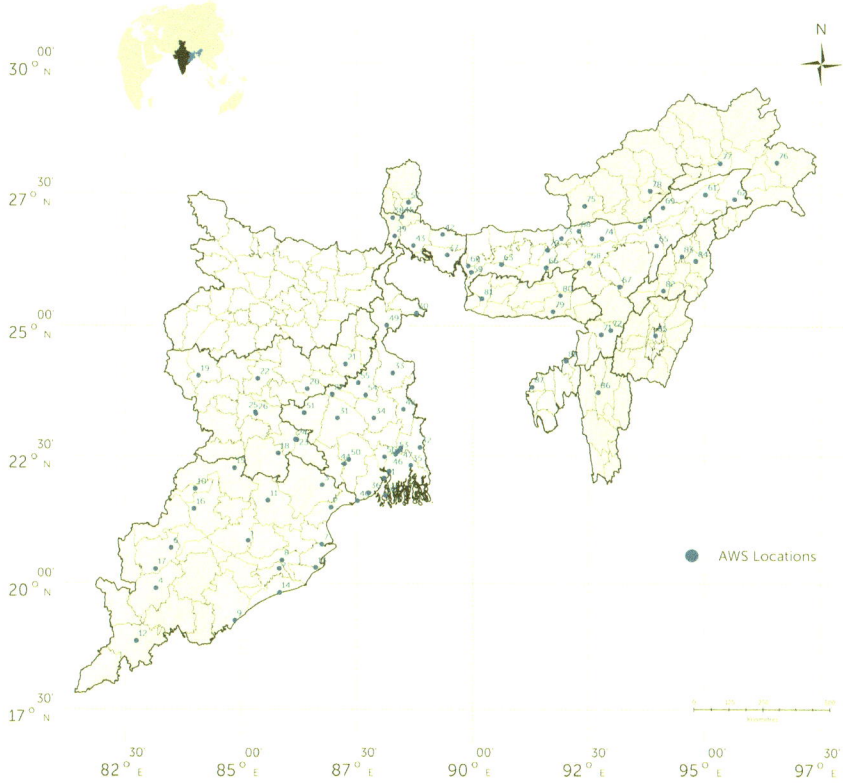

Fig. 7.2 Location of automatic weather stations (AWSs) with surface data available for more than fifty years in eastern and north-eastern India. *Source* IMD Pune

Figure 7.3 depicts such stations in eastern and north-eastern India where the radar and radiosonde observations are operational, and data may be available on request to IMD for research purposes. Some stations are having both radar and radiosonde observatories (e.g. Kolkata, Bhubaneswar, Patna, Ranchi and Gopalpur in eastern India; Mohanbari, Agartala and Guwahati in north-eastern India), while Paradip and Sohra have radar observations only and Jharsuguda, Jalpaiugudi, Shillong, Aizwal, Imphal, Dimapur, Passighat, Gangtok, Bankura stations are listed in IMD upper air rawinsonde/radiosonde sites (http://www.imdpune.gov.in). With the launch of the SAARC-STORM programme by the Department of Science and Technology, Government of India, radiosonde observations were made at the Indian Institute of Technology, Kharagpur for some years (Das et al., 2014a, b), which is not operational currently. Overall, the region has a good number of AWS and radar observations and well-represented radiosonde/upper air observations for understanding the variability in the atmosphere during the genesis or dissipation of any convective activity.

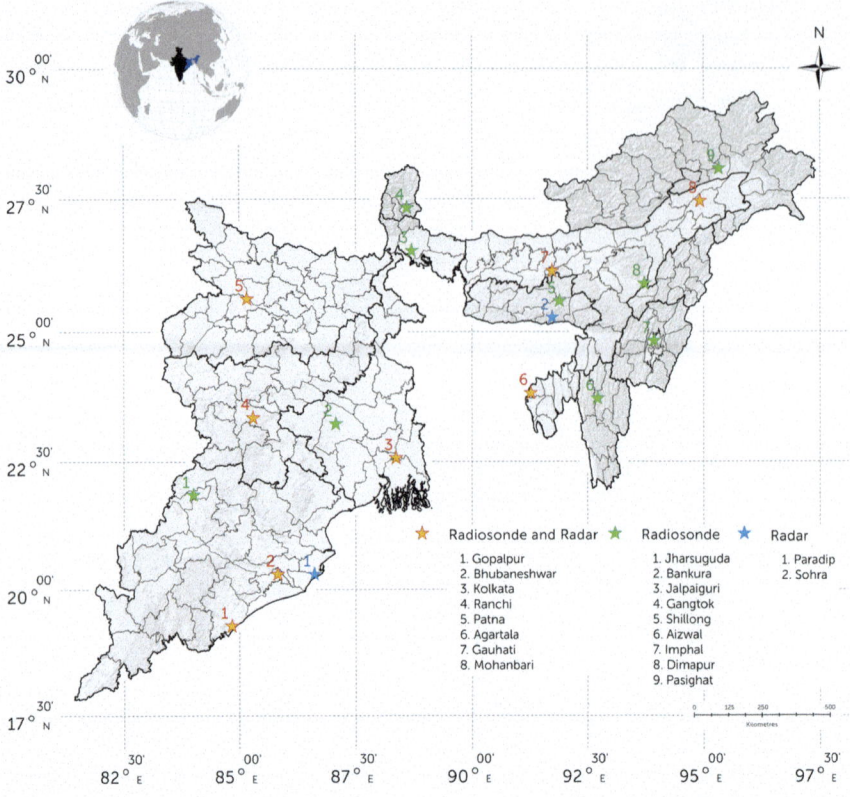

Fig. 7.3 Radiosonde and radar observations in eastern and north-eastern India. *Source* IMD Pune

7.3 Thermodynamic Indices, Skill Scores and Rank Sum Score

Convective activities over any region have different progression rates for the occurrence and associated precipitation, depending upon the existence and strength of several physical and dynamical atmospheric processes. These processes investigate the local and transported water vapour availability and the lower atmosphere's prediction for responsible vertical motions to produce any activity (Peppler 1988). The convective activity development can be understood based on the potential, latent and convective instabilities. Various indices and variables were developed over time, which proved their ability to differentiate the thunderstorm occurrence over the study regions. The most used thermodynamic indices are Boyden Index, Bulk Richardson number, Convective Available Potential Energy (CAPE), Convective Inhibition (CIN), Lifted Index (LI), K Index (KI), Showalter Index (SHOW), Severe Weather Threat Index (SWEAT), Cross Total Index (CTI), Vertical Totals Index (VTI), Total Totals Index (TTI), Humidity Index (HI), Relative Humidity at 700 hPa,

Dew Point Temperature at 850 hPa, Deep Convective Index (DCI) and Normalised CAPE (NCAPE). These indices are calculated mainly by using different levels of temperature/humidity for assessing how the vertical variation of heat and moisture are distributed regions of the lower atmosphere resulting in thunderstorm development. Some indices also combine more than one index for better accountability of atmospheric state and instability progression (e.g. TTI).

As the definitions of these indices are classical, we do not mention the formulations in the main text but provide these leading indices in Appendix 1 of the present article. Apart from these popular indices, many other indices were developed at various regions (e.g. Jefferson Index, SWISS Index, wet bulb zero height, storm-related helicity) and worked well for some regions or types of thunderstorms (Kunz 2007).

After defining the thermodynamic index, the next step involves fixing a threshold value over a particular region, using skill score analysis (Haklander and Delden 2003). The process of skill score development is based on dividing the information related to thunderstorm events into the elements of the contingency table (generally defined as A, B, C and D, for observed and predicted values of event agreement), described as Table 7.4 in Appendix 2. Once we have A, B, C, D values, we can calculate various skill scores. The contingency table method and calculation of skill scores are not limited to thunderstorm research and have been adopted by various meteorological applications to explain the probability of event occurrence based on the available information (Wilks 2001; Hyvärinen 2014). The commonly used skill scores are Probability of Detection (POD), False Alarm Ratio (FAR), Critical Success Index (CSI), True Skill Statistic (TSS), Heidke Skill Scores (HSS), Normalised Skill Score (NSS), Miss rate (MR) and Correct non-occurrence (C-NON), which are defined in Appendix 2. Another critical step after determining the threshold values is calculating the Rank Sum Score (RSS) of computed indices. RSS gives a quantitative assessment for the performance of a particular index for a specific study by assigning it a number indicating the usefulness of a particular index. The higher the value of RSS, the better the index is supposed to perform. The method of calculating RSS has also been briefly explained in Annexure B.

7.4 Atmospheric Thermodynamics Over Eastern and North-Eastern India During Pre-monsoon Thunderstorms

Thermodynamic indices for thunderstorm studies over India are one of the well-explored research techniques for the past many decades. The studies covered pre-monsoon thunderstorms of eastern and north-eastern India, the southern India Peninsular thunderstorms, northern Indian dust storms (Aandhi) and thunderstorms occurring in almost all the seasons over different parts of the country. Though it is challenging to review every study in this aspect, we attempt to briefly summarise the progress of understanding developed by thermodynamic indices in thunderstorm research over eastern and north-eastern India. Kolkata is the most explored city in eastern India for thermodynamic changes during thunderstorms. Bhubaneswar and Ranchi in eastern India and the north-eastern region also reported many studies based on thermodynamics. This section can be explored by categorising studies based on Kolkata and other studies focusing on eastern and north-eastern Indian regions for defining atmospheric thermodynamic changes related to pre-monsoon thunderstorms.

7.4.1 Thermodynamic Changes Related to Pre-monsoon Thunderstorms Over Kolkata

Pre-monsoon thunderstorms over Kolkata got good attention from researchers to study the physical and dynamical behaviour, improve prediction and generate the associated weather data. Many research works deal with different aspects of thunderstorms over Kolkata, but we are discussing only published works with variations in thermodynamic indices. The Nor'westers have different characteristics than mid-latitude thunderstorms, as the lapse rate between 700 and 500 hPa and storm relative helicity over mid-latitude is higher, whereas the lower to middle troposphere is drier compared to tropical regions during thunderstorm occurrences (Chaudhuri and Middey 2014). A combination of various thermodynamic indices (CAPE, CIN, LI, BRN, KI, TTI, SWEAT, wind shear and other indices) has also been used for the data over Kolkata to categorise the types of pre-monsoon thunderstorms for 1997–2008 (Chaudhuri et al. 2014). The results showed that multicellular thunderstorms are the most occurring types over Kolkata, followed by single-cell events and rare appearances of supercell thunderstorms. Another study over pre-monsoon seasons of 2006–2008 over Kolkata reveals that squall lines are the most common thunderstorms occurring over the region (Dalal et al. 2012). These squall line thunderstorms can be further categorised as trailing stratiform, leading stratiform and parallel stratiform. In severe thunderstorms during the pre-monsoon season, overshooting top is a common associated feature. The analysis based on selected severe thunderstorm cases over Kolkata reveals that these overshooting tops thunderstorms have higher instabilities associated with them, defined based on thermodynamic indices, and the lightning is also reportedly higher (Chaudhuri et al. 2020).

Thermodynamic indices for thunderstorm predictions during the pre-monsoon season over Kolkata area significantly explored research area. Basu and Mondal (2002) studied pre-monsoon thunder-squall lines over Kolkata during 1997–1998. The study utilised SHOW, BI, CAPE, CIN, precipitable water content, humidity mixing ratio, propagation speed and vertical wind shear for 26 cases selected over Kolkata Airport. The results showed that KI, CAPE and wind shear are some of the parameters helpful for thunder-squall prediction. Khole and Biswas (2007) used the radiosonde data over Kolkata for the pre-monsoon season of 1991–1996 and verified the accuracy of TTI in prediction. The study found that TTI is one index that can differentiate the thunderstorm occurrences to clear days over the site. To define the state of various thermodynamic indices threshold values related to pre-monsoon thunderstorms over eastern India, Tyagi et al. (2011) analysed multiple indices and parameters over Kolkata during 2006–2008. The study was a significant step forward, as it defined the threshold values of many thermodynamic indices for differentiating pre-monsoon thunderstorms over the region. Thermodynamic indices have been used along with DWR products to formulate a condition support system using rough set theory over Kolkata for pre-monsoon thunderstorms by Chaudhuri and Middey (2012). The study formulated a model with CAPE, CIN, LI, BRN, SWEAT, TTI, KI and SHOW to combine them with DWR products. The model can predict thunderstorm occurrences better than individual thermodynamic indices.

Nayak and Mandal (2014) used radiosonde data over Kolkata for the pre-monsoon season of 2005–2012 to calculate bulk Richardson number, vertical wind shear, energy helicity index and vorticity generation parameter in relation to rainfall. The thunderstorms have been categorised as low precipitation, medium precipitation and high precipitation events and the indices showed different ranges with these events. The thermodynamic indices show different trends (increasing or decreasing) with the frequency and severity of pre-monsoon thunderstorms over Kolkata (Saha et al. 2014). Chakraborty et al. (2016) used in-situ observations of a microwave radiometer at Kolkata for the pre-monsoon season of 2011–2013. The study found that the standard deviation of brightness temperature at 22 GHz and LI, KI and HI can distinguish thunderstorm initiation. The study further developed a now casting algorithm using these parameters, which accurately predicted the thunderstorm events with an accuracy of 80% with about five days in advance.

Atmospheric thermodynamic indices associated with thunderstorm occurrences are also helpful in various models, including mathematical and statistical models, soft computing tools and regional scale atmospheric model simulations over Kolkata. Dasgupta and De (2007) used binary logistic regression for Nor'westers over Kolkata using the difference of pressure and equivalent potential temperature at different levels in the atmosphere as input variables. Chaudhuri (2010) developed an artificial neural network (ANN) model with one hidden layer where CAPE and CIN values were used as input parameters. Using a fuzzy interface for pre-monsoon thunderstorm prediction over Kolkata for 1997–2006 reveals that LI, BRN, CAPE and BI indices as input to a fuzzy interface are better predicting indices (Chaudhuri et al. 2015). Mukhopadhyay (2004) used a regional atmospheric modelling system (RAMS) to simulate a thunderstorm event over Kolkata. He suggested that horizontal resolution improves the thunderstorm prediction for modelling, as the 1 km grid model simulated better than that with a 5 km grid. The Weather Research and Forecasting-Non-hydrostatic Mesoscale Model (WRF-NMM) has been used for simulating the case of the 20th May 2006 thunderstorm over Kolkata (Litta and Mohanty 2008). The model performed well for reproducing thermodynamic indices and 3 km resolution provided good results. Prasad et al. (2014) assimilated DWR observations to Advanced Research Weather Research and Forecasting model (WRF-ARW) for two consecutive thunderstorms on 11th and 12th May 2009 over Kolkata.

The DWR assimilated runs can produce more instability, as shown by the analysis of thermodynamic indices (CAPE, CIN and LI) and improved the model prediction.

7.4.2 Thermodynamic Changes Over Eastern and North-Eastern India (Excluding Kolkata) Related to Pre-monsoon Thunderstorms

Apart from conventional thermodynamic indices, it was observed by Ghosh and De (1997) that the lifting condensation level variation was able to differentiate the thunderstorm possibility over Ranchi and Agartala. The study analysed the equivalent potential temperature and saturated equivalent potential temperature difference along with vertical wind shear and instability below lifting condensation level and discovered that this difference could pick up on the day with possibilities of thunderstorm occurrences. Mukhopadhyay et al. (2003) used radiosonde data over Guwahati, Dibrugarh and Agartala for 1980–1984 for calculating eleven thermodynamic indices (KI, TTI, SLI, DCI, HI, BI, SWEAT, CAPE, NCAPE, SHEAR, RINO) and their respective threshold values using skill score analysis. The work suggested HI as the best index for differentiating thunderstorm days from clear days. Chakrabarti et al. (2008) studied atmospheric thermodynamics over north-eastern India during April–May 2006 by utilising STORM pilot phase experiment over the region. The study explored 62 events that occurred during 1–25th May 2006 over various locations of north-eastern India and analysed surface meteorological variables along with CAPE and satellite imageries. The results suggested that the regions showed inhomogeneous temperature variations with high thermodynamic instability. Although higher CAPE values favour the occurrence of severe thunderstorms, there were some reported cases where even with low values of CAPE, thunderstorms occurred with outbursts in the region. The reason for such episodic behaviour was attributed to topography providing enough triggering to convective cells. Tyagi et al. (2013) analysed radiosonde profiles to understand the thermodynamic structure of the atmosphere over Kharagpur (22.3 °N, 87.2 °E), West Bengal, collected during the STORM programme. The study revealed that although the CAPE values were high for almost all the study days, the CIN played a limiting factor for thunderstorm occurrences over the site. The work also verified that humidity variations are important for differentiating the thunderstorm genesis, and HI is the best index for pre-monsoon thunderstorm now casting over the region.

There were many publications based on case studies of pre-monsoon thunderstorms over the region. Das et al. (2014b) took a case study of a severe thunderstorm over Guwahati (5th April 2010) and analysed the environmental thermodynamics with the help of CAPE, CIN and LI calculations. The study used Kalpana-1 satellite imagery for infrared channels to see the variations of cloud top temperatures and GPS obtained integrated water vapour information to understand the event's genesis, development and dissipation. The study also verified the transportation of moisture by lower-level winds from the Bay of Bengal to the region, helpful for developing severe convection in the pre-monsoon season. Singh et al. (2014) analysed a case study of two consecutive thunderstorm events over Agartala (30th April and 1st May 2012) using radiosonde data, Doppler weather radar imageries, model outputs and infrared satellite imageries of Kalpana-1, along with synoptic and surface observations over the region. The study adopted the threshold values of various thermodynamic indices from literature and analysed the variation of thirteen thermodynamic indices. The findings supported the earlier observations of Chakrabarti et al. (2008), of having lower CAPE values during the thunderstorm occurrence on both the days; however, other indices described favourable conditions for convective activity. Based on the analysis of various thermodynamic indices, Narayanan et al. (2016) attempted to understand the development of pre-monsoon thunderstorms over eastern and north-eastern India for three cases (11th May 2009 and 07th May 2010 over eastern India and 19th April 2013 over north-eastern India). The study utilised satellite, in-situ and re-analysis data to calculate CAPE, CIN, LI and perceptible water in addition to ten other indices, including LCL height and temperature, during pre-monsoon thunderstorm cases.

As the importance of thermodynamic indices analysis paved its way for forecasting the thunderstorm occurrence over eastern and north-eastern Indian regions, some researchers adopted long-term data to understand the trends in these indices and the possible linkages with frequency and intensity thunderstorms occurrences. Chakraborty et al. (2019) analysed long-term variations (1980–2016) of thermodynamic indices over the Indian region by dividing the country into six potential zones. The study took eastern and north-eastern Indian region thunderstorms in the North-East zone and analysed sixteen stability and thermodynamic indices. The study suggested a possible, increasing trend in most of the parameters over the region in association with thunderstorms. Meher and Das (2019) used KI, BI and TTI to analyse pre-monsoon thunderstorms over the Gangetic West Bengal region derived from CMIP5 climate model outputs. The work attempted to relate the trend of these three indices with earlier reported thunderstorm frequency changes of thunderstorm over the region.

Sahu et al. (2020a) analysed radiosonde data for two decades (1987–2006) over Kolkata and Bhubaneswar and for fifteen years (1996–2010) over Ranchi, and various thermodynamic indices have been calculated with threshold values every five years. The surface temperature and humidity show an increasing trend and decreasing thunderstorm frequencies at all three sites. CAPE, CTI, LI and Showalter indices perform better for thunderstorm identifications at all three sites, whereas KI and SWEAT are also suitable for thunderstorm forecasting over Ranchi. The study advocated a change in threshold values of most of the thermodynamic indices; however, the change is different in magnitude for all three sites. Sahu et al. (2020b) used radiosonde observations along with rainfall and surface observations over Agartala and Guwahati from 1987 to 2016 to understand the trends in fourteen thermodynamic indices (BOYD, CAPE, CIN, CTI, VTI, TTI, HI, KI, DPT850, LI, RH700, SWEAT, SHOW, BRN). The study calculated threshold values for five years each in these three decades, and the values for 12 UTC are changing with years. The study revealed a considerable variation in the threshold values and optimum thermodynamic indices: LI, SHOW, CAPE, CTI and BRN for Guwahati, and LI, SHOW, CAPE, DPT and CTI over Agartala. In the thermodynamic indices, CAPE and CIN showed increasing trends, and SWEAT showed a decreasing trend in threshold values for Agartala and Guwahati, which may be associated with the frequency and severity of the thunderstorms.

7.5   Newly Developed Thermodynamic Indices for Nor'westers

As it is clear from the previous section, the use of thermodynamic indices for thunderstorm research has proven its usefulness. Though most of the studies explored already defined thermodynamic indices from the literature, a few newly developed thermodynamic indices are related to pre-monsoon thunderstorm studies over eastern and north-eastern India. The first attempt was made by Mitra et al. (2012) in their study over the Indian region. The study used Moderate Resolution Spectral Radiometer (MODIS) derived moisture and temperature profiles over the region. The profiles have been verified with real-time direct broadcast receiving systems located at three locations: Delhi, Chennai and Guwahati, for validating MODIS profiles for a period of March-June 2011. The receiving system located at Guwahati covers the whole eastern and north-eastern region, beneficial in assessing Nor'westers. The index has been named MODIS profile Index (MPI) and accounts for the vertical integration of specific humidity and pressure from the surface to 300 hPa height. MPI has been compared with traditional indices like LI, TTI and provides good results. The advantage of MPI was its comprehensive coverage over the spatially extended regions, especially over the places with no availability of radiosonde data. It has also been advocated that MPI can detect the initial stage of thunderstorm development. The formulation of MPI has been provided in Table 7.2.

Another attempt in this line was developing the Composite Stability Index (TPI) over Kolkata (Chaudhuri and Middey 2012). The study utilised radiosonde data for the pre-monsoon season (March–May) over Kolkata for 1997–2006. Various stability indices (e.g. LI, CAPE, CIN, BRN, Showalter, Boyden, KI, TTI, SWEAT, Storm-Related Helicity and surface T_d) have been computed and using the rough set theory, many permutations of these indices have been attempted to formulate a better performing index. As LI, CAPE and CIN are the three best performing indices over the study region, the attempt to formulate a new composite index turns out to be successful with the inclusion of these three indices. TPI performed better than any individual index (LI, CAPE and CIN) used in its formulation and resulted in 99.67% accuracy with a lead time of 12 h for thunderstorm prediction. The study also finds that a threshold value of TPI > 1 defines severe thunderstorms and the value of TPI between 0 and 1 indicated ordinary thunderstorm (not severe) occurrence. The TPI formulation is provided in Table 7.2. For predicting pre-monsoon thunderstorms over north-eastern India, locally known as Bordoichila, another composite index

Table 7.2 New indices developed for thunderstorms

Si. no.	Index abbreviation/name	Formula	Area/region of development	References
1	MODIS (Moderate Resolution Imaging Spectroradiometer) profile index (MPI)	$1/2g \sum_{i=1}^{n-1}(q_i + q_{i+1})(p_i - p_{i+1})$	All over India with focus on Delhi, Chennai and Guwahati	Mitra et al. (2012)
2	Composite Stability Index (TPI)	$\frac{CAPE}{2000}\frac{J}{kg} \times \frac{-150}{CIN}\frac{J}{kg} \times \frac{LI}{-5}\,^\circ C$	Kolkata	Chaudhuri and Middey (2012)
3	Bordoichila prediction index (BPI)	$\frac{29}{20.9} \times CTI \times (-14) \times SI$	Guwahati	Chaudhuri et al. (2013a, b)
4	KLURT index	$KLURT = \frac{TTI^2}{KI \times LI}$	Kolkata	Chakraborty et al. (2017)
5	CBHT index	Cloud base height/θe at 850 hPa	Kolkata	Samanta et al. (2020)
6	Modified Instability Indices: MCT, MVT, MTT, MSWI, MKI, MEI	MCT: $T_{d925} - T_{500}$ MVT: $T_{925} - T_{500}$ MTT: $MVT + MCT$ MSWI: 12 $T_{d925} + 20(MTT - 49) + 2f_9 + f_5 + 125(S + 0.2)$ MKI: $(T_{925} + T_{d925}) - (T_{d700} + T_{500})$ MEI: $MSE_{500} - MSE_{925}$	Dhaka and Chittagong (Bangladesh), along with nearby Indian stations	Karmakar and Alam (2011)

[a] where q is specific humidity, p is pressure from the surface to top of the atmosphere (hPa), g = 9.8 m/s^2, T_d: dew point temperature, T: dry bulb temperature, f_5 and f_9: wind speed at 500 and 925 hPa, S: $\sin \alpha_5 - \sin \alpha_9$; MSE: moist static energy, θe: equivalent potential temperature

was formulated with the available pre-monsoon radiosonde data from 1997 to 2006 over Guwahati (Chaudhuri et al. 2013a, b). The study utilised computed thermodynamic indices LI, CAPE, CIN, BRN, Showalter, VTI, CTI, TTI and SWEAT for performing rough set theory analysis for formulating a new index. Unlike the eastern India composite index (TPI), north-eastern India had found a different combination of best performing thermodynamic indices (CTI and Showalter Index) to produce the Bordoichila prediction index (BPI) (formulation given in Table 7.2). BPI also performs well compared to individual indices over the region, and the accuracy was at par with TPI over eastern India.

One more study to be mentioned in this connection is performed using data over Dhaka and Chittagong (Bangladesh) and nearby Indian stations connecting the eastern and north-eastern Indian regions (Karmakar and Alam 2011). The study utilised the CTI, VTI, TTI, SWEAT, KI and energy index (difference of moist static energy at two different levels in the atmosphere) for 00 UTC of the pre-monsoon season of 1990–1995.

With the change in levels for these indices, the study proposed Modified CTI (MCT), Modified VTI (MVT), Modified TTI (MTT), Modified SWEAT (MSWI), Modified KI (MKI) and Modified Energy Index (MEI). These modified indices perform better than traditionally computed indices, and the threshold has been fixed for these indices to differentiate the thunderstorms over Bangladesh and nearby Indian regions, Chakraborty et al. (2017) devised a new non-dimensional index named KLURT Using TTI, KI and LI For the pre-monsoon season of 2005–2015 over Kolkata. KLURT shows a good correlation with meteorological variables and pollutants NO_2 And SO_2 over the study area. The efficiency of KLURT appears to be 75% in predicting thunderstorms over Kolkata. Recently, using cloud base height and equivalent potential temperature at 850 hpa, a new thermodynamic index was developed over eastern India, named CBHT (Samanta et al. 2020). The index utilised radiosonde data over Kolkata for the pre-monsoon season of 2016–2019 and was formulated as a ratio of these two parameters (Table 7.2). The index performs unexpectedly well compared to other traditional indices for individual thunderstorm days over Kolkata. However, The CBHT cannot differentiate the thunderstorm initiation on the days with multiple thunderstorms. Table 7.2 summarises all these newly developed thermodynamic indices for the study region.

7.6   Climate Change and Thunderstorms Activity

With the climate change, the frequency and intensity of extreme events are changing in different parts of the world (IPCC 2021: Summary for Policymakers). For thunderstorms, the intensity and frequency changes reports are relying on the local observations. The observations of frequency for extreme events like thunderstorms are difficult to account in climatic sense over a region. The primary issue comes with spatial fluctuations of extreme events and the coarser density of surface meteorological stations. There are further complications with accounting methods at different stations (Doswell et al. 2005). However, based on local reports, a lot of research has been done on the frequency and intensity changes of thunderstorms globally. There are reports that thunderstorm frequency decreased due to other environmental factors over a region particularly due to pollution (Yang et al. 2013). However, the trends were not same for a long period, and it was observed that from 1950 to 1900, the dust storms were decreased, whereas the dust storms have increased from 1997 to 2002 over northern China (Ding et al. 2005). However, with the temperatures rising continuously over the globe, different parts of world showed evidence that present change in climate will support increased frequency of thunderstorm events in future (Brooks 2013; Allen 2018; Rädler et al. 2019).

The eastern and north-eastern India also face the consequences of global warming. The eastern region has reported a decrease in thunderstorm frequencies and increase in intensities over time. Parts of eastern India showed an increase in night-time thunderstorms over the past few decades (Sahu et al. 2020a). However, for the north-eastern India, the expected frequency may increase over years with reduced severity for nocturnal thunderstorms and enhanced severity for daytime thunderstorms over certain regions (Sahu et al. 2020b). There is a need of more research for understanding the trends of thunderstorm frequency and severity over the eastern and north-eastern region with continuous data and information about thunderstorms at local scales.

7.7 Summary and Conclusions

The present work reviewed the thermodynamic changes in the atmosphere as revealed by thermodynamic indices and parameters for pre-monsoon thunderstorms over eastern and north-eastern India. Thermodynamic indices have proven their utility for prediction improving neural network approach, regional climate models and other mathematical and statistical approaches. The Nor'westers have been studied over a century over the region, and their forecasting was an essential goal in all related studies. Thermodynamic indices bridged this gap between observations and forecasting studies and improved the physical and dynamical understanding of Nor'westers. The understanding of different aspects of Nor'westers based on thermodynamic indices and parameters can be summarised in the following points:

- Kolkata has maximum number of studies related to Nor'westers and the behaviour of thermodynamic indices compared to any other stations/regions of eastern and north-eastern India.
- The radar network of IMD covers the whole eastern and north-eastern region. The upper air observations by radiosonde are available at 9 locations, maintained by IMD. However, not all the radiosonde observatories have long-term or regular data availability in the region, and some parts of the region need better coverage for upper air observations. The in-situ surface observatories in the region have good spatial coverage and long-term data availability, helpful in understanding these convective events.
- The pre-monsoon thunderstorms are mostly multi-cell clusters or squall line thunderstorms in the region of study. There are significant single-cell air mass thunderstorm occurrences and rare events of supercell thunderstorms.
- The eastern and north-eastern regions receive moisture supply from the Bay of Bengal. The pre-monsoon season is characterised by enormous heating of land-mass over the region. Various small hills and plateaus are acting as triggering agents for starting the thunderstorm developments. Chota Nagpur Plateau and Hazaribagh Plateau are the leading triggering platforms for eastern India, whereas the north-eastern India has many hills acting for the trigger. The state of atmospheric thermodynamics may differ at different locations of the region. The best working thermodynamic indices may also change considering this fact, as revealed by various studies.
- The indices based on convective instability and moisture work better than other instability indices for eastern India. In contrast, in north-eastern India, convective instability may not differentiate between thunderstorm days to clear days. The

north-eastern region may experience severe thunderstorms even when the CAPE values are low attributed to triggering.

- The global warming scenario is visible in changing the thermodynamics of the atmosphere, as revealed by the changed Nor'westers frequency and intensity reported by various studies. In turn, the thermodynamic indices threshold values are also found to vary for differentiating these thunderstorm events.
- New thermodynamic indices have been developed for these pre-monsoon thunderstorm predictions over eastern and north-eastern India, which worked better than previously formulated indices and parameters.

Although the studies accounting for the thermodynamic indices and parameters over the region are voluminous, most of them use case studies of thunderstorms or deal with several cases in a couple of years to understand the threshold values and prediction accuracy of a particular index or validating some modelling results. To improve the thunderstorm forecasting, comprehensive mesoscale studies are needed to understand the regional changes for initiation of these thunderstorms. The thermodynamic indices can serve as a good tool for identifying atmospheric changes accountable for thunderstorm occurrence and improving forecasting. More radiosonde observations at strategically defined locations are needed to explore the land–atmosphere interactions. Studies dealing with long-term data and accounting for thermodynamic indices, or their threshold values are limited, and more efforts are needed to identify changing climate scenarios over the region.

Acknowledgements Authors want to acknowledge the Science and Engineering Research Board (SERB), Department of Science and Technology, Government of India for providing the funding [project-funding code: DST/SERB/ECR/2017/001361]. The authors are also thankful to India Meteorology Department for providing the thunderstorm information and data for research. Authors acknowledge the constant efforts of all the public and non-public institutions in tackling the Nor'westers over eastern and north-east India. The authors want to acknowledge the National Institute of Technology Rourkela for providing research facilities for conducting this research. The authors are also thankful to the editors and reviewer for their constructive suggestions to improve the chapter.

Appendix 1

See Table 7.3.

Formulations of Popular Thermodynamic Indices

The brief information of all the thermodynamic parameters calculation is discussed in this section as follows.

Table 7.3 State-wise AWS locations in the eastern and north-eastern India with long-term data availability

State	Id	Station name	State	Id	Station name	State	Id	Station name
Odisha	1	Angul		36	Contai		71	Silchar
	2	Balasore		37	Cooch Behar (A)		72	Silchar/Kumbhigram (A)
	3	Baripada		38	Darjeeling		73	Tangla
	4	Bhawanipatna		39	Diamond Harbour		74	Tezpur
	5	Bhubaneshwar		40	Digha	Arunachal P	75	Bomdila
	6	Bolangir		41	Haldia		76	Hayuliang
	7	Chandbali		42	Hasimara Aero		77	Pasighat (A)
	8	Cuttack		43	Jalpaiguri		78	Ziro
	9	Gopalpur		44	Kalaikonda AF	Meghalaya	79	Cherrapunji
	10	Jharsuguda (A)		45	Kalimpong		80	Shillong (CSO)
	11	Keonjhargarh		46	Kolkata (Alipore)		81	Tura
	12	Koraput		47	Kolkata/Dum Dum (A)	Nagaland	82	Kohima
	13	Paradip (CWR)		48	Krishnanagar		83	Mokokchung
	14	Puri		49	Malda		84	Tuensang
	15	Rourkela		50	Midnapore	Manipur	85	Imphal
	16	Sambalpur		51	Purulia	Mizoram	86	Aizwal
	17	Titlagarh		52	Sagar Island	Tripura	87	Agartala
Jharkhand	18	Chaibasa		53	Sandheads		88	Kailashahar Aero
	19	Daltonganj		54	Santiniketan			
	20	Dhanbad		55	Suri			
	21	Dumka		56	Uluberia			

(continued)

Table 7.3 (continued)

State	Id	Station name	State	Id	Station name	State	Id	Station name
	22	Hazaribagh	Sikkim	57	Gangtok			
	23	Jamshedpur	Assam	58	Chaparmukh			
	24	Jamshedpur (A)		59	Dhubri			
	25	Ranchi		60	Dhubri/Rupsi (A)			
	26	Ranchi (A)		61	Dibrugarh/Mohanbari (A)			
West Bengal	27	Asansol/Panagarh (A)		62	Digboi			
	28	Bagati		63	Goalpara			
	29	Bagdogra Aero		64	Gohpur			
	30	Balurghat		65	Golaghat			
	31	Bankura 1		66	Guwahati/Borjhar (A)			
	32	Basirhat E. M. O		67	Lumding			
	33	Berhampore		68	Majbat			
	34	Burdwan		69	North Lakhimpur (A)/Lilabari			
	35	Canning		70	Rangia			

Source IMD in connection to Fig. 7.2

(a) **Boyden Index (BOYD)**

BOYD is the vertical temperature difference among 700 and 1000 hPa and the temperature at 700 hPa and then subtracted by 200, which reaches a value around 100. It is designed to measure the risk of a frontal thunderstorm; moisture is not considered in this calculation (Boyden 1963).

$$BOYD = 0.1 * (H_{700} - H_{1000}) - T_{700} - 200$$

(b) **Convective Available Potential Energy (CAPE)**

It is the positive buoyancy force of an air parcel rising adiabatically and is vertically integrated (Moncrieff and Miller 1976). CAPE is used for studying the conditional instability of the atmosphere (William and Renno 1993), and normally higher CAPE value signifies more convection.

$$CAPE(J\,kg^{-1}) = \int_{Z_{LFC}}^{Z_{LNB}} g\left(\frac{T_{ve} - T_{vp}}{T_{vp}}\right)dz$$

where T_{ve} is virtual environmental temperature; T_{vp} is virtual parcel temperature; Z_{LNB} is the height of equilibrium level; Z_{LFC} is the height of the level of free convection; and g is the gravity.

(c) **Normalised Convective Available Potential Energy (NCAPE)**

The CAPE divided by the boundary layer height is called as NCAPE, which has units of ms^{-2}. NCAPE is normally defined by terminology of skinny and fat depending on the value ranges of <0.1 and >0.3, indicating lower or higher potential of atmosphere instability.

(d) **Convective Inhibition (CIN)**

It is the energy preventing a parcel of air from rising from surface level to level of free convection (LFC). It is an effective negative buoyancy force of an air parcel which indicates atmospheric instability (Colby 1984).

$$CIN(Jkg^{-1}) = \int_{Z_{bottom}}^{Z_{top}} g\left(\frac{T_{v.parcel} - T_{v.env}}{T_{v.env}}\right)dz$$

where $T_{v.\,parcel}$ is the virtual temperature of the parcel; $T_{v.\,env}$ is virtual environmental temperature; Z_{top} is the height at LFC; and Z_{bottom} is the ground value.

(e) **Bulk Richardson Number**

It is the ratio between CAPE and lifted vertical wind shear of a parcel. BRN specifies the thunderstorm types, i.e. single-cell, multi-cell, supper cell (Weisman and Klemp 1982, 1984).

$$BRN = CAPE/(U^2/2)$$

where U is a shear extent.

(f)   **Lifted Index (LI)**
      It is measured by the difference of temperature within a parcel of air with its environment at the 500 hPa level. It gives the convective stability of the atmosphere at 500 hPa.

$$LI\ (K) = T_{500} - T_{parcel}$$

where T_{parcel} is parcel temperature, and T_{500} is the environmental temperature at 500 hPa. Thunderstorm development is only possible when the values of LI are negative (DeRubertis 2006).

(g)   **K Index (KI)**
      It is measured to determine the air mass thunderstorms and is the combination of the difference of temperature between 850 and 500 hPa and dew point or moisture at 850 and 700 hPa (George 1960) and is given by.

$$KI\ (K) = (T_{850} - T_{500}) + T_{d850} - (T_{700} - T_{d700})$$

(h)   **Showalter Index (SHOW)**
      The Showalter index is the variance among the dry bulb temperature at 500 hPa and the temperature of a parcel of air following the parcel which is lifted from 850 to 500 hPa pseudo-adiabatically (Showalter 1953).

$$SHOW = T_{500} - T_{850hPa-500hPa}.$$

(i)   **Severe Weather Threat Index (SWEAT)**
      It is used for determining the severe weather potential (Miller 1972).

$$SWEAT = 12T_{d850} + 20(TTI - 49) + 2f_{850} + f_{500}$$
$$+ 125[\sin\ (d_{500} - d_{850})] + 0.2$$

where f is wind speed in knots at different levels and d gives the direction of wind.

(j)   **Cross Totals Index (CTI)**
      It is the variation amongst the temperature of dry bulb and dew point by 850 and 500-hPa pressure level (Miller 1967).

$$CTI = T_{d-850} - T_{500}.$$

Here T_d, dew point temperature, and T, dry bulb temperature.

(k)   **Vertical Totals Index (VTI)**

It is the temperature contrast of dry bulb at 850 and 500 hPa level pressure (Miller 1972).

$$\text{VTI} = T_{850} - T_{500}$$

Same as BOYD, VTI is not considering moisture and only assess the conditional instability among 850 and 500 hPa pressure level.

(l) **Total Totals Index (TTI)**

It accommodates the storm strength at diverse levels. However, it did not take latent instability below 850 hPa level (Miller 1967). It is a combination of CTI and VTI and is given as.

$$\text{TTI (K)} = (T_{850} - T_{500}) + (T_{d850} - T_{500})$$

(m) **Humidity Index (HI)**

It is the aggregate temperature of dry bulb and dew point difference on 850, 700, 500 mb, respectively (Litynska et al. 1976).

$$\text{HI (K)} = (T - T_d)_{850} + (T - T_d)_{700} + (T - T_d)_{500}$$

Here T_d and T are the dew point and the dry bulb temperature at 850, 700, 500 mb.

(n) **Relative Humidity at 700 hPa ($RH_{700\ hPa}$)**

The $RH_{700\ hPa}$ is considered as one of the important indexes to identify the moisture availability in the mid-tropospheric region for the development of any convective system (Dhawan et al. 2008; Tyagi et al. 2011).

(o) **Dew Point Temperature at 850 hPa ($DPT_{850\ hPa}$)**

The dew point temperature represents the saturation of air concerning to water vapour in the atmosphere, and the $DPT_{850\ hPa}$ has been identified as a crucial variable to identify the moisture availability in the lower levels of the atmosphere for the development of any convective storm (Dhawan et al. 2008).

(p) **Deep Convective Index (DCI)**

DCI accounts for latent instability of the atmosphere and used temperature and dew point temperature at 850 hPa along with surface LI (Barlow 1993):

$$\text{DCI} = (T + T_d)_{850} - \text{LI}_S$$

where 850 defines 850 hPa and subscript s denotes surface.

Appendix 2

Skill Score Analysis

To analyse the prediction skill and fixing the threshold values of different thermody-namic indices, skill score method is broadly accepted by the scientific community (Lee and Passner 1993; Huntrieser et al. 1997; Haklandar and Delden 2003; Kunz 2007; Tyagi et al. 2011). In this method, each predictor is chosen by certain upper or lower values of the threshold of the thunderstorm event. For example, LI was taken at a definite upper value of threshold which comes from any value of minimum to maximum of LI. Then, the values with 1% increment towards the maximum value are analysed in response to the number of events prediction. There is a total of four cases to analyse this behaviour, viz. A, B, C and D elements formed on forecasted and observed events, described in Table 7.1, known as a contingency table. Based on Table 7.4, different skill scores, i.e. Heidke Skill Score (HSS), the True Skill Statistic (TSS), the probability of detection (POD), the False alarm ratio (FAR), Critical Success Index (CSI), etc., are calculated, as described in Table 7.5 (Kunz 2007) with their formula, symbols and range.

POD and FAR vary from 0 to 1 by which perfect forecast is associated with the value 1 for POD and 0 for FAR. Normalised Skill Score and μ are used in this study, along with these nine skill scores. NSS finds the best standard limits for every constraint. By the definition, NSS and μ reach its maximum values when both HSS and TSS have maximum values. Threshold value has been chosen by taking improved values of all skill scores into account. The HSS (α) signifies HSS for some α, HSS $_{best}$ is the maximum HSS value, and HSS $_{poorest}$ is the minimum in HSS value and likewise for TSS for calculation of μ.

Rank Sum Score (RSS)

RSS is the free, stable distribution that helps to understand the forecast ability of any index (Tyagi et al. 2011). For RSS analysis, we must take a specific index, and by arranging them from higher to lower values, and then for every index, the scores are listed. RSS is calculated by differencing mean rank values of NTD and TD cases and then divided by the total sample size. Higher values of RSS show a specific index can differentiate NTD and TD events. As RSS is independent of prediction expression, one can easily compare the many indices and their execution (Schulz 1989).

Table 7.4 Contingency table for a dichotomous categorical verification of forecasts (after Kunz 2007)

		Observation	
		Yes	No
Forecast	Yes	A Correct event forecast	B False alarms
	No	C Surprise events	D Non-events

Table 7.5 Formulae, symbols and range of different skill scores

Skill scores	Formula	Symbols	Range
Probability of detection	$POD = a/(a + b)$	POD	$0 < POD < 1$
False alarm ratio	$FAR = b/(a + b)$	FAR	$0 < FAR < 1$
Critical success index	$CSI = a/(a + b + c)$	CSI	$0 < CSI < 1$
True skill statistic	$TSS = ((a \times d) - (b \times c))/(a + c) \times (b + d)$	TSS	$-\infty < TSS < 1$
Heidke skill scores	$HSS = (a + d - R)/(a + b + c + d - R)$ with $R = ((a + b) \times (a + c) + (c + d) \times (b + d))/(a + b + c + d)$	HSS	$-\infty < HSS < 1$
Normalised skill score	$\frac{1}{2}(TSS(\times)/TSS_{max} + HSS(\times)/HSS_{max})$	NSS	$0 < NSS < 1$
Miss rate	$MR = c/(c + a)$	MR	$0 < MR < 1$
Correct non-occurrence	$C\text{-}NON = d/(d + b)$	C-NON	$0 < C\text{-}NON < 1$
BIAS	$BIAS = (a + b)/(a + c)$	BIAS	_____
Percentage correct	$PC = [(a + d)/(a + b + c + d)] \times 100$	PC	$0 < PC < 100$
μ	$\mu = 0.5 \times ((TSS(\alpha) - TSS_{poorest})/(TSS_{best} - TSS_{poorest})) + ((HSS(\alpha) - HSS_{poorest})/(HSS_{best} - HSS_{poorest}))$	μ	$0 < \mu < 1$

References

Allen, J. T. 2018. Climate change and severe thunderstorms. In *Oxford Research Encyclopedia of Climate Science*.

Barlow, W. 1993. A new index for the prediction of deep convection. In *Preprints, 17th Conference on Severe Local Storms*, 129–132, St. Louis, MO.

Basu, G.C., and D.K. Mondal. 2002. A forecasting aspect of thundersquall over Calcutta and its parameterisation during pre-monsoon season. *Mausam* 53 (3): 271–280.

Bhardwaj, P., O. Singh, and D. Kumar. 2017. Spatial and temporal variations in thunderstorm casualties over India. *Singapore Journal of Tropical Geography* 38 (3): 293–312.

Bikos, D., J. Finch, and J.L. Case. 2016. The environment associated with significant tornadoes in Bangladesh. *Atmospheric Research* 167: 183–195.

Boyden, C.J. 1963. A simple instability index for use as a synoptic parameter. *Meteorological Magazine* 92: 198–210.

Brooks, C.E.P. 1934. The variation of the annual frequency of thunderstorms in relation to sunspots. *Quarterly Journal of the Royal Meteorological Society* 60 (254): 153–166.

Brooks, H.E. 2013. Severe thunderstorms and climate change. *Atmospheric Research* 123: 129–138.

Chakrabarti, D., H.R. Biswas, G.K. Das, and P.A. Kore. 2008. Observational aspects and analysis of events of severe thunderstorms during April and May 2006 for Assam and adjoining states–A case study on 'Pilot Storm Project.' *Mausam* 59: 461–478.

Chakraborty, R., S. Das, and A. Maitra. 2016. Prediction of convective events using multi-frequency radiometric observations at Kolkata. *Atmospheric Research* 169: 24–31.

Chakraborty, R., U. Saha, A.K. Singh, and A. Maitra. 2017. Association of atmospheric pollution and instability indices: A detailed investigation over an Indian urban metropolis. *Atmospheric Research* 196: 83–96.

Chakraborty, R., M. VenkatRatnam, and S.G. Basha. 2019. Long-term trends of instability and associated parameters over the Indian region obtained using a radiosonde network. *Atmospheric Chemistry and Physics* 19 (6): 3687–3705.

Chaudhuri, S. 2010. Convective energies in forecasting severe thunderstorms with one hidden layer neural net and variable learning rate back propagation algorithm. *Asia-Pacific Journal of Atmospheric Sciences* 46 (2): 173–183.

Chaudhuri, S., and A. Middey. 2014. Comparison of tropical and midlatitude thunderstorm characteristics anchored in thermodynamic and dynamic aspects. *Asia-Pacific Journal of Atmospheric Sciences* 50 (2): 179–189.

Chaudhuri, S., S. Goswami, and A. Middey. 2013a. The coupled influence of instability indices and DWR data in estimating the squall speed of thunderstorms. *Asia-Pacific Journal of Atmospheric Sciences* 49 (4): 451–465.

Chaudhuri, S., J. Pal, A. Middey, and S. Goswami. 2013b. Nowcasting Bordoichila with a composite stability index. *Natural Hazards* 66 (2): 591–607.

Chaudhuri, S., S. Goswami, and A. Middey. 2014. Morphological classification pertaining to validate the climatology and category of thunderstorms over Kolkata, India. *Theoretical and Applied Climatology* 116 (1): 61–74.

Chaudhuri, S., F. Khan, D. Das, P. Mondal, and S. Dey. 2020. Probing for overshooting as extreme event of thunderstorms. *Natural Hazards* 102 (3): 1571–1588.

Chaudhuri, S., and A. Middey. 2012. A composite stability index for dichotomous forecast of thunderstorms. *Theoretical and Applied Climatology* 110(3): 457–469.

Chaudhuri, S., D. Das, and A. Middey. 2015. An investigation on the predictability of thunderstorms over Kolkata, India using fuzzy inference system and graph connectivity. *Natural Hazards* 76(1): 63–81.

Christian, H.J., R.J. Blakeslee, D.J. Boccippio, W.L. Boeck, D.E. Buechler, K.T. Driscoll, S.J. Goodman, J.M. Hall, X.J. Koshak, D.M. Mach, and M.F. Stewart. 2003. MF, Global frequency and distribution of lightning as observed from space by the optical transient detector. *Journal of Geophysical Research* 108: D1.

Colby, F.P., Jr. 1984. Convective inhibition as a predictor of convection during AVE-SESAME II. *Monthly Weather Review* 112 (11): 2239–2252.

da Silva, F.P., O.C. Rotunno Filho, R.J. Sampaio, I.C.D.A.V. Dragaud, A.A.M. de Araújo, M.G.A.J. da Silva, and G.D. Pires. 2019. Evaluation of atmospheric thermodynamics and dynamics during heavy-rainfall and no-rainfall events in the metropolitan area of Rio de Janeiro. *Brazil. Meteorology and Atmospheric Physics* 131 (3): 299–311.

Dalal, S., D. Lohar, S. Sarkar, I. Sadhukhan, and G.C. Debnath. 2012. Organisational modes of squall-type mesoscale convective systems during pre-monsoon season over eastern India. *Atmospheric Research* 106: 120–138.

Das, S., U.C. Mohanty, A. Tyagi, D.R. Sikka, P.V. Joseph, L.S. Rathore, A. Habib, S.K. Baidya, K. Sonam, and A. Sarkar. 2014a. The SAARC STORM: A coordinated field experiment on severe thunderstorm observations and regional modeling over the South Asian Region. *Bulletin of the American Meteorological Society* 95 (4): 603–617.

Das, S., C.S. Tomar, R.K. Giri, K. Bhattacharjee, and B. Barman. 2014b. The severe thunderstorm of 5th April 2010 at Guwahati airport: An observational study. *Mausam* 65(1): 99–102.

Dasgupta, S., and U.K. De. 2007. Binary logistic regression models for short term prediction of pre-monsoon convective developments over Kolkata (India). *International Journal of Climatology* 27 (6): 831–836.

DeRubertis, D. 2006. Recent trends in four common stability indices derived from US radiosonde observations. *Journal of Climate* 19 (3): 309–323.

Desai, B.N. 1950. Mechanism of Nor'wester of Bengal. *Indian Journal of Meteorology and Geophysics* 1 (1): 74–76.

Dhawan, V.B., A. Tyagi, and M.C. Bansal. 2008. Forecasting of thunderstorms in pre-monsoon season over northwest India. *Mausam* 59 (4): 433–444.

Diffenbaugh, N.S., M. Scherer, and R.J. Trapp. 2013. Robust increases in severe thunderstorm environments in response to greenhouse forcing. *Proceedings of the National Academy of Sciences* 110 (41): 16361–16366.

Ding, R., J. Li, S. Wang, and F. Ren. 2005.Decadal change of the spring dust storm in northwest China and the associated atmospheric circulation. *Geophysical Research Letters* 32(2).

Doswell, C.A., III., H.E. Brooks, and M.P. Kay. 2005. Climatological estimates of daily local nontornadic severe thunderstorm probability for the United States. *Weather and Forecasting* 20 (4): 577–595.

Finney, D.L., R.M. Doherty, O. Wild, D.S. Stevenson, I.A. MacKenzie, and A.M. Blyth. 2018. A projected decrease in lightning under climate change. *Nature Climate Change* 8 (3): 210–213.

George, J. 1960. *Weather Forecasting for Aeronautics*, 41. New York: Academic Press.

Ghosh, A., D. Lohar, and J. Das. 2008. Initiation of Nor'wester in relation to mid-upper and low-level water vapor patterns on METEOSAT-5 images. *Atmospheric Research* 87 (2): 116–135.

Ghosh, S., and U.K. De. 1997. A comparative study of the atmospheric layers below first lifting condensation level for instantaneous pre–monsoon thunderstorm occurrence at Agartala (23° 30′ N, 91° 15′ E and Ranchi (23° 14′ N, 85° 14′ E) of India. *Advances in Atmospheric Sciences* 14(1), 93–102.

Guha, A., T. Banik, B.K. De, R. Roy, and A. Choudhury. 2013. Characteristics of severe thunder-storms studied with the aid of VLF atmospherics over North-East India. *Journal of Earth System Science* 122 (4): 1013–1021.

Guhathakurta, P., O.P. Sreejith, and P.A. Menon. 2011. Impact of climate change on extreme rainfall events and flood risk in India. *Journal of Earth System Science* 120 (3): 359–373.

Gupta, P.K.S. 1952. The genesis and movement of the Nor'westers of Bengal. In *Proceedings of the Indian Academy of Sciences-Section A*, vol. 35, no. 6, 303–309. Springer India.

Haklander, A.J., and A. Van Delden. 2003. Thunderstorm predictors and their forecast skill for the Netherlands. *Atmospheric Research* 67: 273–299.

Harel, M., and C. Price. 2020. Thunderstorm trends over Africa. *Journal of Climate* 33 (7): 2741–2755.

Huntrieser, H., H.H. Schiesser, W. Schmid, and A. Waldvogel. 1997. Comparison of traditional and newly developed thunderstorm indices for Switzerland. *Weather and Forecasting* 12 (1): 108–125.

Hyvärinen, O. 2014. A probabilistic derivation of Heidke skill score. *Weather and Forecasting* 29 (1): 177–181.

IMD TN 10. 1944. India Meteorological Department, 1944. *Nor'westers of Bengal, Technical Note, No 10*.

IPCC. 2021. Summary for policymakers. In *Climate Change 2021: The Physical Science Basis. Contribution of Working Group I to the Sixth Assessment Report of the Intergovernmental Panel on Climate Change*, ed. Masson-Delmotte, V., P. Zhai, A. Pirani, S.L. Connors, C. Péan, S. Berger, N. Caud, Y. Chen, L. Goldfarb, M.I. Gomis, M. Huang, K. Leitzell, E. Lonnoy, J.B.R. Matthews,

T.K. Maycock, T. Waterfield, O. Yelekçi, R. Yu, and B. Zhou. Cambridge University Press. In Press.

Karmakar, S., and M.M. Alam. 2011. Modified instability index of the troposphere associated with thunderstorms/nor'westers over Bangladesh during the pre-monsoon season. *Mausam* 62 (2): 205–214.

Khole, M., and H.R. Biswas. 2007. Role of total-totals stability index in forecasting of thunderstorm/non-thunderstorm days over Kolkata during pre-monsoon season. *Mausam* 58 (3): 369.

Koteswaram, P., and V. Srinivasan. 1958. Thunderstorm over Gangetic West Bengal in the pre-monsoon season and the synoptic factors favorable for their formation. *Indian Journalof Meteorology and Geophysics* 9: 301–312.

Kunz, M. 2007. The skill of convective parameters and indices to predict isolated and severe thunderstorms. *Natural Hazards and Earth System Sciences* 7 (2): 327–342.

Kunz, M., J. Sander, and C. Kottmeier. 2009. Recent trends of thunderstorm and hailstorm frequency and their relation to atmospheric characteristics in southwest Germany. *International Journal of Climatology* 29 (15): 2283–2297.

Lee, R.R., and J.E. Passner. 1993. The development and verification of TIPS: An expert system to forecast thunderstorm occurrence. *Weather and Forecasting* 8 (2): 271–280.

Litta, A.J., and U.C. Mohanty. 2008. Simulation of a severe thunderstorm event during the field experiment of STORM programme 2006, using WRF–NMM model. *Current Science* 204–215.

Litynska, Z, J. Parfiniewicz, and H. Piwkowski. 1976. *The Prediction of air Mass-Thunderstorms and Hails*, 128–130. WMO Interpretation of Broad-Scale NWP Prod. for Local Forecasting Purposes (SEE N 77–18671 09–47).

Madala, S., A.N.V. Satyanarayana, C.V. Srinivas, and B. Tyagi. 2016. Performance evaluation of PBL schemes of ARW model in simulating thermo-dynamical structure of pre-monsoon convective episodes over Kharagpur using STORM data sets. *Pure and Applied Geophysics* 173 (5): 1803–1827.

Madala, S., A.N.V. Satyanarayana, and B. Tyagi. 2013. Performance evaluation of convective parameterisation schemes of WRF-ARW model in the simulation of pre-monsoon thunderstorm events over Kharagpur using STORM data sets. *International Journal of Computer Applications* 71(15).

Mahanta, R., and Y. Yamane. 2020. Climatology of local severe convective storms in Assam. *India. International Journal of Climatology* 40 (2): 957–978.

Manohar, G.K., and A.P. Kesarkar. 2003. Climatology of thunderstorm activity over the Indian region: A study of east-west contrast. *Mausam* 54 (4): 819–828.

Manohar, G.K., and A.P. Kesarkar. 2004. Climatology of thunderstorm activity over the Indian region: II Spatial Distribution. *Mausam* 55 (1): 31–40.

Manohar, G.K., and A.P. Kesarkar. 2005. Climatology of thunderstorm activity over the Indian region: III. Latitudinal and Seasonal Variation. *Mausam* 56 (3): 581.

McCann, D.W. 1994. WINDEX—A new index for forecasting microburst potential. *Weather and Forecasting* 9 (4): 532–541.

Meher, J.K., and L. Das. 2019. Skill of CMIP5 climate models to reproduce the stability indices in identifying thunderstorms over the Gangetic West Bengal. *Atmospheric Research* 225: 172–180.

Mezuman, K., C. Price, and E. Galanti. 2014. On the spatial and temporal distribution of global thunderstorm cells. *Environmental Research Letters* 9 (12): 124023.

Miller, R.C. 1967. *Notes on Analysis and Severe-Storm Forecasting Procedures of the Military Weather Warning Center*, vol. 200. Air Weather Service (MAC), United States Air Force.

Miller, R.C. 1972. *Notes on Analysis and Severe-Storm Forecasting Procedures of the Air Force Global Weather Central*, vol. 200. AWS.

Mitra, A.K., A.K. Sharma, I. Bajpai, and P.K. Kundu. 2012. An atmospheric instability derived with MODIS profile using real-time direct broadcast data over the Indian region. *Natural Hazards* 63 (2): 1007–1023.

Mohanty, U.C., D.R. Sikka, O.P. Madan, R.S. Pareek, S. Kiran Prasad, and A.J. Litta. 2007. *Weather Summary Pilot Experiment of Severe Thunderstorms–Observational and Regional Modeling (STORM) Programme–2007*. Department of Science and Technology, Government of India.

Moncrieff, M.W., and M.J. Miller. 1976. The dynamics and simulation of tropical cumulonimbus and squall lines. *Quarterly Journal of the Royal Meteorological Society* 102 (432): 373–394.

Mukhopadhyay, P. 2004. Idealised simulation of a thunderstorm over Kolkata using RAMS. *Journal of Indian Geophysical Union* 8 (4): 253–266.

Mukhopadhyay, P., J. Sanjay, and S.S. Singh. 2003. Objective forecast of thundery/nonthundery days using conventional indices over three northeast Indian stations. *Mausam* 54 (4): 867–880.

Narayanan, S., G. Vishwanathan, and G. Mrudula. 2016. Possible development mechanisms of pre-monsoon thunderstorms over northeast and east India. In *Remote Sensing and Modeling of the Atmosphere, Oceans, and Interactions VI*, vol. 9882, 98821U. International Society for Optics and Photonics.

Nayak, H.P., and M. Mandal. 2014. Analysis of stability parameters in relation to precipitation associated with pre-monsoon thunderstorms over Kolkata, India. *Journal of Earth System Science* 123(4): 689–703.

Peppler, R.A. 1988. A review of static stability indices and related thermodynamic parameters. *ISWS Miscellaneous Publication MP-104*.

Pradhan, D., U.K. De, and U.V. Singh. 2012. Development of nowcasting technique and evaluation of convective indices for thunderstorm prediction in Gangetic West Bengal (India) using Doppler Weather Radar and upper air data. *Mausam* 63 (2): 299–318.

Pramanik, S.K., and C. Alipore. 1939. Forecasting of Nor'westers in Bengal. In *Proceedings of National Institute of Science (India)*, vol. 5, 43.

Prasad, S.K., U.C. Mohanty, A. Routray, K.K. Osuri, S.S.V.S. Ramakrishna, and D. Niyogi. 2014. Impact of Doppler weather radar data on thunderstorm simulation during STORM pilot phase—2009. *Natural Hazards* 74 (3): 1403–1427.

Rädler, A.T., P.H. Groenemeijer, E. Faust, R. Sausen, and T. Púčik. 2019. Frequency of severe thunderstorms across Europe expected to increase in the 21st century due to rising instability. *npj Climate and Atmospheric Science* 2(1): 1–5.

Rao, K.N., and P.K. Raman. 1961. Frequency of days of thunder in India. *Indian Journal of Meteorology and Geophysics* 12: 103–108.

Ray, K., B.K. Bandopadhyay, and S.C. Bhan. 2015. Operational nowcasting of thunderstorms in India and its verification. *Mausam* 66 (3): 595–602.

Robinson, E.D., R.J. Trapp, and M.E. Baldwin. 2013. The geospatial and temporal distributions of severe thunderstorms from high-resolution dynamical downscaling. *Journal of Applied Meteorology and Climatology* 52 (9): 2147–2161.

Romps, D.M., J.T. Seeley, D. Vollaro, and J. Molinari. 2014. Projected increase in lightning strikes in the United States due to global warming. *Science* 346 (6211): 851–854.

Roy, S.C., and G. Chatterji. 1929. Origin of Nor'westers. *Nature* 124 (3126): 481.

Saha, U., A. Maitra, S.K. Midya, and G.K. Das. 2014. Association of thunderstorm frequency with rainfall occurrences over an Indian urban metropolis. *Atmospheric Research* 138: 240–252.

Sahu, R.K., J. Dadich, B. Tyagi, N.K. Vissa, and J. Singh. 2020a. Evaluating the impact of climate change in threshold values of thermodynamic indices during pre-monsoon thunderstorm season over Eastern India. *Natural Hazards* 102: 1541–1569.

Sahu, R.K., J. Dadich, B. Tyagi, and N.K. Vissa. 2020b. Trends of thermodynamic indices thresholds over two tropical stations of north-east India during pre-monsoon thunderstorms. *Journal of Atmospheric and Solar-Terrestrial Physics* 211: 105472.

Samanta, S., B. Tyagi, N.K. Vissa, and R.K. Sahu. 2020. A new thermodynamic index for thunderstorm detection based on cloud base height and equivalent potential temperature. *Journal of Atmospheric and Solar-Terrestrial Physics* 207: 105367.

Sánchez, J.L., J.L. Marcos, J. Dessens, L. López, C. Bustos, and E. García-Ortega. 2009. Assessing sounding-derived parameters as storm predictors in different latitudes. *Atmospheric Research* 93(1–3): 446–456.

Sarkar, S., and R. Maity. 2020. Increase in probable maximum precipitation in a changing climate over India. *Journal of Hydrology* 585: 124806.

Schefczyk, L., and G. Heinemann. 2017. Climate change impact on Thunderstorms: Analysis of thunderstorm indices using high resolution COSMO-CLM simulations. *MeteorologischeZeitschrift* 26.

Schultz, P. 1989. Relationships of several stability indices to convective weather events in northeast Colorado. *Weather and Forecasting* 4 (1): 73–80.

Showalter, A.K. 1953. A convective index as an indicator of cumulonimbus development. *Journal of Applied Meteorology* 5(6), 839–846.

Singh, O., and P. Bhardwaj. 2019. Spatial and temporal variations in the frequency of thunderstorm days over India. *Weather* 74 (4): 138–144.

Singh, C., M. Mohapatra, B.K. Bandyopadhyay, and A. Tyagi. 2011. Thunderstorm climatology over northeast and adjoining east India. *Mausam* 62 (2): 163–170.

Singh, C., B.P. Yadav, S. Das, and D. Saha. 2014. Thunderstorm accompanied with squalls over Agartala for consecutive two days on 30 April and 1 May 2012. *Mausam* 65 (4): 539–552.

Sohoni, V.V. 1928. *Thunderstorms of Calcutta: 1900–1926.*Government of India.

STORM (Severe Thunderstorms—Observations and Regional Modeling) Programme. 2005. *Science plan.* Department of Science and Technology, Government of India.

Tajbakhsh, S., P. Ghafarian, and F. Sahraian. 2012. Instability indices and forecasting thunderstorms: The case of 30 April 2009. *Natural Hazards and Earth System Sciences* 12 (2): 403–413.

The Meteorological Office (Air Ministry). 1920. Climatic conditions on the imperial air routes. *Geographical Journal,* 128–136.

Tinmaker, M.I.R., and K. Ali. 2012. Space time variation of lightning activity over northeast India. *MeteorologischeZeitschrift* 21 (2): 135.

Tyagi, A. 2007. Thunderstorm climatology over Indian region. *Mausam* 58 (2): 189.

Tyagi, B., and A.N.V. Satyanarayana. 2010. Modeling of soil surface temperature and heat flux during pre-monsoon season at two tropical stations. *Journal of Atmospheric and Solar-Terrestrial Physics* 72 (2–3): 224–233.

Tyagi, B., and A.N.V. Satyanarayana. 2013a. Assessment of turbulent kinetic energy budget and boundary layer characteristics during pre-monsoon thunderstorm season over Ranchi. *Asia-Pacific Journal of Atmospheric Sciences* 49 (5): 587–601.

Tyagi, B., and A.N.V. Satyanarayana. 2014a. Coherent structures contributions in fluxes of momentum and heat at two tropical sites during pre-monsoon thunderstorm season. *International Journal of Climatology* 34 (5): 1575–1584.

Tyagi, B., and A.N.V. Satyanarayana. 2014b. Coherent structures contribution to fluxes of momentum and heat during stable conditions for pre monsoon thunderstorm season. *Agricultural and Forest Meteorology* 186: 43–47.

Tyagi, B., and A.N.V. Satyanarayana. 2015. Delineation of surface energy exchanges variations during thunderstorm and non-thunderstorm days during pre-monsoon season. *Journal of Atmospheric and Solar-Terrestrial Physics* 122: 138–144.

Tyagi, B., and A.N.V. Satyanarayana. 2019. Assessment of difference in the atmospheric surface layer turbulence characteristics during thunderstorm and clear weather days over a tropical station. *SN Applied Sciences* 1 (8): 1–9.

Tyagi, B., V.N. Krishna, and A.N.V. Satyanarayana. 2011. Study of thermodynamic indices in forecasting pre-monsoon thunderstorms over Kolkata during STORM pilot phase 2006–2008. *Natural Hazards* 56 (3): 681–698.

Tyagi, B., A.N.V. Satyanarayana, M. Kumar, and N.C. Mahanti. 2012. Surface energy and radiation budget over a tropical station: An observational study. *Asia-Pacific Journal of Atmospheric Sciences* 48 (4): 411–421.

Tyagi, B., A.N.V. Satyanarayana, and N.K. Vissa. 2013. Thermodynamical structure of atmosphere during pre-monsoon thunderstorm season over Kharagpur as revealed by STORM data. *Pure and Applied Geophysics* 170 (4): 675–687.

Tyagi, B., A.N.V. Satyanarayana, R.K. Rajvanshi, and M. Mandal. 2014. Surface energy exchanges during pre-monsoon thunderstorm activity over a tropical station Kharagpur. *Pure and Applied Geophysics* 171 (7): 1445–1459.

Tyagi, B., and A.N.V. Satyanarayana. 2013b. The budget of turbulent kinetic energy during pre-monsoon season over Kharagpur as revealed by storm experimental data. *International Scholarly Research Notices.*

Tyagi, B. 2012. *Surface Energy Exchanges and Thermodynamical Structure of Atmospheric Boundary Layer During Pre-Monsoon Thunderstorm Season Over Two Tropical Stations.* Doctoral dissertation, IIT Kharagpur.

Vujović, D., M. Paskota, N. Todorović, and V. Vučković. 2015. Evaluation of the stability indices for the thunderstorm forecasting in the region of Belgrade, Serbia. *Atmospheric Research* 161: 143–152.

Weisman, M.L., and J.B. Klemp. 1982. The dependence of numerically simulated convective storms on vertical wind shear and buoyancy. *Monthly Weather Review* 110 (6): 504–520.

Weisman, M.L., and J.B. Klemp. 1984. The structure and classification of numerically simulated convective storms in directionally varying wind shears. *Monthly Weather Review* 112 (12): 2479–2498.

Wilks, D.S. 2001. A skill score based on economic value for probability forecasts. *Meteorological Applications* 8 (2): 209–219.

Williams, E.R. 2005. Lightning and climate: A review. *Atmospheric Research* 76 (1–4): 272–287.

Williams, E., and N. Renno. 1993. An analysis of the conditional instability of the tropical atmosphere. *Monthly Weather Review* 121 (1): 21–36.

Yang, X., Z. Yao, Z. Li, and T. Fan. 2013. Heavy air pollution suppresses summer thunderstorms in central China. *Journal of Atmospheric and Solar-Terrestrial Physics* 95: 28–40.

Chapter 8
Real-Time Detection of Tornado-Induced Ionospheric Disturbances by Stand-Alone GNSS Receiver

Batakrushna Senapati, Dibyashakti Panda, Bhaskar Kundu, and Bhishma Tyagi

Abstract Mesoscale atmospheric events, e.g. deep tropical convection, thunderstorms, tropical cyclones, tropospheric jet, typhoons and tornadoes, are generally associated with upward propagating gravity waves and intense variation in electric fields, which may disturb the upper atmosphere. Here, we present a compelling case of variation in the total electron content (TEC) in the ionosphere induced by tornados that occurred over the city of Oklahoma, United States of America. Observations from global navigation satellite system (GNSS) sites indicate a variation in total electron content of about 0.1–0.35 ± 0.024 TECU (1 TECU $= 10^{16}$ electrons/m^2) during the occurrence of tornadoes. The variation in TEC has directly correlated with the intensity of lightning and the severity of the convective system. The TEC anomaly propagated at an apparent speed of ~0.165 m/s. We argued that the rate of change of electron density in the ionosphere is linked with the generation of atmospheric gravity waves, along with the duration and amplitude of lightning, associated with the convective events.

Keywords Tornados · Gravity waves · Total electron content · Global navigation satellite system · Oklahoma

8.1 Introduction

To characterize the state of the Earth's ionosphere, total electron contents (TEC) is the key parameter that has emerged as a new domain in the scientific community. However, it does not remain constant throughout space and time. Rather, it has been influenced by various processes, either directly or indirectly. It includes solar flares (Lee and Reinisch 2006), geomagnetic storm (Dabas et al. 1980; Sun et al. 2012; Park et al. 2010;), tsunamis (Savastano et al. 2017), solar eclipses (Coster et al. 2017; Kundu et al. 2018; Senapati et al. 2020), launching of space shuttles (Bowling et al. 2013), volcanic eruptions (Dautermann et al. 2009; Nakashima et al. 2015;

B. Senapati · D. Panda · B. Kundu (✉) · B. Tyagi
Department of Earth and Atmospheric Sciences, NIT Rourkela, Rourkela 769008, India
e-mail: kundub@nitrkl.ac.in

© The Centre for Science & Technol. of the, Non-aligned and Other Devel. Countries 2022
A. S. Unnikrishnan et al. (eds.), *Extreme Natural Events*,
https://doi.org/10.1007/978-981-19-2511-5_8

199

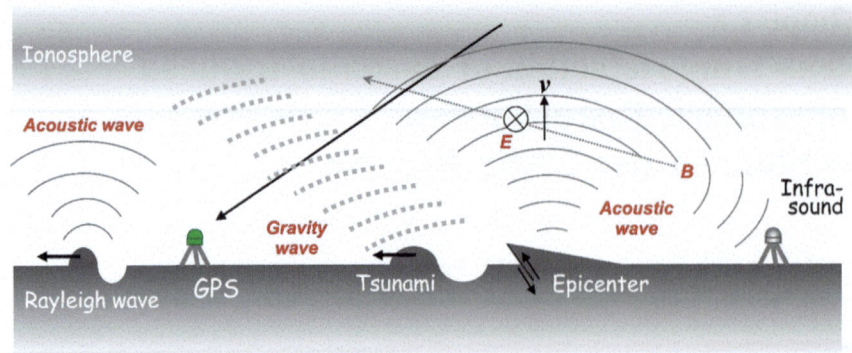

Fig. 8.1 Three kinds of atmospheric waves (e.g. acoustic wave by earthquakes, gravity wave generated by tsunami and acoustic wave excited by Rayleigh surface waves), which can disturb the ionosphere and modify the electron content within it. This variation in ionospheric electron content can be captured by ground-based GPS observations. Part of the direct acoustic wave comes back to the ground by atmospheric refraction and is observed by infrasound sensors. Vertical movements of ionized particles in the geomagnetic field (B) induce current (E) in ionospheric and cause geomagnetic pulsation. Figure is modified after Heki et al. (2006)

Petersen et al. 2012), meteorite showers (Yang et al. 2014), ionospheric irregularities during the post-sunset stage in the low-latitudinal region (Panda et al. 2019), large shallow earthquake (Catherine et al. 2015, 2016; Heki 2011; Cahyadi and Heki 2015; Rolland et al. 2011a, b), nuclear or non-nuclear explosions (Park et al. 2014; Kundu et al. 2021), tropospheric disturbances like thunderstorms (Kumar et al. 2017; Lay et al. 2013), tropical cyclones (Guha et al. 2016; Chou et al. 2017), tornadoes (Nishioka et al. 2013), deep tropical convection (Hagan et al. 2007) and even unknown mechanism (Tsugawa et al. 2007) (Fig. 8.1).

Among all the above process, the disturbances of tropospheric meteorological origin are also considered to cause the significant ionospheric perturbation as travelling ionospheric disturbances (TIDs) (Kelley, 1997; Guha et al., 2016; Xiao et al., 2007; Hung and Kuo 1978). However, understanding the exact coupling mechanism between the ionosphere and underlying troposphere and characterization of TIDs by intense and large-scale tropospheric weather remains equivocal. Hence, we suggest that such an understanding of the complex coupling mechanism deserves much scientific attention. Based on the ionospheric response to the effects of tropical cyclones over the Indian subcontinent, Guha et al. (2016) investigated that aspect. It has been argued that TIDs might be the result of the combined effect of atmospheric gravity waves, acoustic waves, neutral particle dynamics, atmospheric DC electrify fields and lightning-generated electrical field by the underlying large-scale weather system nucleating from the troposphere (Sorokin and Cherny 1999, Forbes et al. 2000; Isaev et al. 2002a, b; Sorokin et al. 2005; Nishioka et al. 2013; Lastovicka 2006; Lay et al. 2013; Shao et al. 2013; Guha et al. 2016).

Further, tropospheric disturbances like tornadoes, which exist in the lowest layer of the troposphere, can mark their prominent signatures in the ionosphere (Lutgens et al. 2001; Nishioka et al. 2013). Tornadoes are violently rotating air columns that blow around an intense low pressure having diameter few metres to more than one kilometre. They are mostly generated by the supercell thunderstorms and are associated with intense lightning along with gravity and acoustic waves (Lutgens et al. 2001), which in turn reaches the ionosphere and changes the TEC. The lightning associated with tornadoes is severe and propagates in all three possible ways (e.g. cloud to cloud, cloud to ground and cloud to atmosphere). From a global perspective, the current generated from the lightning during super cell thunderstorms may vary from 50 to 400 A and peak current up to tens of kilo-ampere. The lightning electric field propagated upward increases the molecular attachment of oxygen and reduces the electron density of the lower ionosphere (Lay et al. 2013; Shao et al. 2012). Similarly, gravity waves (frequency less than Brunt–Vaisala periods, Blanc et al. 2014) produced from a thunderstorm can propagate vertically upward and perturb the ionosphere electron contents (Fig. 8.2). The TEC change in the ionosphere during tornadoes manifestation may be either by vertically upward propagation of gravity

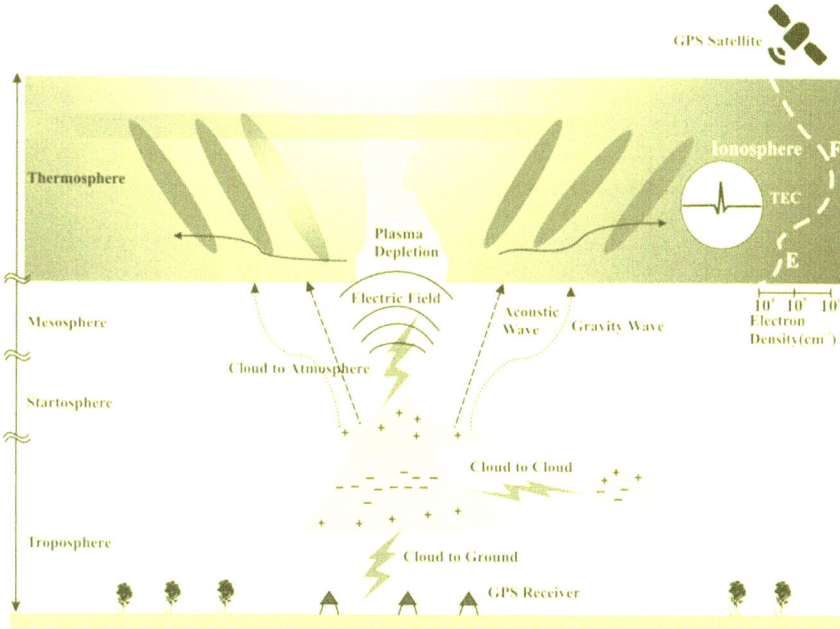

Fig. 8.2 Represents the effect of a tornado on the ionosphere. The white-dashed line indicates the variation of total electron contents (TEC) with the increasing height of the ionosphere. Lightening, acoustic wave and gravity wave generated from tornado propagated upward and change the TEC in the F-layer of the ionosphere by heating and mechanically forcing, respectively. The lightning-induced electric field can increase the attachment of oxygen molecules and decreases the electron density in the lower ionosphere, which is closely related to the time and space of lightning discharge. Figure is modified after Shao et al. (2012)

wave or by the electric field produced from lightning or by combining both mechanisms (Kumar et al. 2017) or by thunderstorm induced Perkins instability (Perkins 1973).

This motivates us to explore the possible troposphere-and-ionosphere coupling process during intense disturbances of three FE-3 to FE-5 categories of tornados that occurred over the city of Oklahoma, United States (also called *"Tornado Alley"*), exploring ground-based stand-alone GNSS receiver. Further, we suggest that in the view of coupling mechanism and associated complexity between the ionosphere and underlying troposphere, the TEC variations related to tornado occurrences have been underexplored (Lay et al. 2013; Nishioka et al. 2013) and will be notable to analyse because of its short lifespan. We hope that the present case studies based on the real-time detection of tornado-induced ionospheric disturbances, exploiting a well-established GNSS-TEC approach, will resolve some of the inherent complexity related to the troposphere-and-ionosphere coupling process.

8.2   Why Oklahoma?

Oklahoma is one of the states in the mid-western United States, receiving most of the tornadoes in the United States and assoiled with the central part of *"Tornado Alley"* (Gagan et al. 2010). Although the United States receives tornados throughout the year, the highest frequency of tornado occurrence over Oklahoma has been reported from March to mid-June every year (National Weather Service 2018). The average number of tornados that occurred in March–June is ~46, with the highest number in May(~23), against the annual average value of ~56, during periods from 1950 to 2017 (according to a report from National Weather Service 2018). The reason for high frequency during these summer months is due to suitable conditions created for the occurrence of these events by a combination of two different air masses: (a) the northward moving air mass into the Northern Great Plains and the Great Lakes area and (b) the southward moving air mass originating from Canada, respectively. However, the frequency decreases in mid-June as the air mass moves further northward (Lutgens et al. 2001).

Therefore, it appears that the city of Oklahoma, United States, is the ideal location to investigate the tornado-induced ionospheric disturbance and TIDs. Moreover, the Earth Scope Plate Boundary Observatory (PBO) has been operating a dense global navigation satellite systems (GNSS) networks over Oklahoma and the surrounding region of the United States (ftp://data-out.unavco.org/pub/rinex/); hence, the availability of the open archive dataset from GNSS network further makes this region suitable location to investigate the tornado-induced ionospheric disturbance and TIDs.

8.3   Datasets

8.3.1   Space Weather Dataset

In the present study, we have considered the Planetary-K (K_P), F10.7, disturbance storm time (DST) indices and variation of vertical (Z) and horizontal (H) components of geomagnetic field to represent geomagnetic storm condition during the time of occurrence of tornados over the city of Oklahoma. The K_P and DST indices described the geomagnetic storm condition, whereas, F10.7 exhibits variation of solar extreme ultraviolet radiation. These datasets have been archived from OMNIWeb (https:// omniweb.gsfc.nasa.gov/form/dx1.html). The components of geomagnetic field data are from the Boulder station, Colorado, USA (BOU, 40.137 °N, 105.237 °W) (https:// www.intermagnet.org/data-donnee/download-eng.php).

8.3.2   GNSS Dataset

To estimate GNSS signal-derived ionospheric TEC variation during tornado-induced ionospheric disturbance and TIDs, we have considered RINEX GNSS data from stations located over the city of Oklahoma and the surrounding region, United States, operated by the Earth Scope Plate Boundary Observatory (PBO) and maintained a dense global navigation satellite systems (GNSS) networks. We have used the standard sampling rate of 15 s in the daily GNSS file to evaluate the TEC perturbations in the ionosphere during the time of respective occurrence of tornados. The GNSS datasets have been archived at both UNAVCO open source (ftp://data-out.unavco. org/pub/rinex/) and Scripps Orbit and Permanent Array Centre (SOPAC, http://sopac-old.ucsd.edu/dataBrowser.shtml).

8.3.3   Other Tropospheric Parameters

We also used near-surface cloud reflectivity data from the Topical Rainfall Measuring Mission (TRMM, https://disc.gsfc.nasa.gov/datasets/TRMM_2A25_7/ summary) provided by NASA and the Japan Aerospace Exploration Agency, to observe the intensity of tornadoes. Further, to quantify the intensity of the considered tornado events, the lightning flash rates also have been used that are archived from the Lightning Imaging Sensor on TRMM (https://fog.nsstc.nasa.gov/pub/lis/trmm/). We have also collected representative thunderstorm intensity regarding information based on the Enhanced Fujita (EF) scale (i.e. considered from EF-3 to EF-5 categories), which are of higher intensity and archived at http://www.tornadohistoryproj ect.com/tornado/Oklahoma/table. It has been presented in Table 8.1.

Table 8.1 Represents the time of tornado manifestation with their length, width and available GNSS sites near to it

Date (dd/mm/yy)	Path	Time (UTC)	Local Time	EF-scale	Length (km)	Width (km)	Fatality	Injured	GNSS site	TEC change
08/05/2003	P1	21.10	16.10	EF-4	27.841	0.64	0	134	SG01, SG04 SG08, SG09 SG10, SG11 SG16, SG18 SG19, SG20	Significant
10/05/2010	P1	21.20	16.20	EF-4	38.624	1.828	2	49	WMOK, SG201, SG08, SG20	Significant
	P2	21.32	16.32	EF-4	35.807	0.804	1	32		
	P3	21.56	16.56	EF-3	59.867	2.01	0	28		
24/05/2011	P2	19.50	14.50	EF-5	101.549	1.609	9	181	WMOK, SG01	Strong
	P3	21.06	16.06	EF-4	53.591	0.804	1	48		
	P4	21.26	16.26	EF-4	37.175	0.804	0	61		

Archived from http://www.tornadohistoryproject.com/tornado/Oklahoma/table

8.4   Methodology

8.4.1   GNSS-TEC Data Processing Strategy

The phase difference between L_1 (~1500 MHz) and L_2 (~1200 MHz) frequency of the microwave signals from GNSS satellites provides important information about the ionospheric electron content by calculating differential code and carrier phase measurements recorded by ground-based GNSS stations. In this tornado-induced ionospheric disturbance and TIDs study, we have used only GNSS satellites. The respective phase difference between these two frequencies provides vital information on the ionospheric electrons content along the line of sight (LOS), called as slant TEC (STEC). The detail methodology for estimating the ionospheric GNSS-TEC has been described by several researchers (e.g. Afraimovich et al. 2001; Calais and Minister 1995; Heki and Ping 2005; Carrano and Groves 2009; Panda et al. 2013; Catherine et al. 2016; Heki 2020; Kundu et al. 2021). In order to estimate the observed STEC, we have eliminated the ambiguities in carrier phase differences by allowing them to align with differential pseudo-ranges (codes). Finally, the observed STEC is a combination of the true STEC and satellite/receiver biases, which can be expressed:

$$STEC_{(observed)} = STEC_{(true)} + Bias_{(satellite)} + Bias_{(receiver)} \qquad (8.1)$$

We have represented the observed STEC in terms of total electron content (TECU), where 1TECU is equal to 10^{16} electrons/m^2. This observed STEC is related to the difference in delay between the L_1 and L_2 frequencies (Δt) and can be expressed as:

$$\Delta t = \left(\frac{40.3 \times TEC}{c} \right) \times \left[\left(\frac{L_1^2 - L_2^2}{L_1^2 L_2^2} \right) \right], \qquad (8.2)$$

where c is the speed of the light.

Here, we have used satellite biases included in the header information of the Global Ionospheric Map files (Mannucci et al. 1998) and computed the receiver bias using the minimum scalloping technique (Rideout and Coster 2006). STEC values have been often converted to absolute vertical TEC (VTEC) values by removing the inter-frequency biases in GNSS receivers and satellites and dividing by the obliquity factor $S(\varnothing)$ (Klobuchar 1996), which depends on the elevation angle of satellite $\varnothing$. The VTEC and $S(\varnothing)$ are represented as follows:

$$VTEC = STEC_{(true)} / S(\varnothing), \qquad (8.3)$$

$$S(\varnothing) = \frac{1}{cos\alpha} = \frac{1}{\sqrt{\left[1 - (R_E cos\varnothing / \{R_E + h\})^2 \right]}}. \qquad (8.4)$$

where α is the incident angle of the LOS with the ionosphere at height h, and R_E is the mean radius of the Earth (i.e. assumed as 6378 km).

To estimate the propagation velocity of the TEC disturbance, we have calculated the distance between the SIP and tornado by considering the location of the occurrence of the tornado in Oklahoma City. Further, we have plotted the distance between sub-ionospheric points (SIP) and the location of the tornado as a function of time. A linear line has been fitted along the location of the maximum disturbance and estimate change in distance and change in time of the maximum disturbance. The propagation velocity of the TEC disturbance is calculated by dividing the change in distance to the change in time of the maximum disturbance.

8.4.2 GNSS-TEC Time Series Analysis

We also have performed wavelet power spectrum analysis of TEC variation using the Morlet wavelet (Torrence and Compo 1998). The wavelet function ($\varphi_0(\eta)$) depends upon a non-dimensional time parameter (η), and Morlet wavelet consists of a plane wave modulated by Gaussian that can be express by:

$$\varphi_0(\eta) = \pi^{-1/4} e^{i\omega_0\eta} e^{-\eta^2/2} \tag{8.5}$$

where ω_0 is the non-dimensional frequency (Torrence and Compo 1998).

8.5 Result and Discussion

8.5.1 Space Weather Condition During Tornado Events

Microwaves undergo a frequency-dependent delay in the ionosphere. The GNSS satellites transmit carrier phases in two frequencies in L-band ($L_1 \sim 1500$ MHz and $L_2 \sim 1200$ MHz). We can separate the ionospheric information by making the phase difference between the two carrier waves expressed in lengths. Further, such phase difference is subsequently converted to TEC (1 TECU $= 10^{16}$ electrons/m^2), and we study the TEC time series from GNSS satellites for various ionospheric disturbances (Heki 2020). However, such ionospheric disturbances can also be caused by the occurrence of solar flares (Lee and Reinisch 2006) or geomagnetic storms (Park et al. 2010; Dabas et al. 1980; Sun et al. 2012). Therefore, it is crucial to assess the geomagnetic condition of the space weather system during occurrence periods of tornadoes in Oklahoma and the surrounding region.

Most of the tornadoes from Oklahoma and the surrounding region have occurred in the evening (Table 8.1). During sunset, strong ionospheric irregularities occurred due to the Rayleigh–Taylor plasma instability known as equatorial plasma bubbles (EPBs) and often masked subtle changes in the ionosphere (Panda et al. 2019; Gentile

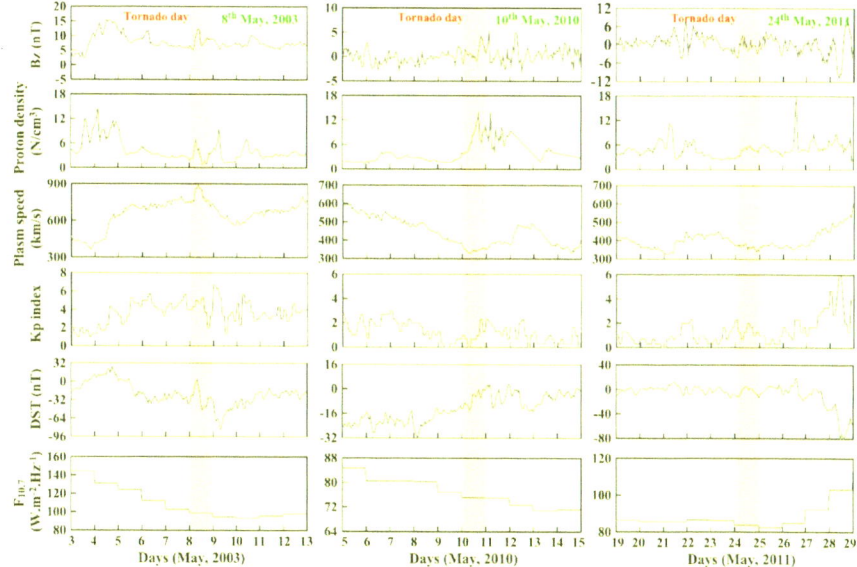

Fig. 8.3 Space weather conditions during tornado events over Oklahoma. Here, we show the geomagnetic activity indices (Magnetic field (Bz), K_P Index, DST index), proton density, plasma speed and F10.7 solar radiation intensity. Note all the tornado events occurred during a low geomagnetic activity day

et al. 2006). Further, Gentile et al. (2006) have argued that EPBs production rates and their occurrence predominantly depend on the regions' seasons and the specific longitude. It has been suggested that high EPB rate concentrates during the spring and autumn time in the worldwide and winter in the America-Atlantic-Africa region. We noticed that the EPB production rate is not high enough at the longitude of the Oklahoma region in May. Further, we have observed that the K_p index (https://www.swpc.noaa.gov/products/planetary-k-index) value of ~2 to 3 for the respective tornado occurrence periods implies that the respective days fall under low geomagnetic storm conditions. We also have noted that respective geomagnetic activity was low around the time of the occurrence of tornadoes, as represented by the other indices related to the space weather dataset (Fig. 8.3). Therefore, from this, it has appeared that our considered cases of investigation related to tornado-induced ionospheric disturbance and associated TIDs are the suitable ones because of the temporal and geographical distribution of the study region.

8.5.2   Observed Changes in Ionosphere: Induced by Tornados

Figure 8.4 represents the sub-ionospheric point (SIP) trajectories of all the GNSS sites at all available PRNs during periods of the respective occurrence of the tornadoes

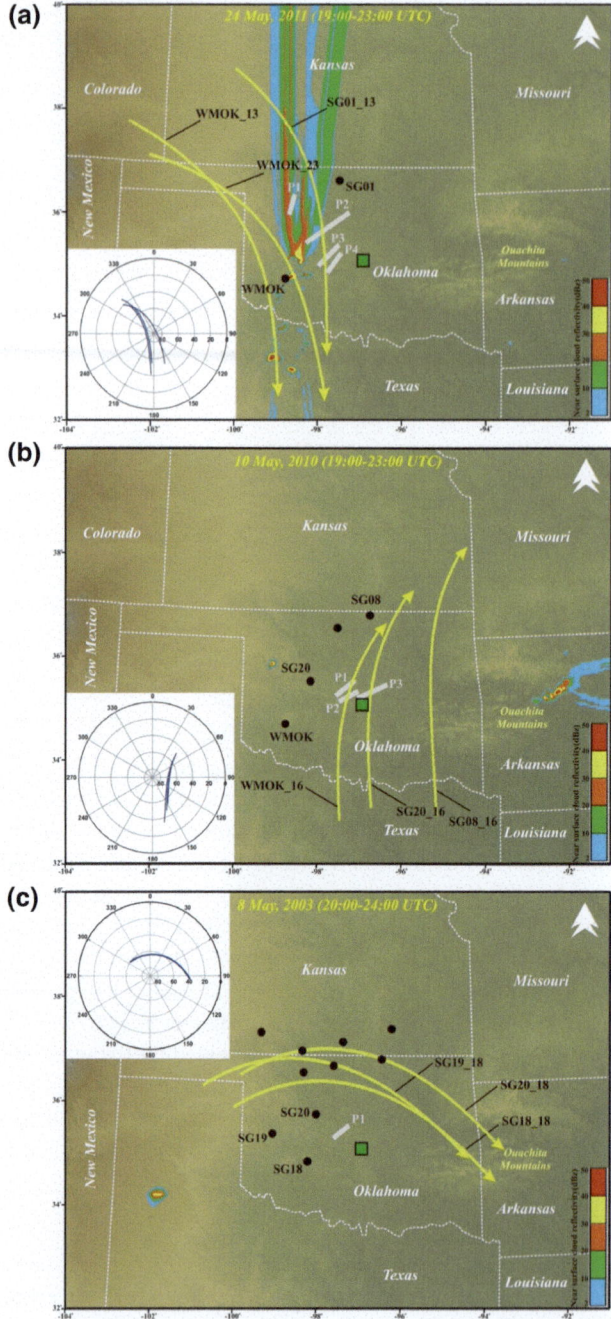

◄**Fig. 8.4**   **a** Location of GNSS station used in the present study. GNSS sites are marked by a black solid circle. The yellow solid lines represent the trajectories of the satellites of PRNs 13 and 23 (between 19:00 to 23:00 UTC hr.) on 24th May, 2011. Grey solid lines (P1, P2, P3 and P4) represent the path of the tornado, and coloured contours indicate the near-surface cloud reflectivity (dBz) which represents the intensity of rainfall during a tornado. The inset figure shows the polar plot of elevation and azimuth angle of satellites of the above PRNs. Green square indicates the location of Oklahoma City. **b** The yellow solid lines represent the trajectories of the satellites of PRN 16 (between 19:00 and 23:00 UTC h.) on 10th May, 2010. Other parameters are the same as presented in Fig. 8.4a. **c** The yellow solid lines represent the trajectories of the satellites of PRN 18 (between 20:00 and 24:00 UTC h.) on 8th May, 2003. Other parameters are the same as presented in Fig. 8.4a

over the city of Oklahoma and adjacent regions. We consider only those satellites, which are passing or near to path of tornadoes having a better line of sight (LOS). It has been noted that SIP trajectories are lying either over or close to the respective paths of the tornado in a synchronous manner (Fig. 8.4a–c). Further, the colour contour shown in Fig. 8.4 indicates near-surface cloud reflectivity (dBz), that has usually used to estimate the intensity of thunderstorm and rainfall activity. Based on near-surface cloud reflectivity (dBz) value, Houze et al. (2007) have classified four types of convective system, comparing the precipitation rate from various storms. Those four categories are as follows: (a) deep convective cores (near-surface cloud reflectivity more than 40 dBz and 10 km in height), (b) wide convective cores (40 dBz km^2 echo volume equal to1000 horizontally), (c) deep and wide convective cores (dBz fall between 1st and 2nd types) and finally (d) board stratiform regions (stratiform echo $\geq$ 50,000 km^2). It has been noted that, out of three considered cases of investigations, the May 24, 2011, tornado event was associated with heavy rainfall, and the near-surface cloud reflectivity (<40 dBz) indicates deep convective cores (i.e. severe convection), along with lightning flash intensity > 100 flashes per minute. Therefore, we can expect relatively higher variations of TEC in the ionosphere for this case, compared to other investigation cases.

To probe that TEC variation and relative comparison due to tornado-induced iono-spheric disturbance and associated TIDs, we have also presented the time series of raw VTEC variation observed by representative GNSS sites (Fig. 8.5) and filtered VTEC response (Fig. 8.6). The raw and filtered VTEC change during respective tornado days is compared with the adjacent day before and after the non-tornado events, respectively (Figs. 8.5 and 8.6). It has been observed that the VTEC varia-tion has magnified by using bandpass filtered of 1–10 min and significantly captured variation in filtered VTEC of about 0.1–0.35 TECU during the occurrence of respec-tive tornadoes (Fig. 8.6). Moreover, it has also been observed that the variation in VTEC is significantly higher in several folds (i.e. ~0.35 to 0.4 TECU) for the May 24, 2011, tornado event, in contrast to the other two events. This indication supports the relatively severe convection system associated with the relatively higher intensity of the lightning flash. However, other tornado events represent relatively lower but significant variation in VTEC of about 0.1–0.2 TECU. This is also consistent with the relative intensity of the respective lightning flashes of respective tornado events that occurred on May 8, 2003 (i.e. 15–25 flashes per minute) and May 10, 2010

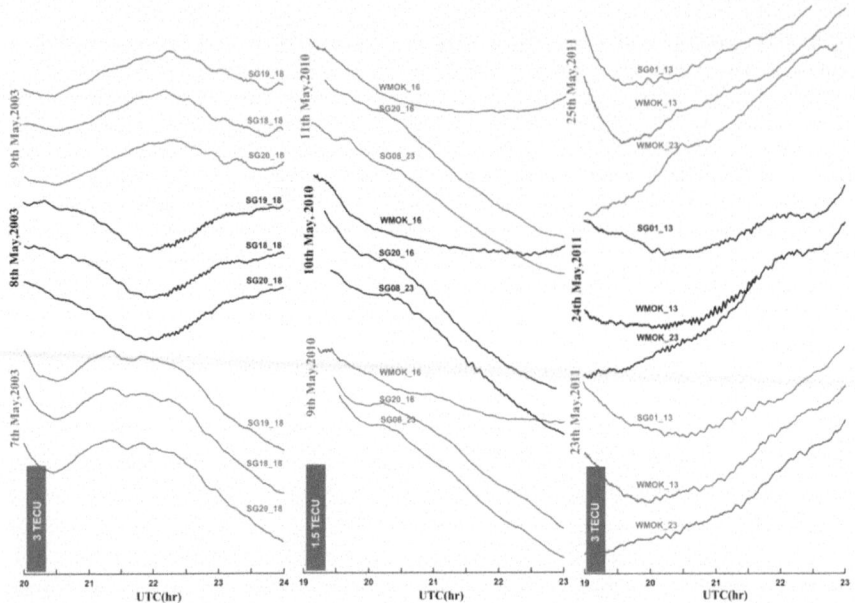

Fig. 8.5 Time series of raw vertical TEC variation observed by representative GNSS sites used in the present study for three tornado events over the Oklahoma region, e.g. 8th May, 2003 (PRN 18), 10th May, 2010 (PRNs 16 and 23) and 24th May, 2011 (PRNs 13 and 23). Note that TEC variation is visible during days of a tornado (black solid line); however, there is no such TEC variation during days before and after the tornado (grey solid lines)

(>25 flashes per minute) over the Oklahoma region. We have also presented wavelet power spectrum of the filtered VTEC variation using Morlet wavelet, normalized by the standard deviation (Torrence and Compo 1998), explicitly for the respective tornado events, and the respective wavelet scalograms that exhibit increasing frequency of power spectrum are exactly coinciding with the time series of TEC variation in all three events (lower panel in Fig. 8.6). However, we have not observed any such TEC variation in the days before and after the respective tornado manifestation. From this analysis and comparison with the intensity of the lightning flash, it has been inferred that observed TEC variation in the ionosphere during tornado events has a strong connection with the intensity of lightning and the intensity of the convective system.

8.5.3 Velocity of Ionospheric Oscillations

In Fig. 8.7, we presented a diagram representing the distance between SIP and occurrence respective tornado events as a function of time with colours indicating the VTEC anomaly. We have used all the available GNSS stations located surrounding

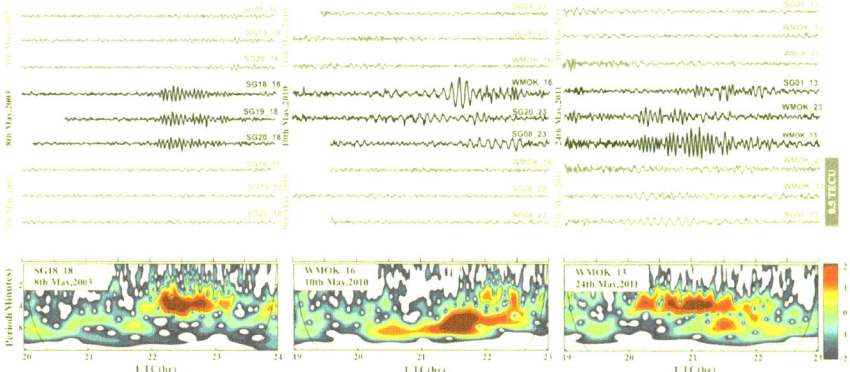

Fig. 8.6 (*Upper panel*) Time series of filter VTEC (bandpass filtered 1–10 min) observed by representative GNSS sites considered in the present study for three tornado events over the Oklahoma region, e.g. 8th May, 2003 (PRN 18), 10th May, 2010 (PRNs 16 and 23) and 24th May, 2011 (PRNs 13 and 23). Note that TEC variation is visible during the days of a tornado (black solid line); however, there is no such TEC variation during days before and after the tornado (grey solid lines). (*Lower panel*) The wavelet power spectrum of the filtered VTEC time series (on the thunderstorm day only) using Morlet wavelet is normalized by standard deviation. The colour bar represents wavelet coefficients, and the black contour represents 95% confidence level of white noise. We can see that the increase of power spectrum exactly coincides with the time series of filtered VTEC variation

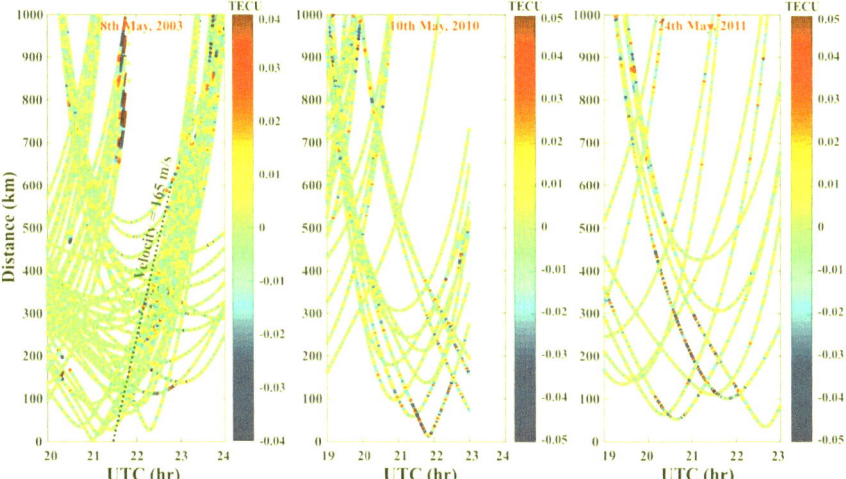

Fig. 8.7 Represent the time and distance from the point of tornadoes occurrence plot of VTEC variation in the response of tornadoes (e.g. 8th May, 2003, 10th May, 2010 and 24th May, 2011) during different observations time for all visible satellite. The colour scale shows the variation of TEC in the ionosphere. The inclined dashed line is the best-fit line of tornadoes generated atmospheric wave's arrival time and indicates the propagation velocity of TEC variation in the ionosphere

the region. We have found that the apparent propagation speed of VTEC anomaly is about ~0.165 m/s, in the case of the May 8, 2003, tornado event. However, the propagation velocity of TEC variation is unable to detect for the other two events due to the availability of a fewer number of SIP trajectories surrounding the Oklahoma region. Interestingly, this velocity is nearly similar to the sound wave velocity in the lower part of the F-region of the ionosphere and much lower than the Rayleigh wave (~3.8 km/s) that often found for co-seismic ionospheric disturbances (Ducic et al. 2003) and relatively comparable with the internal gravity wave (~0.3 km/s) often excited by very large earthquakes (Occhipinti et al. 2006). Moreover, this present estimate is consistent with the reported concentric waves velocity (~170 m/s) and short-period oscillation in the ionosphere induced after 2013, Moore FE-5 tornado over the same region (Nishioka et al. 2013).

8.5.4 *Independent Proxy from Outgoing Long-Wave Radiation (OLR)*

In general conditions with no clouds present above the Earth's surface, the hotter places will show higher OLR values (Zhang et al. 2017). However, the situation will differ in the presence of cloud cover. Especially when supercell thunderstorm clouds are present over any region responsible for tornado generation, one can expect low OLR values. As the OLR values are represented at the top of the atmosphere, they may serve as a good proxy for understanding the TEC variations related to convective events over any place.

The link between convective activities and the generation of gravity waves during the evolution of tornadoes remains equivocal. Ming et al. (2010) have investigated the association between gravity waves and convective activities during the subsequent evolution of tropical cyclone Dina and Faxai. Guha et al. (2016) have also reported converging relations during two tropical cyclones, 2013 Mahasen and 2014 Hudhud, respectively, which originated in the Bay of the Bengal and eventually made landfall over the eastern coast of India. For both works, the OLR maps define the convection and associated deep cloud cover over the study region.

Supporting to our present observations, we have correlated the convective activities associated with respective tornado events using the daily maps of NOAA OLR data (Fig. 8.8). The OLR maps presented in Fig. 8.8 represent successive evolution of large convective zones at the region of Oklahoma and adjacent regions, having energies at ranges about 200–250 W/m^2. It is to be noted here that the OLR spatial map scales are not the same for all the days due to large variations in the values before and after the tornado. For the cases of 8 May 2003 and 10 May 2010, it is clear that before the tornado occurrence, the cloud cover started blocking the OLR values, and to the east of the tornado site, the OLR values were low before the tornado occurrence. On the tornado outbreak day, we can observe lower OLR values in the nearby regions to the site marked by the black square in Fig. 8.8. The very next day shows clear skies

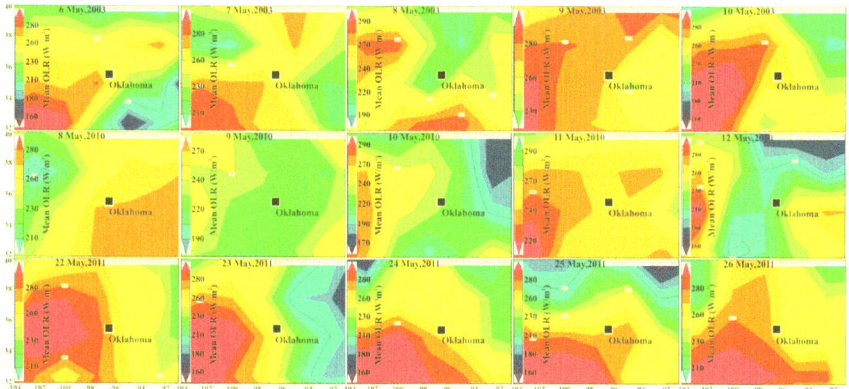

Fig. 8.8 Representative maps of $3° \times 3°$ gridded NOAA interpolated outgoing long-wave radiation (OLR) at Oklahoma surrounding region of the same spatial domain as presented in Fig. 8.4, for three respective tornadoes (e.g. 8th May, 2003, 10th May, 2010 and 24th May, 2011) of corresponding five days (two days before, two days after and during TEC study period, respectively)

over the site, depicted by higher OLR values. However, on 10 May 2010, there was again cloud cover on 12 May 2010, which reduced the OLR values significantly. The third tornado case (i.e. 24 May 2011) was comparatively less violent, and the cloud cover associated with this case was also lower. The low cloud cover allows the passing of OLR into space, and the OLR values in the third case were comparatively higher than the first two cases before and during the tornado outbreak days. These independent observations support the complex evolutionary process of the deep convective system associated with tornadoes. However, with the OLR maps analysis, one cannot conclude that the gravity waves generated by the tornado are not broken below the thermosphere, and these are secondary waves generated therein or waves from other sources like travelling ionospheric disturbances. The OLR data show the convective regime but does not show any signatures of convectively generated gravity waves.

8.5.5 Comparison with Global GNSS-TEC Anomaly Map

We have further focussed on the Calumet–El Reno–Piedmont–Guthrie tornado of FE-5 type, which occurred on May 24, 2011, as it has been associated with much intense convective activity and highest ionospheric TEC variations (Fig. 8.9). During the Calumet–El Reno–Piedmont–Guthrie tornado occurrence, we have also represented two-dimensional maps of the TEC variation (Fig. 8.10) and rate of TEC index (ROTI) map (Fig. 8.11) using the global GNSS-TEC anomaly database. Figures 8.10 and 8.11 are over the entire North American region at about 280 km altitude in the F-region of the ionosphere. From this spatio-temporal pattern of maps, we have observed a prominent concentric waves oscillation, along with the synchronism

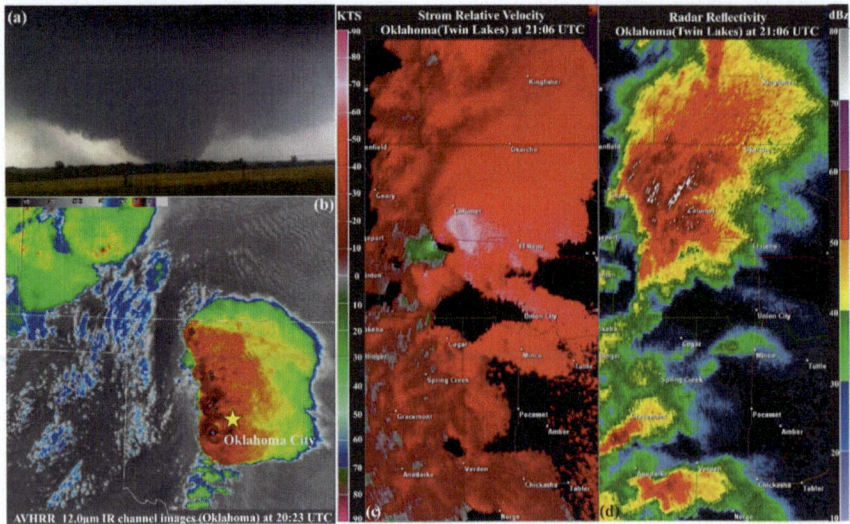

Fig. 8.9 **a** The real image of merged El Reno–Piedmont B$_2$ type tornado (EF-5) on 24 May 2011 at 21:37 UTC over Oklahoma (Photo credit: Roger Edwards); **b** false colour 1-km visible satellite imagery of the tornadic supercells across central Oklahoma at 20:23 UTC; **c** storm-relative velocity and **d** radar reflectivity zoomed over the Twin Lakes (Lat long), Oklahoma at 21:06 UTC. Figures are taken and modified from @https://www.weather.gov/oun/events-20110524-tornado-b2

increase of the rate of TEC index (ROTI) in the ionosphere after the 2011 Calumet–El Reno–Piedmont–Guthrie tornado over the region of Oklahoma. This observation is also converging with other reported cases of a tornado or tropical cyclone-induced ionosphere disturbances.

8.5.6 Tornado-Induced Ionospheric Disturbance: Coupling Mechanism Between Ionosphere and Troposphere

Finally, based on the above observations, we suggest that the coupling mechanism between the ionosphere and underlying troposphere and the TEC variations linked with gravity waves during any intense convective systems (e.g. tornado, tropical cyclone, thunderstorms, mesoscale convective complexes, typhoons, upper-tropospheric jet, etc.) are indeed complex (Fig. 8.2). Moreover, the exact coupling process between the ionosphere and the underlying atmosphere has been proposed to relate through four principal mechanisms: (a) upward propagating waves in the neutral atmosphere, (b) atmospheric DC electric field, (c) lighting-induced electric field and (d) neutral particle dynamics, due to underlying large-scale weather systems originating from the surface or upper troposphere (Sorokin and Cherny 1999, Forbes et al. 2000; Isaev et al. 2002a, b; Sorokin et al. 2005; Lastovicka 2006; Shao et al.

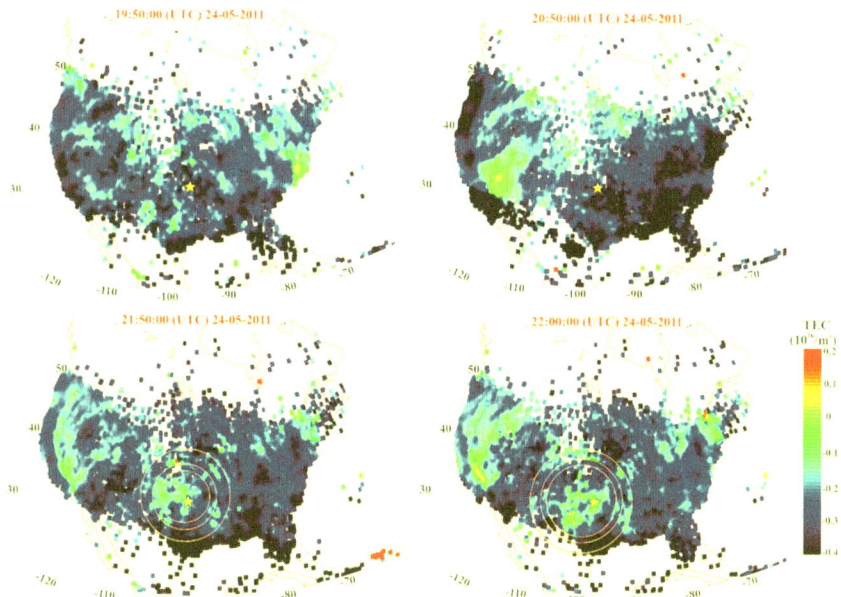

Fig. 8.10 Represents two-dimensional maps (~280 km), of the TEC variation over North America on 24 May 2011 (at the time of 19.50, 20.50, 21.50 and 22.00 UTC, respectively). The colour scale indicates TEC variation components. The yellow star indicates the location of Oklahoma City. The yellow circle indicates concentric wave fronts observed in the TEC variation map. The concentric wave fronts are visible after two hours of tornado occurrence (shown in bottom panels). Each pixel is 0.15 o× 0.15 o in longitude and latitudes and smoothen by using Gaussian filter 5 × 5 pixels, i.e. 0.75 o× 0.75 o in longitude and latitudes. This figure has been modified from the original figure that can be archived at @http://seg-web.nict.go.jp/GPS/N_AMRC/

2013; Guha et al. 2016). Out of those four principle mechanisms, we argue that intense electrical field is originated during lightning activity in a deep convective system, along with the generation of gravity waves responsible for ionosphere TEC variations.

The electric field generated from lightning propagated upward interacts with the ionosphere by heating electrons, affecting electron density by ionization and dissociation (Kumar et al. 2017). It has been argued that in the lower level of the ionosphere, the electron density decreases by increasing the electron attachment of oxygen molecules. However, in the upper ionosphere at F-region, the electric field increases electron density by ionizing nitrogen and oxygen molecules. Further, this rate of electron density change in the ionosphere depends upon the duration and amplitude of lightning (Shao et al. 2013). We suggest that this mode of the coupling mechanism between the ionosphere and underlying troposphere appears to be more suitable in the present case of investigation. We also acknowledge that our present investigation is preliminary and require further improvement by considering more cases in the

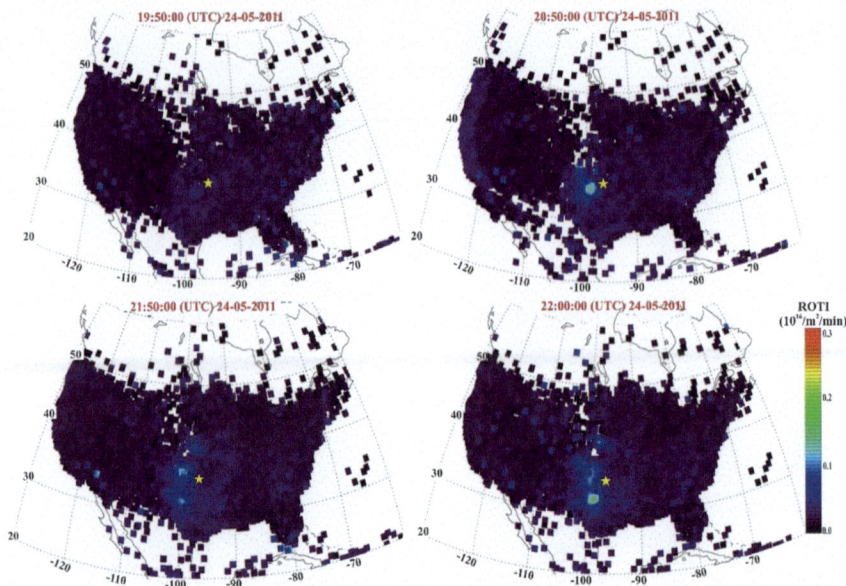

Fig. 8.11 Represents the two-dimensional map (~280 km), of the rate of TEC index (ROTI) map of North America on 24 May 2011 (at the time of 19.50, 20.50, 21.50 and 22.00 UTC, respectively). The colour scale indicates the rate of change of TEC per minute. We can see the TEC change near Oklahoma City in ROTI map after one hour of tornado occurrence (i.e. 20.50, 21.50 and 22.00), whereas there is no such variation detected at the time of tornado occurrence (i.e. 19.50 UTC). This figure has modified from the original figure that can be archived at @ http://seg-web.nict.go.jp/GPS/N_AMRC/RMAP/

future to understand the exact mechanism of ionospheric disturbance caused by any intense convective systems.

8.6 Conclusion

The findings can be summarized as follows:

1. We reported TEC variation of about 0.1–0.35 TECU during the occurrence of three tornadoes over the city of Oklahoma, United States of America.
2. The variation in TEC is directly correlated with the intensity of lightning and the severity of the convective system, where higher variation in electron content (~0.35 to 0.4 TECU) is observed in the case of the May 24, 2011, tornado event than the other two events over the same region.
3. The electron anomaly propagated at an apparent speed ~0.165 m/s during the May 8, 2003, tornado event, which is consistent with the previous observations from the same region.

4.   We suggest that the rate of electron density change in the ionosphere is linked with the generation of atmospheric gravity waves, along with the duration and amplitude of lightning associated with the convective events.

Acknowledgements Geodetic data from the Oklahoma region are maintained by the Earth Scope Plate Boundary Observatory (PBO), available at UNAVCO (ftp://data-out.unavco.org/pub/rinex/). All data supporting the conclusions of this paper are presented in the text and figures. DP and BS are supported by NITR Research Fellowship. Some of the figures were plotted using Generic Mapping Tools. The TEC processing of RINEX datasets have been performed using software GPS-TEC program, Version 2.9.5 (http://seemala.blogspot.com/), developed by Gopi Krishna Seemala, Indian Institute of Geomagnetism (IIG). We thank the editor and anonymous reviewer for their constructive comments, which improved the quality of the manuscript.

References

Afraimovich, E.L., N.P. Perevalova, A.V. Plotnikov, and A.M. Uralov. 2001. The shock acoustic waves generated by the earthquakes. *Advances in Space Research* 19: 395–409.

Blanc, E., T. Farges, A. Le Pichon, and P. Heinrich. 2014. Ten year observations of gravity waves from thunderstorms in western Africa. *Journal of Geophysical Research: Atmospheres* 119: 6409–6418.

Bowling, T., E. Calais, and J.S. Haase. 2013. Detection and modelling of the ionospheric perturbation caused by a Space Shuttle launch using a network of ground-based Global Positioning System stations. *Geophysical Journal International* 192: 1324–1331.

Cahyadi, M.N., and K. Heki. 2015. Coseismic ionospheric disturbance of the large strike-slip earthquakes in North Sumatra in 2012: Mw dependence of the disturbance amplitudes. *Geophysical Journal International* 200: 116–129.

Calais, E., J.B. Minster. 1995. GPS detection of ionospheric perturbations following the January 17, 1994, Northridge earthquakes. *Geophysical research letter* 22: 1045–1048.

Carrano, C., and Groves, K., 2009. *Ionospheric Data Processing and Analysis, Workshop on Satellite Navigation Science and Technology for Africa*, The Abdus Salam ICTP, Trieste, Italy.

Catherine, J.K., M.S.M. Vijayan, U.B. SyedaRabiya, K. Shimna, V.K. Gahalaut, and D.S. Ramesh. 2015. Dichotomy in mode propagation of coseismicionospheric disturbance: Observations from 11 April 2012 Indian Ocean earthquake. *Journal of Geophysical Research Space Physics* 120: 3854–3867.

Catherine, J.K., D.U. Maheshwari, V.K. Gahalaut, P.N.S. Roy, P.K. Khan, and N. Puviarasan. 2016. Ionospheric disturbances triggered by the 25 April, 2015 M7.8 Gorkha earthquake, Nepal: Constraints from GPS TEC measurements. *Journal of Asian Earth Sciences* 133: 80–88.

Chou, M.Y., Charles C. H. Lin, Jia Yue, H.F. Tsai, Y.Y. Sun, J.Y. Liu, and C.H. Chen. 2017. Concentric traveling ionosphere disturbances triggered by Super Typhoon Meranti (2016). *Geophysical Research Letters* 44 (3): 1219–1226.

Coster, A. J., L. Goncharenko, S.-R. Zhang, P.J. Erickson, W. Rideout, and J. Vierinen. 2017. GNSS observations of ionospheric variations during the 21 August 2017 solar eclipse. *Geophysical Research Letters* 44: 12041–12048.

Dabas, R.S., J.B. Lal, T.R. Tyagi, and Y.V. Somayajulu. 1980. Variations in TEC and other parameters associated with magnetic storms. *Indian Journal of Radio & Space Physics* 9: 1–6.

Dautermann, T., E. Calais, P. Lognonne, and G.S. Mattioli. 2009. Lithosphere–atmosphere–ionosphere coupling after the 2003 explosive eruption of the Soufriere Hills Volcano, Montserrat. *Geophysical Journal International* 179: 1537–1546.

Ducic, V., J. Artru, and P. Lognonne. 2003. Ionospheric remote sensing of the Denali earthquake Rayleigh surface waves. *Geophysical Research Letters* 30(18): 1951.

Forbes, J. M., Palo, S. E. and Zhang, X., 2000. Variability of the ionosphere. *Journal of Atmospheric and Solar-Terrestrial Physics* 62: 685–693.

Gagan, J.P., A. Gerard, and J. Gordon. 2010. A historical and statistical comparison of "Tornado Alley" to "Dixie Alley." *National Weather Digest* 34: 145–155.

Gentile, L.C., W.C. Burke, and F.J. Rich. 2006. A global climatology for equatorial plasma bubbles in the topside ionosphere. *Annales Geophysicae* 24: 163–172.

Guha, A., B. Paul, M. Chakraborty, and B.K. De. 2016. Tropical cyclone effects on the equatorial ionosphere first result from the Indian sector. *Journal of Geophysical Research Space Physics* 121: 5764–5777.

Hagan, M.E., A. Maute, R.G. Roble, A.D. Richmond, T.J. Immel, and S.L. England. 2007. Connections between deep tropical clouds and the Earth's ionosphere. *Geophysical Research Letters* 4: L20109.

Heki, K. 2011. Ionospheric electron enhancement preceding the 2011 Tohoku-Oki earthquake. *Geophysical Research Letters* 38: L17312.

Heki, K., and J. Ping. 2005. Directivity and apparent velocity of the coseismicionospheric disturbances observed with a dense GPS array. *Earth and Planetary Science Letters* 236: 845–855.

Heki, K., Y. Otsuka, N. Choosakul, N. Hemmakorn, T. Komolmis, and T. Maruyama. 2006. Detection of ruptures of Andaman fault segments in the 2004 great Sumatra earthquake with coseismic ionospheric disturbances. *Journal of Geophysical Research* 111: B09313.

Heki, K. 2020. Ionospheric disturbances related to earthquakes. In *Advances in Ionospheric Research: Current Understanding and Challenges. Space Physics and Aeronomy*, vol. 3, ed. C. Huang and G. Lu, 320. Wiley/American Geophysical Union, ISBN: 978-1-119-50755-0.

Houze, R.A., D.C. Wilton, D.C. Smull 2007. Monsoon convection in the Himalayan region as seen by the TRMM Precipitation Radar. *Quarterly Journal of the Royal Meteorological Society* 133: 1389–1411.

Hung, R.J., and J. Kuo. 1978. Ionospheric observation of gravity waves associated with hurricane Eloise. *Journal of Geophysics* 45 (1): 67–80.

Isaev, N.V., V.M. Sorokin, V.M. Chmyrev, O.N. Serebryakova, and A.K. Yaschenko. 2002a. Disturbance of the electric field in the ionosphere by sea storms and typhoons. *Cosmic Research* 40: 547–553.

Isaev, N.V., V.M. Sorokin, V.M. Chmyrev, O.N. Serebryakova, and O.Y. Ovcharenko. 2002a. Electric field enhancement in the ionosphere above tropical stormregion. In *Seismo Electromagnetics: Lithosphere–Atmosphere–Ionosphere Coupling*, ed. M. Hayakawa and O. A. Molchanov, 313–315. TERRAPUB, Tokyo.

Kelley, M.C. 1997. In situ ionospheric observations of severe weather-related gravity waves and associated small-scale plasma structures. *Journal of Geophysical Research* 102: 329–335.

Klobuchar, J.A. 1996. Ionospheric effects on GPS. *Global Positioning System: Theory and Application* 164: 485–515.

Kumar, S., W. Chen, M. Chen, Z. Liu, and R.P. Singh. 2017. Thunderstorm-/lightning-induced ionospheric perturbation: An observation from equatorial and low-latitude stations around Hong Kong. *Journal of Geophysical Research Space Physics* 122: 9032–9044.

Kundu, B., D. Panda, V.K. Gahalaut, and J.K. Catherine. 2018. The August 21, 2017 American total solar eclipse through the eyes of GPS. *Geophysical Journal International* 214: 651–655.

Kundu, B., B. Senapati, A.I. Matsushita, and K. Heki. 2021. Atmospheric wave energy of the 2020 August 4 explosion in Beirut, Lebanon, from ionospheric disturbances. *Scientific Report* 11(1): 1–7.

Lastovicka, J. 2006. Forcing of the ionosphere by waves from below. *Journal of Atmospheric and Solar-Terrestrial Physics* 68: 479–497.

Lay, E.H., X.-M. Shao, and C.S. Carrano. 2013. Variation in total electron content above large thunderstorms. *Geophysical Research Letters* 40: 1945–1949.

Lee, C.C., and B.W. Reinisch. 2006. Quiet-condition hmF2, NmF2, and B0 variations at Jica-marca and comparison with IRI-2001during solar maximum. *Journal of Atmospheric and Solar-Terrestrial Physics* 68: 2138–2146.

Lutgens, F.K., E.J. Tarbuck, and D. Tusa. 2001. *The atmosphere*, 484. Upper Saddle River, NJ: Prentice Hall.

Mannucci, A.J., B.D. Wilson, D.N. Yuan, C.H. Ho, U.J. Lindqwister, and T.F. Runge. 1998. A global mapping technique for GPS-derived ionospheric total electron content measurements. *Radio Science* 33: 565–582.

Ming, C.F., Z. Chen, and F. Roux. 2010. Analysis of gravity-waves produced by intense tropical cyclones. *Annales Geophysicae* 28: 531–547.

Nakashima, Y., K. Heki, A. Takeoa, M.N. Cahyadic, and A. Aditiya. 2015. Atmospheric resonant oscillations by the 2014 eruption of the Kelud volcano, Indonesia, observed with the ionospheric total electron contents and seismic signals. *Earth and Planetary Science Letters* 434: 112–116.

National Weather Service. 2018. https://www.weather.gov/oun/tornadodata-ok-monthlyannual.

Nishioka, M., T. Tsugawa, M. Kubota, and M. Ishii. 2013. Concentric waves and short-period oscillations observed in the ionosphere after the 2013 Moore EF5 tornado. *Geophysical Research Letters* 40 (21): 5581–5586.

Occhipinti, G., P. Lognonné, E.A. Kherani, and H. Hébert. 2006. Three-dimensional waveform modelling of ionospheric signature induced by the 2004 Sumatra tsunamis. *Geophysical Research Letters* 33: L20104.

Panda, D., B. Senapati, B. Tyagi, and B. Kundu. 2019. Effects of Rayleigh-Taylor instability and ionospheric plasma bubbles on the global navigation satellite System signal. *Journal of Asian Earth Sciences* 170: 225–233.

Panda, S.K., S.S. Gedam, and G. Rajaram. 2013. GPS derived ionospheric TEC response to annular solar eclipse over Indian region on 15 January 2010. In *International Conference on Space Science and Communication*, 213–218.

Park, Y., Y. Kwak, B. Ahn, Y. Park, and L. Cho. 2010. Ionospheric F2-layer semi-annual variation in middle latitude by solar activity. *Journal of Astronomy and Space Sciences* 27 (4): 319–327.

Park, J., D.A. Grejner-Brzezinska, and R. Von Frese. 2014. GPS discrimination of traveling iono-spheric disturbances from underground nuclear explosions and earthquakes. *NAVIGATION: Japan Institute of Navigation* 62: 125–134.

Perkins, F. 1973. Spread F and ionospheric currents. *Journal of Geophysical Research* 78 (1): 218–226.

Petersen, G.N., H. Bjornsson, and P. Arason. 2012.The impact of the atmosphere on the Eyjafjallajökull 2010 eruption plume. *Journal of Geophysical Research* 117: D00U07.

Rideout, W., and A. Coster. 2006. Automated GPS processing for global total electron content data. *GPS Solutions* 10: 219–228.

Rolland, L.M., P. Lognonné, and H. Munekane. 2011a. Detection and modeling of Rayleigh wave induced patterns in the ionosphere. *Journal of Geophysical Research* 116: A05320.

Rolland, L.M., P. Lognonne, E. Astafyeva, E.A. Kherani, N. Kobayashi, M. Mann, and H. Munekane. 2011b. The resonant response of the ionosphere imaged after the 2011 off the Pacific coast of Tohoku earthquake. *Earth, Planets and Space* 63: 853–857.

Savastano, G., A. Komjathy, O. Verkhoglyadova, A. Mazzoni, M. Crespi, Y. Wei, and A.-J. Mannucci. 2017. Real-time detection of tsunami ionospheric disturbances with a stand-alone GNSS receiver: A preliminary feasibility demonstration. *Scientific Report* 7: 46607.

Senapati, B., J.D. Huba, B. Kundu, V.K. Gahalaut, D. Panda, S.K. Mondal, and J.K. Catherine. 2020. Change in total electron content during the 26 december 2019 solar eclipse: constraints from GNSS observations and comparison with Sami3 model results. *Journal of Geophysical Research: Space Physics* 125: e2020JA028230.

Shao, X.-M., E. Lay, and A.R. Jacobson. 2012. On the behavior of return stroke current and the remotely detected electric field change waveform. *Journal of Geophysical Research* 117: D07105.

Shao, X.-M., E. Lay, and A.R. Jacobson. 2013. Reduction of electrondensity in the night-time lower ionosphere in response to a thunderstorm. *Nature Geoscience* 6(1): 29–33.

Sorokin, V.M., and G.P. Cherny. 1999. It is quite possible to monitor typhoons from outer space. *Aerospace Courier* 3: 84–87.

Sorokin, V.M., N.V. Isaev, A.K. Yaschenko, V.M. Chmyrev, and M. Hayakawac. 2005. Strong DC electric field formation in the low latitude ionosphere over typhoons. *Journal of Atmospheric and Solar-Terrestrial Physics* 67: 1269–1279.

Sun, S.J., P.P. Ban, C. Chen, Z.W. Xu, and Z.W. Zhao. 2012. On the vertical drift of ionospheric F layer during disturbance time: Results from ionosondes. *Journal of Geophysical Research* 117: A01303.

Torrence, C., and G.P. Compo. 1998. *A Practical Guide to Wavelet Analysis.* http://paos.colorado.edu/research/wavelets/.

Tsugawa, T., Y. Otsuka, A.J. Coster, and A. Saito. 2007. Medium-scale traveling ionospheric disturbances detected with dense and wide TEC maps over North America. *Geophysical Research Letters* 34: L22101.

Xiao, Z., S.G. Xiao, and Y.Q. Hao. 2007. Morphological features of ionospheric response to typhoon. *Journal of Geophysical Research* 112: A04304.

Yang, Y.-M., A. Komjathy, R.B. Langley, P. Vergados, M.D. Butala, and A.J. Mannucci. 2014. The 2013 Chelyabinsk meteor ionospheric impact studied using GPS measurements. *Radio Science* 49: 341–350.

Zhang, K., W.J. Randel, and R. Fu. 2017. Relationships between outgoing longwave radiation and diabatic heating in reanalyses. *Climate Dynamics* 49 (7): 2911–2929.

Part III
Extreme Waves, Sea Level Changes and Coastal Inundation

Chapter 9
Extreme Wind-Wave Characteristics in the North Indian Ocean in a Changing Climate

Prasad K. Bhaskaran, S. Neelamani, Khaled Al-Salem, Athira Krishnan, Jiya Albert, and S. Sreelakshmi

Abstract Wind-generated surface gravity waves are the manifestations of sea surface oscillations caused by intense wind stress and momentum transfer acting over the air-sea interface. Understanding the characteristics of wind-wave climate and its spatio-temporal variability over basin scales has significant practical applications in almost all marine-related activities, ocean engineering, coastal zone management, naval applications, etc. In the recent past, the subject of extreme wind-wave activity in a changing climate and its impact on the Indian coast is a topic of immense interest amongst the scientific community having wider socio-economic consequences. Water levels in the nearshore regions due to extreme wind-waves have significant impacts on coastal environment, infrastructure, and dwelling population in the coastal regions. In a broader perspective, extreme waves are a part of the climate system and can be significantly influenced by the natural climate variability. This chapter provides an overview on the generation and dissipation characteristics of wind-waves and the relevance of wind-wave climatology for the North Indian Ocean region. Recent trends observed in the extreme wind-wave activity in a changing climate scenario are a topic of wide interest. Extreme wind-waves and their return periods in a limited-fetch environment for the Arabian Gulf region are also discussed. Observed trends in extreme wind-wave activity for the extra-tropical regions in Indian Ocean and the North Indian Ocean showed an increasing trend at a rate of 3.3 cm/year and 0.27 cm/year, respectively. Also, in the recent decade an increasing trend is observed in the annual distribution of extreme winds and waves over extra-tropical regions having implications on generation of swell wave field that has consequence on local wind-waves in the North Indian Ocean region. Further, the case studies of extreme waves induced by tropical cyclones along with

P. K. Bhaskaran (✉) · A. Krishnan · J. Albert · S. Sreelakshmi
Department of Ocean Engineering and Naval Architecture, Indian Institute of Technology Kharagpur, Kharagpur, West Bengal 721302, India
e-mail: pkbhaskaran@naval.iitkgp.ac.in

S. Neelamani · K. Al-Salem
Coastal Management Program, Environment and Life Sciences Research Centre, Kuwait Institute for Scientific Research, P.O. Box: 24885, 13109 Safat, Kuwait
e-mail: nsubram@kisr.edu.kw

© The Centre for Science & Technol. of the, Non-aligned and Other Devel. Countries 2022
A. S. Unnikrishnan et al. (eds.), *Extreme Natural Events*,
https://doi.org/10.1007/978-981-19-2511-5_9

the recent trends in wind speed and its analysis based on global climate models are also discussed. Finally, a brief overview on the challenges and future directions for more research is also highlighted.

Keywords Extreme wind-waves · Climate change · Satellite data · Numerical model · Indian Ocean

9.1 Introduction

The ocean surface is a dynamic region that plays an important role in the direct transfer of heat, momentum, gas, and particle exchange in the global oceans. The free surface boundary layer is accounted by the surface gravity waves generated by wind stress acting over the air-sea interface. Typical period of surface gravity waves can range between 2 and 30s. Over the past several decades, the physical mechanism governing wave generation, propagation, and dissipation remained as a subject of immense interest amongst the scientific community that has wide practical applications and socio-economic consequences. In the recent past, there has been significant research activity on wind-waves and its variability due to increasing marine and offshore activities. Operational centre, like ESSO-INCOIS (Earth System Science Organization—Indian National Centre for Ocean Information Services) is an organization under the Ministry of Earth Sciences, Government of India, Hyderabad, provides ocean state forecast for the Indian seas. Ocean state forecast is very important for myriad activities such as port and harbour operations, coastal zone management, naval operations, ocean engineering, and efficient ship routing. Locally generated wind-waves are also strongly affected by human-induced activity in coastal and offshore locations. In recent years, there is a growing interest to understand wind-wave climate in the perspective of both historical and futuristic projections (Hemer et al. 2012, 2013; Young and Ribal 2019; Morim et al. 2018, 2019, 2020; Chowdhury et al. 2019; Krishnan et al. 2021). There is ample evidence to show that wind-wave climate is changing over the global oceans based on long-term satellite measurements. Also, there have been coordinated efforts globally in data collection and analysis of model outputs to understand the mean and higher percentile global ocean wind-wave climate (Hemer et al. 2010).

There are also regional studies performed on wind-wave climate projections for the global ocean basins. Morim et al. (2018) made a consensus-based analysis of 91 published wind-wave climate projection assessment to establish the consistent patterns on the impacts of global warming on wind-wave climate across the globe. Their study also discussed on the current limitations and pointed out the opportunities within the existing community ensemble of projections for future scenarios. Hemer et al. (2012) provide more details on advancing wind-wave climate science based on COWCLIP project. Their study highlighted key scientific questions, challenges, and recommendations for wind-waves in a changing climate addressing four themes: historical wind-wave climate variability and change, global wave climate projections, regional wave climate projections, and coupled wind-wave climate modelling. The robustness and uncertainties in global multi-variate wind-wave climate projections were reported by Morim et al. (2019). Their study used a community-driven multi-method ensemble of global wave climate projections to demonstrate regions with robust changes in annual mean significant wave height and mean wave period and shifts in mean wave direction under a high-emission scenario. Recently, Morim et al (2020) provided a global ensemble of ocean wave climate projections from CMIP5-driven models. Wave climate projections along the Indian coast were reported by Chowdhury et al. (2019). Their study analysed regional wave climate along the Indian coast for two time slices, 2011–2040 and 2041–2070, using an ensemble of near-surface winds generated by four different CMIP5 GCMs under RCP4.5 scenario. Their study (Chowdhury et al. 2019) indicates that wave periods at most locations along east coast are expected to increase by almost 20%, whereas the increase would be 10% along the west coast. Also, very recently Krishnan (2021) made an assessment on CMIP5 model performance of significant wave heights for Indian Ocean (IO) using COWSLIP datasets. The study used near-surface wind speed datasets from 8 CMIP5 GCMs to force a spectral wave model. Spatio-temporal variations and projections of mean and extreme wind-wave conditions over the global ocean basins are important for practical needs. However, a proper understanding on the frequency of extreme wind-waves is very important for marine-related activities and coastal engineering applications. It is necessary to have a benchmark to evaluate the design wave parameters for coastal and offshore structures based on the extreme wind-waves. Also, precise knowledge on extreme wind-wave conditions is very crucial to determine the sea levels in nearshore regions and mechanisms of coastal processes and sediment transport. Impact of climate change can aggravate the scenarios of projected wind-wave climate for the global coasts having socio-economic implications (Church et al. 2007; Luijendijk et al. 2018). Therefore, understanding the impact of climate change on extreme wind-waves is very important. The scientific and engineering community has immense interest to understand the associated kinematics and dynamics of surface gravity waves for routine forecast and location-specific studies.

Ocean waves play a significant role in influencing coastal processes in the coastal and nearshore environments. Winds blowing over the ocean surface generate wavelets, and the spectral components eventually develop over time extracting energy from the wind stress. Through nonlinear wave–wave interaction process, the energy within a wave system gets redistributed thereby determining the overall wave energy at a particular location and time and that can be conveniently expressed in the form of a wave spectrum. The free surface boundary is quite dynamic in nature, wherein the exchange of momentum, heat, and gas occurs. Wind stress acting on this boundary layer generates wind-waves or the surface gravity waves. There has been immense interest amongst the scientific community over the past several decades in understanding the characteristics of wind-waves such as their generation mechanism, propagation, and dissipation characteristics having significant practical applications and economic importance. Over the recent decades, there has been significant research on

Table 9.1 Types of natural disaster and its impact due to extreme events in the North Indian Ocean

S. No.	Different types of natural disaster	Impact of extreme events
1	Disasters related to weather and hydrological aspects—tropical cyclones, severe storm events, and floods	Coastal and nearshore inundation
2	Disaster related to geological aspect—plate movement	Extreme coastal with nearshore flooding and inundation
3	Disasters related to global climate change—global warming, depletion of ozone layer, ice-caps melting	Sea level rise, extreme wind-wave activity
4	Tsunami	Widespread coastal and inland inundation

the study of wind-waves and its prediction owing to increasing marine and offshore-related activities. Precise knowledge on prevailing sea-state and its prediction is very vital for many marine-related operations, efficient ship routing, naval operations, port and harbour development activities, and many more. Nevertheless, the scientific and engineering community has significant interest to understand the associated kinematics and dynamics of ocean wind-waves for routine forecast and location-specific case studies. In context to the IO region, there are different types of natural disaster that can significantly affect the vulnerability aspects of coastal and nearshore regions. Table 9.1 illustrates the types, nature, and impact of natural hazards due to extreme events relevant to the IO region.

Engineering community working in related disciplines of ocean engineering, naval architecture, civil and hydraulic engineering requires precise wave information to design, operate, and manage structures in the marine environment. Wave information is also required by coastal engineers in understanding natural processes in the coastal and nearshore environments. Based on existing knowledge, the wind blowing over ocean surface generates wavelets, and the spectral components eventually develop over time extracting energy from wind stress. The nonlinear interaction between waves redistributes the overall energy within the frequency-direction space of the wave spectrum. This is the present state of knowledge acquired despite several years of research in the field of ocean wave modelling. Random nature of waves and its complex interaction in terms of kinematics and dynamics of wave evolution was a major challenge since the past. One can find the fundamental and classical studies on water waves and development of mathematical formulations that dates back to the nineteenth century.

An overview covering the major advances and developments achieved on study of wind-waves during the past few decades is given in Table 9.2. These classical works are testimony and building blocks to the basic research on ocean waves. The following sections in this chapter provide an overview covering the historical aspects of ocean wave studies relevant for the Indian seas, importance of wave energy balance in wave modelling studies, and global perspective of wind-wave modelling studies relevant for the IO, role of Southern Ocean (SO) in wind-wave climate studies, and the impact of extreme wind-waves on coastal inundation.

Table 9.2 Major research advances in the field of ocean surface waves during the past few decades

S. no.	Advancements in ocean wave research studies	1940s	1950s	1960s	1970s	1980s	1990s	2000s
1	Statistical theory	Theory of random noise	Wave statistics and spectral developments	Mathematical developments in wave spectra—nonlinear effects	Similarity form and work on directional spectra	High-frequency wave spectrum	Wave number—frequency spectra	Wave number—frequency spectra
2	Nonlinear theory	Nonlinear theory of regular waves	Nonlinear theory of random waves	Wave instability and wave interaction studies	Computation of dispersion relation	Wave breaking computational works	Wave breaking and energy dissipation	Wave breaking and energy dissipation
3	Experiments (laboratory and field measurements)	Basic studies and visual-based observations	Observations from instrument	Advances in field-based campaigns and planned experiments	Studies on equilibrium—planned ocean experiments	Wave dynamics—use of satellite observations	Microwave remote sensing	Ocean observing systems and satellite-based platforms
4	Air-sea interaction studies and wave projects		Sun glitter project		JONSWAP field experiment	HEXOS	SWADE, RASEX	Coupled atmosphere—ocean models
5	Wave forecasting techniques	Sverdrup and Munk	SMB and PNJ wave forecasting methods	First-generation wave models	Second-generation wave models	3G wave models (WAM)	3G wave models with data assimilation (WAM, WW3)	Third-generation wave models—ensemble modelling

Source Mitsuyasu (2002) from Bhaskaran (2019)

9.1.1 Historical Perspective on Ocean Wave Studies Relevant for the North Indian Ocean

For the north IO region, focused studies on wave research started during the 1980s and using the SMB (Sverdrup-Munk-Bretschneider) method hindcasts were made for wind-seas and swells off Mangalore coast during the south-west monsoon of 1968 and 1969 (Prasada Rao and Durga Prasad 1982). Their study postulated that significant ocean wave characteristics in terms of both wave height and period predicted using this method compared well with the recorded data. In addition, their study (Prasada Rao and Durga Prasad 1982) proposed a bottom friction factor of 0.05 suitable for the study region in evaluating the shallow water wave characteristics off Mangalore coast. In another study, Prasada Rao (1988) reported on the spectral width parameter for wind-generated waves based on wave data analysis using a ship borne wave recorder. Data analysis covered various locations of deep and shallow waters along the east and west coast of India. The study (Prasada Rao 1988) indicated on the bias in the estimation of spectral width parameter using higher order moments. A parametric wave prediction model based on time delay concept was reported by Prasada Rao and Swain (1989). The study used datawell wave rider buoy data recorded from an oceanographic research vessel. Analysis of the data revealed growth and decay phase of sea-states for varying wind speeds. The parametric model used a time delay concept in place of wind duration limit (Prasada Rao and Swain 1989). Studies on wave characteristics and its refraction patterns relating to beach erosion for Kerala coast were reported by Baba et al. (1983).

A significant and pioneering study on ocean wave research for India was made by Baba (1985) that initiated the modern ocean wave research in India. It brought out a concise picture on the latest developments made in the interpretation of ship-based observations, wave hindcasting, and measurements of ocean waves. New approaches on the study of short-term distributions, seasonal and annual climatology, and long-term distributions are discussed (Baba 1985). Importance of nonlinear effects in short-term distributions of wave heights and periods are also highlighted. Developments on ocean wave research in India, wave spectra, numerical methods for wave hindcasting, transformation, tapping of wave energy, and remote sensing techniques are discussed along with recommendations for future research (Baba 1985). In another study, Baba et al. (1989) investigated the wave spectra off Kochi, and the study highlights that spectral shape was multi-peaked and wide banded with high-frequency sides exhibiting similar slopes. The study revealed that the slope was milder than that proposed by Philip's formulation for fully developed conditions. Kurian et al. (1985) reported on the prediction of nearshore wave heights using a refraction programme. In their study, the Dobson wave refraction programme was modified to incorporate the attenuation characteristics due to bottom friction that was verified for prediction of nearshore wave heights. The study focused on the shelf waters off Alleppey coast in Kerala. Swain et al. (1989) used a numerical wave prediction model and performed many case studies for the Arabian Sea (AS) and the Bay of Bengal (BoB). In another study, Ravindran and Koola (1991) investigated the potential for harnessing wave energy emphasizing on the Indian Wave Energy Programme. Using ship-based observations, Chandramohan et al. (1991) developed wave statistics for the Indian coast. In another significant study, Sanil Kumar et al. (1998) estimated the wave direction spread in shallow water utilizing measured wave data for a period of two months at 15 m water depth along the East coast of India. Their study (Sanil Kumar et al. 1998) advocated that shallow water wave directional spread was narrowest at peak frequency and widened towards the lower- and higher-frequency bands. The observed uni-directional spectra were in close resemblance with the Scott wave spectra. Further, Sanil Kumar and Deo (2004) postulated the design wave estimates considering the directional distribution of ocean waves. Based on one-year data measured at three locations along Indian coast and 18 years of ship reported data, the design wave heights were estimated considering the directional distribution of significant wave heights.

Sanil Kumar et al. (2010) investigated the waves in shallow waters off the west coast of India during the onset of summer monsoon period. The study signifies that about 67% of measured waves are attributed due to swells that propagate from south and south-west regions and wind-seas from south-west to north-west directions. Also using measured data, the variations in nearshore wave power for different shallow water locations in the east and west coast of India were reported by Sanil Kumar et al. (2013). Shahul Hameed et al. (2007) using measured data reported on the seasonal and annual variations in wave characteristics off Chavara coast in Kerala. For the Goa coast using measured wave spectral data, Vethamony et al. (2009, 2011) reported superposition of wind-seas with existing swells during the pre-monsoon season. Bhaskaran et al. (2000) investigated the extreme wave conditions for the BoB during severe cyclones with simulations using two spectral models and sea-state hindcast for typical monsoon months (Bhaskaran et al. 2004). Importance of wave models for weather routing of ships in the IO was examined by Padhy et al. (2008). In an operational scenario wave forecasting system and its validation at coastal Puducherry, East coast of India was reported by Sandhya et al. (2013). There were also studies on coastal vulnerability associated with extreme waves (Nayak and Bhaskaran 2014; Sudha Rani et al. 2015; Sahoo and Bhaskaran 2015a, b, 2017a, b; Gayathri et al. 2017). A comprehensive review listing the various challenges and future directions for ocean wave modelling is available in Cavaleri et al. (2007) and Bhaskaran (2019). Studies that examined the trends in wind-wave climate relevant to IO are reported in Bhaskaran et al. (2014), Gupta et al. (2015), Patra and Bhaskaran (2016a, b), and Patra et al. (2017; 2019). The above-discussed review covers some of the seminal studies that are being carried out on ocean wave research for the Indian coast. At present, the ESSO-INCOIS, Hyderabad, provides information on Ocean State Forecast (http://www.incois.gov.in/portal/osf/osf.jsp) for the IO region.

9.1.2   Concept of Energy Balance for Wave Modelling Studies

The concept of energy balance equation was formulated by Gelci et al. (1957) to understand the phenomena of wave evolution. The second- and third-generation wave models used energy balance equation as the governing equation. At present, the third-generation wave models are used for routine wave forecasting of surface gravity waves for the NIO region. The third-generation wave models use sophisticated parameterization of physical processes as compared to the second-generation wave models. Quality of wave forecasts has also drastically improved in the recent years attributed due to tremendous boost in computational power, data acquisition systems, availability of satellite data, and increasing number of in situ observational platforms.

Broadly speaking, the wave models can be classified into phase averaging or phase resolving, wherein the phase-averaged models are expressed in terms of energy balance with appropriate sources and sinks used to represent the relevant physical processes. Phase-resolving models are based on the governing equations of fluid mechanics formulated to obtain the free surface condition. However, the phase-averaging models have no priori restriction on the area to be modelled, whereas the phase-resolving models have an inherent limitation on the spatial dimension of the computational area. The various physical processes that are accounted in phase-averaged models include (i) wave generation by wind accounted due to momentum transfer from atmosphere to ocean, (ii) refraction due to water depth, (iii) shoaling due to shallow water depths, (iv) diffraction due to obstacles, (v) reflection due to impact with solid obstacles, (vi) bottom friction due to heterogeneity of bottom materials, (vii) wave breaking effects when steepness exceeds a critical level, (viii) nonlinear wave–wave interaction due to quadruplets and triads resulting in wave energy redistribution, and (ix) wave–current interaction effects. In deep water environment, the physical processes can result from the combined effects of wave generation by wind, quadruplet wave–wave interaction, and dissipation due to white-capping mechanisms. Deep water waves transform on reaching shallow waters attributed by dominant physical processes like refraction, bottom friction, depth-induced breaking, triad wave–wave interaction, wave–current interaction, diffraction, and reflection (Holthuijsen 2007). Hence, choosing an appropriate wave model for the desired task is very important considering the dominant physical processes relevant to the study area. Wind field and bathymetry are the primary input that governs the dynamic evolution of wind-waves. The spatio-temporal evolution of wave energy is dependent on both wind field and local bathymetry. In addition, the ocean wave spectrum forms an integral part in wave models, as they provide the necessary initial conditions, wherein the energy has a dependence on the wind speed.

9.1.3   Wind-Wave Climate Studies for the Global Oceans

The role of satellites has undoubtedly revolutionized the global ocean observing system during the past two decades. Data obtained from satellites have improved our understanding on the spatio-temporal evolution and variability of meteorological and oceanographic parameters such as wind speed, significant wave height, sea surface temperature (SST), sea surface height anomalies, and surface salinity. Although observed data from voluntary ships of opportunity and in situ measurements at specific locations are important, the role of satellite remote sensing is tremendous in context of understanding the meteorological and oceanographic parameter variability in basin scales. As per the report of IPCC AR4, the long-term record of VOS (1900–2000) signifies negative trends (11 cm/decade) for significant wave height in the SIO (Trenberth et al. 2007). Advent of radar altimeters has made it possible today to measure the maximum significant wave heights and winds across the global ocean basins.

At present, the wealth of data obtained from multi-satellite platforms provides an opportunity to map the spatio-temporal variations in the earth system. Initial studies which utilized the satellite data focused on mapping wind-wave climate and their seasonal variations, and when longer records were available, the inter-annual variability could be determined (Hemer et al. 2010; Young et al. 2011). Practical use of altimeter data to understand the wind-wave climate started with GEOS-3 (Gower 1976) and SEASAT in 1978. The first picture of wind-wave climate for the global oceans was released by Chelton et al. (1981). Further, studies by Challenor et al. (1990) and Carter et al. (1991) used GEOSAT data to determine the seasonal and inter-annual variability of wind-wave climate. Prior studies by Young et al. (2011) examined the global trends in wind-wave climate. Significant wave height projections for the northeast Atlantic in the twenty-first century under different forcing scenarios are reported by Wang et al. (2004), Wang and Swail (2006) indicating an increased tendency during winter and fall seasons. For the Pacific basin, Shimura et al. (2010) examined the extreme values of futuristic significant wave heights and associated wind speed in tropical cyclone zones showing that both had a zonal dependence. Mori et al (2010) examined the projection of extreme wind-wave climate under global warming for the Japan Sea. For the Dutch coast in the North Sea region, Renske et al (2012) reported effect of climate change on extreme waves suggesting a possible increase in the annual wind-wave maxima and their effect on the coastal environment. There are very few studies conducted to determine the trends in wind-wave climate projections, and these were reported for the Atlantic and Pacific Ocean basins (Hemer et al. 2013) and not much reported for the IO region. It was only recent that studies reported on the wind-wave climate variability for the IO region using multi-satellite observations (Bhaskaran et al. 2014; Gupta et al. 2015, 2017; Gupta and Bhaskaran 2016).

In this chapter, the discussions are more focused to analyse and quantify the observed trends in maximum significant wave height and maximum wind speed and identify the zones of maximum variability in the IO basin considering the fair weather conditions. Further, the inter-annual variations in wind-wave climate for the IO basin are reported. Though changing frequency and intensity of deep depressions and cyclones form an integral part of the climate system, the effect of climate change on frequency and cyclone intensity is not covered in this chapter. The analysis from the study is based on data obtained from eight satellite missions covering a period of 21 years. More details are presented in the subsequent sections.

9.1.3.1 Extreme Wave Analysis for the Arabian Gulf

The Arabian Gulf (AG) is a body of water which is an extension of the IO connected through the Strait of Hormuz. It is surrounded by many countries rich in oil reserves both on land and within the Gulf. There are about hundreds of offshore oil drilling platforms that exploit oil and gas reserves, and many more are planned for the future. Many marine structures are being constructed and also in plan such as seawater intake structures, breakwaters for port, harbour, and shore protection, submarine pipelines, open sea loading/unloading terminals, and oil terminals. It is important to note that cost-effective design of all these structures requires precise prediction of extreme waves for different return periods and that is very essential in terms of safety and economic point of view. For example, the armour unit weight of a breakwater depends on the design significant wave height to the power of 3. Hence, selection of 2 or 3 m significant wave height results in an armour unit of weight in the ratio of 8:27. It deciphers the importance to predict design wave heights for different return periods. For the AG waters, as on today, most of the coastal structures appear to be over designed, as there is no systematic extreme wave prediction done so far. Neelamani et al. (2006) had carried out extreme wave analysis especially for the Kuwait territorial waters at 19 different locations, and it is felt this work can be extended to cover the whole Arabian Gulf for the benefit of Gulf Cooperation Council (GCC) countries towards safe and cost-effective design of coastal and offshore structures.

A structure such as seawater intake system designed for the design sea-state of oceans like North Atlantic Ocean or the BoB (that has severe wave climate with frequent cyclones) cannot be adopted for the AG marine environment, primarily this region being fetch limited. In general, the AG is a marginal sea in typical arid zone environment and connected to the IO lying situated between the latitudinal belts $24°–30°$ N. The gulf covers an area of 226,000 km^2. It is about 990 km long, and its width ranges from 56 to 338 km. It has a total volume of 7000 to 8400 km^3 of seawater (Emery 1956; Purser and Seibold 1973; El-Gindy and Hegazi 1996). The average water depth of the AG is about 35 m. But depths more than 107 m occur in some places with water depth increasing in the south-east direction. The Gulf is connected to the Gulf of Oman and the Arabian Sea through the Strait of Hormuz, which is 56 km wide with an average water depth of 107 m that allows free exchange of water between the AG and Arabian Sea. More details on the Oceanographic Atlas of AG can be obtained from Al-Yamani et al. (2004). Dominant wind direction over this region in general is north-westerly (Elshorbagy et al. 2006). It is situated in a strategic location, and most of the countries around Arabian Gulf region rely on seawater for desalination and cooling of power plants. A large number of coastal and offshore project activities are undertaken and being planned in the AG waters like artificial coastal developmental projects such as palm and world-shaped waterfronts in Dubai, Durrat Al-Bahrain—a jewellery-shaped waterfront development in Bahrain and similar projects in Qatar, ultra-modern ports in Kuwait, a number of submarine pipeline and offshore oil and gas platforms, projects for development of tourism industries, etc. Design of all these marine structures requires realistic estimate of design wave height for different return periods, which is not available at present. This chapter discusses on an attempt made to report the extreme waves in AG waters for different return periods. Caires and Sterl (2005) estimated the 100-year return value for significant wave height utilizing ERA-40 data for the global oceans. The spatial resolution of wind data used in this study was $1.5° \times 1.5°$. However, data of this coarse resolution cannot provide valuable information for countries surrounding the AG, due to limited width of the AG. Therefore, finer resolution grid size of $0.5° \times 0.5°$ that was linearly interpolated to a finer size of $0.1° \times 0.1°$ was used to force WAM wave model to hindcast significant wave height and to perform extreme wave analysis.

9.1.4 Role and Influence of Southern Ocean (SO) on Wind-Wave Climate

The Southern Ocean (SO) is the only water body that is freely connected with the major oceans in the world. It plays a major role in governing the wind-wave climate of global ocean basins including the IO sector. It is noteworthy that the extra-tropical belt in the Southern Hemisphere has extremely strong winds and an active potential region for swell generation. A study by Snodgrass et al. (1966) indicated that long waves or 'swells' generated in this region can travel thousands of kilometres circumscribing the hemisphere and influencing locally generated wind-waves elsewhere. Using Empirical Orthogonal Functions (EOFs), the study by Sterl and Caires (2005) analysed the ERA-40 reanalysis significant wave height data and reported that about 15% of global wave activity is contributed by swells generated from the SO region. Hence, this region is extremely important as it can influence the variability of global ocean wave heights. Hemer et al. (2010) used altimeter data to understand the inter-annual climate variability and trends in the directional behaviour of wind-wave climate over the SO region. In a climate perspective, the extreme wind-wave climate variability in the IO sector of SO correlated well with the Southern Annular Mode (Bhaskaran et al. 2014). In another study, Young et al. (2011) reported on the global trends in wind speed and wave heights using altimeter measurements analysing statistical parameters such as monthly mean, 90th and 99th percentile trends for the global oceans. An interesting observation from this study indicated that intensity of extreme weather events increased at a faster rate as compared to the mean conditions. The mean and 90th percentile analysis suggested that wind speed over large areas in the global oceans has increased at a rate of 0.25–0.5% per year. Young and Ribal (2019) using significant wave height data obtained from 31 satellite missions for the period (1985–2018) comprising altimeters, radiometers, and scatterometers witnessed an increase in extreme wave heights (90th percentiles) than the mean significant wave height.

9.1.5 Impact of Extreme Wind-Waves on Coastal Inundation

One of the biggest threats to coastal communities is severe inundation and its frequency associated with climate change on the extreme water levels in the nearshore regions. The immediate threat is on coastal communities and the islands (Nicholls et al. 2007; Seneviratne et al. 2012). The impact resulting from coastal inundation can significantly affect the shoreline configuration, damage to infrastructure, saltwater intrusion into groundwater, destruction of crops, and affecting the human population having wide socio-economic consequences. It is imperative to understand the processes that cause extreme water levels having paramount significance to adaptation strategies in a changing climate scenario. Prior studies by Munk et al. (1963) and Snodgrass et al. (1966) indicated that distant swells can propagate basin scale. Swells can influence the local wind-generated waves as well modulate and modify the resultant wave energy in nearshore regions (Nayak et al. 2012). Breaking wind-waves can influence the wave set-up that can reach approximately one-third of the incident wave height along coasts in the tropical and subtropical islands (Munk and Sargent 1948; Vetter et al. 2010). Therefore, wave-induced set-up due to breaking of extreme wind-waves is a significant environmental driver that can affect the coastlines.

Arrival of extreme swells can trigger inundation events along major coastlines of the world. In the literature, the causative factors attributed to more commonly reported extreme nearshore water levels are astronomical tides, tropical cyclone-induced storm surges, and regional sea-level variability due to El Nino and Southern Oscillations (ENSO) phenomena (Church et al. 2006; Menendez and Woodworth 2010; Walsh et al. 2012). Perhaps one of the reasons addressing swells as the cause of extreme water levels could be remoteness of island communities, poor reporting networks (Kruke and Olsen 2012), and scarcity of in situ observations and surface waves (Lowe et al. 2010). Widespread major inundation events during December 2008 for the Pacific islands that was triggered by distant swells were reported by Hoeke et al. (2013). Reports suggested that inundation was significant and that occurred over several consecutive days during high tide conditions with additional impacts from wave run-up and infra-gravity bores impacting the low-lying island's locations. Widespread damages were also reported at islands such as Micronesia, the Marshall Islands, Kiribati, Papua New Guinea, and Solomon Islands (Hoeke et al. 2013). The timing of inundation was clearly correlated with the arrival of extra-tropical storm swells, wherein the potential of inundation from swells was accounted due to steep bathymetric slopes at all affected locations. It is important to note that steep bathymetry can result in high dissipation gradient along the coast causing extreme wave set-up and run-up conditions (Kennedy et al. 2012). Changing wind-wave climate and extreme waves can have large impact on the future frequency and magnitude of coastal inundation events. Though still uncertainty prevails in the changes of these events, the current knowledge on storm wave heights and frequency in particular for the mid- and high-latitude regions shows a positive trend (Young et al. 2011; Aucan et al. 2012; Hemer et al. 2013). It is warranted to perform more rigorous studies on the frequency and magnitude of coastal inundation due to extreme wave events under sea-level rise projections for the future.

In the IO perspective, the acronym '*Kallakadal*' that literally means '*sea thief*' is a phenomenon reported for the Kerala coast in south-western coast of India during the pre-monsoon season and during the periods of monsoon breaks (Kurian et al. 2009). This acronym was adopted by UNESCO referring to sea creeping into the coastal region attributed due to swells generated by storms in the SO. This phenomenon mostly occurs during the pre-monsoon season and sometimes during the monsoon breaks and continues for a few days inundating the low-lying coastal regions. During high tide condition, the water levels can reach about 3–4 m above the maximum record. A recent study by Remya et al. (2016) has documented on the teleconnection between NIO high swell events and the prevailing meteorological conditions in the SO. Their study used combination of in situ measurements and model generated simulations for the year 2005 in addressing the flooding associated with high swell activity or *Kallakadal* event in the Kerala coast. During the year 2005, there were about ten high swell events reported in the NIO basin, and the study (Remya et al. 2016) confirmed these events were triggered by distant swells propagating from south of 30S. A severe low-pressure system, also termed as '*Cut-Off-Low*' quasi-stationary in nature, occurred in the SO about 3–5 days prior to high swell events attacking the south-west coast of India. Their study (Remya et al. 2016) reported that strong surface winds (about 25 m/s) sustained with longer duration (about 3 days) over a large fetch were the essential conditions for the generation of long-period swell waves that prevails.

Coastal inundation associated with *Kallakadal* depends on the onshore topography, and it appears to be more severe and frequent on the south-western Indian coast compared to the northern coast. Though not well documented in the literature, reported sources from fishermen state this flash flood event occurs almost every year (Kurian et al. 2009). As per available literature, the *Kallakadal* event of May 2005 was quite intense and documented (Narayana and Tatavarti 2005; Baba 2005; Murty and Kurian 2006). The associated wave characteristics are typical swells having moderate heights between 2–3 m and long periods (about 15 s). At present, there is no operational forecasting system for *Kallakadal* events being a remotely forced event. There are several studies over the recent past that investigated the coastal inundation characteristics attributed by tropical cyclone-induced storm surges using stand-alone ADCIRC (Advanced Circulation) model and coupled wave-hydrodynamic (ADCIRC + SWAN) models. For example, the performance of coupled ADCIRC + SWAN model for *Thane* cyclone event and coastal inundation along with validation was reported by Bhaskaran et al. (2013a, b, 2014); for *Phailin*event (Murty et al. 2014), for *Aila* event (Gayathri et al. 2016), for *Hudhud* event (Murty et al. 2016; Dhana Lakshmi et al. 2017; Samiksha et al. 2017), for 1999 Odisha Super cyclone (Sahoo and Bhaskaran 2019a, b). Jismy et al. (2017) reported the role of continental shelf on the nonlinear interaction mechanism between storm surge, tides, and wind-waves. Gayathri et al. (2019) investigated the role of river–tide–storm surge interaction characteristics for the Hooghly estuary located in the East coast of India. Murty et al. (2019) investigated the effect of wave radiation stress in storm surge-induced indication for the East coast of India. A comprehensive overview on tropical cyclone-induced storm surges and wind-waves for the BoB region is available in Bhaskaran et al. (2019). Influence of wave set-up along Indian coasts was investigated by Nayak et al. (2012). However, with an increasing trend seen of extreme wind-wave activity (Bhaskaran et al. 2014) in the SO region, the severity and frequency of flash flood events such as *Kallakadal* and its impact on the Indian coast are a research topic of immense interest having socio-economic implications.

9.2   Data and Methodology

The recent trends in wind-wave climate for the IO region in a changing climate was reported by Bhaskaran et al. (2014) and Gupta et al. (2015). The study used altimeter data for maximum significant wave height using daily datasets from eight satellite missions covering a period of 21 years (1992–2012). Measured data from satellite used advanced microwave techniques that enabled data coverage irrespective of cloud and sunlight conditions. As mentioned above, the eight satellite missions are from ERS-1/2, TOPEX/Poseidon, GEOSAT Follow-On (GFO), JASON-1/2, ENVISAT, and CRYOSAT. Quality checked altimeter data were used in this study (Queffeulou et al. 2011), and the corrected altimeter data available as daily data files were used for analysis. Details of eight satellite missions covering various aspects on the satellite passes and data collection time are given in Table 9.3.

Data from these eight satellite missions were available for the global ocean basins. In this study, the IO basin (30 E–120 E; 30 N–60 S) is the region of interest (Fig. 9.1a). The Basic Radar Altimetry Toolbox (BRAT) version 3.1.0 was used to convert the binary data into text files. There are three different filters in BRAT such as smooth, extrapolate, and loess with functionality of smoothing, filling data gaps and combinations of both, respectively.

Relevant parameters, such as maximum significant wave height and maximum wind speed for the domain as shown in Fig. 9.1a, were derived from the BRAT processed daily data. The study examined the variations of these parameters along the two transect lines (Fig. 9.1a) covering the western and eastern sectors of the IO basin. There are two points each marked along these transects and that represents the extra-tropical, equatorial, and tropical belts of the IO. More precisely, these ten locations cover south IO, south subtropical, south trade wind, and equatorial and tropical north-east and north-west IO sectors. Recent trends in the variability of maximum wind speed and maximum significant wave heights were analysed at these

Table 9.3 Data record of satellite products and respective time coverage used in the study

S. No.	Satellite	Product	Cycles	Time Period	Remarks
1	ERS-1	OPR	Not defined	01 August 1991 to 30 March 1992	Phases A and B 3-days
			83 to 101	14 April 1992 to 20 December 1993	Phase C 35-days
			Not defined	24 December 1993 to 10 April 1994	Phase D 3-days
			Not defined	10 April 1994 to 21 March 1995	Phases E and F 168-days
			144 to 156	24 March 1995 to 02 June 1996	Phase G 35-days
2	ERS-2	OPR	1 to 169	15 May 1995 to 04 July 2011	
3	ENVISAT	GDR v2.1	6 to 113	14 May 2002 to 08 April 2012	
4	TOPEX/ POSEIDON	M-GDR	1 to 481	25 September 1992 to 08 October 2005	
5	JASON-1	GDR	1 to 525	15 January 2002 to 15 February 2013	Mission going on
6	JASON-2	GDR	0 to 168	04 July 2008 to 01 February 2013	Mission going on
7	GEOSAT Follow-On (GFO)	GDR	37 to 222	07 January 2000 to 07 September 2008	
8	CRYOSAT-2	IGDR	11 to 580	28 January 2011 to 08 April 2013	Mission going on

From Bhaskaran et al. (2014)

ten locations. As the study focused on understanding the trends during fair weather conditions, cases of extreme weather events such as tropical cyclones that had short durations were excluded from the analysis.

Satellite-derived products used in this study have undergone thorough calibration and validation checks with other in situ observations leading to the development of homogeneous research quality data. Therefore, the altimeter derived maximum significant wave height is a quality product beyond doubt that aids one to deduce meaningful conclusions. To determine the long-term trends, inter-annual, and intra-seasonal variability, a recent study analysed 41 years of ERA-5 wind-wave data covering the period from 1979 to 2019 (Sreelakshmi and Bhaskaran 2020). Figure 9.1b shows the study area in the IO region grouped into six sub-domains. Parameters such as significant wave heights of combined wind-seas and swells, total swells, and wind-seas were analysed with data retrieved at $0.5° \times 0.5°$ spatial resolution on a monthly averaged time frequency. Further, the performance evaluation of reanalysis and model products were done using various statistical measures such

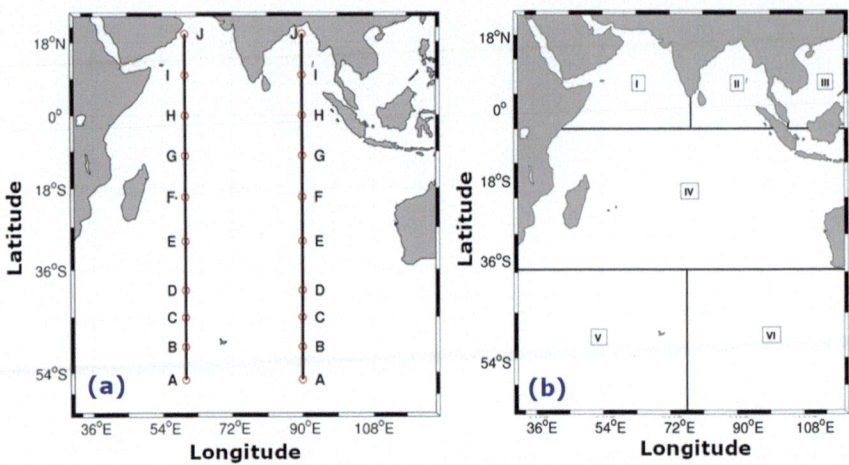

Fig. 9.1 (a) Study region and transects for analysis in the Arabian Sea and Bay of Bengal (from Bhaskaran et al. 2014), and (b) Indian Ocean region grouped into six sub-domains (from Sreelakshmi and Bhaskaran 2020)

as correlation coefficient, root mean square error, bias, average absolute error, and percentage of bias (Sreelakshmi and Bhaskaran 2020).

To assess the climatology and annual trend analysis of extreme waves, the seasons were classified based on the India Meteorological Department (IMD) nomenclature such as winter (January–February; season 1) pre-monsoon (March–May; season 2), monsoon (June–September; season 3), and post-monsoon (October-December; season 4). Linear regression analysis using poly-fitting function was used to investigate the trend in wave heights quantifying the changes per year by employing a suitable statistical significance test for the study domain. Statistical significance of trends was determined using the Mann-Kendal trend test for 95% confidence limit (Sreelakshmi and Bhaskaran 2020). To examine the various modes in the spatio-temporal variability, the widely used principal component analysis/empirical orthogonal function (EOF) was employed. This analysis helps to represent the data according to the variance by performing Eigen value decomposition, wherein the principal component analysis provides the measure of temporal variation, and the EOF attributes to the spatial variability. It is noteworthy that this method can essentially capture the nonlinearity and high-dimensional characteristics for a given dataset preserving significant patterns and their variability, thereby aiding the users to derive meaningful information for data interpretation and analysis. Analysis was carried out separately for swells and wind-seas for seasonal and annual scales to understand the variability in the individual characteristics.

From the variability analysis, twelve locations were identified from the domain that exhibited more significant variability in the EOF first three modes. Another powerful technique is the wavelet transform that analyses the frequency components of a given signal in the dataset. Techniques using Fourier transform can also be used to analyse the frequency components; however, if this technique is applied for the entire time series dataset, one cannot interpret at what time instant a particular frequency rises. A short-time Fourier transform uses sliding window for spectrogram that provides information in both time and frequency scales. However, there is a constraint with the window length that can limit the frequency resolution. This limitation is taken care in the wavelet transform which is a powerful technique to provide the solution. Wavelet transforms work on the principle based on small wavelets with limited duration. The first type of wavelet transform is the orthogonal Haar wavelet (Addison 2018). A second type of orthogonal wavelet is the Meyer wavelet that was formulated in 1985. Wavelet transform algorithm had demonstrated numerous applications in the field of signal processing in diverse disciplines. Complex wavelets have Fourier transforms that is zero for negative frequencies. Advantages of using a complex wavelet are its capability in separating the phase and amplitude components of the signal. The Morlet wavelet is the most commonly used complex wavelet, which is used in this study (Sreelakshmi and Bhaskaran 2020).

There are several studies carried out on extreme value analysis of wind and waves. Gumbel (1958) developed the statistical technique to analyse the extreme values of natural random events like wind speed. Recorded annual maximum wind speed covering the time series of many years is the input for this method. It is important to note that Gumbel extreme value distribution is being widely used by the wind engineering community across the globe, as this method is simple and robust. St. Denis (1969, 1973) had discussed on Gumbel distribution for extreme wave prediction. More information pertaining to data sample collection for extreme value analysis is available in Nolte (1973), Cardone et al. (1976), Petrauskas and Aagaard (1971), and Jahns and Wheeler (1973). Detailed information on the plotting formula used for extreme wave predictions is available in Kimball (1960), Gringorten (1963), and Petrauskas and Aagaard (1971). Also, the procedure for extreme wave height predictions is explained in Sarpkaya and Isaacson (1981) and Kamphuis (2000). Extreme value analysis for waves is discussed in Mathiesen et al. (1994), Goda et al. (1993), and Goda (1992). In addition, Coles (2001) provides more statistical details on extreme value prediction based on the annual maximum data points and peak over threshold (POT) method. Additional information on POT and its application is provided in Ferreira and Guedes Soares (1998) and Leadbetter (1991). All these literatures provide the information and knowledge for carrying out a detailed extreme value analysis and used for the present study. For the AG region, the wave data was hindcasted using WAM wave model covering a total period of 12 years (1 January 1993 till 31 December 2004). The output from WAM model comprises significant wave height and the mean wave period at every one-hour interval. The hindcasted wave data for the entire AG waters covers a grid resolution of $0.1° \times 0.1°$. Model was validated using measured data as provided in Al-Salem et al. (2005). Extreme wave analysis was carried out for a total of 38 different locations in the Gulf region shown in Fig. 9.2. Each location had a total of 1, 05,192 data points. More details of each location are provided in Table 9.4.

Maximum and mean significant wave heights at 38 locations based on the 12 year hindcasted data are provided in Fig. 9.3. The highest maximum significant wave

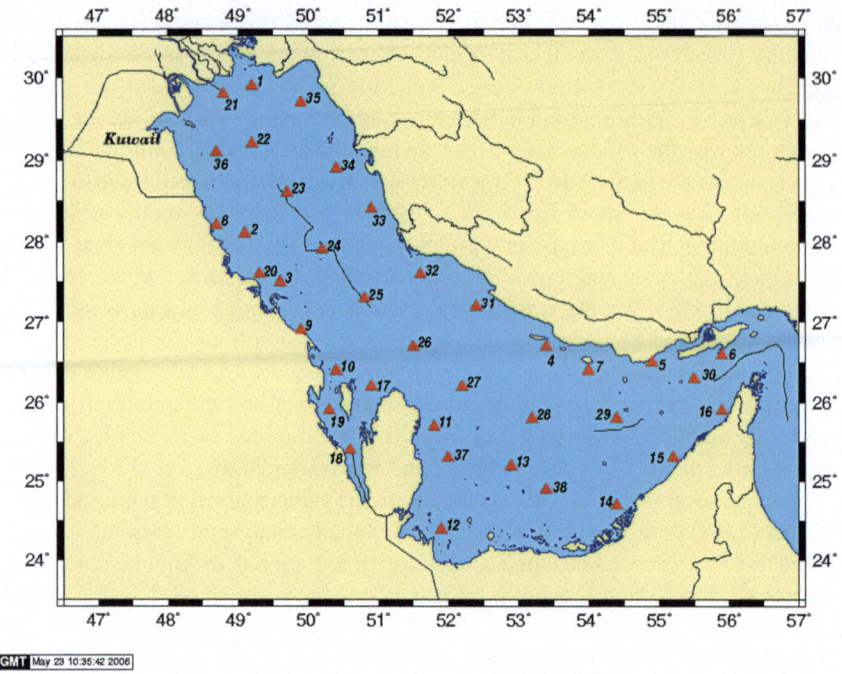

Fig. 9.2 Locations in the Arabian Gulf waters for extreme wave analysis

height is hindcasted at location 28 ($H_s = 5.33$ m), and the lowest maximum significant wave height is hindcasted at location 8 ($H_s = 1.82$ m). Similarly, the maximum mean wave height for 12-year period is at location 27 (Fig. 9.2) with $H_s = 0.77$ m, and the minimum average wave height is at location 5 (Fig. 9.2) with $H_s = 0.21$ m.

The Gumbel and Weibull distributions were used for the extreme value analysis. Statistics of long-term wave prediction require that the individual data used in the statistical analysis be statistically independent. Hence, the hourly wave height depends very much on the wave height of the previous hours, fulfilling that the condition of statistical independence is not met. Hence, in order to produce independent data points, only the storms should be considered. The commonly used method to separate wave heights into storms is called the peak over threshold (POT) analysis (Coles 2001). A threshold wave height of 1.0 m is selected for the present analysis. Figure 9.4 illustrates the number of storm events/year with threshold wave height of 1.0 m for different locations.

As seen from Fig. 9.4, there are nine locations with more than 80 number of storm events/year and 14 locations had 60–80 storm events/year with threshold significant wave heights of 1.0 m. There are six locations amongst the selected 38 locations with less than 40 storm events/year with threshold significant wave height of 1.0 m. This important information is vital for marine operations around these locations. The data points used in the POT analysis are the peaks occurring during each storm with

Table 9.4 Details of geographical locations and respective local water depths at 38 different locations in the Arabian Gulf waters

Location	Longitude (°E)	Latitude (°N)	Water depth (m)	Remarks/nearest country
1	49.2	29.9	15	Iran
2	49.1	28.1	15	Saudi Arabia
3	49.6	27.5	15	Saudi Arabia
4	53.4	26.7	61	Iran
5	54.9	26.5	11	Iran
6	55.9	26.6	31	Iran
7	54.0	26.4	55	Iran
8	48.7	28.2	9	Saudi Arabia
9	49.9	26.9	16	Saudi Arabia
10	50.8	26.4	12	Bahrain
11	51.8	25.7	19	Qatar
12	51.9	24.4	10	UAE
13	52.9	25.2	16	UAE
14	54.4	24.7	10	UAE
15	55.2	25.3	16	UAE
16	55.9	25.9	20	UAE
17	50.9	26.2	9	North-west of Qatar
18	50.6	25.4	17	South-west of Qatar
19	50.3	25.9	20	In between Saudi Arabia and Bahrain
20	49.3	27.6	9	Saudi Arabia
21	48.8	29.8	10	In between Kuwait and Iran
22	49.2	29.2	33	In between Saudi Arabia and Iran
23	49.7	28.6	45	In between Saudi Arabia and Iran
24	50.2	27.9	48	In between Saudi Arabia and Iran
25	50.8	27.3	62	In between Bahrain and Iran
26	51.5	26.7	39	In between Qatar and Iran
27	52.2	26.2	44	In between Qatar and Iran
28	53.2	25.8	54	In between UAE and Iran
29	54.4	25.8	59	In between UAE and Iran
30	55.5	26.3	57	In between UAE and Iran
31	52.4	27.2	79	Iran
32	51.6	27.6	22	Iran
33	50.9	28.4	42	Iran
34	50.4	28.9	44	Iran
35	49.9	29.7	24	Iran

(continued)

Table 9.4 (continued)

Location	Longitude (°E)	Latitude (°N)	Water depth (m)	Remarks/nearest country
36	48.7	29.1	19	Kuwait
37	52.0	25.3	15	East of Qatar
38	53.4	24.9	20	UAE

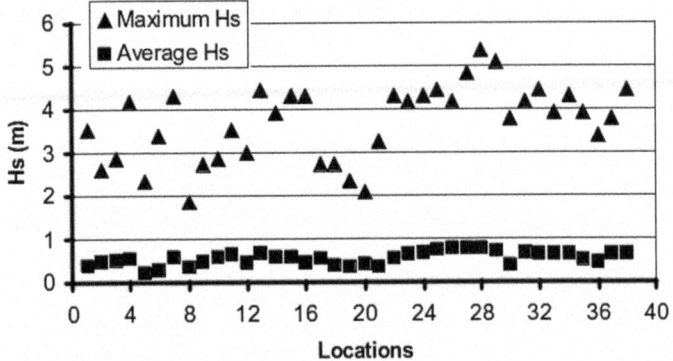

Fig. 9.3 Maximum and mean hindcasted significant wave height for Arabian Gulf waters during the period January 1993–December 2004

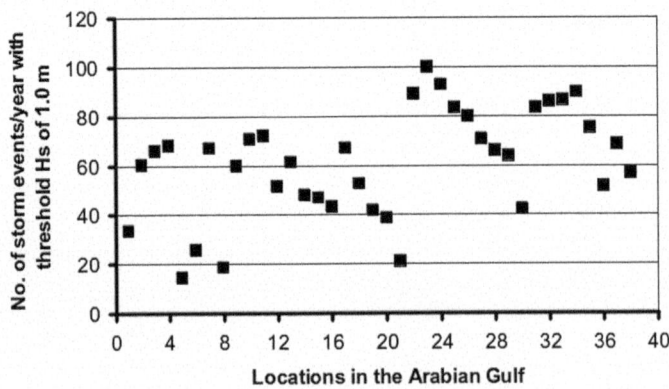

Fig. 9.4 Number of storm events/year with threshold significant wave height of 1.0 m in the Arabian Gulf waters

threshold wave height of 1.0 m. The total number of data points used for the extreme wave analysis is hence 12 times the number of storm events/year with threshold H_s value of 1.0 m as shown in Fig. 9.4. Data points for each location are arranged in the descending order, and the probability of exceedance (Q) is calculated using the formula:

$$Q = \frac{(i - c_1)}{(N + c_2)} \qquad (9.1)$$

where the symbols 'i' represents the rank; 'N' represents the total number of data points; $c_1 = 0.44$ and $c_2 = 0.12$ for Gumbel distribution, and $c_1 = 0.20 + (0.27/\alpha)$ and $c_2 = 0.20 + (0.23/\alpha)$ for Weibull distribution, where α is the shape parameter. The value of α is varied from 0.8 to 1.3 with an increment of 0.05, and the value of α, which gives best fit for the dataset, is selected. More details on the Gumbel and Weibull distributions are available, for example in Kamphuis (2000).

Prediction of wave height for the selected return period (T_R) and the probability of exceedance are linked by the following expression:

$$Q = 1/(1 T_R) \qquad (9.2)$$

where 1 represents the number of event/year. In this case, we know the total number of storm events exceeding threshold value of $H_s = 1.0$ m for each location in the Arabian Gulf waters. As the data used are for a total duration of 12 years, the value of 1 can be determined from Fig. 9.4. According to Gumbel distribution, the wave height expected for a selected return period H_{TR} can be estimated using the formula:

$$H_{TR} = g - bln[\ln(1/P] \qquad (9.3)$$

that becomes

$$H_{TR} = g - bln[ln\{(1 T_R)/(1 T_R - 1)\}] \qquad (9.4)$$

According to the Weibull distribution, the wave height expected for a selected return period H_{TR} can be estimated from the following formula:

$$H_{TR} = g + b[\ln(1/Q)]^{1/\alpha} \qquad (9.5)$$

i.e.

$$H_{TR} = g + b[\ln(1 T_R)]^{1/\alpha} \qquad (9.6)$$

It is now possible to obtain the extreme wave height for any selected return period.

Wave height and wave periods are independent parameters. However, as the wave height increases, it is also likely that wave period increases. On the other hand, the probability of occurrence of high waves and long periods is more pronounced than the probability of occurrence of high waves and short periods. Joint probability of wave height and wave period is used for predicting the wave period for a given wave height of any desired return periods (Kamphuis 2000). An example of the joint wave period–wave height distribution for location 23 (Fig. 9.2) in the AG waters is illustrated in Table 9.5.

The joint distribution is simplified by relating the wave period to wave height using the combinations of greatest frequency. For example, in Table 9.5, the interpolation gives $T_{mean} = 5.627$ s corresponding to $H_s = 2.625$ m. Now the significant wave height and mean wave period are related by the following equation:

$$T_{mean} = C_3(H_s)^{C_4} \tag{9.7}$$

The values of C_3, C_4 and the corresponding coefficient of correlation R^2 are obtained for all the 38 locations in the Arabian Gulf waters and shown in Fig. 9.5a–c, respectively.

As seen from Fig. 9.5, the average coefficient of correlation is 0.948, which is an acceptable value for using C_3 and C_4 values to obtain the mean wave period for a given H_s value. It is recommended to use the respective C_3 and C_4 values for the chosen locations to estimate T_{mean}. The average value of C_3 and C_4 is 4.398 and 0.2648, respectively, and it can be used to estimate the approximate value for the mean wave period in Arabian Gulf waters for a given significant wave height corresponding to the desired return period of the event.

9.2.1 Extreme Wind-Wave Analysis for the North Indian Ocean

Extreme wind-waves are also generated by extreme weather events like tropical cyclones. The sea-state is quite rough during cyclone events, and significant wave height is extremely high in both open sea and in coastal regions. In the recent past, studies were attempted using coupled wave-hydrodynamic models (ADCIRC + SWAN) that provides a clear picture on the extreme water levels near coastal regions during tropical cyclone events (Bhaskaran et al. 2013a, b; Murty et al. 2014, 2016). Radiation stress obtained from wave model is dynamically exchanged with hydrodynamic models modifying the water level elevation and further mutually exchanged with wave model to update the radiation stress at coupled time steps. The contribution from wave models is the wave set-up that contributes to the extreme water levels. Readers can refer to the above-mentioned references for more details on studies conducted for different severe cyclone cases. This chapter discusses on extreme waves computed using the coupled model for very severe cyclonic storm *Hudhud* that made landfall in Andhra Pradesh, East coast of India (Murty et al. 2016).

Figure 9.6a is the satellite imageries of this cyclone, and the corresponding track is shown in Fig. 9.6b. The bathymetry and finite element mesh for the computational domain was generated using the surface modelling system (SMS) as shown in

Table 9.5 Joint distribution of significant wave height and mean wave period for Location 23 in the Arabian Gulf waters (Number of occurrences over 12 years)

H_s (m)	Mean wave period, T_{mean} (s)									Total no. of occurrence	Mean value (s)
	0–1.0	1.01–2.0	2.01–3.0	3.01–4.0	4.01–5.0	5.01–6.0	6.01–7.0	7.01–8.0	8.01–9.0		
0–0.25	0	2519	21,308	6202	350	35	0	0	0	30,414	2.648
0.251–0.5	0	91	7930	21,796	1308	102	0	0	0	31,227	3.289
0.51–0.75	0	0	42	13,349	2028	214	2	0	0	15,635	3.654
0.751–1.0	0	0	0	3225	2360	144	1	0	0	5730	3.963
1.01–1.25	0	0	0	1391	3153	135	2	0	0	4681	4.233
1.251–1.5	0	0	0	276	6775	287	13	1	0	7352	4.505
1.51–1.75	0	0	0	6	4465	322	36	4	0	4833	4.583
1.751–2.0	0	0	0	0	1563	1145	26	4	0	2738	4.942
2.01–2.25	0	0	0	0	105	1288	31	2	0	1426	5.451
2.251–2.5	0	0	0	0	3	542	22	5	0	572	5.551
2.51–2.75	0	0	0	0	0	217	25	3	0	245	5.627
2.751–3.0	0	0	0	0	0	72	67	1	0	140	5.993
3.01–3.25	0	0	0	0	0	7	96	1	0	104	6.442
3.251–3.5	0	0	0	0	0	0	28	1	0	29	6.534
3.51–3.75	0	0	0	0	0	0	48	0	0	48	6.5
3.751–4.0	0	0	0	0	0	0	12	1	0	13	6.577
4.01–4.25	0	0	0	0	0	0	5	0	0	5	6.5
4.251–4.5	0	0	0	0	0	0	0	0	0	0	–

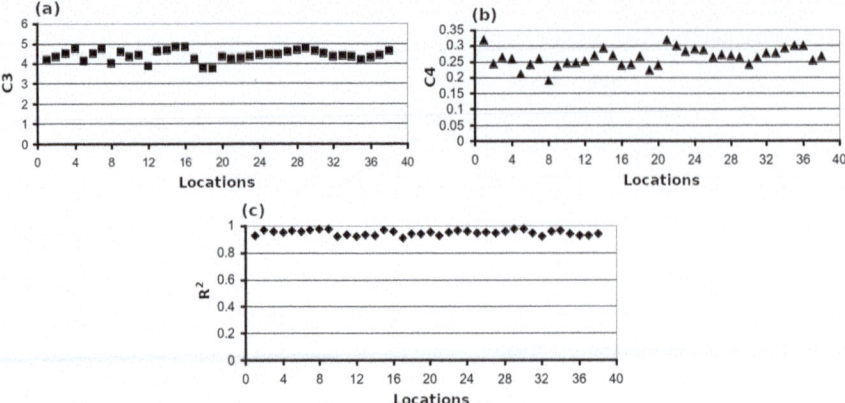

Fig. 9.5 Values of **a** C_3, **b** C_4, and **c** R^2 for different locations in the Arabian Gulf waters

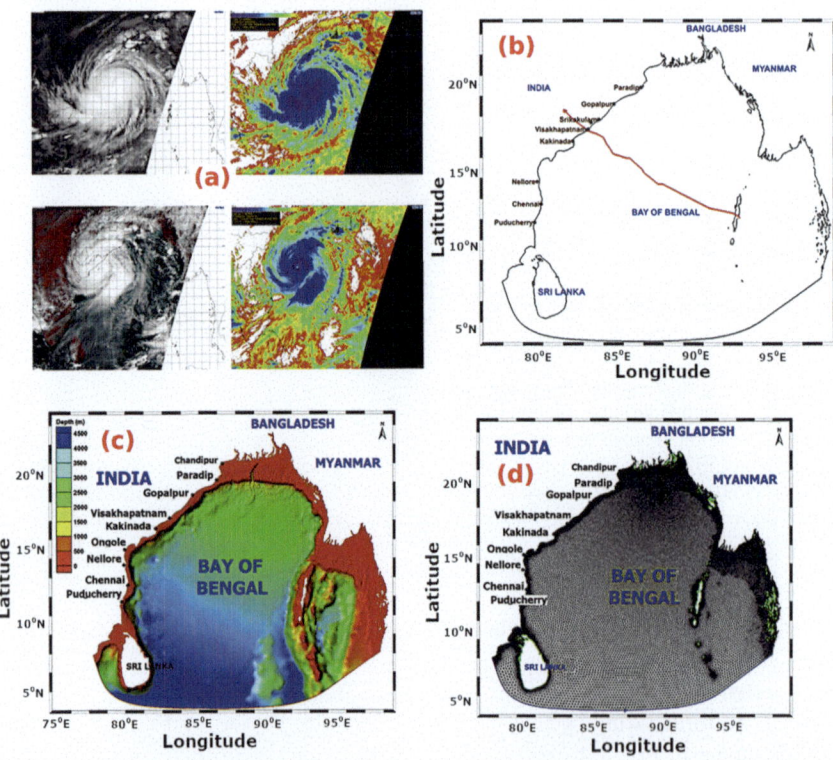

Fig. 9.6 **a** Satellite imageries of Hudhud cyclones, **b** track of Hudhud cyclone from IMD best track estimates, **c** bathymetry for the Bay of Bengal region, and **d** finite element mesh of the study region (from Murty et al. 2016)

Fig. 9.6c, d. The finite element mesh shown in Fig. 9.6d resolves well sharp bathymetric gradients in the coastal environment. The bathymetric data General Bathymetric Chart of the Oceans (GEBCO) having a grid spacing of 30 arc seconds maintained by the British Oceanographic Data Centre (BODC) were used in this study. The unstructured grid that is optimized in context to computational time and shown in Fig. 9.6d comprises 123,594 vertices and 235,952 triangular elements. Flexibility in grid structure provides allowance to relax in deep waters and refines accordingly based on the specified bathymetric features in the nearshore areas. Spatial grid resolution is 500 m along coast in the nearshore regions and relaxing to 30 km in the offshore boundary at deep ocean. Study by Bhaskaran et al. (2013a, b) signifies that a high-resolution flexible mesh in the nearshore areas better resolves the complex bathymetric features reflected in the wave transformation. Also, the criteria in fixing the grid size not exceeding 1 km for nearshore regions are justified in the study by Rao et al. (2009) indicating the optimum size for precise computation of storm surge height for the East coast of India.

Bottom friction coefficient used in ADCIRC model is 0.0028 with a time step size of 10s and that suits well for the sandy bottom environment prevailing at Andhra Pradesh coast (Murty et al. 2014). Model run was executed for the period from 8 October 2014 (00 h) when *Hudhud* was in deep waters, until the landfall time (forenoon of 12 October 2014) with total simulation length of 120 h including the ramp function of one day. Computation was carried out in the high-performance computing system at INCOIS utilizing 320 processors. The coupling time step in wave-hydrodynamic (ADCIRC + SWAN) was specified as 600s. Study by Bhaskaran et al. (2013a, b) advocates that this prescribed time step is good enough to understand the nonlinear interaction effects from changing water levels in the presence of wave field. The wave model (SWAN) configuration was prescribed having 36 directional and 35 frequency bins that can optimally resolve the spectral distribution of wave energy propagation, as well capture realistically the evolution of spectral wave energy in both geographic space and time. Wave frequencies used logarithmic frequency bins ranging between 0.04 and 1.0 Hz with angular resolution of 10 degrees. Quadruplet nonlinear wave–wave interaction using discrete interaction approximation technique was used in the wave model configuration along with Madsen et al. (1988) formulation for the bottom resistance. Wave rider buoy located off Visakhapatnam was used for verification of the model computed results.

9.2.2   Analysis of GCM Results for the North Indian Ocean

Wave models are sensitive to input wind fields, and therefore identification of the best available wind field serves to provide better quality wave forecast of extreme waves. Reliability of wind forcing produced by GCMs directly influences the quality of wave outputs (Bricheno and Wolf 2018). Also, to evaluate the futuristic projections of extreme wave datasets, the wind fields generated from general circulation models (GCMs) is a necessity. Climate models that use GCMs and their ensemble can provide simulated data for historical, near-, and futuristic projections to force wave models. It is therefore imperative to determine the best-performing GCMs for the IO region. The readers can refer to the studies by Krishnan and Bhaskaran (2019a, b, 2020) that deals with CMIP5/CMIP6 wind speed comparison between satellite altimeters and reanalysis products, global climate models for the BoB region and its projection.

The best-performing climate models under CMIP5 category verified for the BoB region are available in Krishnan and Bhaskaran (2019a, b). Utilizing this knowledge, the available 20 GCMs under CMIP6 category were subjected to performance evaluation. Models employed under CMIP5 and CMIP6 family belong to ensemble 'r1i1p1' and 'r1i1p1f1', respectively. Historical simulations from CMIP5 and CMIP6 datasets span the period 1850–2005 and 1850–2014, respectively. Monthly near-surface wind speed data simulated by GCMs are extracted for the historical and projection analysis. CMIP5 future projections are characterized as Representative Concentration Pathways (RCP) scenarios comprising of RCP2.6, RCP4.5, RCP6.0, and RCP8.5 radiative forcing of 2.6 W/m^2, 4.5 W/m^2, 6 W/m^2, and 8.5 W/m^2, respectively. Four shared socio-economic pathways (SSP) scenarios under the Tier-1 experiment such as SSP1-2.6, SSP2-4.5, SSP3-7.0, and SSP5-8.5 were considered for evaluating the future changes in wind speed. These emission scenarios correspond to the low-end future category indicating the end-century temperature rise to be less than 2° to a high-end future with a temperature rise of 5° (Gidden et al. 2019). More details pertaining to different SSP scenarios are presented in Table 9.6.

Skill level of simulated near-surface wind speed from models under the CMIP5 and CMIP6 family is evaluated against merged scatterometer data (Sreelakshmi and Bhaskaran 2020), the ERA-Interim Reanalysis product, and in situ observations from Research Moored Array for African–Asian–Australian Monsoon Analysis and

Table 9.6 Shared socio-economic pathways (SSPs)

Emission scenario	Scenario description
SSP1-2.6	Strong economic growth via sustainable pathways
SSP2-4.5	Middle of the road scenario with moderate population growth and slower convergence of income levels across countries Intermediate vulnerability and climate forcing and its median positioning of land use and aerosol emissions
SSP3-7.0	Futures with high inequality between countries (i.e. 'regional rivalry') and within countries Quantification of avoided impacts (e.g. relative to SSP2) has significant emissions from near-term climate forcing (NTCF) species such as aerosols and methane (also referred to as short-lived climate forcers, or SLCF)
SS5P-8.5	Strong economic growth via fossil fuel pathways, delayed climate action End of the century (EOC) temperature outcomes span a large range, from 1.4 °C at the lower end to 4.9 °C for SSP5-8.5

From Gidden et al. (2019)

Prediction (RAMA) buoys. Monthly wind speed data retrieved from satellites, ERS-1 (1992–1996), ERS-2 (1997–1999), QuikSCAT (1999–2007), and ASCAT (2008–2014) are merged to form a continuous time series of 23 years and used as primary reference dataset in the study. Statistical evaluation of climate models is performed using the Taylor diagram (Taylor 2001). It is an advanced method to express the skill level of models by representing the correlation coefficient, standard deviation, and root mean square error between the models and reference datasets. Further, the wind speed obtained from CMIP6 models was extracted at the in situ RAMA buoy locations and comparison carried out by estimating various statistical measures such as correlation coefficient, bias error, root mean square error, Nash–Sutcliffe efficiency, and index of agreement. Based on the statistical analyses performed, the best-performing models were selected and employed to construct a Multi-Model Mean (MMM) to understand the future changes. Futuristic changes in the wind speeds from CMIP5 ensemble for the near future (2026–2050), mid-century (2051–2075), and end-century (2076–2100) are calculated as the respective change from historical period (1980–2014).

9.3 Results and Discussion

Extreme waves that coincide with high spring tide conditions longer fetch and strong winds are catastrophic in particular for coastal regions that are highly populated and industrialized. Higher waves that are superimposed on extreme water levels can instantaneously lead to flash floods near coast and eventually cause run-up of large volumes of water in short time period. Mean overtopping discharges that exceed 0.03 l/s per m as function of wave height, steepness, and water depth can pose significant hazard to public safety (Allsop et al. 2005; Burcharth and Hughes 2006). This chapter discusses on important aspects related to recent trends in maximum wind speed and significant wave height for the IO region utilizing multi-satellite datasets, long-term trends, inter-annual and inter-seasonal variability of total wind-generated waves, wind-seas and swells using ERA-5 datasets (41 years), extreme wind-waves associated with *Hudhud* cyclone, and best-performing GCMs for the IO region for futuristic prediction of extreme waves.

9.3.1 Recent Trends in Maximum Wind Speed and Significant Wave Heights for Indian Ocean

The Intergovernmental Panel on Climate Change (IPCC 2007, 2012) report clearly indicates on the effect of climate change noticed across the globe. Projected results also mention that in the future, the frequency and intensity of extreme weather events are likely to increase. Studies that investigated on the basin-scale variability of maximum winds and wave heights for the IO region were only recent (Bhaskaran et al. 2014; Gupta et al. 2015). The Hovmoller diagram is commonly used in the field of meteorology and oceanographic applications to handle data that vary on space–time scales. The decadal variation of daily averaged maximum wind speed as a Hovmoller diagram for the zonal belt between 40° and 60° S along the meridian (30°–120° E) is shown below in Fig. 9.7.

This figure clearly demonstrates that wind speed in the Southern Ocean (SO) belt of IO sector has increased in the past years. The decadal variability of maximum wind speed from 2002 is higher than the variability seen during the period from 1992 until 2001. The conspicuous feature noticed is regarding wind speed maxima that extend all along the meridian during the past one decade (2002 until present). This wind speed maxima (core of maximum winds) show an increased activity during the current decade along the meridian. It clearly signifies that the extreme winds have increased with time for the SO belt. It has practical implications concerning the NIO basin. It is worthwhile to mention here on the recent study by Nayak et al (2013) that highlights on swells generated from SO sector crossing the hemisphere and

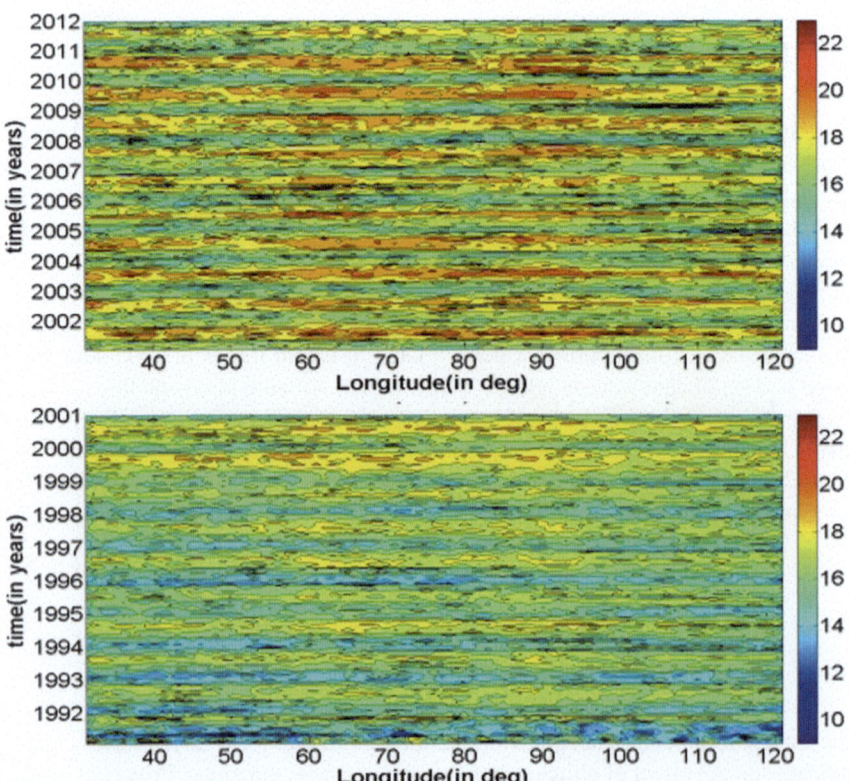

Fig. 9.7 Decadal variation of zonally averaged maximum wind speed between the geographic coordinates 40° S–60° S in the Southern belt of the Indian Ocean region (from Gupta et al. 2015)

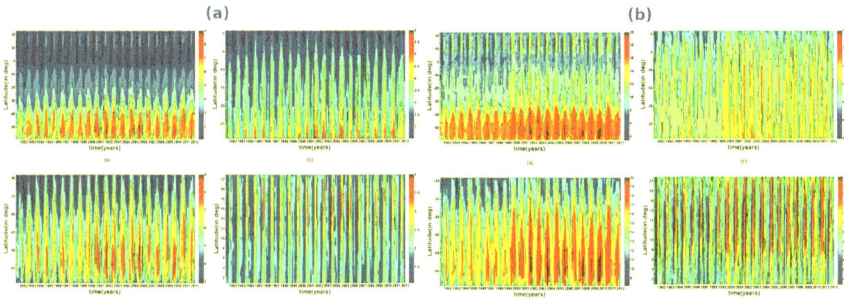

Fig. 9.8 Hovmoller diagram of basin-scale meridional averaged (30° E–120° E) **a** maximum significant wave height and **b** maximum wind speed for whole Indian Ocean basin, between 60° S and 30° S, between 30° S and Equator, between Equator and 23° N (from Gupta et al. 2015)

reaching NIO basin (in a period of ~4 days). These swells modify as well modulate the local wind-waves during their propagation to East coast of India in the BoB region (Nayak et al. 2013). Hence, this analysis signifies expectation of higher swell activity observed from the recent increasing trends of maximum wind speed in the SO basin. The increased swell activity and its long-distance propagation confine not only to the IO basin, but influence other ocean basins as well. The consequences that result from an increased wind magnitude in the current decade particularly for the SO basin are vital in terms of wave climatology for tropical NIO basin. It means an increased wave activity in NIO has direct implications on the nearshore physical oceanographic processes such as coastal erosion and sediment transport mechanisms.

Zonal distribution of daily averaged meridional (30°–120° E) maximum significant wave height is shown in Fig. 9.8a. Maximum significant wave height shows a steady rise in wave activity for the past two decades. The core of maximum significant wave height as well as the contour slopes in the latitudinal band between 60° S and 30° S signifies higher wave activity spread over larger regions in the SO belt during the recent years. Findings for the current decade are analogous with increased wind speed activity over the SO region (Gupta et al. 2015).

Trends in the maximum wind speed distribution between 10° S and 20° S have also increased. In general, the magnitude in wind speed for regions north of 10° N in the NIO basin has increased by about 2.5 m s^{-1} in the past two decades. There is a paradigm shift in the distribution of wind speed (Fig. 9.8b (d)) for the NIO basin. Increased wind magnitudes are evident for the equatorial regions (Fig. 9.8b (c, d)) covering the Inter-Tropical Convergence Zone (ITCZ) during the current decade (from 2001 until present). Highest impact of climate change is apparent in the SO region (band extending from 40° S to 60° S). Hemer et al. (2013) used five independent wave models to show that wave heights have increased in the seas off Indonesia, Antarctica, and east coast of Australia. In context to the SO, one can expect a shift in the Southern Annual Mode (SAM) that strengthens the westerly wind patterns in the SO sector. Therefore, increased wave activity over this region influences swell propagation in the northward direction that reaches other ocean basins. In context to the IO sector, increased wave activity due to climate change has implications on fishing industry and coastal mitigation measures. Figure 9.9 illustrates the trend distribution of meridional averaged maximum significant wave heights.

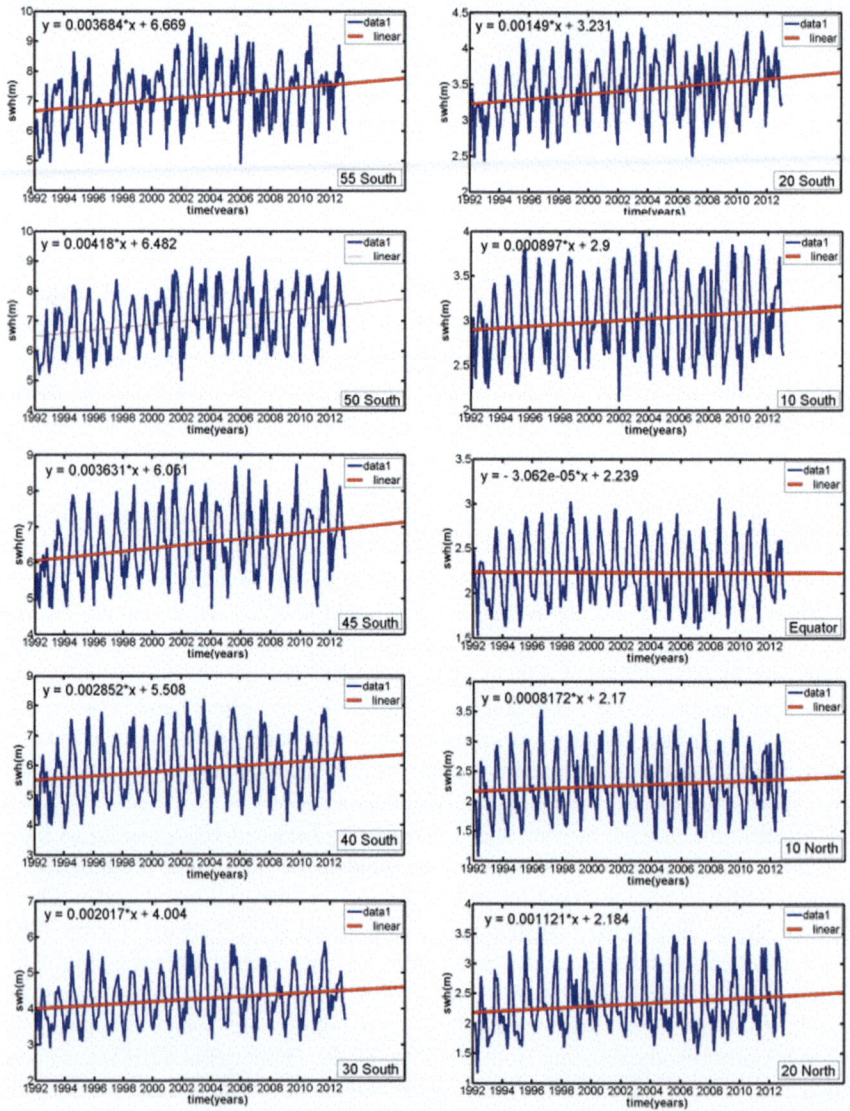

Fig. 9.9 Zonal distribution of meridional averaged maximum significant wave height (in m) between 55° S and 20° N (from Gupta et al. 2015)

The solid linear line in each panel (Fig. 9.9) represents the best-fit regression equations pertaining to maximum significant wave height for the IO basin. In the SO (55° S latitude) basin, there is an overall increased wave activity of almost $+0.93$ m in the past two decades. The trend line (Fig. 9.9) indicates that the highest maximum significant wave height was about 6.6 m during 1992 that increased to 7.6 m during 2012. On an average, there is a steady rise of about $+4.5$ cm per year in the maximum significant wave height for this zone. There is an overall increase of 1.524 m and 2.38 m s^{-1} for maximum significant wave height and maximum wind speed (Gupta et al. 2015). Along the eastern side of the IO basin (transect in the BoB corresponding to 50° S), the maximum significant wave height has increased by 1.427 m and wind speed by 3.16 m s^{-1} in the last two decades. Though swells circumscribe crossing the equator, the equatorial regions exhibit insignificant variation in maximum significant wave height, whereas the wind speed maxima showed a rise compared to the tropical south IO. For regions in the NIO basin, wind speeds have increased by about 1.8 m s^{-1} in the last two decades.

9.3.2   Trends in Extreme Waves Analysed Using ERA5 for the Indian Ocean

Analysis was carried out using ERA5 wind-wave data covering a period of 41 years (1979–2019) to determine the long-term trends, inter-annual, and inter-seasonal variability of total wind-generated waves, wind-seas, and swells in the IO region (Sreelakshmi and Bhaskaran 2020). More details are provided on the validation aspect of the reanalysis product (ERA5) with altimeter data, annual and seasonal climatology and trends in wind-seas and swells, inter-annual and inter-seasonal spatio-temporal variability, and wavelet spectrum analysis of selected potential locations that experienced significant variability in the IO basin.

9.3.2.1   Validation with Altimeter Data

Details on the validation aspects of ERA5 combined significant wave height with altimeter dataset are discussed. Study by Vinoth and Young (2011) pointed out that by using altimeter data, one can estimate 100-year return period of extreme significant wave height within 5% error of buoy data. The altimeter wave record is available from 1985, and in this study, 25 years (1992–2017) of data were used to evaluate the significant wave heights obtained from ERA5. Statistical measures such as average absolute error, bias, bias percentage, root mean square error, correlation coefficient, and standard deviation were performed with significant wave height in this study (Sreelakshmi and Bhaskaran 2020). These calculations were performed separately for all the six sub-domains shown in Fig. 9.1b. Study revealed that the correlation coefficient of altimeter waves with ERA5 is about 0.97 considering the entire IO region, and the overall agreement between ERA5 and altimeter annual averaged significant wave height is excellent. The root mean square error in significant wave height from ERA5 is about 0.29 m for the entire IO, found higher in the extra-tropical south IO region (37 cm), and comparatively less over the AS and BoB domains (21 cm). The average absolute error is ~ 30 cm for the extra-tropical south IO region (shown as regions 5 and 6 in Fig. 9.1b), whereas for the NIO and tropical south IO regions, the absolute average bias ranged between 15 and 18 cm. In general, it is observed that ERA5 underestimates the altimeter satellite observation. The correlation coefficient in all the six sub-domains is above 0.9 and that is higher (>0.94) for the NIO and tropical SIO regions. Keeping in view the quality of significant wave height data from ERA5 for the entire IO sector, further analysis has been carried out.

9.3.2.2 Annual, Seasonal Climatology, and Trends in Indian Ocean Extreme Waves

As compared to other ocean basins, the IO is quite unique due to the reversal of the monsoon wind system and that plays a major role in the wind-wave climate. In terms of variability in wind speed, it is higher over the NIO as compared to the SIO. However, the climatological ranges in wind speed and total significant wave heights are higher over the SIO region. Keeping in view the superposition of locally and remotely generated waves, it is very essential to understand the climatology and variability of wind-seas and swells separately. The long-term annual distribution of swells and winds-seas over IO region along with their trends for 41 years with ERA5 is shown in Fig. 9.10.

Annual climatological significant wave height for combined wind-seas and swells varied between 0 and 4.5 m. The total swell heights are higher than the wind-seas in most of the regions except the AS. At most of the locations in BoB and tropical SIO, the swell height varied between 1.5 and 2.5 m. In context to wind-seas, the maximum range of climatological significant wave heights is higher over the extra-tropical SIO (3 m) as compared to the other regions. Wind-seas over the central tropical SIO and regions off Somalia coast are in the range of about 1.5 m. Spatial trend (Fig. 9.10) for swells is positive (0–1.5 cm/yr.) in the IO, specifically north of 60° S. Interestingly, the east coast of Australia, the southern African coast, and central extra-tropical SIO (60° E–110° E) showed a noticeable increasing trend of swell wave activity. Along the 40° S belt, the trend in wind-seas appears to be decreasing, whereas the complete Southern Ocean westerly belt exhibited a growing wind-seas trend. Both these locations exhibited the highest wind-seas and swell activity with an increasing trend (0.5 cm/yr). The study reveals that the trend in total swells is found increasing in the Arabian Sea sector (Sreelakshmi and Bhaskaran 2020). Irrespective of the seasons, the swell activity exhibited a rise in the IO region and noticeably in the AS (0.9–1 cm/yr) during the monsoon period. Northern Arabian Sea and off Oman coast an increasing trend is noticed around 0.8 cm/yr. Also, the eastern coast of Australia and southern African coast showed an increased swell activity in all seasons at a rate of 0.6–0.9 cm/yr.

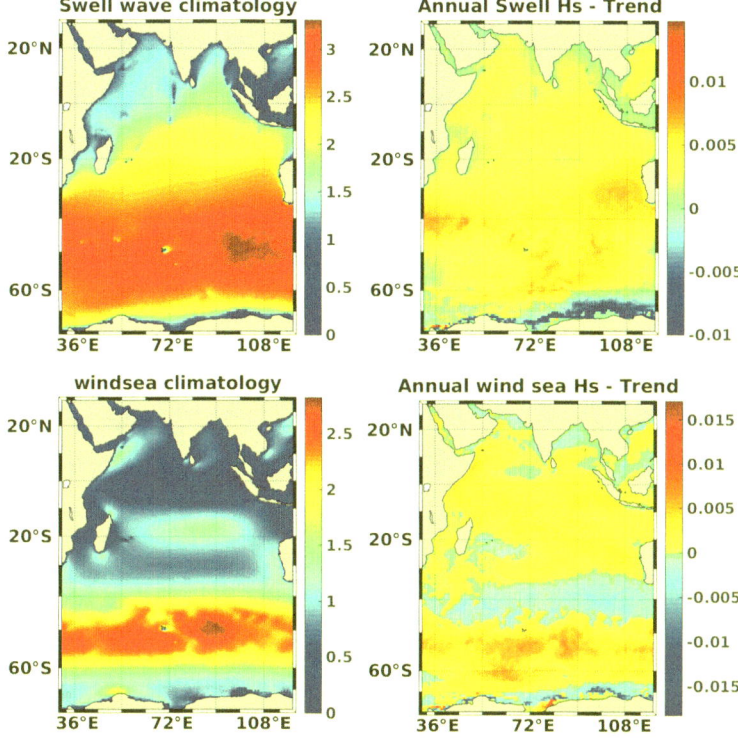

Fig. 9.10 Spatial distribution of wave climatology and trend. Top row indicates the swell climatology and bottom row for wind-seas (from Sreelakshmi and Bhaskaran 2020)

9.3.2.3 Inter-Annual Variability of Extreme Waves

The geographical locations that experienced high inter-annual variability obtained using different modes of Empirical Orthogonal Function (EOF) in terms of Eigen vectors are discussed in this chapter. Analysis using the principal component analysis (PCA) aims to investigate the relative contribution of wind-seas and swells on the total significant wave height variability. Figure 9.11 illustrates the inter-annual variability of SWH, SWH_{SW}, and SWH_{WS} in terms of Eigen vectors (EOF and PCA) using 41 years of ERA5 data.

As seen from Fig. 9.11, the first mode of variability represents 80%, 82%, and 70% of the total variability for the above-mentioned waves, respectively. The highest variability is noticed in the extra-tropical SIO region. Over this region, the highest variability amongst the first EOF modes is observed for the total significant wave height and partitioned wind-seas. At the same time, the first principal component (PC1) of total significant wave height is seen to be synchronized well and in-phase with that of swells indicating the influence and role of swells on the total significant wave height.

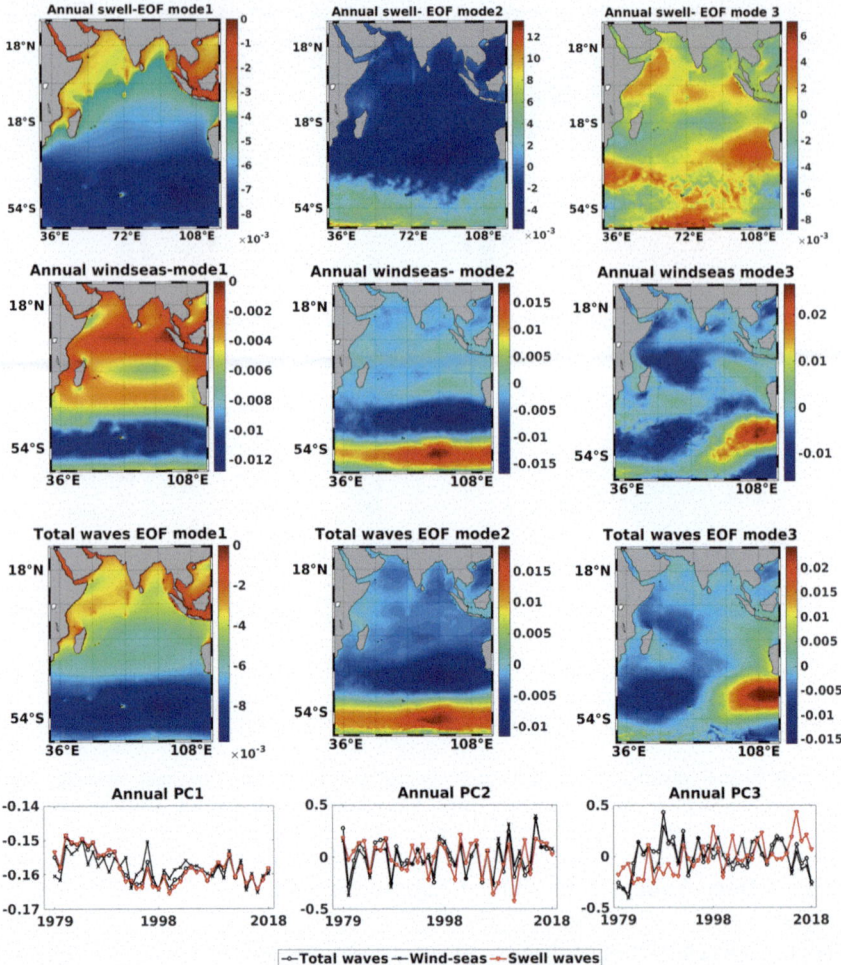

Fig. 9.11 Spatial distribution of inter-annual variability of significant wave height in terms of first three modes of Eigen vectors (left to right), for swells (first row), wind-seas (second row), and total significant wave height (third row) (from Sreelakshmi and Bhaskaran 2020)

Higher modes such as PC2 and PC3 of total wave heights are in-phase with the wind-seas. Second mode for significant wave height (1.6%) and that for wind-seas (2.7%) exhibited a zonal dipole variability over the extra-tropical SIO (40° S–63° S), indicating the influence of Southern Annular Mode (SAM). The correlation of PC2 of significant wave height as well as wind-seas with the Southern Annular Mode Index (SAMI) (http://www.nerc-bas.ac.uk/icd/gjma/sam.html) is observed to be moderate (0.61, 0.8) and significant (95% confidence level). The first mode of annual wind-seas (Fig. 9.11) illustrates a noticeable variability in the Gulf of Mannar and the south-eastern tip of the Sri Lanka. Over the AS, it is concentrated over northeast (off

Gujarat and Maharashtra coast) and south-east (off Kerala and Mangalore coast). Dominant mode of inter-annual variability of significant wave height in Arabian Sea has a mixed pattern representing the active contributions both from wind-seas and swells. In the BoB sector, the variability due to swells and total significant wave heights is comparable. More details are available in Sreelakshmi and Bhaskaran (2020).

9.3.2.4   Inter-Seasonal Variability of Extreme Waves

Spatial distribution of significant wave height in a region differs with seasons (intra-season), and the inter-seasonal variability attributes to the inter-annual variation. For seasonal analysis in the Northern and Southern hemispheres, the seasons considered are season 1 (October–March) and season 2 (April–September). Figures 9.12 and 9.13 show the first three dominant modes (EOF and PCs) corresponding to season 1 and season 2.

The primary mode of EOF (EOF1) and PC1 shown in Figs. 9.12 and 9.13 indicates that the variability of SWH (92.7% of the total variance in season 1 and 89% in season 2) and SWH$_{SW}$ (93% in season 1 and 92% in season 2) is synchronous and in-phase with time (Sreelakshmi and Bhaskaran 2020). The variation of SWH$_{SW}$ in both the seasons has a major contribution to the total wave field. In both seasons, the highest variability is identified in the AS, South China Sea, and extra-tropical SIO sectors. In the AS, EOF1 patterns for significant wave heights are a mixture of SWH$_{SW}$ from the north AS, and SWH$_{WS}$ (off Somalia jet) in both seasons. The second and third modes of variability are due to SWH$_{SW}$ waves that propagate from the north AS. The north–south shift of Southern Ocean westerly belt and the Australian summer

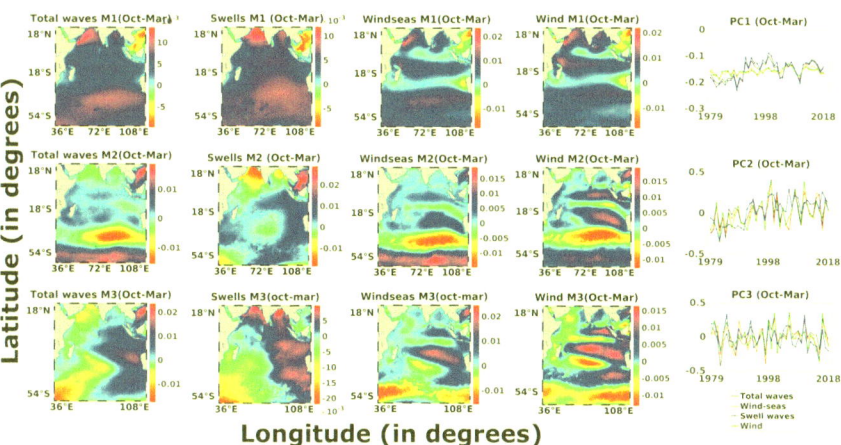

Fig. 9.12 Spatial distribution of inter-annual seasonal anomaly (October–March) of total significant wave height, swells, wind-seas, and wind in terms of EOF and PCA mode 1 (first row), mode 2 (second row), and mode 3 (third row) (from Sreelakshmi and Bhaskaran 2020)

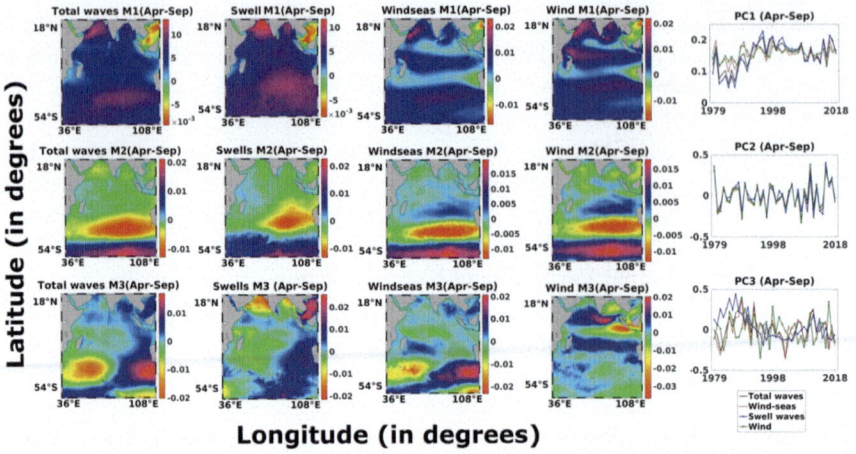

Fig. 9.13 Spatial distribution of inter-annual seasonal anomaly (April–September) of total significant wave height, swells, wind-seas, and wind in terms of EOF and PCA mode 1 (first row), mode 2 (second row), and mode 3 (third row) (from Sreelakshmi and Bhaskaran 2020)

monsoon (north-westerly wind) are responsible for the second mode variability of total significant wave height in the SIO. The estimated coefficient values indicated that PC2 of SWH, SWH$_{WS}$, and SWH$_{SW}$ for season 1 (October–March) have significant correlation with SAMI in austral summer (0.8, 0.8, 0.34) and, for season 2, the correlations are significant in austral winter (0.79, 0.8, and 0.7). These correlation coefficients are statistically significant corresponding to 95% confidence level. Significant influences of SAM in autumn, as well as winter waves, are consistent with a previous study conducted by Hemer et al. (2010).

The third and fourth modes of EOF in season 2 (Fig. 9.13) explain a bimodal (dipole) signature in the west and east of the extra-tropical SIO, resembling an elongated 'S' curve. This observation is consistent with the inter-annual variability (EOF3) of total significant wave height; however, the correlation of SWH or SWH$_{WS}$ is not significant in the seasonal analyses. In contrast, PC3 of the SWH$_{SW}$ is moderately correlated in both the seasons (0.45 and 0.42) with a 95% confidence level. The inconsistency of the third mode for SWH$_{SW}$ variability with that of SWH$_{WS}$ and SWH indicates the propagation of SWH$_{SW}$ from the near SO sectors to the IO sector. It is interesting to note that the PC2 of inter-annual variability for significant wave heights in region 6 is synchronous and in-phase with SWH$_{WS}$, whereas it was SWH$_{SW}$ for the region 5 (Fig. 9.1b). A separate PCA analysis for region 6 concerning inter-seasonal variability of significant wave height showed that season 2 is more consistent with the SAMI with a correlation coefficient of 0.63, and for season 1, the correlation is about 0.4 (Sreelakshmi and Bhaskaran 2020). Similarly, the region 5 showed a significant correlation between SWH and SAMI (0.33 and 0.6) in seasons 1 and 2, respectively. The study identified a number of locations that experienced significant variability, such as off the Somalia coast, the North AS, north and the central BoB, South China Sea, off Australia, and different locations in the extra-tropical SIO, wherein the frequency of dominant variability is examined through wavelet analysis.

9.3.2.5   Analysis of Wavelet Spectrum

This section deals with a detailed investigation on the wavelet spectrum analysis at those locations where the variability is found to be significant. Total of six sites were selected from the NIO using spatial variability gradients pertaining to the first three modes of EOF. Continuous Morlet wave transform is mostly used in the atmospheric and oceanic studies which consist of a plane wave modified by Gaussian envelope and proven with high precision. This study utilized the Torrence's code for the Morlet continuous wavelet transform.

In the mid-latitudinal belt of south Asia, strong northerly low-level jets are influenced by presence of Makran mountain range and the wind system blow predominantly in the west–north-west and northeast directions. These wind systems influence the surface gravity waves over the Arabian Sea region that is also referred as 'Makran Swells'. They are prevalent during October to May. High wave activity associated with Makran events can influence the marine operations and coastal processes in the west coast of India. Monthly time series of selected locations are subjected to wavelet spectrum analysis, and the wavelet chosen for the study is Morlet at 90% confidence level for the red-noise process. Figure 9.14 shows the six locations situated in the NIO region. Each location is described with four subplots a, b, c, and d that represent the wavelet power spectrum for the significant wave height; global wavelet power spectrum which is time averaged along the x-axis; the power-averaged time series for 2–8 (semiannual), and 8–16 (annual) monthly components. In subplot, the area above the red-dashed line corresponds to 90% confidence level, and the thick black contour in subplot 'a' highlights the time–frequency region higher than the specified limit. Figure 9.14 represents the six points selected in the NIO, and the first two locations are near the west of Somalia coast and the north AS. These are the only two regions in the IO, where the continuous significant power is perceived along with the annual and semi-annular frequency. The subplots a and b showed an increasing annual as well as semi-annular variance indicating the increasing trend of Makran wave activity in the north Arabian Sea. Markers P9, P10, and P11 are the locations situated in the BoB, precisely, North Head Bay, central Bay, and south of Sri Lanka coast. These locations exhibited a dominant and continuous annular variation. For P9 and P10, the annular variance is found to have decadal breaks. Similarly, there is a decadal break for average variance in the semi-annular variance off the Somalia coast. At the same time, the annular variance at P11 was uniform throughout the past 41 years. The South China Sea (P12) is detected with vigorous wave activity comparable to that of the extra-tropical SIO.

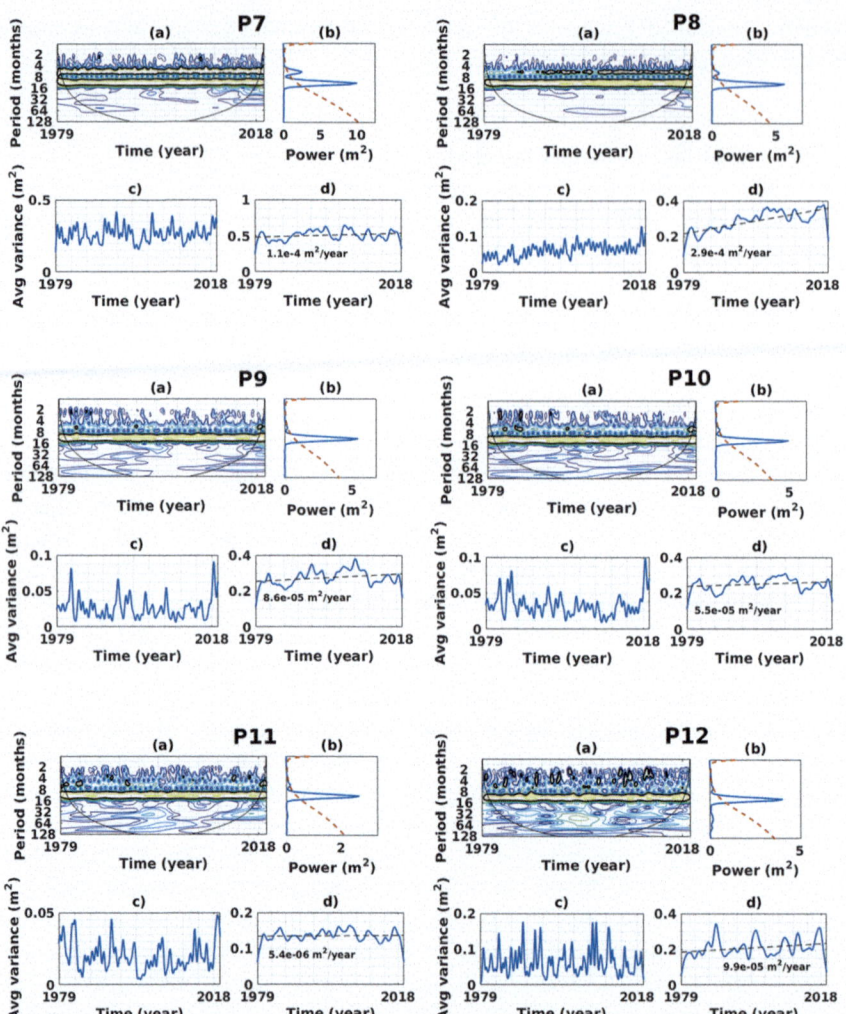

Fig. 9.14 Wavelet spectrum for locations P7-P12 in the NIO. For each location, four subplots are included, **a** wavelet power spectrum, **b** time-averaged global wavelet power spectrum, **c** 2–8 months averaged time series, and **d** 8–16 months averaged time series. Red-dashed line represents the 90% confidence level and black-dashed line is the best fitting linear regression line

9.3.3 Extreme Waves for Different Return Periods in the Arabian Gulf

To determine the long-term prediction of extreme waves in the AG region, the dataset for each location was obtained based on POT value of significant wave height of 1.0 m at all 38 locations for the hindcasted period January 1993–December 2004. Further,

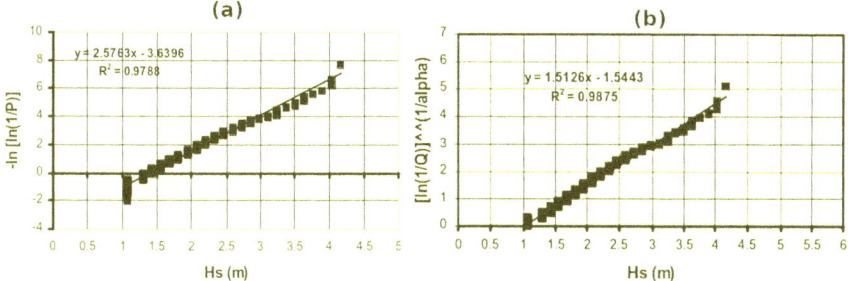

Fig. 9.15 **a** Gumbel distribution and **b** Weibull distribution plots for the location 23 in the Arabian Gulf

the wave heights obtained were arranged in the descending order, and the plotting formula shown in Eq. (9.1) was used to reduce wave height data to a set of points expressing the probability of exceedance of wave height (Q). Wave heights are then represented against the reduced variate of Gumbel and Weibull distributions. The least square technique deciphers the trend using the straight line passing through the set of points, wherein the slope and intercept values are obtained and hence the probability distribution. Further using Eqs. (9.4) and (9.6), the wave heights can be predicted for a chosen return period (12 yrs., 25 yrs., 50 yrs., 100 yrs., 200 yrs. etc.) for Gumbel and Weibull distributions, respectively. The typical Gumbel and Weibull distribution plots for the location 23 (Fig. 9.2) are shown in Fig. 9.15a, b, respectively. The equations corresponding to the best line fit and the correlation coefficients are provided. Similar plots are prepared for all the 38 locations. It is found that Weibull distribution is better than Gumbel distribution for all the 38 locations.

The location parameter, scale parameter, and the coefficient of regression obtained for all the 38 locations based on the Gumbel distribution are shown in Fig. 9.16a–c, respectively. These parameters can be used in Eq. (9.4) in order to obtain the wave

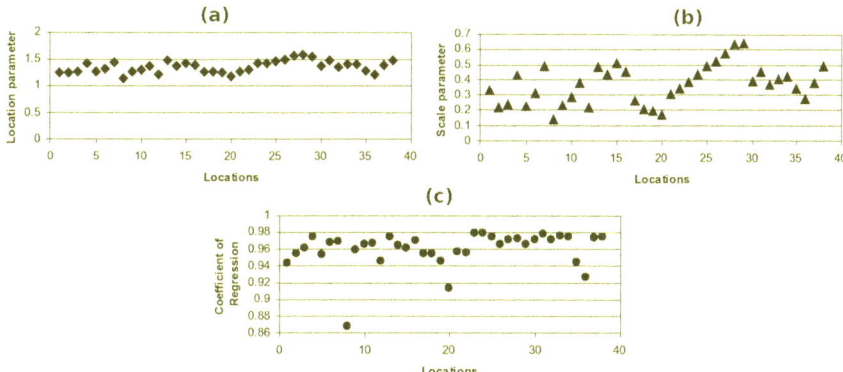

Fig. 9.16 **a** Location parameter, **b** scale parameter, and **c** coefficient of regression for the best-fit line based on Gumbel distribution for 38 locations in the Arabian Gulf

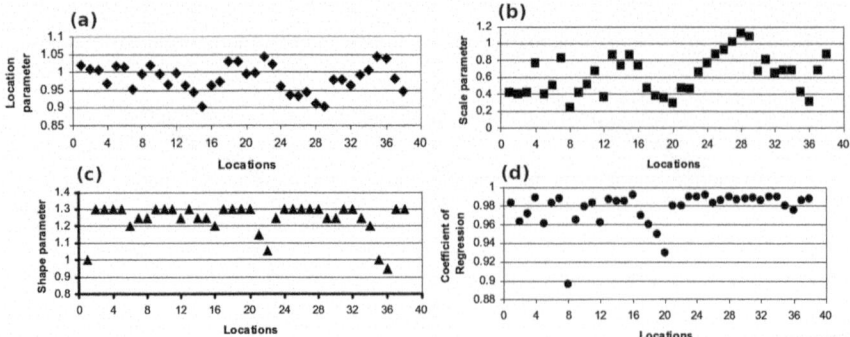

Fig. 9.17 **a** Location parameter, **b** scale parameter, **c** shape parameter, and **d** coefficient of regression for best-fit line based corresponding to Weibull distribution for 38 locations in the Arabian Gulf

heights corresponding to the required return period. The location parameter varies from 1.18 to 1.59, and the scale parameter varies from 0.14 to 0.64. The coefficient of regression for most of the locations is closer to 0.95 (except location 8 where the total number of data points with POT significant wave height of 1.0 is small compared to other locations), and hence, one can have confidence in the best line fit for the data.

Similarly, the location parameter, scale parameter, shape parameter, and the coefficient of regression obtained for all the 38 locations based on the Weibull distribution are shown in Fig. 9.17a–d. These parameters can be used in equation [6] to obtain wave heights corresponding to any required return period. It is seen that the location parameter varies from 0.904 to 1.044 and scale parameter from 0.24 to 1.126. The shape parameter for the best-fit line also varies from 0.95 to 1.3. The coefficient of regression for the best line fit for most of the locations fit is closer to 1.0 and is better than the corresponding Gumbel distribution fits for different locations. Hence, it is recommended to use Weibull distribution for extreme wave height prediction in the AG waters.

The predicted wave heights for different locations based on Gumbel distribution for return periods of 12 years, 25 years, 50 years, 100 years, and 200 years are shown in Fig. 9.18a. Similar plot based on the Weibull distribution is shown in Fig. 9.18b. In addition, the predicted extreme significant wave height for 100-year return period in the AG at different locations is shown in Fig. 9.18c.

In general, the extreme waves in the territorial waters of Kuwait, Saudi Arabia, Bahrain, Qatar, and UAE are smaller compared to the Iran's territorial waters and in the Arabian Gulf mid-way between the longitudinal boundaries of both the sides. By considering the entire AG, the predicted 100-year significant wave varies from 2.2 m (in the Saudi Arabian territorial waters) to 7.0 m (midway between UAE and Iran). Even on the longitudinal direction of the AG along its midway, the 100-year return period waves are about 5 m in the northern part and about 6.0–7.0 m in the southern part of the Gulf. This could be due to the higher water depths and longer fetch length available for the southern part of the gulf for the north-west winds.

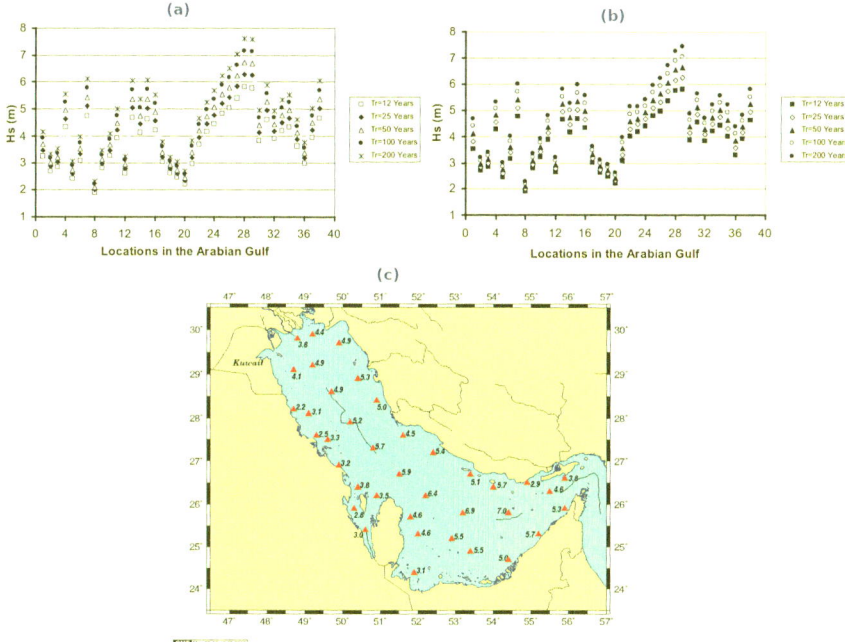

Fig. 9.18 Predicted extreme significant wave heights in the Arabian Gulf waters for different return periods based on **a** Gumbel distribution, **b** Weibull distribution, and **c** predicted extreme significant wave heights for 100-year return periods in the Arabian Gulf waters based on Weibull distribution

Design of any marine structures in these locations needs to consider this point for safety and economic designs. The territorial waters off UAE coast, where a large number of artificial coastal development projects are being undertaken, and the 100-year return period significant waves are in the order of 5.0–5.5 m. The complete picture of the predicted extreme waves for different return periods in the AG can be used for economic and safe design of the projects proposed for the near future and also for assessing the reserve strengths of different ocean structures functioning at present in these waters.

The relationship between significant wave height and mean wave period for location 23 (Fig. 9.2) is shown in Fig. 9.19a. The best-fit polynomial equation and the coefficient of regression are also shown in this figure.

For all the 38 locations in AG, similar exercise was carried out to obtain the mean wave periods for different return periods, viz. 12, 25, 50, 100, and 200 years, and the mean wave period obtained is shown in Fig. 9.19b. It is seen that the mean wave period ranged between 4.5 s and 8.1 s when all the locations and all return period ranging from 12 to 200 years are considered together. In fact, for a selected location, the difference between the mean wave period for 12-year return period and 200-year return period is only of the order of 0.5 s, whereas the location has very significant effect on change of mean wave period. For example, for location 8 (Saudi Arabian

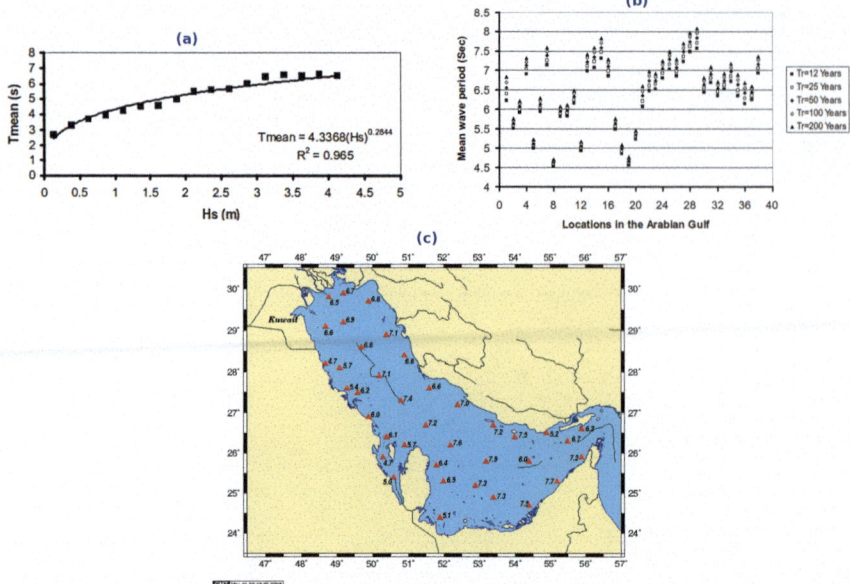

Fig. 9.19 **a** Relationship between the mean wave period and significant wave height for location 23 in the Arabian Gulf, **b** predicted mean wave period for significant wave height obtained based on Weibull for 12, 25-, 50-, 100-, and 200-year return periods, and **c** mean wave period in the Arabian Gulf for 100-year return periods

territorial waters), the mean wave period is in between 4.5 and 5 s for return periods ranging from 12 to 200 years, whereas for location 28 (offshore in between UAE and Iran), the mean wave period is in between 7.5 and 8 s for return periods in the range of 12 to 200 years. This aspect is very important in the design of ocean structures which are very sensitive for wave periods (such as wave transmission characteristics of floating breakwater, which is very sensitive for wave period). Figure 9.19c illustrates the mean wave period at different locations in the AG for 100-year return period event.

9.3.4 *Tropical Cyclone Induced Extreme Waves*

This section deals with extreme waves that are generated by tropical cyclones. Massive flooding along coastlines due to storm surges coupled with extreme wave activity is a major threat to human life, property, and damages to the ecosystem and infrastructure. Potential of destruction depends mainly on cyclone intensity, maximum radius of curvature of high winds, and landfall location. Total water level near the coast is a cumulative effect resulting from astronomical tides, storm surges, wind-waves, wave induced set-up/set-down, and sea-level rise. A review on the

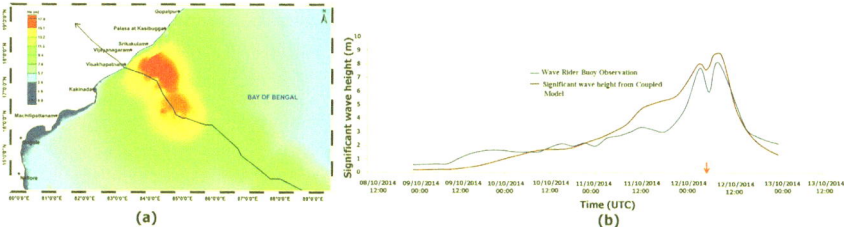

(a)   (b)

Fig. 9.20 **a** Model computed maximum significant wave height (in m) for *Hudhud* event, **b** significant wave height validation between model and wave rider buoy off Visakhapatnam (arrow indicates the landfall time) (from Murty et al. 2016)

studies carried out on tropical cyclone-induced storm surges and extreme waves is referred in Sect. 9.3. The discussion in this chapter is with reference to *Hudhud* cyclone of 2014 that made landfall in Andhra Pradesh, located in the East coast of India. Figure 9.20 illustrates the computed maximum significant wave height using the coupled ADCIRC + SWAN model.

Extreme waves are strong along the coastal belt north of Visakhapatnam (exceeding 8.0 m) facing the right side of the cyclone track (Fig. 9.20a) attributed to strong onshore winds. Wave heights on the lee side of the track experienced relatively lower wave heights (less than 4.0 m) as the predominant winds were in the offshore direction. Validation of significant wave height is shown in Fig. 9.20b between the coupled model and wave rider buoy recorded at Gangavaram, south of Visakhapatnam. It is seen that the coupled model performed very well in representing the variations of extreme waves in the nearshore regions off Visakhapatnam quite satisfactorily. The wave induced set-up at two locations Visakhapatnam and Bheemunipatnam is shown in Fig. 9.21.

There was a gradual increase in wave set-up during the approach of *Hudhud* cyclone and that diminished rapidly after the landfall (Fig. 9.21). In contrast to the

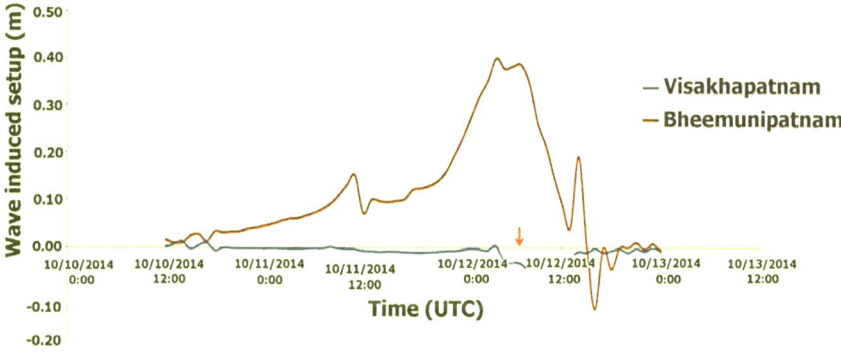

Fig. 9.21 Comparison of wave-induced set-up at Visakhapatnam and Bheemunipatnam

observed wave set-up phenomena at Bheemunipatnam, the location off Visakhapatnam experienced set-down. The wave set-up remained almost invariant at Visakhapatnam followed by set-down during the landfall event. On the other hand, at Bheemunipatnam the wave set-up increased during the approach of *Hudhud*. There is a steady
increase seen until the landfall time, and thereafter the wave set-down attributes
from predominant offshore winds at this location. Wave set-up characteristics have a
dependence on the coastal geomorphic features. For the Bheemunipatnam location,
the bottom features have Karstic pinnacle features both along the mid and shelf edge
regions retarding wave propagation towards the nearshore areas, causing piling up
of water during extreme weather events. It is unlike the bottom comprising of dome-
shaped features observed off Visakhapatnam with reef structures and higher gradient
in beach slopes (Murty et al. 2016).

9.3.5 Projections and Validation of Wind Speed from GCMs in the Indian Ocean Region

Over the past half century, the IO basin has been warming throughout, and there
are a few studies that examined the causative factors and effects of basin-scale IO
warming (Klein et al. 1999; Dong et al. 2014; Swapna et al. 2014). A study by
Roxy et al. (2014) indicates that the western IO has been warming over a century.
Sea surface temperature rise can lead to manifold effects such as variations in sea
surface pressure distribution leads to changes in wind speed, sea-level rise, and other
related consequences. Occurrence of tropical cyclones in the NIO during pre- and
post-monsoon seasons also leads to the rise in wind speed (Shanas and Kumar 2015).
It is considered that significant wave height is approximately proportional to wind
speed (Young et al. 2011). Therefore, understanding the wind speed variability has
profound importance in the development of futuristic wind-wave climate projections.
There are a few studies that investigated the influence of climate change on wind
and wave characteristics (Semedo et al. 2011; Kumar et al. 2016; Wang et al. 2014;
Young and Ribal 2019). Mean wind speed over the global oceans has increased by
0.25–0.5% per year (Young et al. 2011). Hemer et al. (2010) proposed a futuristic
projection of increased wave height owing to increased wind speeds and mid-latitude
storms foreseen by statistical global wave climate projections. Wind speed simulated
by CMIP5 models validated against reanalysis products, satellite data, and in situ
observations have resulted in identifying the best-performing models that can be used
as a tool to project future with better accuracy (Krishnan and Bhaskaran 2019a, b).
Very recently, the CMIP6 models were released, and in this study, a comprehensive
assessment based on inter-comparison experiments between CMIP5 and CMIP6
models was evaluated that would be beneficial in understanding extreme waves.

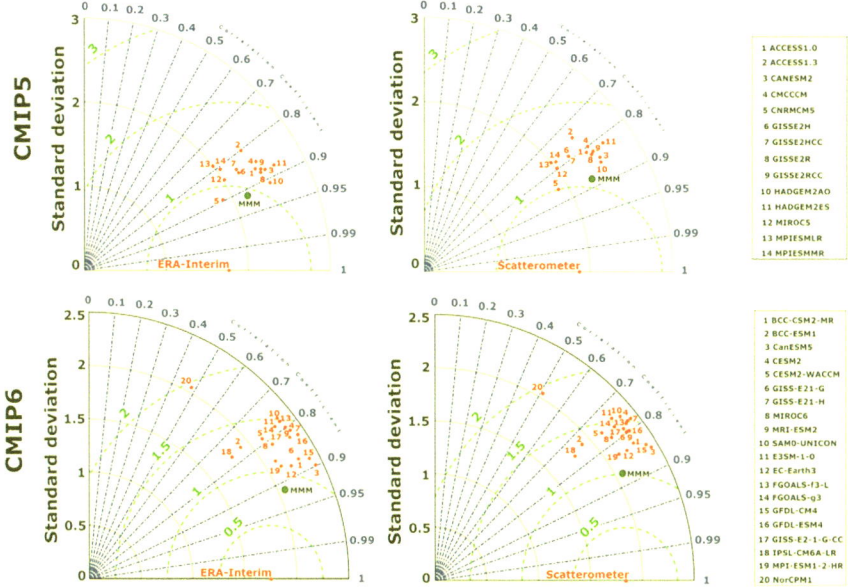

Fig. 9.22 Taylor diagram of monthly mean wind speed in the Bay of Bengal based on CMIP5 models (top panels) and CMIP6 models (bottom panels) along with the Multi-Model Mean (MMM) and its comparison against ERA-Interim Reanalysis and Scatterometer for the historical period (from Krishnan and Bhaskaran 2020)

9.3.5.1 Comparison of CMIP5 and CMIP6 Models with Reference Data Sets

Taylor diagram shown in Fig. 9.22 represents the inter-comparison of wind speed obtained from various models in CMIP5 and CMIP6 family validated against ERA-Interim and Scatterometer datasets covering the historical time period (1992–2005). This inter-comparison exercise considered output of 14 GCMs from CMIP5 (Krishnan and Bhaskaran 2019a, b) and 20 GCMs from CMIP6 family.

In addition, the best-performing models from CMIP5 showed a good correlation between 0.7–0.9 and RMSE between 0.8 and 1.7 m/s in wind speed as compared against ERA-Interim and Scatterometer datasets. Also, a greater number of CMIP6 models fit into correlation range 0.8–0.9 and RMSE of 1.0–1.5 m/s compared to CMIP5 models. The best-performing models from Taylor's skill are chosen for spatial analysis. They are ACCESS-1.0, CanESM2, CNRM-CM5, GISS-E2R, GISS-E2RCC, and HadGEM2-ES from CMIP5 family and BCC-CSM2-MR, CanESM5, EC-Earth3, IPSL-CM6A-LR, and MPI-ESM-1-2-HR from the CMIP6 family. The Multi-Model Means (MMM) constructed using CMIP5 and CMIP6 in general correlated well with both ERA-Interim and Scatterometer datasets (Fig. 9.23).

Correlation Coefficient   **Bias**

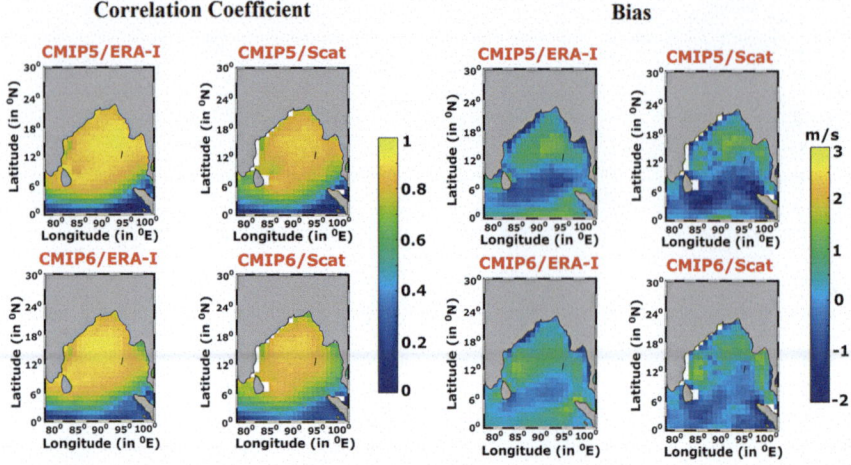

Fig. 9.23 Spatial distribution of correlation coefficient and bias error of Multi-Model Mean (MMM) wind speed compared against ERA-Interim and Scatterometer data (from Krishnan and Bhaskaran 2020)

Over the equatorial region and in the central BoB, the correlation and bias estimates of CMIP6 MMM are relatively better than CMIP5 MMM. The overall inference in the spatial bias distribution for CMIP5 MMM and CMIP6 MMM is analogous with the spatial correlation distribution. In order to verify the effectiveness of individual models and MMM, the study was extended to evaluate the performance of scores by comparing them with ERA-Interim and Scatterometer datasets. More details on the statistical measures are available in Krishnan and Bhaskaran (2020). Figure 9.24 shows the time series comparison of in situ wind speed data obtained from individual CMIP6 models, Multi-Model Mean (MMM), and the error bars show the difference of MMM from RAMA buoy observations.

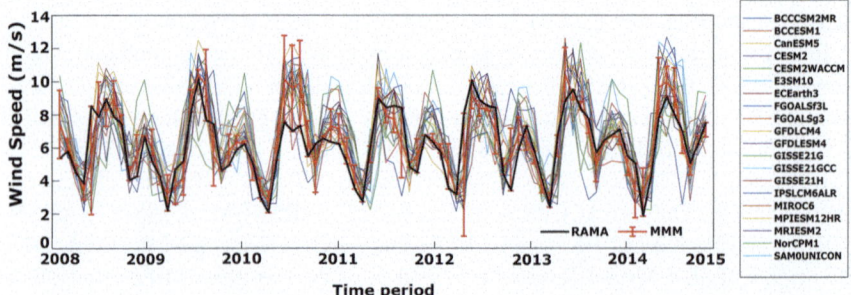

Fig. 9.24 Time series comparison of in situ wind speed data obtained from individual CMIP6 models (thin lines), Multi-Model Mean MMM (thick solid red line), and the error bars show the difference of MMM from RAMA buoy observations (from Krishnan and Bhaskaran 2020)

The comparison exercise showed that few of the individual models are either overestimating or underestimating the observations. However, the MMM established a moderate agreement with the RAMA buoy time series distribution except for a few months. Analysis reveals that maximum deviation in the GCMs is observed during the summer months of 2010, and thereafter the MMM closely follows the observations reasonably well. For the winter months with low wind speeds, the models are found to slightly overestimate the RAMA buoy observations during 2009, 2010, and 2014. Models CanESM5, GFDL-CM3, MPI-ESM12-HR, EC-EARTH3, and BCC-CSM2-MR belong to category having the highest correlation amongst the 20 models from CMIP6 family when compared against RAMA buoy data. Models such as MPI-ESM12-HR, GISS-E21G-CC, GISS-E21G, and EC-Earth have a lower bias within the range of 0.1–0.2 m/s. Study also signifies that the models with higher correlation show low *RMSE* values of 1.3–1.7 m/s with RAMA buoy observations. Evaluating other statistical measures for the goodness-of-fit such as Nash–Sutcliffe efficiency (*NSE*) and *IA*, the study clearly reveals on the superior performance of MMM. An index of agreement (*IA*) value close to 1 shows the best match for the datasets. The models, with exception for NOR-CPM1, FGOALS-F3, BCC-ESM1, IPSL-CM6A-LR, CESM2-WACCM, and SAM0-UNICON, shows an *IA* value ranging between 0.8 and 1.0. While the maximum agreement obtained for models MPI-ESM12-HR, EC-EARTH3, and CanESM5 is close to 0.9, the MMM shows an *IA* value of 0.91 at the higher side. Based on the statistical estimates, the study recommends that EC-EARTH3 and MPI-ESM12-HR are the two models in CMIP6 family that performed the best amongst the GCMs. From these several proven methods, CMIP6 models and MMM constructed from the best-performing models are suitable for investigating the projected wind speed patterns over the NIO region. Superior performance of CMIP6 models is owing to the improvement in resolution, model physics, accounting for responses to aerosols and short-term forcing agents in the simulations.

9.4 Summary and Conclusions

Recent trends in extreme wave heights and maximum wind speeds over the IO region based on multi-satellite platform observational datasets were examined. Inferences were based on a comprehensive analysis of all available altimeter data from eight satellite missions such as ERS-1/2, TOPEX-POSEIDON, GEOSAT Follow-On (GFO), JASON-1/2, ENVISAT, and CRYOSAT products. Preliminary analysis of the altimeter data indicates that extreme wave activity showed a pronounced increase in the extra-tropical southern IO. High waves generated over this region can circumscribe the southern hemisphere and affect the locally generated wind-seas over the north IO. An increasing trend is also observed in the wind magnitudes especially over the Southern Ocean basin, and this rise is substantial in the present decade. Analysis covered two transects, extending from the Southern Ocean to north IO basin, one along the Arabian Sea and other in the BoB sector. An increasing trend has been observed along these transects during the past two decades. The highest variability in extreme waves was seen in the southern IO (between 45°S and 55° S) indicating a steady rise of about 7.2 cm yr^{-1} and 0.12 m s^{-1} yr^{-1} for maximum significant wave height and maximum wind speed, respectively. Observed variability is quite marginal over the equatorial regions, whereas the maximum significant wave height and maximum wind speeds for the north IO basin showed an increasing trend of + 1.5 cm yr^{-1} and 0.09 m s^{-1} yr^{-1}, respectively. The study has practical relevance for scientists and engineers, and the increasing trend observed in extreme wave activity has potential for design aspects of coastal and offshore structures as well flash floods due to high wave activity in coastal regions.

A comprehensive study was also carried out to investigate the climatological aspects of extreme wave activity pertaining to combined total waves, wind-seas, and swells separately utilizing ERA5 datasets for the IO region. Study used 40 years of ERA5 data to examine the trends and spatio-temporal variations in extreme wave activity over the IO region. Keeping in view the observed variability, the entire IO region was subdivided into six sectors that covered the AS, BoB, South China Sea, Tropical south IO, western, and eastern sectors of extra-tropical south IO, and the Southern Ocean regions. Detailed validation exercise was carried out using multi-platform altimeter derived wave heights for about 25 years (1992–2016). The correlation obtained was about 0.97 for the entire IO, and it ranged between 0.94 and 0.97 at different geographical locations in the study region. Highest wave activity with increased trends in both wind-seas and swells (ranging between 0.5 and 1 cm/yr) is evident over the extra-tropical south IO. The north–south movement of westerly wind system resulted in pole-ward increasing and equator-ward decreasing trend over this region in context to wind-seas. Over the north Arabian Sea and off the Oman coast, the swells showed an increasing trend (0.8 cm/yr) during the monsoon season. However, the activity of wind-seas in the AS and head BoB is seen decreasing. In addition, the inter-annual and inter-seasonal variability of extreme waves were evaluated using Empirical Orthogonal Functions, and the potential regions of high variability were identified. Study clearly brings out the fact that swells play a very important role and a major contributor to the dominant mode of total wave variability in all regions of the IO basin. It is seen that over the north IO, the variability in wind-seas has a direct linkage with ENSO years, whereas the total combined wave system is least affected. Second dominant mode of inter-annual variability in the extra-tropical regions is connected with the Southern Annular Mode (SAM) seen more prominent in the eastern extra-tropical south IO. Seasonal variability of extreme swells is also consistent with SAM index. Based on the seasonal and annual variability analyses, there are 12 identified spots in the IO that experienced the highest variability. For these 12 locations, the wave spectrum analysis was carried out for the combined significant wave heights. It is seen that the Arabian Sea exhibited dominant variability both in the annular and semi-annular frequency scales, and the south IO experiences enhanced extreme wave activity on intra-annular scales.

Gumbel and Weibull extreme value distributions are used in order to obtain significant wave heights at 38 different marine locations in the Arabian Gulf waters. Data obtained based on WAM model for 12 years were used. Peak over threshold of 1.0 m is used for synthesizing the raw data. Based on joint probability analysis, the mean wave period for the significant wave heights of intended return periods was obtained based on polynomial fit between the mean period and significant wave period. Study signifies that the Weibull distribution is very suitable for extreme wave height prediction in the Arabian Gulf waters. Though the Arabian Gulf covers an area of about 2, 26,000 square km, the extreme significant wave height varies between 2.2 and 7.0 m for 100-year return period, amongst these 38 locations. This spatial variation of the wave height must be considered for design of marine structures at these locations. In general, the value of extreme significant wave heights is smaller in the territorial waters of Kuwait, Saudi Arabia, Bahrain, and Qatar compared to the territorial waters of UAE and Iran. The maximum value for the 100-year return period significant wave height is 7.0 m, and it is expected to occur in deeper waters in between UAE and Iran. It is found that the mean wave period in the Arabian Gulf waters ranged between 4.5 and 8.1 s for the significant wave heights corresponding to 100-year return period. The mean wave periods are more sensitive for spatial variations rather than the return periods. Variation of return period from 12 to 200 years has increased the mean wave period by 0.5 s only, whereas variation of space from coastal waters of Saudi Arabia to the offshore waters of UAE has changed the 100-year return period mean wave period from 4.7 s to 8.0 s. A large number of coastal and offshore projects are in progress, and many new projects are planned for the near future in the Arabian Gulf waters. The results from this study will be highly useful for optimal design of different types of ocean structures in the Arabian Gulf.

In order to evaluate the impact of tropical cyclone-induced extreme waves, a study was carried out using the coupled wave-hydrodynamic (ADCIRC + SWAN) model for an extremely severe cyclonic storm *Hudhud* that made landfall in October 2014 near Visakhapatnam, Andhra Pradesh, East coast of India. The IMD best-track data were used for this study. The coupled model simulation for extreme waves and wave-induced set-up were investigated. Modelling system utilized a flexible unstructured finite element grid with a resolution of about 30 km in the open ocean boundary refining to less than 1 km along the nearshore regions. Overall performance of the coupled model with modified winds showed a good match with available observations. Extreme waves simulated for this event were validated with the wave ride buoy observation located near Visakhapatnam. In addition, a comprehensive analysis was carried out to understand the variations in wave-induced set-up for the coastal stretch between Visakhapatnam and Bheemunipatnam separated by a distance of about 24 km. Study revealed that the wave set-up at both these locations was different and that is directly linked with the coastal geomorphic features and beach slope geometry. Importance of wave set-up and its role in determining the total water level elevations along the coastal region of Andhra Pradesh has been examined.

This chapter also discussed on the impact of climate change on extreme wind speed characteristics in the IO region utilizing general circulation models (GCMs) developed under the CMIP project. Historical wind speed datasets in CMIP5 and CMIP6 models were validated against scatterometer, ERA-Interim Reanalysis, and in situ RAMA buoy observations. Study revealed the improved capability of CMIP6 over CMIP5 models in simulating wind speed based on inter-comparison exercises and statistical estimates. Models such as BCC-CSM2-MR, CanESM5, EC-EARTH3, IPSL-CM6A-LR, and MPIESM12-HR are identified as the best-performing models under CMIP6 family in simulating wind speed in the BoB. Also, the Multi-Model Means (MMM) constructed from best-performing CMIP6 models were used to investigate the projected changes in wind speed under each emission scenarios. Over the BoB region, futuristic changes in CMIP6 wind speed under different emission scenarios such as near-future (2026–2050), mid-century (2051–2075), and end-century (2076–2100) relative to the historical period of 1980–2014 were examined. Projected MMM wind speed for the near-future, mid-century, and end-century showed a rising trend in northern BoB regions under all forcing scenarios along with a decline in the southern Bay regions under SSP5-8.5. Study signifies that southwest monsoon winds in the head Bay region are projected to change up to 15% in the near-century intensifying to 25% during the end-century from historical period (1980–2014) under SSP3-7.0 and SSP5-8.5 scenarios, respectively. Maximum rise of 25% (0.5–1 m/s) in wind speed is observed for the SSP3-7.0 scenario over BoB in the near future. However, the simulated outputs of future projections are highly dependent on various factors such as underlying physics, parameterization of convection schemes, inclusion of carbon cyclone dynamics, and model resolution representing the mesoscale and sub-mesoscale topography.

From the above discussions and as far as the climate change adaptation and effective disaster mitigation strategies are considered, it is crucial to understand the role and influence of extreme waves in net water level elevation in the nearshore regions. However, there are number of scientific questions that can emerge in this topic such as: What is the possible track and intensity characteristics that can lead to widespread inundation along coastal and island locations? What combinations of extreme waves, astronomical tides, and sea-level anomaly can lead to severe impacts? What are the possible scenarios on the severity and frequency of extreme waves that can change over time at a given location? It is warranted to have high density of in situ observations along coastal regions for monitoring extreme waves, sea levels, bathymetry, and topographic details for better mapping of coastal inundation. Also, more frequent and severe inundation events from extreme waves can have longer time implications along coastal regions such as impact on coastal groundwater resources, shifts in erosion and sedimentation rates and patterns, coastal ecosystem shifts, and synergistic effects between the processes.

The study has significant applications where the impact is greater for developing countries exposed to frequent events. In the recent past, there has been greater attention paid to adaptation strategies and mitigation efforts to impact of extreme waves in the IO region by undertaking coastal vulnerability assessment studies. Understanding and reducing coastal vulnerability is a multi-disciplinary task. Understanding the dynamics of coastal geomorphology has contributed significantly in understanding the assessment of flooding associated with extreme waves. Historically, excessive human interventions in coastal zones and upstream of catchment areas have led to non-sustainable exploitation of resources. In a global warming scenario, accelerated rise of sea level and higher occurrence frequency of severe storm events pose additional stress to coastal regions. Climate change-related impacts on extreme waves can have significant socio-economic implications affecting the tourism sector, fisheries, agriculture, and human life. The feedback between human intervention and environmental drivers on climate change can have significant environmental impacts in the coastal zone, and many of these feedbacks are positive leading to increased vulnerability. For example, there has been significant degradation of natural buffers that protects the coast from storm events and extreme waves such as salt marshes, mangroves, dunes, and coral reefs. Increased population pressure has also led to planned and unplanned development in coastal regions. Adaptation strategies to impact from climate change require taking advantage of new opportunities or to cope with the consequences.

References

Addison, P.S. 2018. Introduction to redundancy rules: The continuous wavelet transform comes of age. *Philosophical Transactions of the Royal Society a: Mathematical, Physical and Engineering Sciences* 376: 2126. https://doi.org/10.1098/rsta.2017.0258.

Allsop, W., B. Phillip, and B. Tom. 2005. Wave overtopping at vertical and steep seawalls. *Proceedings of The Institution of Civil Engineers-Maritime Engineering* 158: 103–114. https://doi.org/10.1680/maen.2005.158.3.103.

Al-Salem, K., K. Rakha, W. Sulisz, and W. Al-Nassar. 2005. Verification of a WAM Model for the Arabian Gulf. In *Arabian Coast Conference*, Dubai, November 27–29, 2005.

Al-Yamani, F.Y., J. Bishop, E. Ramadhan, M. Al-Husaini, and A. Al-Ghadban. 2004. *Oceanographic Atlas of Kuwait's Waters*, 203. Kuwait: Kuwait Institute for Scientific Research.

Aucan, J., R. Hoeke, and M.A. Merrifield. 2012. Wave-driven sea level anomalies at the Midway tide gauge as an index of North Pacific storminess over the past 60 years. *Geophysical Research Letters* 39(17). https://doi.org/10.1029/2012GL052993.

Baba, M. 2005. Occurrence of "Swell Waves" along the southwest coast of India from southern Indian Ocean storm. *Journal of the Geological Society of India* 66: 248–249.

Baba, M., J. Dattatri, and S. Abraham. 1989. Ocean wave spectra off Cochin, west coast of India. *Indian Journal of Marine Sciences* 18: 106–112.

Baba, M., N.P. Kurian, K.V. Thomas, M. Prasannakumar, T.S.S. Hameed, and C.M. Harish. 1983. *Study of the Waves and their Refraction in Relation to Beach Erosion along the Kerala Coast*, 28. Centre for Earth Science Studies, Technical Report, No. 29.

Baba, M. 1985. New trends in Ocean Wave Research in India. *Mahasagar—Bulletin of the National Institute of Oceanography* 18(2): 231–248.

Bhaskaran, P.K. 2019. Challenges and future directions in ocean wave modeling—a review. *Journal of Extreme Events* 6 (2): 1950004. https://doi.org/10.1142/S2345737619500040.

Bhaskaran, P.K., R. Kalra, S.K. Dube, P.C. Sinha, A.D. Rao, R. Kumar, and A. Sarkar. 2000. Extreme wave conditions over the Bay of Bengal during a severe cyclone—simulation experiment with two spectral wave models. *Marine Geodesy* 23 (2): 91–102.

Bhaskaran, P.K., R. Kalra, S.K. Dube, P.C. Sinha, and A.D. Rao. 2004. Sea State Hindcast with ECMWF data using a Spectral wave model for typical monsoon months. *Natural Hazards* 31 (2): 537–548.

Bhaskaran, P.K., S. Nayak, B. SubbaReddy, P.L.N. Murty, and D. Sen. 2013a. Performance and validation of a coupled parallel ADCIRC-SWAN model for THANE cyclone in the Bay of Bengal. *Environmental Fluid Mechanics* 13: 601–623.

Bhaskaran, P.K., R. Gayathri, P.L.N. Murty, B. SubbaReddy, and D. Sen. 2013b. A numerical study of coastal inundation and its validation for Thane Cyclone in the Bay of Bengal. *Coastal Engineering* 83: 108–118.

Bhaskaran, P.K., N. Gupta, and M.K. Dash. 2014. Wind-wave climate projections for the Indian Ocean from Satellite observations. *Journal of Marine Science Research & Development* S11: 005. https://doi.org/10.4172/2155-9910.S11-005.

Bricheno, L.M., and J. Wolf. 2018. Future wave conditions of Europe, in response to high-end climate change scenarios. *Journal of Geophysical Research Oceans* 123 (12): 8762–8791.

Burcharth, H.F., and S.A. Hughes. 2006. Design of coastal project elements. In *Fundamentals of Design. Coastal Engineering Manual 4*, ed. by S. A. Hughes. Engineer Manual 1110-2-100. Washington, DC: U.S. Army Corps of Engineers.

Caires, S., and A. Sterl. 2005. 100-year return value estimates for ocean wind speed and significant wave height from the ERA-40 data. *Journal of Climate* 18: 1032–1048.

Cardone, V.J., W.J. Pierson, and E.G. Ward. 1976. Hindcasting the directional spectra of hurricane generated waves. *Journal of Petroleum and Technology* 28: 385–394.

Carter, D.J.T., S. Foale, and D.J. Webb. 1991. Variations in global wave climate throughout the year. *International Journal of Remote Sensing* 12: 1687–1697.

Cavaleri, L., J.H.G.M. Alves, F. Ardhuin, A. Babanin, M. Banner, K. Belibassakis, M. Benoit, M. Donelan, J. Groeneweg, T.H.C. Herbers, P. Hwang, P.A.E.M. Janssen, T. Janssen, I.V. Lavrenov, R. Magne, J. Monbaliu, M. Onorato, V. Polnikov, D. Resio, W.E. Rogers, A. Sheremet, J. McKee Smith, H.L. Tolman, G. van Vledder, J. Wolf, and I. Young. 2007. Wave modelling–The state of the art. *Progress in Oceanography* 75: 603–674.

Challenor, P.G., S. Foale, and D.J. Webb. 1990. Seasonal changes in the global wave climate measured by the GEOSAT altimeter. *International Journal of Remote Sensing* 11: 2205–2213.

Chandramohan, P., V. Sanil Kumar, and B.U. Nayak. 1991. Wave statistics around the Indian coast based on ship observed data. *Indian Journal of Marine Sciences* 20: 87–92.

Chelton, D.B., K.J. Hussey, and M.E. Parke. 1981. Global satellite measurements of water vapor, wind speed and wave height. *Nature* 294: 529–532.

Chowdhury, P., M.R. Behera, and D.E. Reeve. 2019. Wave climate projections along the Indian coast. *International Journal of Climatology*. https://doi.org/10.1002/joc.6096.

Church, J.A., N.J. White, and J.R. Hunter. 2006. Sea-level rise at tropical Pacific and Indian Ocean islands. *Glob. Planetary Change* 53: 155–168.

Church, J., S. Wilson, P. Woodworth, and T. Aarup. 2007. Understanding sea level rise and variability. *EOS Transactions of American Geophysical Union* 88: 43.

Coles, S. 2001. *An Introduction to Statistical Modeling of Extreme Values*, 208. Springer.

Dhana Lakshmi, D., P.L.N. Murty, P.K. Bhaskaran, B. Sahoo, T. Srinivasa Kumar, S.S.C. Shenoi, and A.S. Srikanth. 2017. Performance of WRF-ARW winds on computed storm surge using hydrodynamic model for Phailin and Hudhud cyclones. *Ocean Engineering* 131: 135–148.

Dong, L., T. Zhou, and B. Wu. 2014. Indian Ocean warming during 1958–2004 simulated by a climate system model and its mechanism. *Climate Dynamics* 42: 203–217. https://doi.org/10.1007/s00382-013-1722-z.

El-Gindy, A., and M. Hegazi.1996. *Atlas on Hydrographic Conditions in the Arabian Gulf and the Upper Layer of the Gulf of Oman*, 170. University of Qatar.

Elshorbagy, W., M.H. Azam, and K. Taguchi. 2006. Hydrodynamic characterization and modeling of the Arabian Gulf. *Journal of Waterway, Port, Coastal and Ocean Engineering, ASCE* 132 (1): 47–56.

Emery, K.O. 1956. Sediments and water of the Persian Gulf. *Bulletin of the American Association Petroleum Geologists* 40 (10): 2354–2383.

Ferreira, J.A., and C. Guedes Soares. 1998. An application of the peaks over threshold method to predict extremes of significant wave height. *Journal of Offshore Mechanics and Arctic Engineering* 120: 165–176.

Gayathri, R., P.L.N. Murty, P.K. Bhaskaran, and T. Srinivasa Kumar. 2016. A numerical study of hypothetical storm surge and coastal inundation for AILA cyclone in the Bay of Bengal. *Environmental Fluid Mechanics* 16 (2): 429–452.

Gayathri, R., P.K. Bhaskaran, and F. Jose. 2017. Coastal inundation research: an overview of the processes. *Current Science* 112 (2): 267–278.

Gayathri, R., P.K. Bhaskaran, and P.L.N. Murty. 2019. River-tide-storm surge interaction characteristics for the Hooghly estuary, East coast of India. https://doi.org/10.1080/09715010.2019.160
3087.

Gelci, R., H. Cazalé, and J. Vassal. 1957. Prévision de la Houle. La Méthode des DensitésSpec-
troangulaires, Bulletin Inform. *Comité Central Oceanography, d'étude Cotes* 9: 416.

Gidden, M.J., K. Riahi, S.J. Smith, S. Fujimori, G. Luderer, E. Kriegler, D.P. van Vuuren, M. van
den Berg, L. Feng, D. Klein, K. Calcin, J.C. Doelman, S. Frank, O. Fircko, M. Harmsen, T.
Hasegawa, P. Havlik, J. Hilaire, R. Hoesly, J. Horing, A. Popp, E. Stehfest, and K. Takahashi.
2019. Global emissions pathways under different socioeconomic scenarios for use in CMIP6: A
dataset of harmonized emissions trajectories through the end of the century. *Geoscientific Model
Development* 12 (4): 1443–1475.

Goda, Y. 1992. Uncertainty of design parameters from viewpoint of extreme statistics. *Journal of
Offshore Mechanics and Arctic Engineering, ASME* 114: 76–82.

Goda, Y., P. Hawkes, E. Mansard, M.J. Martin, E. Mathiesen, E. Peltier, E. Thompson, and G. Van
Vledder. 1993. Intercomparison of extremal wave analysis methods using numerically simulated
data. In *Proceedings of. 2nd International Symposium On Ocean Wave Measurement and Analysis*,
963–977, ASCE, New Orleans.

Gower, J.F.R. 1976. GEOS-3 Ocean wave measurements in Northeast Pacific. *Transactions
American Geophysical Union* 57 (12): 944–944.

Gringorten, I.I. 1963. A plotting rule for extreme probability paper. *Journal of Geophysical Research*
68: 813–814.

Gumbel, E.J. 1958. *Statistics of Extremes*. New York: Columbia University Press.

Gupta, N., and P.K. Bhaskaran. 2016. Inter-dependency of wave parameters and directional analysis
of ocean wind-wave climate for the Indian Ocean. *International Journal of Climatology* 37:
3036–3043. https://doi.org/10.1002/joc.4898.

Gupta, N., P.K. Bhaskaran, and M.K. Dash. 2015. Recent trends in wind-wave climate for the Indian
Ocean. *Current Science* 108 (12): 2191–2201.

Gupta, N., P.K. Bhaskaran, and M.K. Dash. 2017. Dipole behaviour in maximum significant wave
height over the Southern Indian Ocean. *International Journal of Climatology* 37: 4925–4937.
https://doi.org/10.1002/joc.5133.

Hemer, M.A., J.A. Church, and J.R. Hunter. 2010. Variability and trends in the directional wave
climate of the Southern Hemisphere. *International Journal of Climatology* 30 (4): 475–491.
https://doi.org/10.1002/joc.1900.

Hemer, M.A., X.L. Wang, R. Weisse, and V.R. Swail. 2012. Advancing wind-waves climate
science—the COWCLIP project. *Bulletin of the American Meteorological Society* 93 (6):
791–796.

Hemer, M.A., Y. Fan, N. Mori, A. Semedo, and X.L. Wang. 2013. Projected changes in wave climate
from multi-model ensemble. *Nature Climate Change* 3: 471–476.

Hoeke, R.K., K.L. McInnes, J.C. Kruger, R.J. McNaught, J.R. Hunter, and S.G. Smithers. 2013.
Widespread inundation of Pacific Islands triggered by distant-source wind-waves. *Global and
Planetary Change* 108: 128–138.

Holthuijsen, L.H. 2007. *Waves in Oceanic and Coastal Waters*, 387. Cambridge, UK: Cambridge
University Press.

Jahns, H.O., and J.D. Wheeler. 1973. Long-term wave probabilities based on Hindcasting of severe
storms. *Journal of Petroleum and Technology* 25: 473–486.

Jismy, P., A.D. Rao, and P.K. Bhaskaran. 2017. Role of continental shelf on non-linear interaction
of storm surges, tides and wind waves: An idealized study representing the west coast of India.
Estuarine, Coastal and Shelf Science. https://doi.org/10.1016/j.ecss.2017.06.007.

Kamphuis, J.W. 2000. Long term wave analysis. In *Introduction to Coastal Engineering and
Management, Advanced Series on Ocean Engineering*, vol.16, 81–102. Singapore: World
Scientific.

Kennedy, A.B., J.J. Westerink, J.M. Smith, M.E. Hope, M. Hartman, A.A. Taflanidis, S. Tanaka,
H. Westerink, K.F. Cheung, T. Smith, M. Hamann, M. Minamide, A. Ota, and C. Dawson.

2012. Tropical cyclone inundation potential on the Hawaiian Islands of Oahu and Kauai. *Ocean Modelling* 52: 54–68.

Kimball, B.F. 1960. On the choice of plotting positions on probability paper. *Journal of the American Statistical Association* 55: 546–560.

Klein, S.A., B.J. Soden, and N.-C. Lau. 1999. Remote sea surface temperature variations during ENSO: Evidence for a tropical atmospheric bridge. *Journal of Climate* 12: 917–932.

Krishnan, A., and P.K. Bhaskaran. 2019a. Performance of CMIP5 wind speed from global climate models for the Bay of Bengal Region. *International Journal of Climatology*. https://doi.org/10.1002/joc.6404.

Krishnan, A., and P.K. Bhaskaran. 2020. Skill assessment of global climate model wind speed from CMIP5 and CMIP6 and evaluation of projections for the Bay of Bengal. *Climate Dynamics*. https://doi.org/10.1007/s00382-020-05406-z.

Krishnan, A., P.K. Bhaskaran, and P. Kumar. 2021. CMIP5 model performance of significant wave heights over the Indian Ocean using COWCLIP datasets. *Theoretical and Applied Climatology*. https://doi.org/10.1007/s00704-021-03642-9.

Krishnan, A., and P.K. Bhaskaran. 2019b. *CMIP5 Wind Speed Comparison between Satellite Altimeter and Reanalysis Products for the Bay of Bengal*. https://doi.org/10.1007/s10661-019-7729-0.

Kruke, B.I., and O.E. Olsen. 2012. Knowledge creation and reliable decision-making in complex emergencies. *Disasters* 36 (2): 212–232.

Kumar, P., S.K. Min, E. Weller, H. Lee, and X.L. Wang. 2016. Influence of climate variability on extreme ocean surface wave heights assessed from ERA-interim and ERA-20C. *Journal of Climate* 29 (11): 4031–4046. https://doi.org/10.1175/JCLI-D-15-0580.1.

Kurian, N.P., M. Baba, and T.S. Shahul Hameed. 1985. Prediction of nearshore wave heights using a wave refraction programme. *Coastal Engineering* 9 (4): 347–356.

Kurian, N.P., M. Nirupama, M. Baba, and K.V. Thomas. 2009. Coastal flooding due to synoptic scale, meso-scale and remote forcings. *Natural Hazards* 48: 259–273.

Leadbetter, M.R. 1991. On a basis for "peak over threshold" modeling. *Statistics & Probability Letter* 12: 357–362.

Lowe, J.A., P.L. Woodworth, T. Knutson, R.E. McDonald, K.L. McInnes, K. Woth, H. von Storch, J. Wolf, V. Swail, N.B. Bernier, S. Gulev, K.J. Horsburgh, A.S. Unnikrishnan, J.R. Hunter, and R. Weisse. 2010. *Past and Future Changes in Extreme sea Levels and Waves, Understanding Sea-Level Rise and Variability*, 326–375. Wiley-Blackwell.

Luijendijk, A., G. Hagenaars, R. Ranasinghe, F. Baart, G. Donchyts, and S. Aarninkhof. 2018. The state of the world's beaches. *Science and Reports* 8: 6641. https://doi.org/10.1038/S41598-018-24630-6.

Madsen, O.S., Y.-K. Poon, and H.C. Graber. 1988. Spectral wave attenuation by bottom friction: theory. In *Proceedings of 21st International Conference on Coastal Engineering*, 492–504, Malaga.

Mathiesen, M., P. Hawkes, M.J. Martin, E. Thompson, Y. Goda, E. Mansard, E. Peltier, and G. Van Vledder. 1994. Recommended practice for extreme wave analysis. *Journal of Hydraulic Research, IAHR* 32: 803–814.

Menendez, M., and P.L. Woodworth. 2010. Changes in extreme high water levels based on a quasi-global tide-gauge data set. *Journal of Geophysical Research: Oceans* 115 (C10): C10011.

Mitsuyasu, H. 2002. A historical note on the study of ocean surface waves. *Journal of Oceanography* 58: 109–120. https://doi.org/10.1023/A:1015880802272.

Mori, N., T. Yasuda, H. Mase, T. Tom, and Y. Oku. 2010. Projection of extreme wave climate change over global warming. *Hydrological Research Letters* 4: 15–19.

Morim, J., M. Hemer, N. Cartwright, D. Strauss, and F. Andutta. 2018. On the concordance of 21st century wind-wave climate projections. *Global and Planetary Change* 167: 160–171.

Morim, J., C. Trenham, M. Hemer, X.L. Wang, N. Mori, M. Casas-Prat, A. Semedo, T. Shimura, B. Timmermans, P. Camus, L. Bricheno, L. Mentaschi, M. Dobrynin, Y. Feng, and L. Erikson. 2020.

A global ensemble of ocean wave climate projections from CMIP5-driven models. *Scientific Data* 7: 105. https://doi.org/10.1038/s41597-020-0446-2.

Morim, J., M. Hemer, X.L. Wang, N. Cartwright, C. Trenham, A. Semedo, I. Young, L. Bricheno, P. Camus, M. Casas-Prat, L. Erikson, L. Mentaschi, N. Mori, T. Shimura, B. Timmermans, O. Aarnes, O. Breivik, A. Behrens, M. Dobrynin, M. Menendez, J. Staneva, M. Wehner, J. Wolf, B. Kamranzad, A. Webb, J. Stopa, and F. Andutta. 2019. Robustness and uncertainties in global multivariate wind-wave climate projections. *Nature Climate Change* 9: 711–718.

Munk, W.H., G.R. Miller, F.E. Snodgrass, and N.F. Barber. 1963. Directional recording of swell from distant storms. *Philosophical Transactions of the Royal Society of London. Series A, Mathematical and Physical Sciences* 255(1062): 505–584.

Munk, W., and M. Sargent. 1948. Adjustment of Bikini Atoll to waves. *Transactions of American Geophysical Union* 29 (6): 855–860.

Murty, T.S., and N.P. Kurian. 2006. A possible explanation for the flooding several times in 2005 on the coast of Kerala and Tamil Nadu. *Journal of the Geological Society of India* 67: 535–536.

Murty, P.L.N., K.G. Sandhya, P.K. Bhaskaran, F. Jose, R. Gayathri, T.M. Balakrishnan Nair, T. Srini-vasa Kumar, and S.S.C. Shenoi. 2014. A coupled hydrodynamic modeling system for PHAILIN cyclone in the Bay of Bengal. *Coastal Engineering* 93: 71–81.

Murty, P.L.N., P.K. Bhaskaran, R. Gayathri, B. Sahoo, T. Srinivasa Kumar, and B. SubbaReddy. 2016. Numerical study of coastal hydrodynamics using a coupled model for Hudhud cyclone in the Bay of Bengal. *Estuarine, Coastal and Shelf Science* 183: 13–27.

Murty, P.L.N., A.D. Rao, K. Siva Srinivas, E.P. Rama Rao, and P.K. Bhaskaran. 2019. Effect of wave radiation stress in storm surge-induced inundation: A case study for the East Coast of India. *Pure and Applied Geophysics*. https://doi.org/10.1007/s00024-019-02379-x.

Narayana, A.C., and R. Tatavarti. 2005. High wave activity on the Kerala coast. *Journal of the Geological Society of India* 66: 249–250.

Nayak, S., and P.K. Bhaskaran. 2014. Coastal vulnerability due to extreme waves at Kalpakkam based on historical tropical cyclones in the Bay of Bengal. *International Journal of Climatology* 34: 1460–1471.

Nayak, S., P.K. Bhaskaran, and R. Venkatesan. 2012. Nearshorewave induced setup along Kalpakkam coast during an extreme cyclone event in the Bay of Bengal. *Ocean Engineering* 55: 52–61.

Nayak, S., P.K. Bhaskaran, R. Venkatesan, and S. Dasgupta. 2013. Modulation of local wind-waves at Kalpakkam from remote forcing effects of Southern Ocean swells. *Ocean Engineering* 64: 23–35. https://doi.org/10.1016/j.oceaneng.2013.02.010.

Neelamani, S., K. Al-Salem, and K. Rakha. 2006. Extreme waves for Kuwaiti territorial waters. In *Communicated to Ocean Engineering*. U.K.: Pergaman Press.

Nicholls, R.J., P.P. Wong, V.R. Burkett, J.O. Codignotto, J.E. Hay, R.F. NcLean, S. Ragoonaden, and C.D. Woodroffe. 2007. Coastal systems and low-lying areas. In *Climate Change 2007. Impacts, Adaptation and Vulnerability. Contribution of Working Group II to the Fourth Assessment Report of the Intergovernmental Panel on Climate Change*, 315–356, ed. by Parry, M.L., O.F. Canziani, J.P. Palutikof, P.J.V.D. Linden, and C.E. Hanson. Cambridge, UK: Cambridge University Press.

Nolte, K.G. 1973. Statistical methods for determining extreme sea states. In *Proceedings of 2nd International Conference on Port and Ocean Engineering Under Arctic Conditions*, 705–742. University of Iceland.

Padhy, C.P., D. Sen, and P.K. Bhaskaran. 2008. Application of wave model for weather routing of ships in the North Indian Ocean. *Natural Hazards* 44: 373–385.

Patra, A., and P.K. Bhaskaran. 2016a. Trends in wind-wave climate over the Head Bay of Bengal Region. *International Journal of Climatology* 36 (13): 4222–4240.

Patra, A., and P.K. Bhaskaran. 2016b. Temporal variability in wind-wave climate and its validation with ESSO-NIOT wave atlas for the Head Bay of Bengal. *Climate Dynamics* 49 (4): 1271–1288.

Patra, A., P.K. Bhaskaran, and F. Jose. 2017. Time evolution of atmospheric parameters and their influence on sea level pressure over the head Bay of Bengal. *Climate Dynamics* 50 (11–12): 4583–4598.

Patra, A., P.K. Bhaskaran, and R. Maity. 2019. Spectral wave characteristics over the Head Bay of Bengal: A modeling study. *Pure and Applied Geophysics.* https://doi.org/10.1007/s00024-019-02292-3.

Petrauskas, C., and P. Aagaard. 1971. Extrapolation of historical storm data for estimating design wave height. *Society of Petroleum Engineers Journal* 11: 23–37.

Prasada Rao, C.V.K. 1988. Spectral width parameter for wind generated ocean waves. *Proceedings of the Indian Academy of Sciences (earth and Planetary Sciences)* 97: 173–181.

Prasada Rao, C.V.K., and N. Durga Prasad. 1982. Analysis of hindcasting wind waves and swell Mangalore. *Indian Journal of Marine Sciences* 11: 21–25.

Prasada Rao, C.V.K., and J. Swain. 1989. A parametric wave prediction model based on time-delay concept. *Mausam* 40: 381–388.

Purser, B.H., and E. Seibold. 1973. The principal environmental factors influencing holocene sedimentation and diagenesis in the Persian Gulf. In *Persian Gulf*, 1–9, ed. by Purser, B.H. Berlin.

Queffeulou, P., F. Ardhuin, and J.M. Lefevre. 2011. Wave height measurements from altimeters: validation status and applications. In *OSTST Meeting*, San Diego, October 19–21, 2011.

Rao, A.D., I. Jain, M.R. Murthy, T.S. Murty, and S.K. Dube. 2009. Impact of cyclonic wind field on interaction of surge–wave computations using finite-element and finite-difference models. *Natural Hazards* 49 (2): 225–239.

Ravindran, M., and P.M. Koola. 1991. Energy from sea waves—The Indian wave energy programme. *Current Science* 60 (12): 676–680.

Remya, P.G., S. Vishnu, B. Praveen Kumar, T.M. Balakrishnan Nair, and B. Rohith. 2016. Teleconnection between the North Indian Ocean high swell events and meteorological conditions over the Southern Ocean. *Journal of Geophysical Research: Oceans* 121 (10): 7476–7494.

Renske, C.W., A. Sterl, J.W. deVries, S.L. Weber, and G. Ruessink. 2012. The effect of climate change on extreme waves in front of the Dutch coast. *Ocean Dynamics.* https://doi.org/10.1007/s10236-012-0551-7.

Roxy, M.K., K. Ritika, P. Terray, and S. Masson. 2014. The curious case of Indian Ocean warming. *Journal of Climate* 27 (22): 8501–8509.

Sahoo, B., and P.K. Bhaskaran. 2015a. Assessment on historical cyclone tracks in the Bay of Bengal, east coast of India. *International Journal of Climatology* 36 (1): 95–109.

Sahoo, B., and P.K. Bhaskaran. 2015b. Synthesis of tropical cyclone tracks in a risk evaluation perspective for the east coast of India. *Aquatic Procedia* 4: 389–396.

Sahoo, B., and P.K. Bhaskaran. 2017a. A comprehensive data set for tropical cyclone storm surge induced inundation for the east coast of India. *International Journal of Climatology.* https://doi.org/10.1002/joc.5184.

Sahoo, B., and P.K. Bhaskaran. 2017b. Multi-hazard risk assessment of coastal vulnerability from tropical cyclones—a GIS based approach for the Odisha coast. *Journal of Environmental Management.* https://doi.org/10.1016/j.jenvman.2017.10.075.

Sahoo, B., and P.K. Bhaskaran. 2019a. Prediction of storm surge and coastal inundation using Artificial Neural Network—a case study for 1999 Odisha Super Cyclone. *Weather and Climate Extremes.* https://doi.org/10.1016/j.wace.2019.100196.

Sahoo, B., and P.K. Bhaskaran. 2019b. Prediction of storm surge and inundation using climatological datasets for the Indian coast using soft computing techniques. *Soft Computing.* https://doi.org/10.1007/s00500-019-03775-0(0123456789(),-volV)(0123456789,-().volV).

Samiksha, V., P. Vethamony, C. Antony, P.K. Bhaskaran, and T.M. Balakrishnan Nair. 2017. Wave-current interaction during Hudhud cyclone in the Bay of Bengal. *Natural Hazards and Earth System Sciences* 17: 2059–2074.

Sandhya, K.G., T.M. Bala Krishnan Nair, P.K. Bhaskaran, L. Sabique, N. Arun, and K. Jeykumar. 2013. Wave forecasting system for operational use and its validation at coastal Puducherry, East Coast of India. *Ocean Engineering* 80: 64–72. https://doi.org/10.1016/j.oceaneng.2014.01.009.

Sanil Kumar, V., and M.C. Deo. 2004. Design wave estimation considering directional distribution of waves. *Ocean Engineering* 31: 2343–2352.

Sanil Kumar, V., M.C. Deo, N.M. Anand, and P. Chandramohan. 1998. Estimation of wave directional spreading in shallow water. *Ocean Engineering* 26 (1): 83–98.

Sanil Kumar, V., C.S. Philip, and T.M. Balakrishnan Nair. 2010. Waves in shallow water off west coast of India during the onset of summer monsoon. *Annals of Geophysics* 28: 817–824.

Sanil Kumar, V., K.K. Dubhashi, T.M. Balakrishnan Nair, and J. Singh. 2013. Wave power potential at a few shallow water locations around Indian coast. *Current Science* 104 (9): 1219–1224.

Sarpkaya, T., M. de St., and Q. Isaacson. 1981. *Mechanics of Wave Forces on Offshore Structures*. New York, USA: Van Nostrand Reinhold Company.

Semedo, A., K. Sušelj, A. Rutgersson, and A. Sterl. 2011. A global view on the Wind Sea and swell climate and variability from ERA-40. *Journal of Climate* 24 (5): 1461–1479.

Seneviratne, S.I., N. Nicholls, D. Easterling, C.M. Goodess, S. Kanae, J. Kossin, Y. Luo, J. Marengo, K. McInnes, M. Rahimi, M. Reichstein, A. Sorteberg, C. Vera, and X. Zhang. 2012. Changes in climate extremes and their impacts on the natural physical environment. In *Intergovernmental Panel on Climate Change Special Report on Managing the Risks of Extreme Events and Disasters to Advance Climate Change Adaptation*, ed. by Field, C.B., V. Barros, T.F. Stocker, D. Qin, D. Dokken, K.L. Ebi, M.D. Mastrandrea, K.J. Mach, G.-K. Plattner, S.K. Allen, M. Tignor, P.M. Midgley. Cambridge University Press, Cambridge, United Kingdom and New York, NY, USA.

Shahul Hameed, T.S., N.P. Kurian, K.V. Thomas, K. Rajith, and T.N. Prakash. 2007. Wave and current regime off the southwest coast of India. *Journal of Coastal Research* 23 (5): 1167–1174.

Shimura, T., N. Mori, T. Tom, T. Yasuda, and H. Mase. 2010. Wave climate change projection at the end of 21st century. In *Proceedings of the International Conference on Coastal Engineering*, Shanghai, China.

Snodgrass, F.E., G.W. Groves, K. Hasselmann, G.R. Miller, W.H. Munk, and W.H. Powers. 1966. Propagation of ocean swell across the Pacific. *Philosophical Transactions of the Royal Society of London* 249A: 431–497.

Sreelakshmi, S., and P.K. Bhaskaran. 2020. Regional wise characteristic study of significant wave height for the Indian Ocean. *Climate Dynamics* 54: 3405–3423. https://doi.org/10.1007/s00382-020-05186-6.

St. Denis, M. 1969. On wind generated waves. In *Topics in Ocean Engineering*, vol. 1, 37–41, ed. by Bretschneider, C.L. Gulf Publishing Co., Texas.

St. Denis, M. 1973. Some cautions on the employment of the spectral technique to describe waves of the sea and the response thereto of oceanic systems. In *Proceedings of Offshore Technology Conference*, 827–837. Houston, Paper No. OTC 1819.

Sterl, A., and S. Caires. 2005. Climatology, variability and extrema of ocean waves: The web-based KNMI/ERA-40 wave atlas. *International Journal of Climatology* 25: 963–977. https://doi.org/10.1002/joc.1175.

Sudha Rani, N.N.V., A.N.V. Satyanarayana, and P.K. Bhaskaran. 2015. Coastal vulnerability assessment studies over India: A review. *Natural Hazards* 77: 405–428.

Swain, J., C.V.K. Prasada Rao, and P.N. Ananth. 1989. Numerical wave prediction—some case studies for Arabian Sea and Bay of Bengal. In *Ocean Applications*, 33–43, ed. by Baba, M., and T.S. Shahul Hameed. CESS, Trivandrum.

Swapna, P., R. Krishnan, and J.M. Wallace. 2014. Indian Ocean and monsoon coupled interactions in a warming environment. *Climate Dynamics* 42: 2439–2454. https://doi.org/10.1007/s00382-013-1787-8.

Taylor, K.E. 2001. Summarizing multiple aspects of model performance in a single diagram. *Journal of Geophysical Research Atmospheres* 106 (D7): 7183–7192.

Trenberth, K., et al. 2007. Observations: Surface and atmospheric climate change. In *Climate Change 2007: The Physical Science Basis. Contribution of Working Group I to the Fourth Assessment Report of the Intergovernmental Panel on Climate Change*, edited by S. Solomon. Cambridge, U.K., and New York: Cambridge Univ. Press.

Vethamony, P., V.M. Aboobacker, K. Sudheesh, M.T. Babu, and K. Ashok Kumar. 2009. Demarcation of inland vessels' limit off Mormugao Port region, India: A pilot study for the safety for inland vessels using wave modelling. *Natural Hazards* 49: 411–420.

Vethamony, P., V.M. Aboobacker, H.B. Menon, K. Ashok Kumar, and L. Cavaleri. 2011. Super-imposition of windseas on pre-existing swells off Goa coast. *Journal of Marine Systems* 87 (1): 47–54.

Vetter, O., J.M. Becker, M.A. Merrifield, A.C. Pequignet, J. Aucan, S.J. Boc, and C.E. Pollock. 2010. Wave setup over a Pacific Island fringing reef. *Journal of Geophysical Research* 115 (C12): C12066.

Vinoth, J., and I.R. Young. 2011. Global estimates of extreme wind speed and wave height. *Journal of Climate* 24 (6): 1647–1665. https://doi.org/10.1175/2010JCLI3680.1.

Walsh, K.J.E., K.L McInnes, and J.L. McBride. 2012. Climate change impacts on tropical cyclones and extreme sea levels in the South PacifiC—a regional assessment. *Global and Planetary Change* 0: 149–164.

Wang, X.L., and V.R. Swail. 2006. Climate change signal and uncertainty in projections of ocean wave heights. *Climate Dynamics* 26: 109–126.

Wang, X.L., F.W. Zwiers, and V.R. Swail. 2004. North Atlantic Ocean wave climate change scenarios for the twenty-first century. *Jour. of Climate* 17: 2368–2383.

Wang, X.L., Y. Feng, and V.R. Swail. 2014. Change in global ocean wave heights as projected using multimodel CMIP5 simulations. *Geophysical Research Letters* 41: 1026–1034. https://doi.org/10.1002/2013GL058650.

Young, I.R., and A. Ribal. 2019. Multiplatform evaluation of global trends in wind speed and wave height. *Science* 364: 548–552.

Zieger, I.R.S., and A.V. Babanin. 2011. Global trends in wind speed and wave height. *Science* 332 (6028): 451–455. https://doi.org/10.1126/science.1197219.

Chapter 10
Changes in Extreme Sea-Level in the North Indian Ocean

A. S. Unnikrishnan and Charls Antony

Abstract A review of studies on past changes and future projections in extreme sea level in the tropical oceans with a focus on the north Indian Ocean is made. Studies based on historical tide-gauge data for different oceanic basins show that changes in extremes are caused primarily due to changes in mean sea level. Analysis of data from tide gauges along the east coast of India and the head of the Bay of Bengal indicates similar results consistent with the studies elsewhere. Trends in extreme sea level are large (~5.0 mm/year) at Hiron Point, in the Ganga–Brahmaputra delta, pointing out that subsidence of the delta contributes considerably to changes in extremes through changes in mean sea level. Accurate assessments of the past changes and future projections of extreme sea level need data on rates of land motion. Besides, excess groundwater extraction in some regions is found to affect mean sea-level changes and its impacts on extremes are not fully understood. Extreme sea level along the east coast of India and the head of the Bay is found to have variability at different time scales such as seasonal and inter annual, with the latter having a correlation with modes of climate variability, such as the Indian Ocean Dipole mode and El Niño–Southern Oscillation. Region-specific modelling studies on projections of extreme sea level indicate that the increase in extreme sea level is associated with an increase in mean sea level, while a few studies show an increase with changes in storminess. The north Indian Ocean is one of the regions having the greatest threats from extreme sea levels, at the same time it is also a region that lacks sufficient observations, both in space and time, to facilitate climate studies related to extreme sea levels. The changing climate and increasing coastal population necessitate a systematic monitoring of extreme sea level through a dense network of observations along with Global Navigation Satellite System stations to monitor the land motions.

Keywords Extreme sea level · Storm surge · Tide · Mean sea-level rise · Coastal inundation · North Indian Ocean

A. S. Unnikrishnan (✉)
Former Chief Scientist, CSIR-National Institute of Oceanography, Dona Paula, Goa 403004, India
e-mail: as.unnikrishnan@gmail.com

C. Antony
Indian National Centre for Ocean Information Services, Hyderabad 500090, India

© The Centre for Science & Technol. of the, Non-aligned and Other
Devel. Countries 2022
A. S. Unnikrishnan et al. (eds.), *Extreme Natural Events*,
https://doi.org/10.1007/978-981-19-2511-5_10

10.1   Introduction

Coastal regions are among highly populated regions on the globe. The low elevation coastal zone (LECZ; land below 10 m height above sea level and contiguous along the coast) accounts for nearly 2% of the total land area and about 10% of the total human population (McGranahan et al. 2007). The LECZ population density (241 people/km^2) is about five times the global average (47 people/km^2), and nearly 83% of the LECZ population is in the less developed nations (Neumann et al. 2015). The countries which share a significant portion (together they account for 56%) of the LECZ population are from Asia, namely, China, India, Bangladesh, Indonesia and Vietnam (Neumann et al. 2015). Large and fast-growing human population in coastal regions are exposed to threats from natural hazards (storm surges, tsunamis and flooding associated with sea-level rise, etc.). Climate change exacerbates the impacts in the coastal zone already having an increasing population resulting in more people at the risk of extreme events. Some of the events in the recent decades like Hurricane Katrina, Sandy, Cyclone Nargis and the 2004 Indian Ocean tsunami indicate the vulnerability of the coastal population to extreme events.

Considering various natural hazards, storm surge disasters are noted to cause significant losses of both life and property (Needham et al. 2015). Some regions of the world's coastlines are affected by storm surges, which are caused by tropical and extratropical cyclones. Tropical and extratropical cyclones have different characteristics, and surges generated by them also show variations. Extratropical cyclones are much larger systems compared to those in tropical regions, but tropical cyclones remain most intense. The spatial extent of extratropical cyclones reaches to more than 1000 km, whereas it is only a few hundred kilometres for tropical cyclones. Storm surges due to tropical cyclones are generally larger in amplitude and smaller in spatial extent and duration, compared to those generated by extratropical cyclones (von Storch and Woth 2008). Disasters caused by storm surges are a major concern for the countries in the north Indian Ocean compared to other storm surge risk regions, as the majority of the deadliest events occurred in the north Indian Ocean, whereas the events with largest economic loss were reported in the United States of America. The deadliest cyclone disaster recorded was the one associated with the Bhola cyclone, which hit the coast of Bangladesh in 1970. About 300,000 people lost their lives during that event (Frank and Husain 1971). A very severe cyclonic event in Bangladesh in 1991 and the Odisha super cyclone (India) in 1999 were other notable events in the north Indian Ocean during the 1990s. There were instances of severe events during the 2000s in the north Indian Ocean, such as cyclones Sidr (2007), Nargis (2008) and Aila (2009) with considerable loss of human lives. In the last decade, during events like cyclones Phailin (2013) and Hudhud (2014), human fatalities were well reduced, mainly due to the success of early warning and evacuation. Most recently, Cyclone Amphan (2020), which hit the coast of West Bengal, in India was the first super cyclone since 1999 in the Bay of Bengal (LeComte 2021).

Figure 10.1 shows a schematic of various processes that cause changes in mean sea level (MSL) and extreme sea level (ESL). MSL changes are caused primarily by

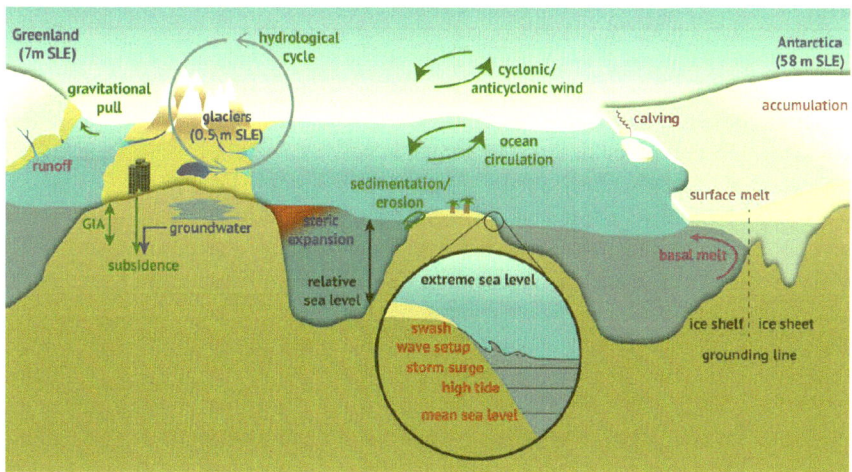

Fig. 10.1 Processes of mean sea level and extreme level changes (Source: Oppenheimer et al. (2019), Chap. 4, SROCC)

thermal expansion due to ocean warming and melting of ice sheets and glaciers. ESL is the total water level occurring during an episodic event, such as a storm surge. Total water level includes MSL, tide and surge components. Gregory et al. (2019) suggested that the term 'extreme sea level' may be used for referring to coastal sea level during an episodic event. We follow this mostly in the present paper. Tsunamis also result in ESL; however, they are non-climatic and not included in the present article. Modes of climate variability also affect ESL by MSL anomalies associated with climate modes and mode-related changes in storminess. And lastly, local MSL rises at that location also affect the ESL. ESL is an important parameter for coastal zone management. Planners are interested to know the maximum sea level attained during an episodic event such as a storm surge.

Many studies have reported changes in ESL in different ocean basins, however, only a few in the Indian Ocean. In the present paper, we review the available literature in tropical oceans, particularly focussing on the north Indian Ocean to assess the past changes in ESL and their future projections. We also discuss the gaps in knowledge and future perspectives. A few studies on ESL in the Red Sea and the Persian Gulf, marginal seas of the north Indian Ocean, are also discussed.

10.2   Past Changes in Extreme Sea Level

ESL changes can occur mainly due to two processes: (i) Changes in storminess (wind and surface pressure fields in the atmosphere) and (ii) Changes in MSL. In addition, other factors such as land subsidence and waves can also change ESL particularly, in deltaic regions and low-lying islands, respectively.

Studies that focus on variability and long-term changes in MSL started in the 1980s, whereas there is paucity of literature on variability and changes in ESL prior to 2000. From the beginning of the twenty-first century, attention has been given to ESL and the research outcomes on the topic have started increasing. The Third Assessment Report of the Intergovernmental Panel on Climate Change (TAR, IPCC) (Church et al. 2001) has only a mere mention of ESL but in the IPCC AR4 (Fourth Assessment Report) (Bindoff et al. 2007), a few papers were used for the assessment. AR4 concluded that the highest water levels have been increasing since the 1950s in most regions of the world, caused mainly by increasing MSL. Lowe et al. (2010) presented more literature on the topic and their general conclusion is that there is little evidence for ESL changing worldwide by amounts significantly different to changes in MSL. Similar conclusions can be found in Woodworth et al. (2011), who also presented a short review of ESL studies. In the special report on managing the risks of extreme events and disasters to advance climate change adaptation (Seneviratne et al. 2012) and in the Fifth Assessment Report (AR5) of the IPCC (Church et al. 2013; Rhein et al. 2013), a review of ESL is presented. The report concluded that the recent increase in observed extremes worldwide has been caused primarily by an increase in MSL, although the dominant modes of climate variability (particularly, the El Niño–Southern Oscillation (ENSO) but including the Indian Ocean Dipole (IOD), North Atlantic Oscillation (NAO) and other modes) also have a measurable influence on extremes in many regions, via variability in MSL and/or mode-related changes in storminess. Oppenheimer et al. (2019) noted that many extreme events that are rare in the present scenario, will become more frequent by 2100. They also concluded that in some of the tropical islands, for example, the Maldives in the Indian Ocean, waves play a role in the total sea level during an extreme event. In deltaic regions, for instance, Ganga-Brahmaputra delta, subsidence considerably enhances MSL change thereby changing ESL. In the next paragraphs, we discuss a few studies on ESL based on observations in the tropical oceans, adjoining many developing countries.

In the tropical Atlantic, Torres and Tsimplis (2014) analysed tide-gauge data in stations in the Caribbean Sea. In five stations, having a record of more than 20 years, they found a significant increase in extremes, which is found to be associated with MSL rise, with no evidence of secular changes in the storm activity. Feng and Tsimplis (2014) studied trends in extremes between 1954 and 2012 at 20 tide gauges along the coast of China. They concluded that the trends are primarily driven by changes in MSL but also linked with increases in tidal amplitudes at three stations. Neither the Pacific Decadal Oscillation (PDO) nor ENSO was found to be indicators of changes in the size of extremes, but ENSO appears to regulate the number of tropical cyclones that reach the Chinese coasts.

Le Cozannet et al. (2021) found that flooding in coastal areas of Petit Cul-de-sac in Guadeloupe (West Indies) is increasing with sea-level rise. Using total water levels including tides, rates of land motion (subsidence) and MSL rise for the regional projections for Representative Concentration Pathways (RCP) 8.5 scenario, the number of flood days is found to become once in every two days within two decades. Antony et al. (2016) examined ESL changes along the Bay of Bengal coast using selected tide gauges. The major findings are summarised in Sect. 6.

Most of the above studies were on ESL at the coast. In a study for the open ocean using satellite-altimeter data, Woodworth and Menéndez (2015) found that long-term change in the standard deviation of sea surface height (SSH) to be small throughout the global ocean (typically ~0.5 per cent/year) and the trends of positive and negative SSH are similar to those of MSL, with a few regional exceptions.

In brief, relative role of MSL and storminess on ESL changes is still not fully understood, the available evidence presented in AR5 and other studies indicate that the role of MSL is certain, while uncertainties exist over the role of change in storminess.

10.3   Role of Tides in Extreme Sea-Level Changes

Tides have magnitudes comparable to or even higher than those in storm surges in some regions. They occur mainly with a periodicity of semi-diurnal or diurnal, with longer periods (fortnightly, monthly and longer) also contributing to a smaller extent. In addition to any long-term changes in ESL observed by a tide gauge, there will be changes on inter-annual and decadal timescales because of the nodal and perigean contributions to the astronomical tide (Pugh and Woodworth 2014). However, these variations in extremes are predictable once one has adequate knowledge of the ocean tide at a location, if several years of good quality data are available.

However, it may be noted that since tide gauges measure sea level at the coastal stations. Observed changes in tidal heights in shallow waters could occur due to changes in propagation characteristics of tides. Since tides propagate as shallow-water waves, their phase speed and propagation characteristics depend on depth, which can vary with MSL rise. However, these changes are small and less certain, because some of the anthropogenic activities such as dredging in harbour areas could also make changes in local bathymetry, thereby changing characteristics of tidal propagation. A few studies available on this topic are region-specific and more work will be needed to get a more definite picture on changes in tides in shallow waters. Using a global tidal model, Pickering et al. (2017) investigated the effect of MSL rise on tides in the shelf seas. They noted a significant change in M_2 (principal lunar constituent) and S_2 (principal solar constituent) in shelf seas, but K_1 (luni-solar diurnal) and O_1 (lunar diurnal) are found to increase in Asian seas only.

10.4   Future Projections

In this section, global studies on ESL projections and regional studies on the same in the tropical oceans are discussed. Projections of ESL are made using two approaches (i) using hydrodynamical models and (ii) statistical methods.

10.4.1 Hydrodynamical Approach

Hydrodynamical models that simulate tides and storm surges have been used. The models are in general two-dimensional type that use vertically integrated hydrodynamic equations. The models are regional and driven by surface atmospheric wind fields and MSL pressure obtained from regional climate models for different climate scenarios. Tides are forced along the open boundaries, which are obtained from a global tidal model. Models also introduce a MSL rise to get future projections of ESL due to a MSL rise. Recently, some studies made use of global models for getting future climate projections.

A global tide and surge model with a grid resolution of 2.5 km was used to simulate ESL for the present scenario (1979–2017) and a future climate scenario from 2040 to 2100 (Muis et al. 2020). It is found that one in ten year, water levels increase by 0.34 m, while some regions show increases up to 0.5 m for RCP 4.5 scenario. Using a global model for tides, surges and wave set-up, Kirezci et al. (2020) showed that projections of ESL for 2100 under RCP 8.5 scenario could bring nearly 50% of the coastal population under risk for flooding resulting from episodic events. They identified many hotspots for ESL events by 2100, which include the coasts surrounding the Bay of Bengal.

Recent regional studies on ESL projections in tropical oceans can be briefly discussed as follows. Feng et al. (2018) analysed data from 15 tide gauges along the Chinese coast and used different Coupled Model Intercomparison Project Phase 5 (CMIP5) models for 1.5 and 2.0 °C warming to estimate return levels of ESL. They argued that there is a significant difference in return levels between the two scenarios with return levels in 2 °C warming scenario higher than those of 1.5 °C. Widlansky et al. (2015) made projections using climate models forced with increased greenhouse gases and removing the effects of MSL rise; they found that occurrence of low sea level (Taimasa) is found to increase by two times in the tropical southwest Pacific. In the entire tropical Pacific, inundation caused due to extreme La Niña events is found to increase. McInnes et al. (2014) modelled ESL in Fiji Island in the Pacific and found that higher storm surges found in southwest of Viti Levu and Vanua Levu islands during El Niño and La Niña years than normal years, but in other regions of these islands, storm surges are found to be lower during La Niña years. They also found that future projections of ESL will be changed primarily due to MSL rise.

10.4.2 Statistical Projections

Statistical approach is used for analysis of ESL by using past tide gauge data. In general, a generalised extreme value distribution or Gumbel distribution is used for the analysis.

In the statistical approach, it is assumed that changes in extremes occur only through the MSL rise, with storminess not changing. Hunter (2012) used hourly tide gauge data for 198 stations over the globe and fitted a Gumbel extreme value distribution to determine the return periods of extreme events in the present scenario.

For making the projections, it is assumed that tide and surge statistics do not change with time. Then increasing the values of sea level with a uniform MSL rise and taking into account its uncertainty, Hunter determined a multiplication factor by which the frequency of occurrence will change by 2100. For regions having high tidal ranges and/or surges, the multiplication factor is small and vice versa. The method allows to obtain a reasonably good estimate of return levels of future ESL for designing of coastal structures (Hunter et al. 2013), assuming that increases occur primarily through MSL changes.

It is found that frequencies of extreme events vary depending also on tidal range in a region, with low tidal range regions experiencing large increases. Indirectly, this helps to understand that frequency of 100 years can become 10 or even 1 year depending on the characteristics of tides and surges in a given location, as reported in AR5 (Church et al. 2013). Oppenheimer et al. (2019) reported these projections for the mid-century and towards the end of the century for different RCP scenarios in the special report on the Ocean and Cryosphere in a Changing Climate (SROCC). Figure 10.2 shows statistical projections of ESL for future scenarios at tide gauge stations in the Global Extreme Sea Level Analysis Version 2 (GESLA-2) database (Woodworth et al. 2017a). MSL rises projections for climate scenarios, RCP 2.6, RCP 4.5 and RCP 8.5 for mid-century and towards the end of the twenty-first century were used for making the projections. These analyses showed that the frequency of an ESL occurring in 100 years now will occur by once in 10 year or 1 year depending on the location (Oppenheimer et al. 2019).

10.5 Mean Sea-Level Variability and Changes in the North Indian Ocean

The Indian Ocean, though smallest among the three ocean basins, has unique characteristics (Fig. 10.3). Landlocked in the north, the north Indian Ocean is characterised by the monsoon. In the north Indian Ocean, during summer monsoon (June and September), winds are southwesterlies and during northeast monsoon (November to February), winds are northeasterlies. Recent studies indicate that the Indian Ocean has been warming rapidly (Roxy et al. 2014; Beal et al. 2020). Swapna et al. (2017) found a relationship between weakening of the Indian summer monsoon since the 1950s causing an increase in MSL in the north Indian Ocean.

The locations of tide gauges, having more than 40 years of data, are shown in Fig. 10.3. The Mumbai (along the west coast of India) record has more than 100 years duration. MSL data are provided by the Permanent Service for Mean Sea Level

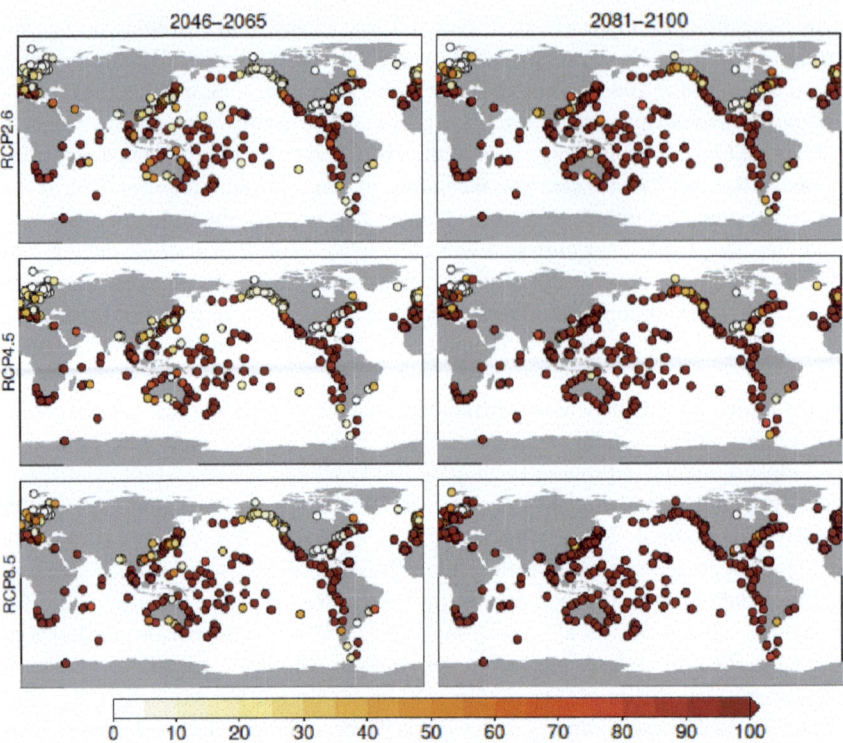

Fig. 10.2 Future projections of extreme sea level for tide gauge stations in GESLA-2 database. The colours of the dots indicate the factor by which the frequency of extreme events having a 100-year return period increases. The results are shown for RCP 2.6, 4.5 and 8.5, respectively, for the mid-century and towards the end of the century (Adapted from Oppenheimer et al. (2019), Chap. 4, SROCC, IPCC)

(PSMSL, http://www.psmsl.org). The data are available as monthly mean and annual mean, which can be used for determining sea-level rise trends. Real-time data can be obtained from the Intergovernmental Oceanographic Commission's Sea Level Station Monitoring Facility (http://www.ioc-sealevelmonitoring.org), operated by the Flanders Marine Institute (VLIZ, Belgium). Since the 2004 tsunami in the Indian Ocean, real-time stations have increased. Hourly tide-gauge data for the stations along the Indian coast are obtained from Survey of India, Dehradun.

Tide-gauge data need to be corrected for land movements for obtaining accurate sea-level rise trends. Earlier, Ice-model (for instance, Peltier 2004) results have been widely used to correct for glacial isostatic adjustment. Recently, Global Navigation Satellite System (GNSS) stations co-located with tide gauges have been installed in many regions over the globe and processed data are being stored and distributed through the Service d'Observation du Niveau des Eaux Littorales (SONEL) data centre (http://www.sonel.org). Woodworth et al. (2017b) pointed out the need for having GNSS stations, co-located with tide gauges. King (2014) pointed out the lack

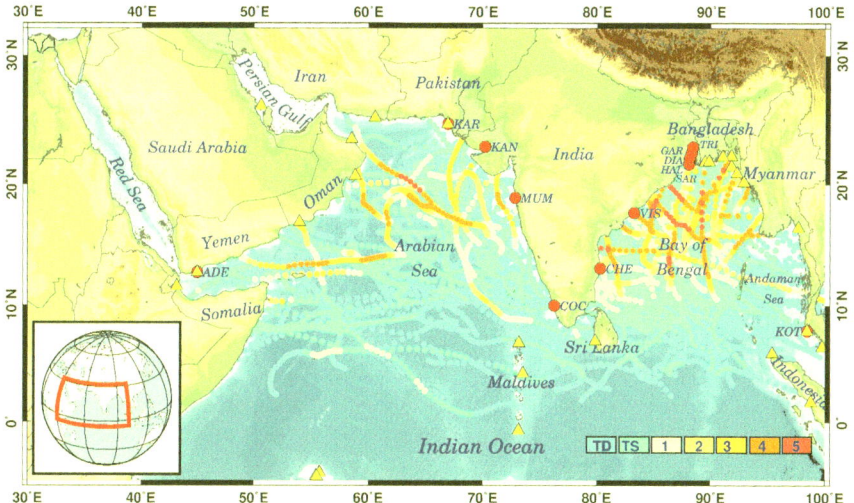

Fig. 10.3 Map of the north Indian Ocean with tracks of the tropical cyclones from 1980 to 2020. The colour bar represents the Saffir–Simpson Hurricane wind scale (TD = Tropical Depression, TS = Tropical Storm and 1 to 5 = Category 1 to 5 storms). Locations of Aden (ADE), Karachi (KAR), Kandla (KAN), Mumbai (MUM), Cochin (COC), Chennai (CHE), Visakhapatnam (VIS), Sagar (SAR), Haldia (HAL), Diamond Harbour (DIA), Garden Reach (GAR), Tribeni (TRI) and Ko Taphao Noi (KOT) tide gauges, where long-term (> 40 years) mean sea-level data are available, are marked with circles. Locations at which research quality sea-level data at hourly intervals are available are marked with triangles

of GNSS stations in some of the tide-gauge stations having long records in the Indian Ocean.

Since 1992, a series of satellites, TOPEX/Poseidon, Jason-1 and Jason-2 provide continuous sea-level data using altimeters covering the globe, except high latitudes. The analysis of altimeter data provides spatial variations in MSL rise trends. Altimetry data were also used, for example, Nerem et al. (2018) to determine acceleration in global MSL, mainly because the data of about 25 years are available without any gaps. Sea-level rise trends obtained from altimetry and tide gauge data analysis are complementary to assess MSL rise trends due to global warming. An assessment of the various studies is given in Church et al. (2013) and Oppenheimer et al. (2019).

Mean sea level in the north Indian Ocean undergoes variability at different time scales (Shankar et al. 2010). Aparna et al. (2012) showed using tide-gauge data and altimeter record that during the occurrence of IOD and ENSO, sea-level anomalies in the Bay of Bengal become negative and positive depending on the phase of the modes. Nidheesh et al. (2013) found that decadal variabilities, though smaller than interannual variabilities in the tropical Indo-Pacific regions, are associated with decadal variations in wind stress. The study was done using various sea-level data sets available and steric sea level computed from an ocean model. Srinivasu et al. (2017) found a reversal in MSL rise trends in the north Indian Ocean during the last two decades. A detailed description of MSL variability in the Indian Ocean is beyond the scope of the present article.

The long records in the north Indian Ocean were used in earlier studies to determine long-term sea-level-rise trends (Unnikrishnan and Shankar 2007; Unnikrishnan et al. 2015) and also causes of sea-level change (Han et al. 2010). They analysed long records in the north Indian Ocean, Aden, Karachi, Mumbai, Kochi, Visakhapatnam, Diamond Harbour and Ko Taphoi Noi (Fig. 10.3) and reported MSL-rise trends varying from 1.06 to 1.75 mm/year along the coast, except at Diamond Harbour, where trends exceeding 5.0 mm/year are found. These trends are consistent with global estimates. In the region of Ganga–Brahmaputra delta, for example, at Diamond Harbour and Hiron Point, the trends are above 5.00 mm/year, which are partly due to the subsidence, as reported in earlier studies. Goodbred and Kuehl (2000) reported subsidence rates of 4 mm/year in Ganga–Brahmaputra delta using sedimentological evidences, but recent assessments point towards higher rates of subsidence (Oppenheimer et al. 2019). Jyoti et al. (2019) reported accelerated MSL rise in the south Indian Ocean during the recent decades. In the recent climate change assessment for regions surrounding the Indian region, Swapna et al. (2020) assessed various studies on MSL changes and variability in the Indian Ocean.

Excess groundwater extraction is found to alter MSL rise trends in some regions. Veit and Conrad (2016) showed using a global model that excess groundwater pumping depresses sea level in the region of south Asia. Rates of sea-level rise, as found from Mumbai tide gauge data record, are about 1.2 mm/year, which is slightly lower than global MSL rise trends. Veit and Conrad (2016) attributed the slightly reduced sea-level rise trends than the global mean due to excess groundwater pumping in the region.

Figure 10.4 shows spatial variations in MSL rise trends, estimated using altimeter data, in the north Indian Ocean for the recent period. The spatial average of trend is 3.37 mm/year for the period 1993–2018, which is higher than those found during the historical period. Relatively large trends in some regions, such as the head of the Bay, exceeding 5.00 mm/year are partly due to variability at different time scales, as pointed out in earlier studies. The average trends in the north Indian Ocean are consistent with global mean sea-level rise trends for this period but higher than those estimated from tide gauges for the last century.

Global MSL projections for different scenarios are given in the SROCC (Oppenheimer et al. 2019). Sea-level projections for the south Asia region (Harrison 2020) for 2100 show slightly lower values in the Arabian Sea and the Bay of Bengal compared to global MSL projections for all the RCP scenarios.

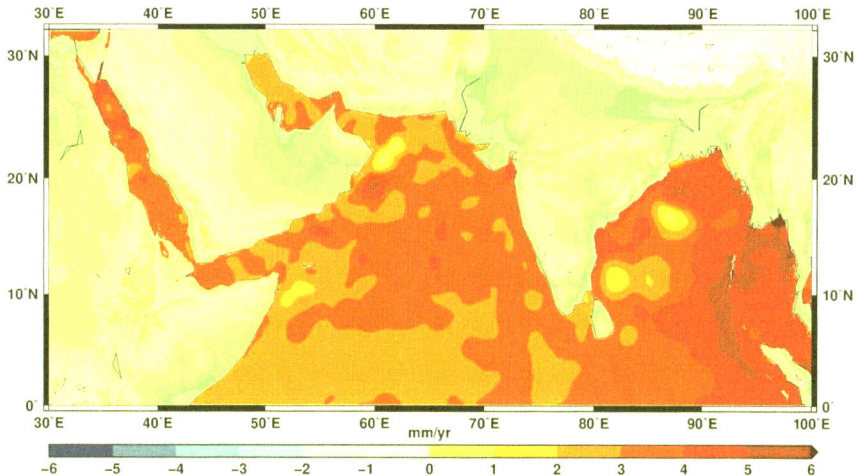

Fig. 10.4 Trends in mean sea-level rise estimated from altimeter data in the north Indian Ocean for the period 1993–2018

10.6 Extreme Sea-Level Changes and Projections in the North Indian Ocean

The north Indian Ocean (Fig. 10.3) is one among the cyclone risk regions over the globe. Though average annual number of cyclones in this basin are the smallest among various cyclone risk basins, the past events show that the impact of tropical cyclones is higher along the coast of the north Indian Ocean, especially in the northern Bay of Bengal due to high population in this region as well due to the topographical features that amplify storm surges. The Arabian Sea and the Bay of Bengal are the two sub-basins of the north Indian Ocean in the west and east divided by the Indian subcontinent. The annual distribution of tropical cyclones is bimodal in nature for the north Indian Ocean with tropical cyclones occurring in the post and pre-monsoon seasons. The Bay of Bengal has a greater number of cyclones occurring in a year on an average compared to the Arabian sea, the ratio being 4:1. Most of the cyclones formed in the Bay of Bengal move west, northwest and north directions and cross the east coast of India and the coast of Bangladesh (Fig. 10.3). However, there are some cyclones that cross the Myanmar coast, for instance, the Cyclone Nargis crossed the Myanmar coast in 2007. In the Arabian Sea, northern and western coastal areas have frequent landfall of cyclones. While tropical cyclones related storm surges are the frequent contributor to the ESL in the region, there are also other events such as tsunamis, meteotsunamis, seiches and swells which contribute to extremes and coastal flooding. Though the basin had experienced much larger-scale impact from the 2004 tsunamis, occurrence of such events is rare. In this section, we summarise various studies related to ESL changes in the north Indian Ocean.

Studies on the ESLs and storm surges, identified in past tide gauge records, are not many in the case of the north Indian Ocean. Availability of good quality data is the major reason for this. Figure 10.3 presents locations of tide gauges (marked with triangles) where research quality sea-level data are available at hourly intervals. These data are freely provided by the University of Hawaii Sea Level Center (https:// uhslc.soest.hawaii.edu/). The freely available data with decent space–time coverage are limited to the Bangladesh coast, where 20–30 years of data are available at a few stations. Although several tide gauge stations are available along the Indian coast, hourly sea-level data at these stations are not publicly available but real-time data are available at a few stations through the VLIZ website. Another issue is that very often surges do not get recorded in gauges, mainly because gauges are not closely enough located to capture the signals. However, some studies have documented the characteristics of ESLs and surges along the east coast of India and the head of the Bay by analysing many years of data. (Unnikrishnan et al. 2004; Antony et al. 2016). These data have come from Survey of India and Indian National Centre for Ocean Information Services.

An example of an ESL event resulting from a tropical cyclone is presented in Fig. 10.5. Figure 5a depicts the track of the Cyclone Aila of May 2009, which hits the northern Bay of Bengal coast. The cyclone attained the strength of a Category 1 storm before its landfall. The reported maximum wind speed was 65 kt and minimum pressure at the centre of the cyclone was 970 mb. Time series of observed sea level at Hiron Point (located on the right side and about 130 km away from cyclone landfall point) and its components predicted tide and residual sea level are shown in Fig. 5b. The storm surge attained a height of 1.7 m and the maximum sea level was nearly 0.8 m higher than the highest astronomical tide for the year.

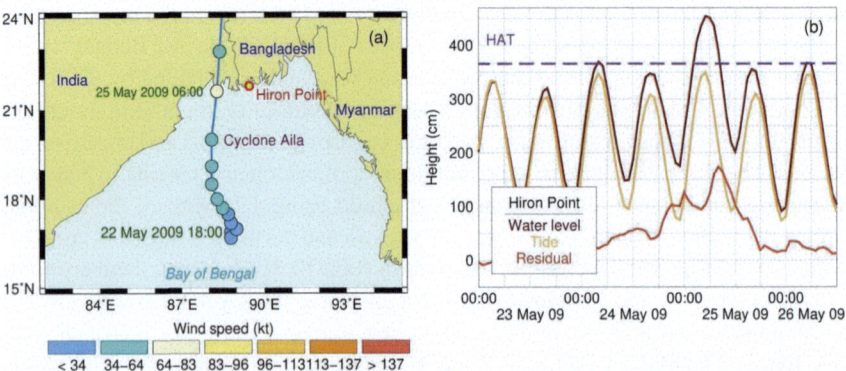

Fig. 10.5 Extreme sea level and storm surge during Cyclone Aila. Map showing the track of Cyclone Aila (**a**). Total water level, astronomical tide and residual sea level from Hiron Point tide gauge during the passage of cyclone Aila (**b**). Highest astronomical tide for the year is also marked [Modified from Antony et al. (2020)]

10.6.1 *Variability and Trends in Extreme Sea Level*

Even though there had been a number of numerical modelling studies on ESL in the north Indian Ocean, only a few studies based on observations exist. The study by Pethick and Orford (2013) showed a rapid rise in high water maxima in southwest Bangladesh during the last few decades. The average trend in the monthly mean high water in the region was about 14 mm/year. The authors also showed that this high rate was principally as a result of increased tidal range in estuarine channels recently constricted by embankments and also include a combination of deltaic subsidence, including sediment compaction, and sea-level rise.

Antony et al. (2016) studied spatial variability, seasonal, inter-annual and long-term changes in extreme high waters along the east coast of India and at the head of the Bay of Bengal. This study was conducted using long-term hourly sea-level observations from 1974 to 2007 at five tide gauges stations in the Bay, namely Chennai, Visakhapatnam, Paradip, Hiron Point and Cox's Bazaar (Fig. 10.6). The highest water levels are reported towards the head of the Bay of Bengal, which are a combination of large tides and moderate or small surges. Strong seasonality is also identified in the extreme high waters, which also shows higher values towards the head of the Bay. Variability at the northernmost stations is characterised by a

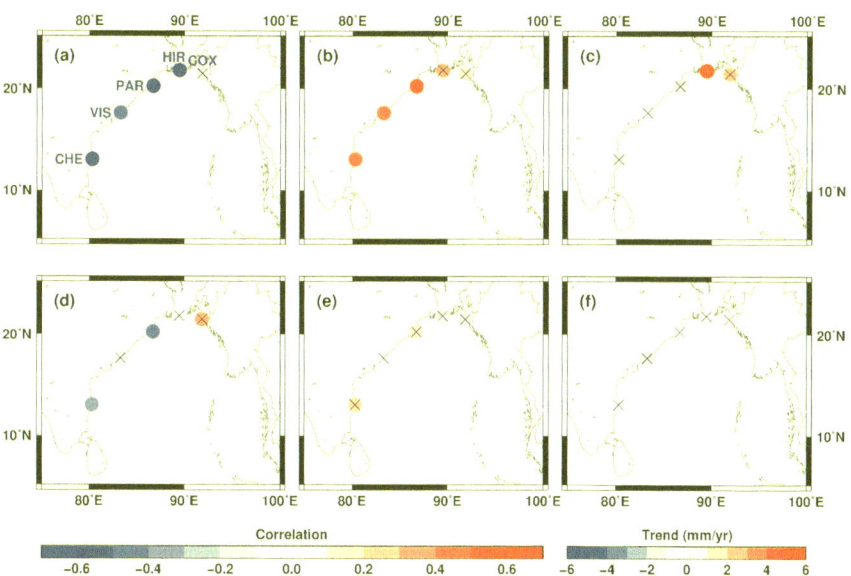

Fig. 10.6 Correlations between 99th percentile of sea level and dipole mode index (DMI) (**a**) and southern oscillation index (SOI) (**b**) at Chennai (CHE), Visakhapatnam (VIS), Paradip (PAR), Hiron Point (HIR) and Cox's Bazaar (COX). Trends in 99th percentile of sea level (**c**). Correlations of DMI (**d**) and SOI (**e**) with 99th percentile of sea level after mean sea-level (50th percentile) subtracted. Trends after mean sea level subtracted (**f**). Correlations and trends which are not statistically significant are marked with a cross sign

strong annual cycle, whereas towards the south-western part it is dominated by a semi-annual cycle. There is also consistency in the seasonality of extreme and MSL. The study further showed that interannual variations in the extreme high waters are correlated with regional climate phenomena such as, the IOD and ENSO (Fig. 10.6). The extremes showed negative correlations with dipole mode index indicating lower (higher) extremes during positive (negative) IOD events. A direct relationship (positive correlations) was observed between extremes and the southern oscillation index indicating lower (higher) ESL during El Niño (La Niña) events. A trend analysis of extreme high waters showed larger trends in the head Bay region compared to the rest of the Bay of Bengal and only Hiron Point tide gauge showed statistically significant increasing trends of about 5 mm/year (Figs. 10.6 and 10.7). Further, the study also showed that when MSL is subtracted from the extremes the reported correlations are reduced. Similarly, the high positive trend at Hiron Point also reduced to small negative values, which are insignificant. The study concluded that changes in extremes

Fig. 10.7 Trends in various sea-level percentiles at Hiron Point, a station in the Ganga–Brahmaputra delta. The 50th percentile represents the mean sea level [Adapted from Antony et al. (2016)]

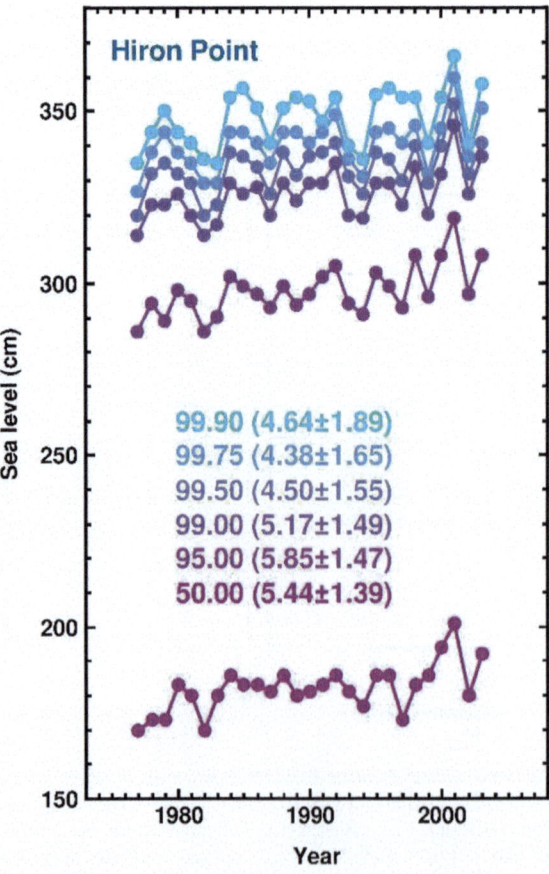

are in line with changes in MSL, consistent with the global study of Woodworth and Blackman (2004) and Menéndez and Woodworth (2010).

10.6.2 Future Projections of Extreme Sea Level

Most of the studies on future projections of ESL in the north Indian Ocean have been done for the Bay of Bengal. A few studies on ESL changes in the Arabian Sea and the marginal sea, the Red Sea, are also discussed briefly.

Unnikrishnan et al. (2011) carried out a study to understand future changes in ESL along the east coast of India using a hydrodynamic model forced with MSL pressure and wind fields from a regional climate model. They considered 1961–1990 and 2071–2100 as periods for current and future climate scenarios. The results indicate 100-year return levels in the future climate are 15–20% higher than those in the current climate, particularly in the region of low tidal ranges (southeast coast of India). In the region of high tidal ranges (northeast coast of India, Sagar and Kolkata), the changes are below 5%.

The study by Kay et al. (2015), similar in approach to that of Unnikrishnan et al. (2011), showed an increased likelihood of ESL events in the Ganga–Brahma-putra–Meghna delta through the twenty-first century, with the frequency of events increasing greatly in the second half of the century: water levels that occurred at decadal time intervals under current climate occurred in most years by the middle of the twenty-first century and 3–15 times per year by 2100. In a recent study by Rahman et al. (2019) showed that combined with sea-level rise of 0.5, 1 and 1.5 m, a Category 4 cyclone such as Cyclone Sidr would inundate 2.6%, 3.67% and 5.84% of the area of Bangladesh and 21.0%, 42.1% and 65.1%, respectively, of the Sundarbans mangrove forest.

Another study by Lee (2013) using tide gauge data at Hiron Point from 1977 to 2009 estimated the 100-year return level of storm surge at the station as 1.75 m. The author also estimated the sea-level rise trend for the station as 4.46 mm/year and the resulting sea-level rise in 2050 as 0.34 m. Further, he pointed out that in 2050 the 100-year event due to sea-level rise and storm surge would be 2.09 m.

Khan et al. (2020) used a regional tidal model for the Bay of Bengal to simulate the effects of MSL rise under various scenarios on modulating tidal characteristics in the Ganga–Brahmaputra–Meghna delta. They showed significant flooding could occur for a MSL rise over 0.5 m, with the area of flooding found to vary in different parts of the delta.

A few other studies on climate change and storm surges along the Indian coast include those of Rao et al. (2020a), Rao et al. (2020b) and Poulose et al. (2020). Rao et al. (2020a) showed an average increase of about 20% and 30% in maximum water elevations for the moderate and extreme climate change scenarios (wind speeds increased by 7% and 11%, respectively). Numerical experiments by enhancing wind speeds by 7% and 11% showed additional water levels of 0.5 to 1.0 m and 1.0 to

2.0 m in the northern regions (Gulf of Kutch and Khambhat) of the west coast of India, for the two scenarios considered (Poulose et al. 2020).

10.6.3   Extreme Sea Levels in the Marginal Seas and Low-Lying Islands

The Red Sea and the Persian Gulf are two marginal water bodies in the western part of the north Indian Ocean. Unlike the Arabian Sea, which is at risk for the occurrence of tropical cyclones, the Red Sea and the Persian Gulf are considered to be safe zones. In the Red Sea, sea-level observations are limited and a few numerical modelling studies can be found on the ESL. A numerical modelling study for long term by Antony et al. (2021) reported extremes may reach up to 0.75 m in the central eastern Red Sea. They also noted high-surge events are correlated with winds over the Southern Red Sea. The study also reported that projected sea-level rise would reduce the average recurrence intervals of extreme water levels in the region (Fig. 10.8). A recent study by Antony et al. (2022) carried out for the entire Red Sea reported that the extreme water levels simulated using a barotropic model, forced with meteorological fields generated by a 5 km assimilative Weather Research and Forecasting (WRF) model, have no significant trend over the period 1980–2016.

The Persian Gulf is mainly affected by extratropical weather systems (El-Sabh and Murty 1989) and some of the strongest winds in the region are associated with the shamal winds, the seasonal north-westerly winds which occur during winter as well as summer. Storm surges are reported in the Persian Gulf (Khalilabadi 2016) and more recently, Heidarzadeh et al. (2019) and Kazeminez had et al. (2020) reported occurrence of 'meteotsunami' in the region. The meteotsunami event had maximum

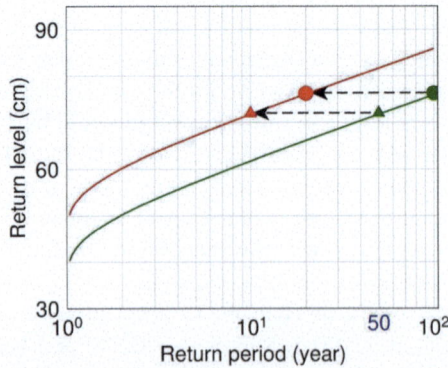

Fig. 10.8 Return period curves at a location in the central Red Sea coast of Saudi Arabia under current climate (green) and future sea-level rise scenario (red). Green triangle and circle represent 50- and 100-year return levels under current climate, and red triangle and circle represent positions of current 50- and 100-year return levels on the future return curve

surveyed run-up of about 3 m at Dayyer in southern Iran, where waves inundated the land for a distance of ~1 km. Besides a few local studies, the region lacks detailed analysis of ESL processes and their long-term changes.

Low-lying small island nations have been assessed to be the most vulnerable to sea-level rise. Maldives Islands in the north Indian Ocean, which is one of the lowest lying nations in the world with an average height above MSL of about 2 m and nearly 80% of its land is below 1 m. This makes the Maldives highly vulnerable to sea-level rise. Wadey et al. (2017) showed that coastal flooding in the Maldives occurs due to multiple interacting sources. These include long-period (up to 20 s) energetic waves generated in the Southern Ocean and propagating northward combined with spring tides.

10.7   Coastal Flooding

The impacts of flooding in the coastal regions are felt through the occurrence of ESL events rather than the gradual MSL rise. Coastal flooding depends mainly on onshore topographic gradients. Estimates of area of inundations were made using geographic information system (GIS) techniques and simulations from fine-resolution storm surge and tide models, coupled with inundation models.

A recent study on coastal flooding for a projected sea-level rise was made by Kulp and Strauss (2019). They identified particular hotspots in Asia and reported that, the area of coastal inundation by 2050 for an annual maximum flood event in the region surrounding the megacity of Mumbai, India could be large. Using the newly available coastal digital elevation data, Kulp and Strauss (2019) used a global model for inundation estimates and showed that people affected globally for annual maximum flood line could be three times more than the earlier estimates made using the data from the digital elevation model, Shuttle Radar Topography Mission. However, caution is needed while interpreting these results, as these estimates largely depend on the accuracies of the digital elevation model (DEM) used, besides uncertainties in model projections of MSL. Hasan et al. (2020) estimated MSL rise projections using temperature projections for different RCP scenarios and making a semi-empirical approach for obtaining sea-level rise as 1 m for 2100. They estimated the area of inundation of about 2098 km^2 for Bangladesh coasts for 2100. Using GIS techniques, there have been many region-specific studies on coastal inundation, for example, along the east coast of India (Pramanik et al. 2015) and Kochi, the west coast of India (Mani Murali and Dinesh Kumar 2015), etc.

Impacts of mean sea-level rise and ESL are felt in low-lying areas. Deltas, which are undergoing subsidence in this region, are also sensitive to MSL rise and extreme events.

10.8 Conclusions and Future Perspectives

10.8.1 Conclusions

Most of the studies on past changes using tide-gauge data indicate that changes in ESL are in line with those of MSL. However, a few limitations of the studies are briefly discussed below.

Although there have been many developments in the construction of a global tide gauge network that can sample most of the world coastline (Merrifield et al. 2009), one has to recognise that the presently available sea-level data set has many gaps, both in space and time and so conclusions about extremes may not be as globally representative as one would wish. The research depends critically on the availability of suitable data sets. However, access to high-frequency (hourly or more rapid sampling) tide-gauge data is restricted in some countries for various reasons, while in others, there are many years of original records (tide-gauge charts or tabulations) that await reprocessing to modern standards in a process called 'data archaeology' that is followed in some developed countries. As a result, there are gaps in the data sets available for research even where the data themselves exist.

In addition, large gaps in locations of adjacent tide gauges may mean that the largest extremes that lead to the most catastrophic flooding, and which are of most practical interest, might not always be represented adequately in the data sets. This reservation applies particularly for the ESLs that are caused by the storm surges during tropical storms. By definition, these are rare events, and when they do happen the highest sea level may not occur exactly at the gauge, but at some distance along the coast. Therefore, these large events may not be sampled at all even in a record of many decades. Moreover, the technologies used at some tide gauge stations may not allow the highest sea levels to be recorded, and during the most energetic tropical cyclones during the historical period. Moreover, some events do not get recorded, for instance, when the super cyclone that crossed at Paradip, along the east coast of India, in 1999, the tide gauge was not functional and the sea level was not recorded during the event.

Tides have magnitudes comparable or even higher than surges in many regions. They occur in a variety of time scales varying from semi-diurnal, diurnal, fortnightly spring-neap cycle and longer. So far, no study exists that documents the occurrence of various events and phase of the tide. Occurrence of storm surges, especially large surges, is a low-probability event than the occurrence of high probability events of high tides (spring tides). Moreover, surges occurring during low tide or neap tide do not cause much flooding. High tides, particularly during spring tide, when modulated with MSL rise can cause increased flooding in future. This also means that in spite of the difficulties in getting accurate regional projections of wind fields that drive storm surge models, a reasonably good picture of future coastal inundation can be obtained by using high resolution regional tidal models and storm surge models coupled with inundation models that incorporate MSL rise.

10.8.2   Future Perspectives

Future inundation due to the occurrence of ESL events is projected to increase. This would require long-term adaptation measures for coastal infrastructure development and for people residing in coastal areas.

Recommendations for ESL monitoring and analysis can be listed as follows.

(i)   A close network of tide gauges in the regions prone to cyclones will help in better monitoring of storm surges and other ESL events.

(ii) Documentation and reporting of ESL events along the coast.

(iii) Return period estimates including allowance for MSL rise could improve design heights of coastal and offshore structures.

(iv) Inundation maps for annual maximum flood levels need to be prepared with the fine resolution DEM for accurate assessment for future projections.

Acknowledgements   We thank Survey of India, Dehradun, and Indian National Ocean Information Services, Hyderabad, for providing hourly sea-level data along the Indian coast. The authors thank Permanent Service for Mean Sea Level, UK, for making available mean sea-level data.

Charls Antony was supported by the Science and Engineering Research Board (SERB) vide project PDF/2019/001415.

References

Antony, C., A.S. Unnikrishnan, and P.L. Woodworth. 2016. Evolution of extreme high waters along the east coast of India and at the head of the Bay of Bengal. *Global and Planetary Change* 140, 59–67. https://doi.org/10.1016/j.gloplacha.2016.03.008.

Antony, C., A.S. Unnikrishnan, Y. Krien, P.L.N. Murty, S.V. Samiksha, and A.K.M.S. Islam. 2020. Tide–surge interaction at the head of the Bay of Bengal during Cyclone Aila. *Regional Studies in Marine Science* 35. https://doi.org/10.1016/j.rsma.2020.101133.

Antony, C., S. Langodan, H.P. Dasari, O. Knio, and I. Hoteit. 2021. Extreme water levels along the central Red Sea coast of Saudi Arabia: Processes and frequency analysis. *Natural Hazards and Earth System Sciences* 105: 1797–1814.

Antony, C., S. Langodan, H.P. Dasari, Y. Abualnaja, and I. Hoteit. 2022. Sea level extremes of meteorological origin in the Red Sea. *Weather and Climate Extremes* 35. https://doi.org/10.1016/wace.2022.100409.

Aparna, S.G., J.P. McCreary, D. Shankar, and P.N. Vinayachandran. 2012. Signatures of Indian Ocean Dipole and El Niño-Southern Oscillations events in the Bay of Bengal. *Journal of Geophysical Research C: Oceans* 117: C10012, https://doi.org/10.1029/2012JC008055.

Beal, et al. 2020. A roadmap to Indoos-2: Better observations of the rapidly warming Indian Ocean. *Bulletin of the American Meteorological Society*. https://doi.org/10.1175/BAMS-D-19-0209.1.

Bindoff, N.L., J. Willebrand, V. Artale, A. Cazenave, J.M. Gregory, S. Gulev, K. Hanawa, C. Le Quéré, S. Levitus, Y. Nojiri, C.K. Shum, L.D. Talley, and A.S. Unnikrishnan. 2007. Observations: Oceanic climate change and sea level. In *Climate Change 2007: The Physical Science Basis. Contribution of Working Group I to the Fourth Assessment Report of the Intergovernmental Panel on Climate Change*, ed. S. Solomon, D. Qin, M. Manning, Z. Chen, M. Marquis, K.B. Averyt, M. Tignor, and H.L. Miller. Cambridge: Cambridge University Press.

Church, J.A., J.M. Gregory, P. Huybrechts, M. Kuhn, K. Lambeck, M.T. Nhuan, D. Qin, and P.L. Woodworth. 2001. Changes in sea level. In *Climate Change 2001: The Scientific Basis: Contribution of Working Group I to the Third Assessment Report of the Intergovernmental Panel*, ed. J.T. Houghton, Y. Ding, D.J. Griggs, M. Noguer, P.J. van der Linden, X. Dai, K. Maskell, and C.A. Johnson, 639–694.

Church, J.A., P.U. Clark, A. Cazenave, J.M. Gregory, S. Jevrejeva, A. Levermann, M.A. Merrifield, G.A. Milne, R.S. Nerem, P.D. Nunn, A.J. Payne, W.T. Pfeffer, D. Stammer, and A.S. Unnikrishnan. 2013. Sea level change. In *Climate change 2013: The Physical Science basis. Working Group I Contributions to the Fifth Assessment Report of the Intergovernmental Panel on Climate Change*, ed. T.F. Stocker, D. Qin, G.K. Plattner, M. Tignor, S.K. Allen, J. Boschung, A. Nauels, Y. Xia, V. Bex, and P.M. Midgley. Cambridge: Cambridge University Press.

El-Sabh, M.I., and T.S. Murty. 1989. Storm surges in the Arabian Gulf. *Natural Hazards and Earth System Sciences* 1: 371–385.

Feng, X., and M.N. Tsimplis. 2014. Sea level extremes at the coasts of China. *Journal of Geophysical Research, c: Oceans* 119: 1593–1608. https://doi.org/10.1002/2013JC009607.

Feng, J., H. Li, D. Li, Q. Liu, H. Wang, and K. Liu. 2018. Changes of extreme sea level in 1.5 and 2.0 °C warmer climate along the Coast of China. *Frontiers in Earth Science* 6. https://doi.org/10.3389/feart2018.00216.

Frank, N.L., and S.A. Husain. 1971. The deadliest tropical cyclone in history? *Bulletin of the American Meteorological Society* 52 (6): 438–444.

Goodbred, S.L., and S.A. Kuehl. 2000. The significance of large sediment supply, active tectonism, and eustasy on margin sequence development: Late Quaternary stratigraphy and evolution of the Ganges-Brahmaputra delta. *Sedimentary Geology* 133: 227–248.

Gregory, J.M., S.M. Griffies, C.W. Hughes, J.A. Lowe, J.A. Church, I. Fukimori, N. Gomez, R.E. Kopp, F. Landerer, G. Le Cozannet, and R.M. Ponte. 2019. Concepts and terminology for sea level: Mean, variability and change, both local and global. *Surveys in Geophysics* 40(6), 1251–1289. https://doi.org/10.1007/s10712-019-09525-z.

Han, W., et al. 2010. Patterns of Indian Ocean sea-level change in a warming climate. *Nature Geoscience* 3, 546–550.https://doi.org/10.1338/NGEO901.

Harrison, B. 2020. Sea level projections for South Asia (2020). Report on main findings. Met Office, United Kingdom.

Hasan, K., L. Kumar, and T. Gopalakrishnan. 2020. Inundation modelling for Bangladeshi coasts using downscaled and bias-corrected temperature. *Climate Risk Management* 27 (2020): 100207. https://doi.org/10.1016/j.crm.2019.100207.

Heidarzadeh, M., J. Šepić, A. Rabinovich, M. Allahyar, A. Soltanpour, and F. Tavakoli. 2019. Meteorological tsunami of 19 March 2017 in the Persian Gulf: Observations and analyses. *Pure and Applied Geophysics* 177 (3): 1231–1259. https://doi.org/10.1007/s00024-019-02263-8.

Hunter, J. 2012. A simple technique for estimating an allowance for uncertain sea-level rise. *Climate Change* 113, 239–252. https://doi.org/10.1007/s10584-011-0332-1.

Hunter, J.R., J.A. Church, N.J. White, and X. Zhang. 2013. Towards a global regionally varying allowance for sea-level rise. *Ocean Engineering* 71, 17–27. https://doi.org/10.1016/j.oceaneng.2012.12.041.

Jyoti, J., P. Swapna, R. Krishnan, and C.V. Naidu. 2019. Pacific modulation of accelerated sea level rise during the early 21st century. *Climate Dynamics*. https://doi.org/10.1007/s00382-019-04795-0.

Kay, S., J. Caesar, J. Wolf, L. Bricheno, R.J. Nicholls, A.S. Islam, A. Haque, A. Pardaens, and J.A. Lowe. 2015. Modelling the increased frequency of extreme sea levels in the Ganges–Brahmaputra–Meghna delta due to sea level rise and other effects of climate change. *Environmental Science: Processes & Impacts* 17 (7): 1311–1322.

Kazeminezhad, M.H., I. Vilibić, C. Denamiel, P. Ghafarian, and S. Negah. 2020. Weather radar and ancillary observations of the convective system causing the northern Persian Gulf meteotsunami on 19 March 2017. *Natural Hazards*, pp. 1–23.

Khalilabadi, M.R. 2016. Tide–surge interaction in the Persian Gulf, Strait of Hormuz and the Gulf of Oman. *Weather* 71 (10): 256–261. https://doi.org/10.1002/wea.2773.

Khan, M.U.J., F. Durand, L. Testut, Y. Krien, and I.M. Saiful. 2020. Sea level rise inducing tidal modulation along the coasts of Bengal delta. *Continental Shelf Research* 211.https://doi.org/10.1016/j.csr.2020.104289.

King, M. 2014. Priorities for installation of continuous Global Navigation Satellite System (GNSS) near to tide gauges. Technical Report. University of Tasmania. October 2014. https://doi.org/10.13140/RG.2.1.1781.7049.

Kirezci, E., I.R. Young, R. Ranasinghe, S. Muis, R.J. Nicholls, D. Lincke, and J. Hinkel. 2020. Projections of global-scale extreme sea levels and resulting episodic coastal flooding over the 21st century. Science Report. 10-11629. https://doi.org/10.1038/s41598-020-67736-6.

Kulp, S.A., and B.H. Strauss. 2019. New elevation data triple estimates of global vulnerability to sea-level rise and coastal flooding (2020). *Nature Communications.* https://doi.org/10.1038/s41467-019-12808-z.

LeComte, D. 2021. International weather highlights 2020: Record Atlantic tropical season, historic flooding in Asia and Africa. *Weatherwise* 74 (3): 26–35.

Le Cozannet, G., D. Idier, M. de Michele, Y. Legendre, M. Moisan, R. Pedreros, R. Thiéblemont, G. Spada, D. Raucoules, and Y. de la Torre. 2021. Timescales of emergence of chronic flooding in the major economic center of Guadeloupe. *Natural Hazards and Earth System Sciences Discussions*, 21: 703–722. https://doi.org/10.5194/nhess-21-703-2021.

Lee, H.S. 2013. Estimation of extreme sea levels along the Bangladesh coast due to storm surge and sea level rise using EEMD and EVA. *Journal of Geophysical Research, c: Oceans* 118 (9): 4273–4285.

Lowe, J.A., P.L. Woodworth, T. Knutson, R.E. McDonald, K. McInnes, K. Woth, H. Von Storch, J. Wolf, V. Swail, N. Bernier, S. Gulev, K. Horsburgh, A.S. Unnikrishnan, J. Hunter, and R. Weisse. 2010. Past and future changes in extreme sea levels and waves. In *Understanding Sea-Level Rise and Variability* (Chap. 11), *ed.* J.A. Church, P.L. Woodworth, T. Aarup, and W.S. Wilson. London: Wiley-Blackwell.

Mani Murali, R., and P.K. Dinesh Kumar. 2015. Implications of sea level rise scenarios on land use/land cover classes of the coastal zones of Cochin, India. *Journal of Environmental Management* 148: 124–133.

McGranahan, G., D. Balk, and B. Anderson. 2007. The rising tide: Assessing the risks of climate change and human settlements in low elevation coastal zones. *Environment and Urbanisation* 19 (1): 17–37.

McInnes, K.L., K.J.E. Walsh, R.K. Hoeke, J.G. O'Grady, F. Colberg, and G.D. Hubbert. 2014. Quantifying storm tide risk in Fiji due to climate variability and change. *Global and Planetary Change* 116: 115–129.

Menéndez, M., and P.L. Woodworth. 2010. Changes in extreme high water levels based on a quasi-global tide-gauge dataset. *Journal of Geophysical Research, c: Oceans* 115: C10011. https://doi.org/10.1029/2009JC005997.

Merrifield, M., et al. 2009. The global sea level observing system (GLOSS). In *Proceedings of Ocean Obs'09: Sustained Ocean Observations and Information for Society* (Vol. 2), Venice, Italy, 21–25 September 2009, ed. J. Hall, D.E. Harrison, and D. Stammer European Space Agency Publication WPP-306. https://doi.org/10.5270/OceanObs09.cwp.63. http://www.oceanobs09.net/cwp/.

Muis, S., M.I. Apecechea, J. Dullaart, J.D.L. Rego, K.S. Madsen, J. Su, K. Yan, and M. Ferlaan. 2020. A high-resolution global dataset of extreme sea levels, tides, and storm surges, including future projections. *Frontiers in Marine Science* https://doi.org/10.3389/fmars.2020.00263.

Needham, H.F., B.D. Keim, and D. Sathiaraj. 2015. A review of tropical cyclone-generated storm surges: Global data sources, observations, and impacts. *Reviews of Geophysics* 53 (2): 545–591.

Nerem, R.S., B.D. Beckley, J.T. Fasullo, et al. 2018. Climate-change driven accelerated sea-level rise detected in the altimeter era. *PNAS* 115: 2022–2205.

Neumann, B., A.T. Vafeidis, J. Zimmermann, and R.J. Nicholls. 2015. Future coastal population growth and exposure to sea-level rise and coastal flooding—A global assessment. *PLoS ONE* 10 (3): e0118571.

Nidheesh, A.G., M. Lengaigne, J. Vialard, A.S. Unnikrishnan, and H. Dayan. 2013. Decadal and long-term sea level variability in the tropical Indo-Pacific Ocean. *Climate Dynamics* 41: 381–402.

Oppenheimer, M., B.C. Glavovic, J. Hinkel, R. van de Wal, A.K. Magnan, A. Abd-Elgawad, R. Cai, M. CifuentesJara, R.M. DeConto, T. Ghosh, J. Hay, F. Isla, B. Marzeion, B. Meyssignac, and Z. Sebesvari. 2019. Sea level rise and implications for low-lying Islands, coasts and communities. In *IPCC Special Report on the Ocean and Cryosphere in a Changing Climate*, ed. H.-O. Pörtner, D.C. Roberts, V. Masson-Delmotte, P. Zhai, M. Tignor, E. Poloczanska, K. Mintenbeck, A. Alegría, M. Nicolai, A. Okem, J. Petzold, B. Rama, and N.M. Weyer. Cambridge: Cambridge University Press.

Peltier, W.R. 2004. Global Isostacy and the surface of the ice-age earth: The ICE-5G (VM2) model and GRACE. *Annual Review of Earth and Planetary Sciences* 32: 111–149.

Pethick, J., and J.D. Orford. 2013. Rapid rise in effective sea-level in southwest Bangladesh: Its causes and contemporary rates. *Global and Planetary Change* 111: 237–245.

Pickering, M.D., K.J. Horsburgh, J.R. Blundell, J.J.M. Hirschi, R.J. Nicholls, M. Verlaan, and N.C. Wells. 2017. The impact of future sea-level rise on the global tides. *Continental Shelf Research* 142: 52–68. https://doi.org/10.1016/j.csr.2017.02.004.June.

Poulose, J., A.D. Rao, and S.K. Dube. 2020. Mapping of cylcone induced water levels along Gujarat and Maharashtra coasts: A climate change perspective. *Climate Dynamics*. https://doi.org/10.1007/s00382-020-05463-4.

Pramanik, M.K., S.S. Biswas, T. Mukherjee, A.K. Roy, R. Pai, and B. Mondal. 2015. Sea level rise and coastal vulnerability along the eastern coast of India through geo-spatial technologies. *Journal of Geophysics & Remote Sensing* 4: 2. https://doi.org/10.4172/2169-0049.1000145.

Pugh, D.T., and P.L. Woodworth. 2014. *Sea-Level Science: Understanding Tides, Surges, Tsunamis and Mean Sea-Level Changes*, 408p. Cambridge: Cambridge University Press. ISBN 9781107028197.

Rahman, S., A.S. Islam, P. Saha, A.R. Tazkia, Y. Krien, F. Durand, L. Testut, G.T. Islam, and S.K. Bala. 2019. Projected changes of inundation of cyclonic storms in the Ganges–Brahmaputra–Meghna delta of Bangladesh due to SLR by 2100. *Journal of Earth System Science* 128 (6): 1–11.

Rao, A.D., P. Upadhaya, S. Pandey, and J. Poulose. 2020a. Simulation of extreme water levels in response to tropical cyclones along the Indian coast: A climate change perspective. *Natural Hazards and Earth System Sciences* 100 (1): 151–172.

Rao, A.D., P. Upadhaya, H. Ali, S. Pandey, and V. Warrier. 2020b. Coastal inundation due to tropical cyclones along the east coast of India: An influence of climate change impact. *Natural Hazards and Earth System Sciences* 101 (1): 39–57.

Rhein, M., S.R. Rintoul, S. Aoki, E. Campos, D. Chambers, R.A. Feely, S. Gulev, G.C. Johnson, S.A. Josey, A. Kostianoy, C. Mauritzen, D. Roemmich, L. Talley, and F. Wang. 2013. In *Observations: Ocean*. In *Climate Change 2013: The Physical Science Basis. Contribution of Working Group I to the Fifth Assessment Report of the Intergovernmental Panel on Climate Change*, ed. T.F. Stocker, D. Qin, G.-K. Plattner, M. Tignor, S.K. Allen, J. Boschung, A. Nauels, Y. Xia, V., Bex, and P.M. Midgley, 255–315. Cambridge: Cambridge University Press.

Roxy, M.K., K. Ritika, P. Terray, and S. Masson. 2014. The curious case of Indian Ocean warming. *Journal of Climate* 27 (22): 8501–8509. https://doi.org/10.1175/JCLI-D-14-00471.1.

Seneviratne, S., et al. 2012. Changes in climate extremes and their impacts on the natural physical environment. Managing the risks of extreme events and disasters to advance climate change adaptation (SREX). In *A special report of Working Groups I and II of the Intergovernmental Panel on Climate Change*, ed. C.B. Field, V. Barros, T.F. Stocker, D. Qin, D.J. Dokken, K.L. Ebi, M.D. Mastrandrea, K.J. Mach, G.-K.Plattner, S.K. Allen, M. Tignor and P.M. Midgley, 582p. Cambridge: Cambridge University Press.

Shankar, D., S.G. Aparna, J.P. McCreary, I. Suresh, S. Neetu, F. Durand, S.S.C. Shenoi, and M.A. Al Saafani. 2010. Minima of interannual sea-level variability in the Indian Ocean. *Progress in Oceanography* 84 (3–4): 225–241.

Srinivasu, U., M. Ravichandran, W. Han, S. Sivareddy, H. Rahman, Y. Li, and S. Nayak. 2017. Causes for the reversal of North Indian Ocean decadal sea level trend in recent two decades. *Climate Dynamics*. https://doi.org/10.1007/s00382-017-3551-y.

Swapna, P., P. Jyoti, R. Krishnan, N. Sandeep, and S.M. Griffes. 2017. Multidecadal weakening of Indian summer monsoon circulation induces an increasing northern Indian Ocean Sea level. *Geophysical Research Letters*. https://doi.org/10.1002/2017GL074706.

Swapna, P., M. Ravichandran, G. Nidheesh, J. Jyothi, N. Sandeep, and J.S. Deepa. 2020. Sea-level rise, 2020. *Assessment of Climate Change over the Indian Region*. https://doi.org/10.1007/978-981-15-4327-2_9.

Torres, R.R., and M.N. Tsimplis. 2014. Sea level extremes in the Carrebean Sea. *Journal of Geophysical Research*. https://doi.org/10.1002/2014/C00929.

Unnikrishnan, A.S., D. Sundar, and D. Blackman. 2004. Analysis of extreme sea level along the east coast of India. *Journal of Geophysical Research. C: Oceans* 109: C06023.https://doi.org/10.1029/2003JC002217.

Unnikrishnan, A.S., and D. Shankar. 2007. Are sea-level-rise trends along the coasts of the north Indian Ocean consistent with global estimates? Global and Planet. *Change* 57: 301–307.

Unnikrishnan, A.S., M.R. Ramesh Kumar, and B. Sindhu. 2011. Tropical cyclones in the Bay of Bengal and extreme sea level projections along the east coast of India in a future climate scenario. *Current Science* 101 (3): 327–331.

Unnikrishnan, A.S., A.G. Nidheesh, and M. Lengaigne. 2015. Sea-level-rise trends off the Indian coasts during last two decades. *Current Science* 108 (5): 966–971.

Veit, E., and C.P. Conrad. 2016. The impact of groundwater depletion on spatial variations in sea level change during the past century. *Geophysical Research Letters*. https://doi.org/10.1002/2016GL068118.

Von Storch, H., and K. Woth. 2008. Storm surges: Perspectives and options. *Sustainability Science* 3 (1): 33–43.

Wadey, M., S. Brown, R.J. Nicholls, and I. Haigh. 2017. Coastal flooding in the Maldives: An assessment of historic events and their implications. *Natural Hazards and Earth System Sciences* 89 (1): 131–159.

Widlansky, M.J., A. Timmermann, and W. Cai. 2015. Future extreme sea level seesaws in the tropical Pacific. *Science Advances* 2015 (1): e1500560.

Woodworth, P.L., and D.L. Blackman. 2004. Evidence for systematic changes in extreme high waters since the mid-1970s. *Journal of Climate* 17: 1190–1197.

Woodworth, P.L., M. Menéndez, and W.R. Gehrels. 2011. Evidence for century-timescale acceleration in mean sea levels and for recent changes in extreme sea levels. *Surveys in Geophysics* 32 (4–5): 603–618.

Woodworth, P.L., and M. Menendez. 2015. Changes in the mesoscale variability and in extreme sea levels over two decades as observed by satellite altimetry. *Journal of Geophysical Research, c: Oceans*. https://doi.org/10.1002/2014JC10363.

Woodworth, P.L., J.R. Hunter, M. Marcos, P. Caldwell, M. Menendez, and I. Haigh. 2017a. Towards a global higher-frequency sea level dataset. *Geoscience Data Journal* 3: 50–59. https://doi.org/10.1002/gdj3.42.

Woodworth, P.L., G. Wöppelmann, M. Marcos, M. Gravelle, and R.M. Bingley. 2017b. Why we must tie satellite positioning to tide gauge data. *Eos* 98 (4): 13–15.

Chapter 11
Mapping of Coastal Inundation Due to Tropical Cyclones: A Numerical Study for the Indian Coasts

A. D. Rao and Smita Pandey

Abstract The Indian coastline is often prone to tropical cyclones that lead to high rise in water elevations and associated flooding. This chapter reviews studies on inter-action of storm surges and tides in the North Indian Ocean and the recent develop-ments in the numerical simulation of associated coastal inundation along the Indian coasts. The entire coast is mapped with maximum possible water elevations and water levels of the landward inundation as a combined effect of the storm surges, tides and wind waves. The simulations are made using a standalone ADCIRC to compute combined effect of storm surges and tides and coupled ADCIRC + SWAN model to include wind-wave effects well. These computations use climate change projections by enhancing the cyclonic wind intensity by 7 and 11% over the present scenario (no climate change) according to the IPCC fifth assessment report. The most probabilistic cyclone tracks for each coastal district and the intensity of the cyclone based on a 100-year return period are considered for the simulations. The study reveals that maximum water levels of about 10–12 m are found along with the northern coastal districts of Odisha. West Bengal districts are most vulnerable with reference to coastal flooding in the present scenario; however, they are least affected by 5–6% in climate change scenario (CCS). The districts in the Godavari river deltaic region in Andhra Pradesh will be inundated by more than 50% in CCS. Ramanathapuram district in Tamil Nadu will get water levels of more than 8 m in any scenario, while it is unaffected by the inundation due to local high topography. Along the west coast of India, the Gulfs of Khambhat and Kutch have maximum elevations of about 9–9.5 m in the present scenario; however, it will be increased to 10–11 m under climate change projections. The low-lying areas of Great and Little Rann of Kutch, Mumbai and high-tide mudflats of Bhavnagar are maximum affected by coastal inundation.

Keywords Tropical cyclone · Storm surge · Storm tide · Coastal inundation · Climate change scenario · Arabian Sea and Bay of Bengal

A. D. Rao (✉) · S. Pandey
Centre for Atmospheric Sciences, Indian Institute of Technology Delhi, Hauz Khas, New Delhi 110016, India
e-mail: adrao@cas.iitd.ac.in

© The Centre for Science & Technol. of the, Non-aligned and Other Devel. Countries 2022
A. S. Unnikrishnan et al. (eds.), *Extreme Natural Events*,
https://doi.org/10.1007/978-981-19-2511-5_11

Abbreviations

ACE	Accumulated Cyclone Energy
ADCIRC	Advanced Circulation Model
AP	Andhra Pradesh
AS	Arabian Sea
BMTPC	Building Materials and Technology Promotion Council
BoB	Bay of Bengal
CCS	Climate Change Scenario
CVI	Coastal Vulnerability Index
GEBCO	General Bathymetric Chart of the Oceans
GWCE	Generalized Wave Continuity Equation
IDW	Inverse Distance Weighted
IMD	India Meteorological Department
IPCC	Intergovernmental Panel for Climate Change
MWE	Maximum Water Elevation
MWL	Maximum Water Level
NIO	North Indian Ocean
PD	Pressure Drop
RSG	Radiation Stress Gradient
SRTM	Shuttle Radar Topography Mission
SS	Storm Surge
ST	Storm Tide
SWAN	Simulating Wave Nearshore
TC	Tropical cyclone
TN	Tamil Nadu
WB	West Bengal
WDS	Wetting and Drying Scheme

11.1 Introduction

Tropical cyclones (TC) are one of the significant hazards faced globally by coastal communities. The impact of an impinging TC includes strong winds, heavy rains, storm surges and inland flooding, leading to heavy destruction in the coastal regions. A storm surge (SS) is an increase in water levels near the coast in response to strong onshore cyclonic winds. The coastal inundation due to SS also causes heavy human casualties and disturbs man-made structures and derails the country's economy. The inland flooding due to cyclonic storm leads to the intrusion of saline water into inland coastal areas, which affects the agricultural productivity of the region, and in return disturb the livelihood of the local people. Land falling TC frequently threatens the North Indian Ocean (NIO) that encompasses the Arabian Sea (AS) in the north-west and the Bay of Bengal (BoB) in the northeast. The annual global average of cyclones in the basin is about 7% (Grey 1985), with higher cyclone activity in the BoB compared to the Arabian Sea (AS). The cyclone season in the NIO has two peaks, one during the pre-monsoon season (March to May) and another one during post-monsoon (October–December). On an average, about 5–6 tropical cyclones form in the BoB and the AS every year, out of which 2 to 3 may be severe. Figure 11.1 provides the monthly frequency of cyclonic storms and severe cyclonic storms together for the BoB and AS based on the cyclone data during 1891–2020. In the post-monsoon period, peak storm activity reaches in the second half of October and the first half of November. Compared to the pre-monsoon season, the months of October and November are known for severe storms. District-wise land falling cyclonic storms for both the east coast and west coasts of India are shown in Fig. 11.2.

The cyclones that develop in the BoB mainly make landfall along the east coast of India as the movement of storms from their location of genesis is mostly westward or north-westward (Holland 1983, 1995; Carr and Elsberry 1994, 1995). In the AS, contrary to this, a significant share of cyclones makes landfall along the coasts of the Persian Gulf and east African countries. The 1970 Bhola cyclone, the 1977 Machilipatnam cyclone, the 1999 Super cyclone and the recent Fani cyclone in 2019 are some of the most intense storms that made landfall at different locations along the east coast. Although the frequency of cyclones is relatively less in the AS, the coastal regions of north Maharashtra and Gujarat are more prone to tropical cyclones. Moreover, the low-lying areas having topography less than 10 m in the Great Rann of Kutch and Gulf of Khambhat are at higher risk due to cyclones. The cyclones in 1982 and 1998 are some of the notable ones, which caused higher surges and human causalities along the Gujrat coast.

When a cyclone passes through shallow oceans, the water levels keep on increasing and reach maximum along the coast during the landfall time (Murty et al. 1986).

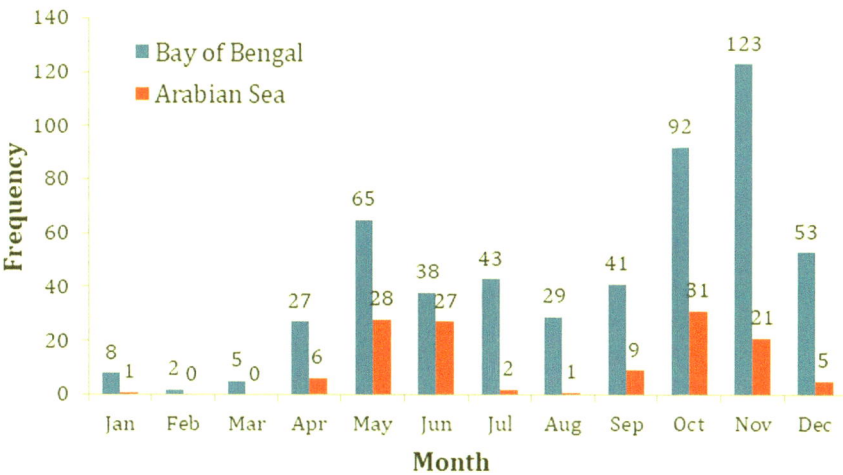

Fig. 11.1 Monthly frequency of cyclonic storms and severe cyclonic storms during 1891–2020 in the Bay of Bengal & Arabian Sea

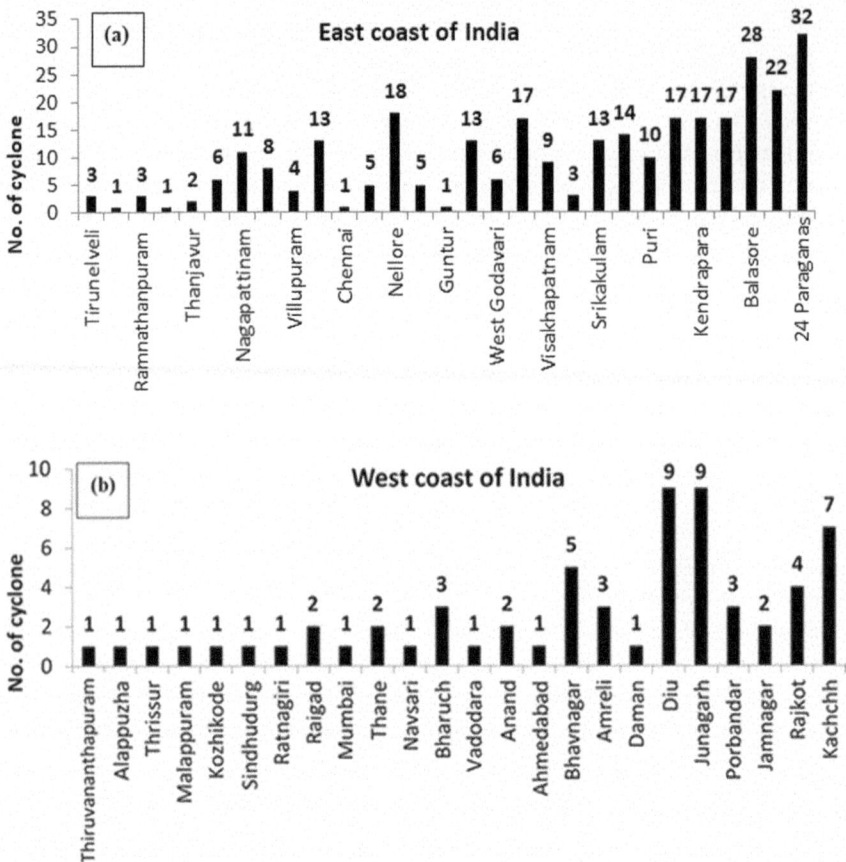

Fig. 11.2 Land falling cyclonic storms in district-wise during 1891–2020. **a** East coast, **b** West coast

The generation of SS depends upon many factors that include the intensity of the cyclone, its size, forward speed and the cyclone's approach angle. Rego and Li (2010) demonstrated that a slow-moving cyclone generates higher surges in the coastal bay/estuaries. Pandey and Rao (2019) carried out numerical experiments by taking idealized cyclone tracks and bathymetry to understand the role of the approach angle of a cyclone on the generation of surges and its interaction with tides along the coast. They inferred that cyclones making landfall perpendicular to the coast generate maximum surges, while the cyclones that travel parallel to the coast affect the maximum extent of the coastal stretch. The computation of SS is also sensitive to the coastal geometry and offshore bathymetry. The role of varying shelf width on the nonlinear tide-surge interaction is studied by Poulose et al. (2018). It is inferred that the interaction is maximum at low tide and minimum at flood tide on a broad shelf region with the maximum interaction of 15–20% observed for any tidal phase. The combined effect of tide and surge is referred to as storm tide (ST). Validation of the cyclonic wind distribution and its direction in the BoB using buoy data at different locations for the recent cyclonic cases is carried out for storm tides by Pandey and Rao (2018). Based on this study, a suitable modification is suggested to the Jelenianski's cyclonic wind distribution (Jelesnianski and Taylor 1973).

The coast along the western BoB is dominated by the semi-diurnal tide and the tidal range increases from south to north with a range exceeding 4 m at few locations in the head-bay region (Sindhu and Unnikrishnan 2013; Antony and Unnikrishnan 2013; Rao et al. 2020a). On the west coast of India, it is a mixed tide, while the southern part of the coast is dominated by the mixed semi-diurnal tide. The increase in tidal range from 0.8 m in the southern tip of Kerala to 5 m near north of Bombay is mainly due to an increase in the shelf width from 45 to 300 km, respectively. This is further amplified to about 11–12 m in the Gulf of Khambhat. Antony et al. (2014) demonstrated using satellite altimeter and tide-gauge data sets that the altimeter data can be used as a complementary data for the study of storm surges. In the recent paper of Antony et al. (2020), the tide-surge interaction during the Aila cyclone is studied using the ADCIRC model. It is displayed that shallow bathymetry, high tidal range and complex coastal configuration in the head-bay region result in strong tide-surge interaction. An investigation is also made by Antony et al. (2021) on extreme water levels along the central Red Sea coast using extreme value analysis based on annual maxima of hindcast water level data. It is inferred that the sea-level rise would reduce the average recurrence intervals of extreme water levels along the coastline.

Lewis et al. (2013) developed a regional inexpensive inundation model for the northern BoB using freely available Shuttle Radar Topography Mission (SRTM) topography data to compute flood extent during the 2007 Sidr cyclone. Coastal inundated area in response to a cyclone depends upon the surge levels, tidal amplitude and its phase and the nonlinear interaction of tides and surges. A particular storm may lead to significant coastal inundation when the storm surge coincides with the occurrence of the local high tide. Recently, Pinheiro et al. (2020) assessed the sensitivity of the flooded area due to tide-surge interaction in the Ria de Aveiro lagoon and found that it leads to higher flooding in the adjacent areas as the non-linearity increases from its mouth to the upper reaches. Therefore, for better understanding and estimation of extreme water elevations and coastal flooding, it is essential to consider the tide-surge interaction, which is mainly governed by the amplitude and phase of the tide (Horsburgh and Wilson 2007).

Among various processes, wind-wave setup also plays a vital role in the computation of ST and coastal flooding, particularly during a storm-approaching period, when the waves are gigantic at the coast. In shallow waters, the depth-induced breaking of waves releases radiation stress gradients (RSG) (Longuett-Higgins and Stewart 1964), which is an additional momentum to the water column causing wave setup of 10–20% of the breaking wave height and raising water levels adjacent to the coast (Holthuijsen 2010). This increase may be of tens of centimetres on the water elevation and is also affected by the tidal phase (Olabarrieta et al. 2011). For the Indian coast, Bhaskaran et al. (2013) simulated extreme waves along the Tamil Nadu (TN) coast during the Thane cyclone. They highlighted the significance of including the wind waves in the simulation of water levels at the coast. Murty et al. (2014)

used a coupled surge, tide and wave hydrodynamic modelling system to compute wave-induced setup for the Phailin cyclone. They inferred that there is an additional 23–36% increase in the peak surge compared to that of an uncoupled simulation. Recently, Murty et al. (2019) investigated the contribution of wave setup on the storm surge-induced coastal inundation using recent severe cyclonic storms, which made landfall along the east coast of India. The combined effect of storm surge, tide and wind wave is referred to as total water elevation (TWE).

The presence of river deltas in the cyclone-affected areas further enhances the flood risk in the coastal region. The increase in the population near the estuaries and deltas has further added to the coastal flood disaster. The rise of cyclone-generated water elevations at the coast penetrates through the river system and overflows from its banks, contributing to riverine flooding. This landward intrusion of seawater depends upon the local topography, upstream river discharge and precipitation associated with a cyclone. The distribution of precipitation around the cyclone varies and heavy precipitation occurs during landfall of the cyclone; hence, it is essential to understand their interaction and quantify its contribution. The intense rainfall during any cyclonic event causes sudden upstream runoff that leads to flash floods (Jonkman et al. 2009). Such flooding leads to severe damage and fatalities in the coastal regions. The river systems that join the sea transport accumulated inland precipitation towards the coast and the TWE that is contributed by storm surges, tides and wind waves from the downstream of the river interacts with the freshwater discharge and enhances the flood risk along the river banks (Bacopoloulus et al. 2017). The Krishna and Godavari rivers in Andhra Pradesh (AP), Mahanadi, Brahmani and Baitarani rivers in Odisha and Hooghly in West Bengal (WB) make the coastal districts more vulnerable to flooding. The coastal regions of Kutch, Jamnagar, Rajkot and Porbandar are vulnerable to inland flooding due to ST, and rivers like Sabarmati, Tapi, Mahi, Narmada and Dhadhar, enhance further the disaster due to cyclones. Ikeuchi et al. (2017) studied SS influence on fluvial flood by considering the Sidr cyclone as a case study in the Ganga, Brahmaputra, Meghna river delta regions. They found that SS causes an increase of about 3 m in water levels near the river mouth and about 0.7 m increase around 200 km away from the river mouth. It is also inferred that including sea-level dynamics in the river model has considerable effects in predicting the inundation. Gayathri et al. (2019) investigated the interaction of surge, tide and upstream discharge in the Hooghly River by taking an idealized river bed.

Future climate change projections reveal that intense cyclones with high frequency are more likely to happen in the global warming scenario. Singh et al. (2001) considered historical cyclone data during 1877–1998 and showed an increase in the frequency of intense cyclones in the NIO. In addition, the effect of climate change (CC) increases the risk of coastal flooding. The IPCC (2014a) report projects an increase in the frequency of intense cyclones in response to a rise in sea surface temperature. Furthermore, various studies are conducted regarding the increase in the number and intensity of cyclones in the AS. A recent study of Deo and Ganer (2014) indicates an increase in the intensity of tropical cyclones in the NIO over the past 15 years. Anthropogenic global warming increases the probability of post-monsoon extreme severe cyclonic storms over the AS compared to the BoB (Murakami et al.

2017). A study by Evan et al. (2011) reveals that weakening of vertical wind shear in monsoon circulation due to increase in anthropogenic emissions of aerosols favours pre-monsoon and post-monsoon TC intensification in the AS. As per the supplementary material of AR5 (Fifth Assessment Report), IPCC (2014b), the expected percentage change in mean Lifetime Maximum Intensity of TCs over the period 2081–2100 relative to 2000–2019 ranges between −10 and +10% for the North Indian Ocean. The projected tropical cyclone wind speed increment for the end century (RCP8.5, 2081–2100) in the Arabian Sea is ~ 2 to 8 m/s (IPCC 2014b). Based on the studies of Knutson et al. (2010, 2013), it is reported that the average intensity of the cyclones globally will increase 2–11% by 2100.

Coastal flooding due to variability in future cyclones and sea-level rise (SLR) is one of the main concerns for the coastal populations today. Moreover, SLR is expected to continuously escalate in the twenty-first century. The studies of Nicholls et al. (1999), Church et al. (2006), Irish and Resio (2013), Lin et al. (2012) have focussed on the contribution of SLR and increase number of intense storms in cyclone flooding. Karim and Mimura (2008) demonstrated that rise in sea surface temperature by 2 °C and sea-level rise of 0.3 m leads to storm surge flooding in the western Bangladesh. Using geospatial technique, Hoque et al. (2018) developed cyclone risk modelling approach for the present and future climate change scenarios and tested it in Sarankhola Upazila of coastal Bangladesh. They integrated local sea-level rise of 0.34 m for the year 2050 with surge models developed for 100-year return periods and showed that climate change scenario intensifies the cyclone risk area by 5–10% in every return period.

Rao et al. (2020a) adopted the same 2–11% increment in the cyclonic winds, representing 7% (moderate scenario) and 11% (extreme scenario), which also falls in the range of 2–8 m/s as reported by IPCC. In this study, Maximum Water Elevations (MWE) at the coast are computed for the return period of 10, 50 and 100 years for both east and west coast of India. These are computed as a combined effect of storm surges and local high tide. It is demonstrated that MWE in the moderate and extreme scenarios will increase by 20% and 30%, respectively. In the climate change scenario (CCS), Rao et al. (2020b) carried out computation of coastal inundation due to TC in the present and climate change scenarios. They inferred that the river deltaic regions of Krishna, Godavari, Mahanadi and Hooghly along the east coast are highly prone to high water levels and associated coastal flooding. Poulose et al. (2020) carried out coastal flood mapping along the Gujrat and north Maharashtra coasts by enhancing the present cyclonic wind intensity by 7% and 11%. It is found that the regions will be more affected by the depth of inundation, while it will be a negligible impact on the extent of inundation in the CCS. The focus of this chapter is to highlight and consolidate all the studies in the past on coastal inundation along the Indian coasts as result of ST and wind waves generated by TCs. However, this study does not take into account of mean sea-level rise.

11.1.1 Brief Description of ADCIRC and SWAN

In recent studies, numerical modelling has included a finite-element based hydro-dynamic model, the Advanced Circulation Model (ADCIRC) developed by the US Army Corps of Engineers, Engineering Research and Development Centre, University of Notre Dame and University of North Carolina jointly in 1990 (Luettich et al. 1992; Luettich and Westerink 1999). To minimize the damage, the prediction of inland inundation is as important as the prediction of ST along the coast. With the implementation of a wetting and drying scheme (WDS), the model can simulate coastal flooding. The ADCIRC, the widely used circulation model, implemented the WDS to compute coastal inundation associated with the cyclone. According to the scheme, the particular element is activated or deactivated after checking it against the minimum threshold depth such that if the depth of the node is more than the threshold value, then it is activated and included in the computations; otherwise, it remains dry and deactivated. To avoid spurious noise, the current stage of the node remains active for some time before changing the stage.

Moreover, to include the effect of wind waves on the computation of SS, the coupled version of the ADCIRC+ Simulating Wave Nearshore (SWAN) model (Dietrich et al. 2011, 2012) is used. The storm tides and associated currents generated in the ADCIRC model interact with the wave-induced setup from the SWAN model through RSG. The coupled ADCIRC + SWAN runs on the same unstructured mesh and the information from both the models are exchanged at the coupling time interval that helps to capture the physical mechanism in the wave-current information. These models use numerous information like cyclonic wind distribution, tides, wind waves, coastal topography and its geometry. The MWE computed near the coast is the combined effect of the nonlinear interaction of surges, tides and wind waves.

11.1.2 Computation of Storm Surges and Coastal Inundation

Rao et al. (2013) configured a finite-element-based ADCIRC model for the east coast of India to compute coastal inundation due to 1989, 1996 and 2000 cyclones crossing different parts of the east coast of India. Bhaskaran et al. (2014) computed the extent of coastal inundation along the TN coast for the Thane cyclone. The model computed peak surge and inland inundation are found in good agreement with the field measurements. They emphasized that coastal beach slope primarily affects the inundation extent. Later many researchers (Srinivasa et al. 2015; Gayathri et al. 2016; Murty et al. 2017; Mandal et al. 2020) carried out numerical simulations to compute coastal inundation associated with cyclones. Recently, river-surge-tide interaction is illustrated by Gayathri et al. (2019) by incorporating idealized Hooghly River in the ADCIRC model. They also investigated the role of cyclonic wind stress forcing along with river discharge on storm surge propagation and inferred that lower discharge with higher stress increases the penetration of surge inside the river. Sahoo and

Bhaskaran (2018) investigated coastal vulnerability index (CVI) associated with TC for the Odisha coast in a CCS. They included surge height and coastal inundation along with eight weather parameters to determine the physical vulnerability index. Later, they combined social, economic and environmental vulnerability to compute the overall CVI. Moreover, it is also essential to assess the coastal vulnerability in terms of maximum elevation and extent of inundation for the entire Indian coast for long-term planning due to climate change.

11.2 Model Domain and Computation of Tides

For computation of storm tides, a high-resolution mesh is configured for the east coast and the west coast of India by using the ADCIRC model. The model bathymetry is derived from 30 s General Bathymetric Chart of the Oceans (GEBCO) and topography from the 90 m resolution SRTM data. While the model bathymetry for the Gulf of Khambhat region is considered from the field observations of National Institute of Ocean Technology (Giardino et al. 2014). Unstructured triangular mesh for both the coasts is generated using the above bathymetry and topography data, as shown in Figs. 11.3 and 11.4. The finite-element-based mesh is generated using Surface Modelling System. The east coast model has a minimum resolution of 200 m near the shallow regions and a maximum of about 20 km in deeper waters, while the minimum resolution for the west coast model is 20 m and 35 km near the nearshore and open boundary, respectively.

The ADCIRC model is used to compute tides along the Indian coastline by prescribing tidal elevations at the open boundary of the east coast mesh, extracted from 'Le Provost' tidal database using 13 major tidal constituents (K1, M2, N2, O1, P1, S2, K2, L2, 2N2, MU2, NU2, Q1 and T2) (Le Provost et al. 1995, 1998). However, the major tidal constituents are extracted from the TPXO model (Egbert and Erofeeva 2002) and provided as the open boundary forcing in the west coast model. Bottom friction is an essential factor in deciding the amplitude of modelled tide. In the ADCIRC, a hybrid bottom friction formulation with a minimum drag coefficient of 0.0015 and 0.0022 is used for the east and west coast, respectively. Time discretization using an explicit scheme is maintained at a model time-step of 2 s for the east coast and 0.5 s for the west coast. The models are initially spun up for 60 days using tidal amplitude and corresponding phase of different constituents till it reaches a steady state. The tidal solution for the last 15 days is validated at different tide-gauge locations as discussed in Rao et al. (2020a). In Fig. 11.5, the maximum tidal range is shown at every 10 km along the east and west coasts of India. Along the west coast, the tidal range increases from 1 to 9 m from the southern part of the coast to the Gulf of Khambhat and about 5 m in the Gulf of Kutch. However, a monotonic increase in tidal range is noticed along the east coast from south to north with a maximum of about 3.6 m in the Hooghly region.

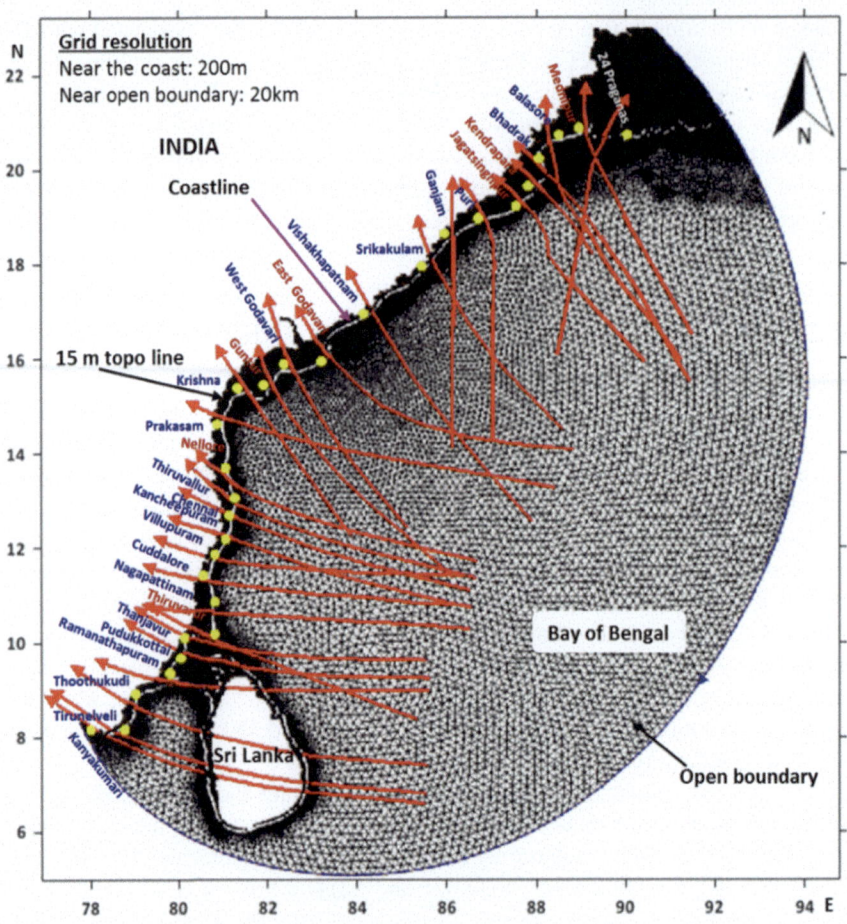

Fig. 11.3 Model mesh covering the east coast of India along with synthetic cyclone tracks

11.3   Coastal Inundation Along the East Coast of India: A Climate Change Perspective

11.3.1   Data and Methodology

The inventory of the past cyclone tracks is used to compute synthetic tracks for each district along the east coast of India. The return period of pressure drop (PD) for different time periods is also calculated based on the historical cyclone data. The experiments are performed for each return period of 10, 50 and 100 years by considering the cyclonic wind enhancement of 7% and 11% over the present. Here, 7% refers to the moderate scenario and 11% is an extreme one. The ADCIRC model is used for computing maximum possible water elevations as a combined effect of

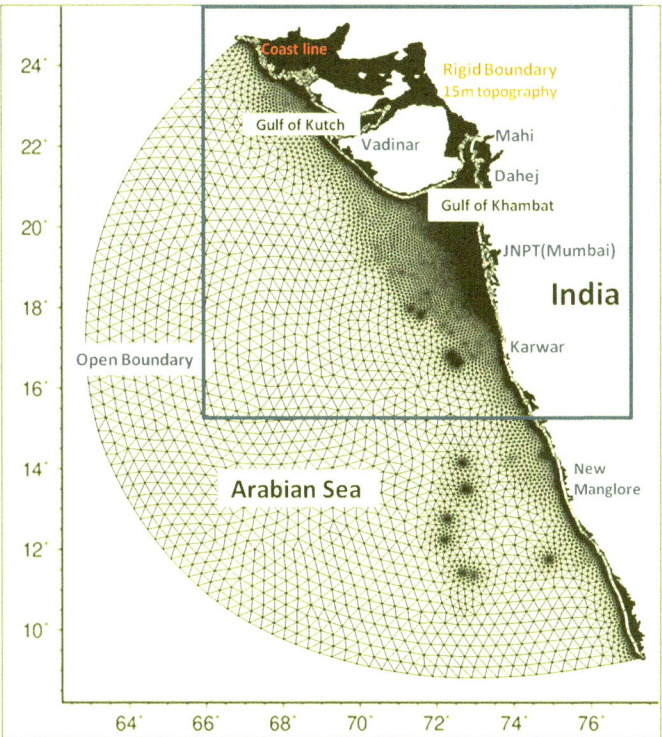

Fig. 11.4 Model grid for the west coast of India

surge and tide, known as ST as mentioned earlier and the extent of inland inundation with water levels along the coast. The ST is computed at every 10 km along the coast for different return periods and CCS. While the computations are made using the PD for a 100-year return period to compute the maximum possible impact on the ST and inundation. As shown in Fig. 11.3, the rigid boundary of the domain is kept at 15 m topography contour line to simulate coastal inundation.

There are four coastal marine states along the east coast of India, viz., TN, AP, Odisha and WB in the north. These states consist of total of 31 coastal districts having 13 in TN, 9 in AP, 7 in Odisha and 2 in WB. Some of the coastal districts become more vulnerable due to prevailing major river systems like Krishna and Godavari in AP, Mahanadi, Brahmani and Baitarani Rivers in Odisha and Hooghly in WB. The low-lying areas in the river deltaic regions are prone to higher flooding during any cyclonic event. The cyclone track information and its intensity are collected from India Meteorological Department (IMD) for the period 1891–2018 and other sources like vulnerability atlas (BMTPC 2006). A uniform master data set is prepared after the reconciliation of all the data available. Synthetic tracks, the most probable tracks, are constructed based on the dataset of cyclone tracks. The tracks are generated using the inverse distance weighted (IDW) method using actual cyclone tracks (Weber

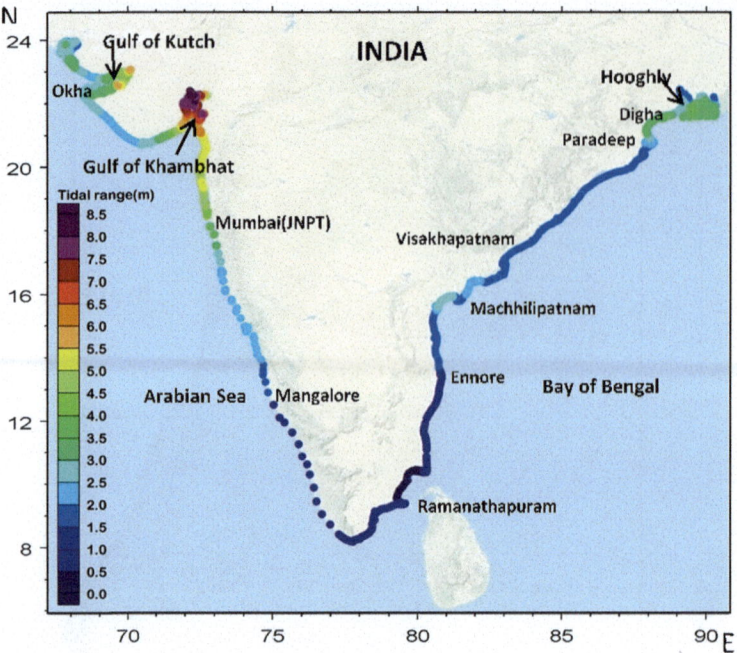

Fig. 11.5 Tidal range (m) along the Indian coasts

and Englund 1994; Nayak and Bhaskaran 2014; Sahoo and Bhaskaran 2016). The return periods are also computed using statistical projection based on extreme value analysis (Gumbel 1958) of the PD and the details of the distribution are given in Rupp and Lander (1996). In Fig. 11.6, all the major river systems along the east coast are represented in our model and the detailed land topography is shown in Fig. 11.7, which is used to explain the extent of inland flooding and corresponding water levels.

For the computation of MWE and inland inundation due to cyclonic storms, one of the essential input parameters is cyclonic surface wind distribution. In the present simulations, it is computed using a dynamic wind model of Jelesnianski and Taylor (1973). The only input required by the wind model is the maximum PD, the radius of maximum winds and the position of the cyclone (latitude and longitude) to compute the wind distribution. In the present simulations, an average value of 30 km for R_{max} is considered for all the simulations based on the observations of the cyclones in this region (Mohapatra and Sharma 2015). The ramp function for all the simulations is about one day. The weighting factor (τ_0) in Generalized Wave Continuity Equation (GWCE) and eddy viscosity are considered as 0.05 and 2 m^2/s (Cyriac et al. 2018), respectively. The wetting and drying algorithm is activated in these simulations, and Garratt's (1977) wind drag formulation is applied.

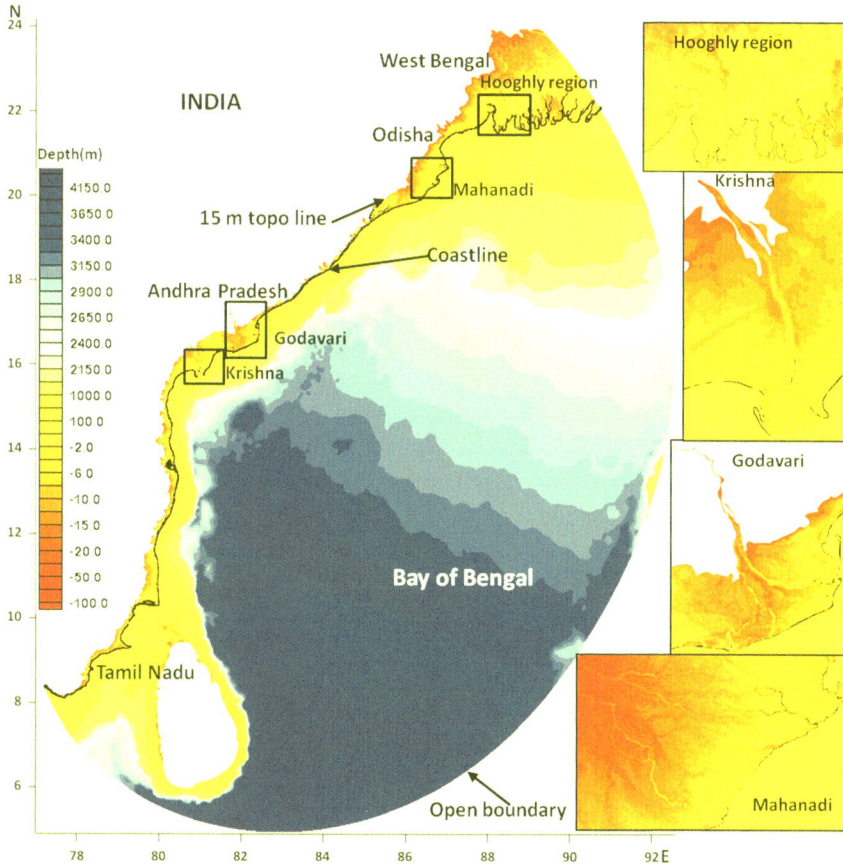

Fig. 11.6 Bathymetry and onshore topography for the model domain

11.3.2   Results and Discussions

11.3.2.1   Odisha

Odisha experienced a super cyclonic storm in 1999 with a maximum PD of 98 hPa, which represents a 100-year return period (Table 11.1). There are six coastal districts in Odisha, having a total extent of about 480 km coastline. Six synthetic tracks are constructed for these districts and moved to every 10 km within the district along the coastline, resulting in 48 tracks to compute maximum elevations and coastal inundation. Figure 11.8 depicts the MWE along the coast and inland inundation with maximum water levels (MWL) above the local topography due to the land falling cyclones within Odisha. Based on the past cyclone data, 28 cyclones made landfall in the Balasore district along the Odisha coast. The cyclones in this district generate high water elevations along the coast and the extent of inundation is more

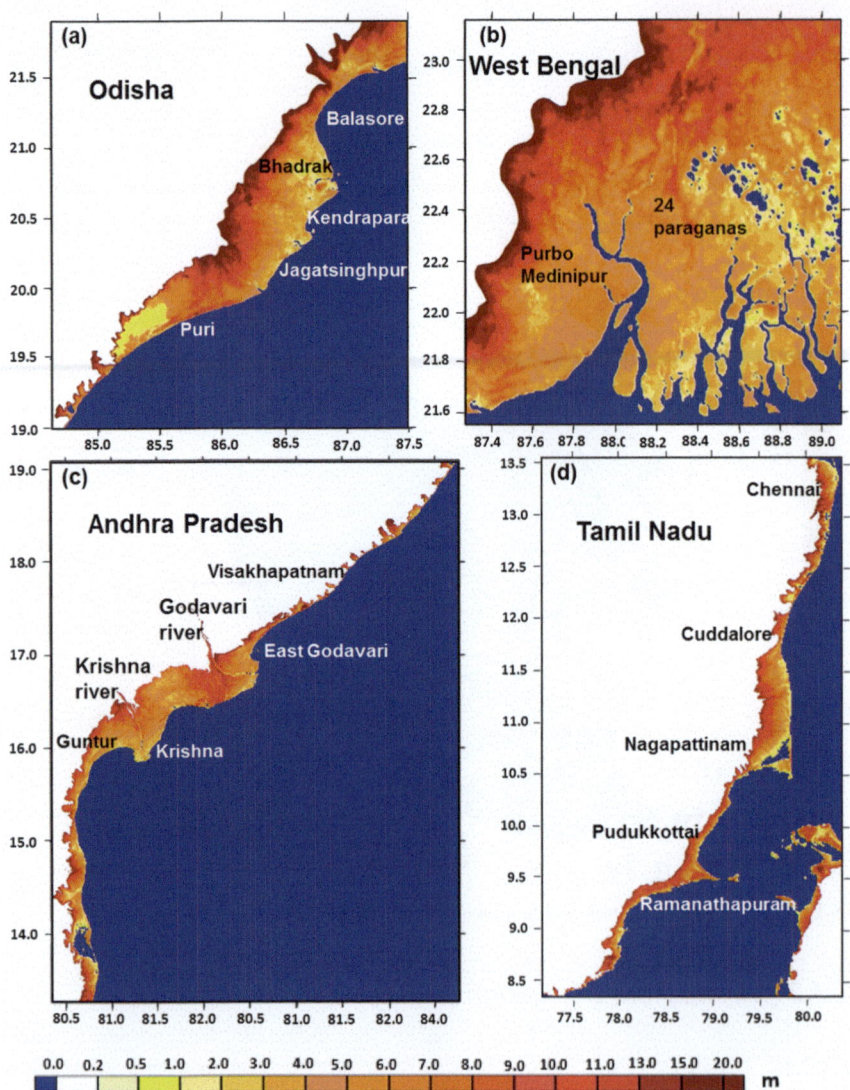

Fig. 11.7 Detailed land topography up to 15 m for all maritime states along the east coast of India: **a** Odisha, **b** WB, **c** AP and **d** TN

Table 11.1 Values of PD by return period for the east coast of India

S. No.	Return period (year)	PD (hPa)			
		WB	Odisha	AP	TN
1	10	25	45	48	26
2	50	51	78	70	47
3	100	75	98	82	66

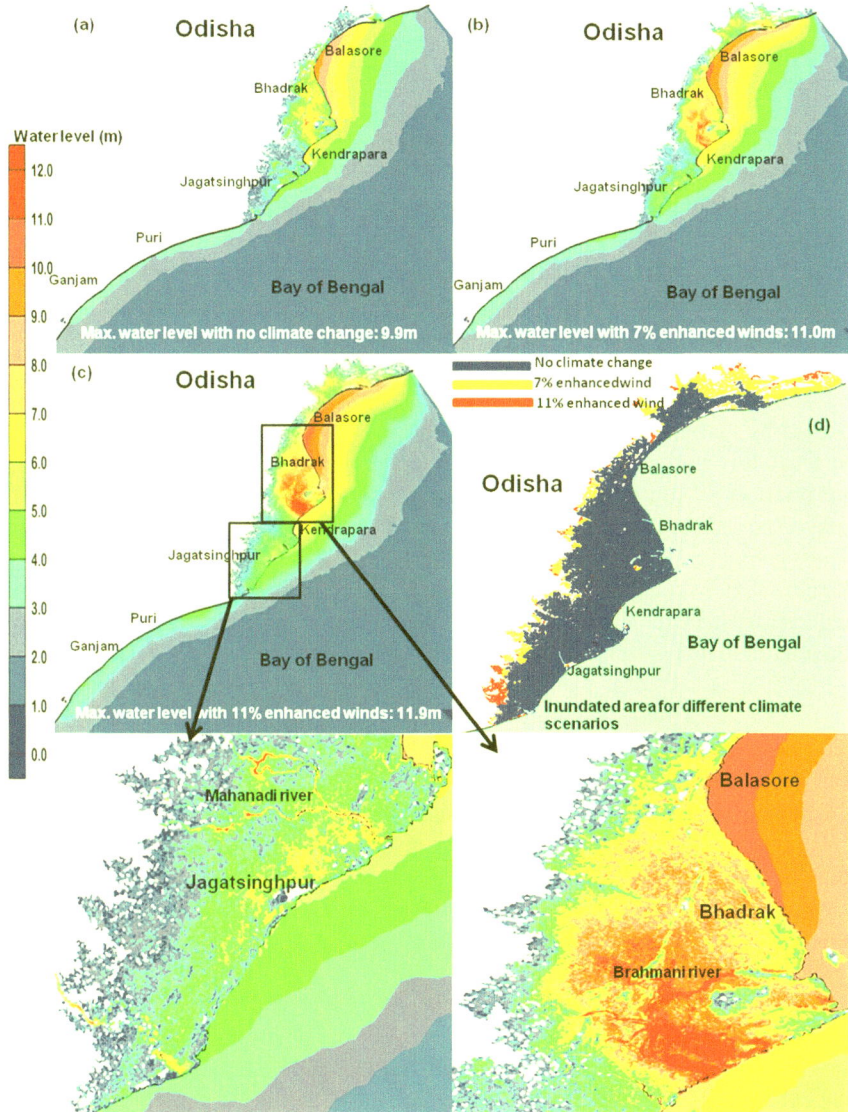

Fig. 11.8   Composite depiction of MWE and MWL due to all possible cyclones crossing Odisha coast with **a** no climate change, **b** 7% increment in winds, **c** 11% increment in winds and **d** Inundated area for all climate scenarios

in the southern districts of Bhadrak and Kendrapara. This may be attributed to the fact that these districts experience onshore cyclonic winds, leading to higher water elevations and inundation before the landfall of the cyclone. The MWL of about 10 m is simulated in the Dhamara region near the Brahmani River's tributaries (Fig. 11.8a). The maximum extent of inundation is found up to 40 km interior as the storm waters penetrate through the river systems.

In the CCS, the MWL is increased by about 1 m and 2 m for 7% (Fig. 11.8b) and 11% (Fig. 11.8c), respectively. However, the associated inundation is not much affected due to the local topography (Fig. 11.7a). Figure 11.8d shows the possible inundated area superimposed from all different climate scenarios for the Balasore district, including no climate change (present scenario). It is important to note that south of Balasore district is more affected with the present scenario itself. The MWE and associated inundation for all the scenarios reveal that Puri and Ganjam are not affected by the inundation. However, a storm tide of about 4 m is simulated along the coast. It is due to the presence of high topography, which varies from 8 to 15 m. The presence of river deltaic regions, local coastline geometry and lower topography generates higher storm tides and inundation in the districts of Bhadrak and Jagatsinghpur in all scenarios. The northern districts are affected by the maximum inundated extent of about 50 km inside in the case of no climate change, while it will be the least affected by the CCS. This is mainly due to the high local topography of about more than 5 m beyond the inundated topo line in the no CCS.

11.3.2.2   West Bengal

WB has two coastal districts, Medinipur and 24-Paraganas, with about 150 km coastline. In this case, 15 synthetic tracks are considered for every 10 km interval. It has the most irregular and complicated coastline geometry with many river systems and its tributaries extending much interior of the coast. As a result, it leads to higher inundation due to cyclones making landfall in this area. The MWE along the coast for the 10-year return period is about 3 m. While the MWL is about 8–9 m observed for the 100-year return period in Medinipur and 24 Parganas, causing vast inundated area in all the scenarios as shown in Fig. 11.9a–c. This may be attributed to the presence of a large river deltaic region with topography varying between 1 and 6 m (Fig. 11.7b). Most of the coastal districts are inundated in the present scenario up to the extent of about 130 km interior, which is the highest among all the coastal districts (Fig. 11.9d). It is interesting to note that not much additional area of this region will be affected due to any CCS over the present scenario.

11.3.2.3   Andhra Pradesh and Tamil Nadu

For the state of AP, experiments are carried out for eight coastal districts. It is the second-largest state in terms of coastline of about 972 km and is the most affected by TC after Odisha. Simulations carried out for MWE and coastal inundation suggest

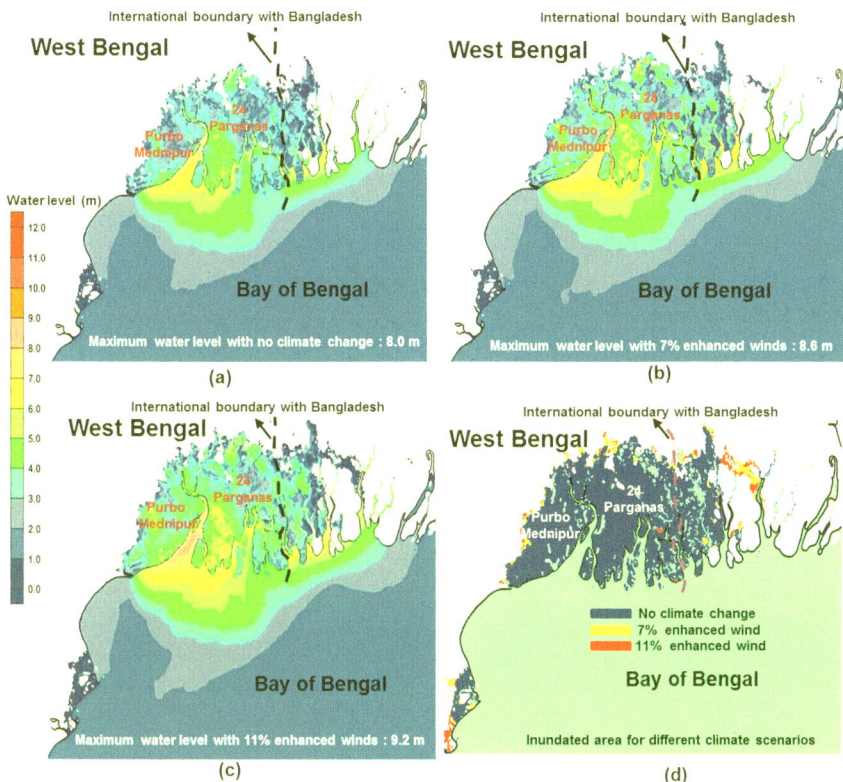

Fig. 11.9 Composite depiction of MWE and MWL due to all possible cyclones crossing WB coast with **a** no climate change, **b** 7% increment in winds, **c** 11% increment in winds and **d** Inundated area for all climate scenario

that East and West Godavari and Krishna districts are prone to higher water elevations due to the presence of large river systems, as shown in Fig. 11.10. A maximum coastal flooding extent of about 60 km is computed in Krishna and Guntur districts as they are located in the low-lying river deltaic regions (Fig. 11.7c). While the MWL of about 6–9 m is computed along the rivers due to enhanced wind stress in response to CCS. Moreover, it is also noticed that all the southern districts are more affected for any return period compared to northern districts.

In TN, there are 13 coastal districts with a total coastline length of about 1076 km. For the model simulations, 107 synthetic tracks are used for every 10 km. In terms of maximum elevation and coastal inundation, the region between Nagapattinam and Ramanathapuram is more affected (Fig. 11.11). From the simulations of 10-year return period for all scenarios, it is observed that the Ramanathapuram is more affected by an elevation of about 3 m. In comparison, the MWE for 50 and 100-year return periods has a similar pattern. Around Nagapattinam, the MWL of 6 m is computed. The coastal inundation extent is about 45 km as the topography varies

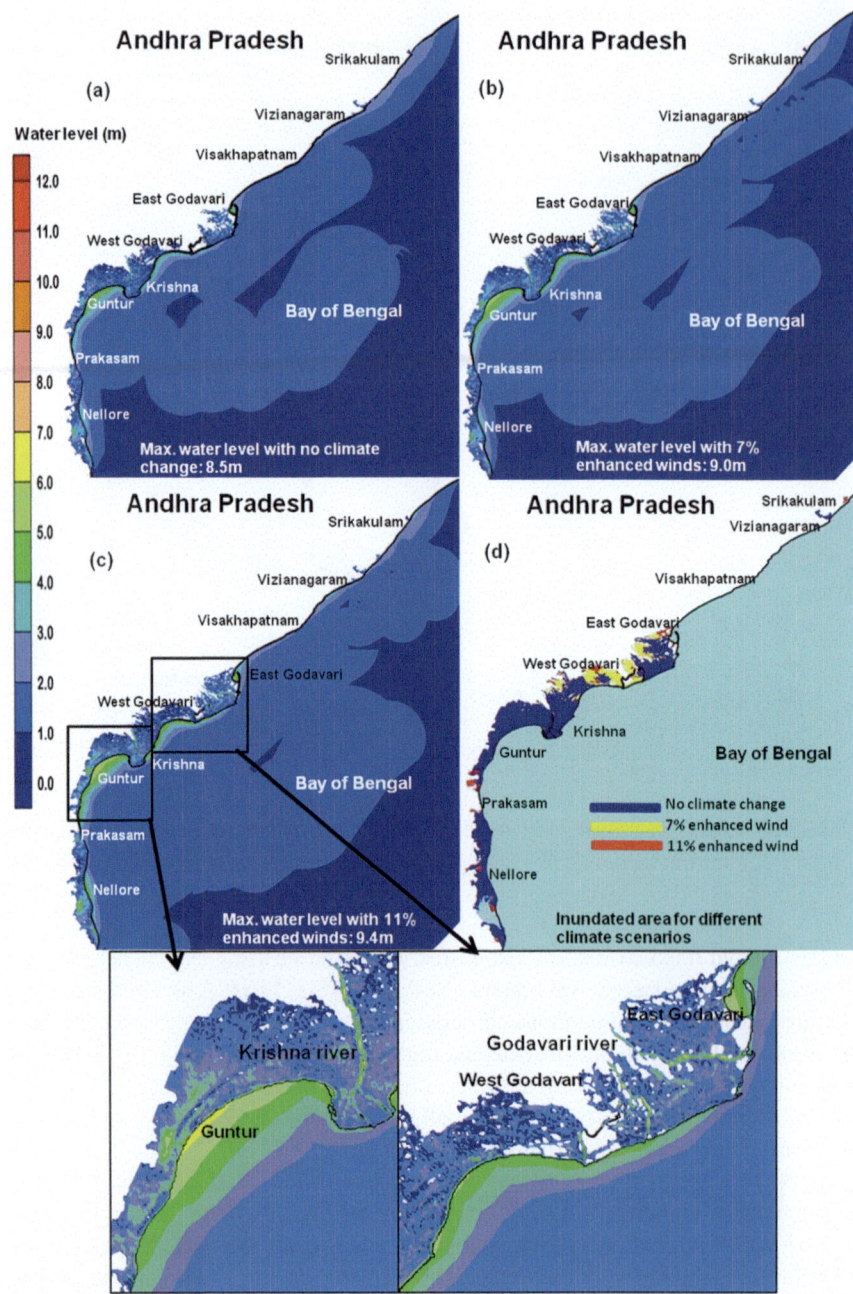

Fig. 11.10 Composite depiction of maximum water levels due to all possible cyclones crossing AP coast with **a** no climate change, **b** 7% increment in winds, **c** 11% increment in winds and **d** Inundated area for all climate scenarios

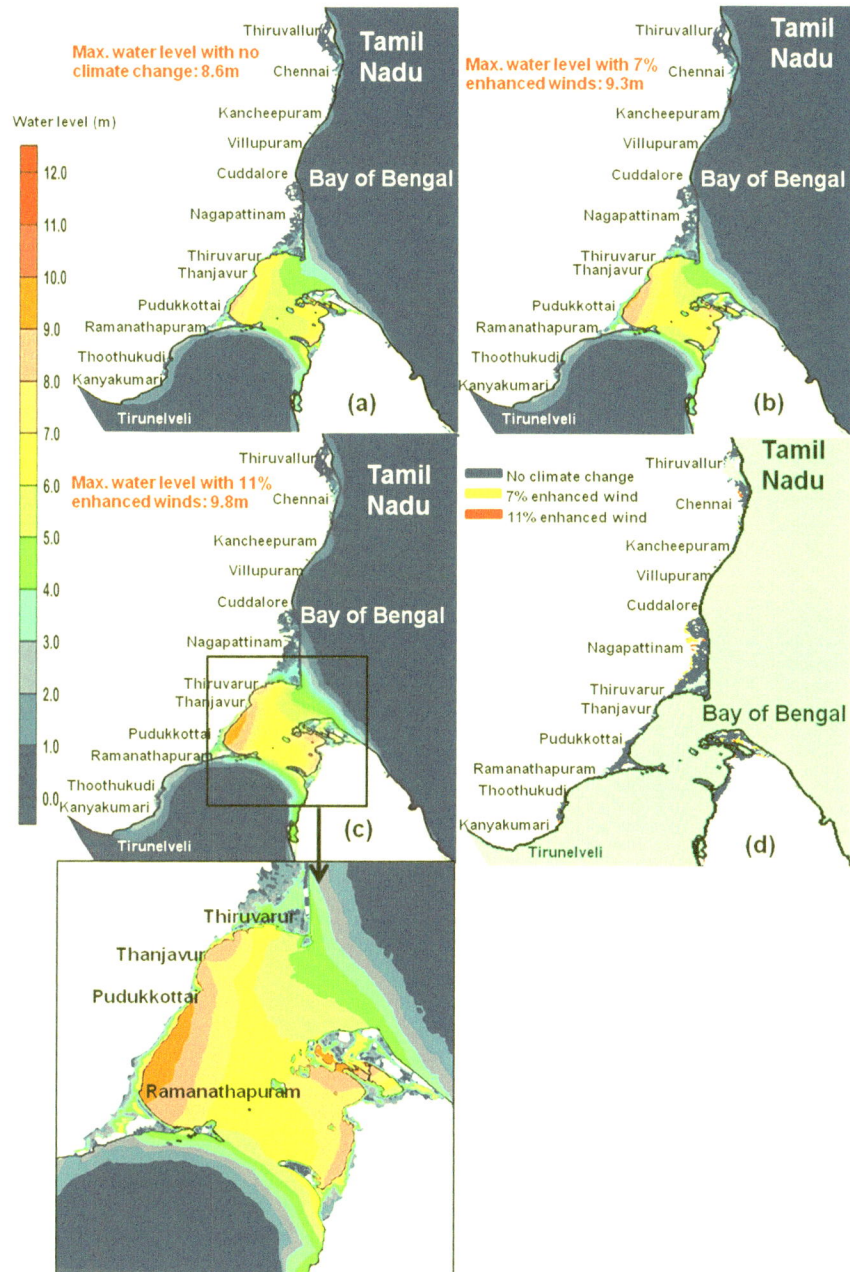

Fig. 11.11 Composite depiction of MWE and MWL due to all possible cyclones crossing TN coast with **a** no climate change, **b** 7% increment in winds, **c** 11% increment in winds and **d** Inundated area for all climate scenarios

from 2 to 5 m (Fig. 11.7d). Though the regions of Ramanathapuram experience 8-10 m of water elevation and the extent of inundation is only about 7 km with no flooding is seen interior (Fig. 11.11d) as the topography varies from 9 to 12 m. The higher elevation in the region is due to the coastline's local curvature and also it is bounded by Sri Lanka on the eastern side. As a result, storm waters are trapped, causing higher elevations on either side of the coast.

It is observed that all the coastal districts experience a different kind of water elevations for all possible scenarios, mainly varying between 2 and 12 m. This may be caused due to different intensities of cyclones, local topography and curvature of the coastline. It is also noticed that higher water elevations near the coast may not increase inundated areas of a particular coastal stretch. Figure 11.12a shows how much total area is inundated district-wise for each scenario. Among all the districts,

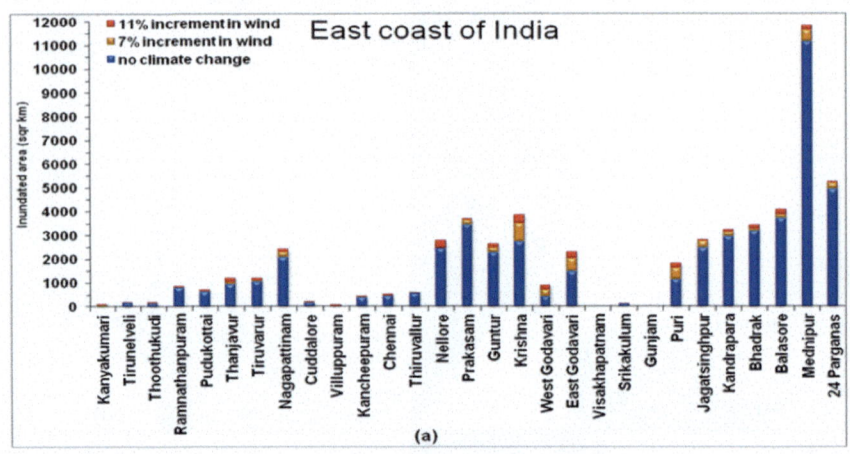

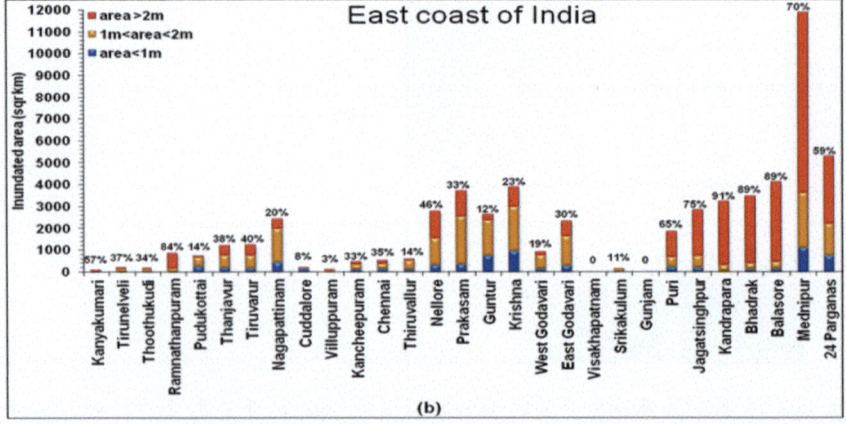

Fig. 11.12 **a** District-wise inundated area along the east coast of India for all possible climate change scenarios **b** Inundated area at different water levels

Medinipur in WB has the maximum inundated area of about 11,500 m^2 in a present scenario, mainly due to complex coastal geometry. To understand the inundated area's vulnerability/severity, the total area with elevation less than 1 m, between 1 and 2 m and more than 2 m in the case of extreme CCS are shown in Fig. 11.12b. It is observed that most of the coastal areas in AP experience water elevations only up to 2 m. Coastal areas of Odisha and WB are mostly inundated by more than 2 m with a maximum inundated area in Medinipur. This study is useful for stakeholders to prepare a long-term development plan.

11.4 Coastal Inundation for the West Coast of India: A Climate Change Perspective

11.4.1 Data and Methodology

In this section, the simulations of MWE and associated inland inundation are discussed for the west coast of India. Here, the computation of MWE and MWL are limited to the North of Maharashtra and Gujarat as it is the most cyclone-prone area compared to the southern part of west coast of India (refer to square box in Fig. 11.4). The model domain is used in a coupled version of the ADCIRC + SWAN model to compute maximum elevation and inland flooding as a combined effect of the surge, tide and wind wave. For the west coast of India, the nonlinear interaction of storm surges, tides and wind waves is considered and hence MWE represents in this section the combined effect of all. The construction of synthetic tropical cyclone tracks for the study region is carried out using the IDW method as discussed in the earlier section utilizing the past cyclone information. The TC tracks information is collected from the best-track data (www.rsmcnewdelhi.imd.gov.in, www.metoc.navy.mil/jtwc/jtwc.html?best-tracks) and cyclone eAtlas (www.rmcche nnaiatlas.tn.nic.in), produced by IMD. As in the case of the east coast of India, the rigid boundary of the domain for the west coast (refer Fig. 11.4) is also kept at 15 m topography contour line for simulation of coastal inundation.

Based on the past cyclone tracks' approach angle that made landfall along the coast, the region of interest is divided into five different zones as Zone1, Zone2, Zone3, Zone4 and Zone5, as shown in Fig. 11.13. Zone1 is considered from the northern tip of Gujarat coast to Porbandar, including the Gulf of Kutch. Zone2 covers from Porbandar to Diu. The entire Gulf of Khambhat is comprised within Zone3 and extended up to Valsad. Zone4 is a small stretch of coast from Valsad to the north of Mumbai (Nandgaon). Mumbai region is enclosed in Zone5 till Dapoli in Maharashtra. The bathymetry and topography of all five zones are shown in Fig. 11.14. A synthetic track in the category of Severe Cyclonic Storm to Extremely Severe Cyclonic Storm is constructed for each zone separately. The total number of synthetic tracks considered for each zone is 25, 13, 13, 9 and 14 for Zone1 to Zone5, respectively. In Table 11.2, the maximum wind speed in the present scenario from Zone5 to Zone1 varies from

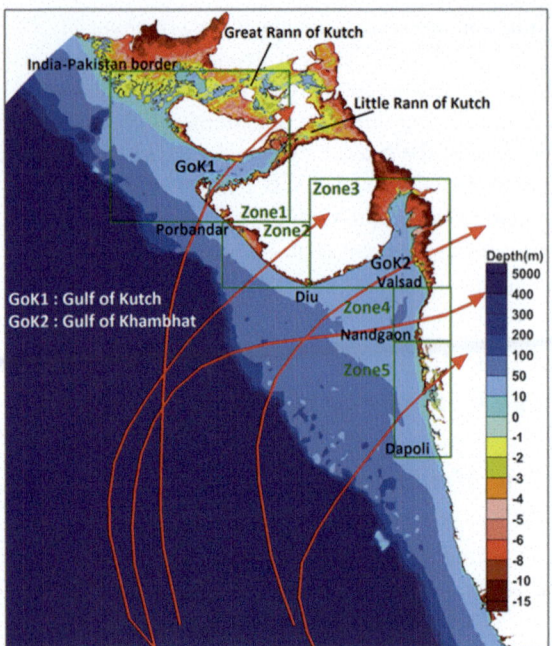

Fig. 11.13 Analysis area for computation of coastal inundation along with synthetic cyclone tracks for Zone1, Zone2, Zone3, Zone4 and Zone5

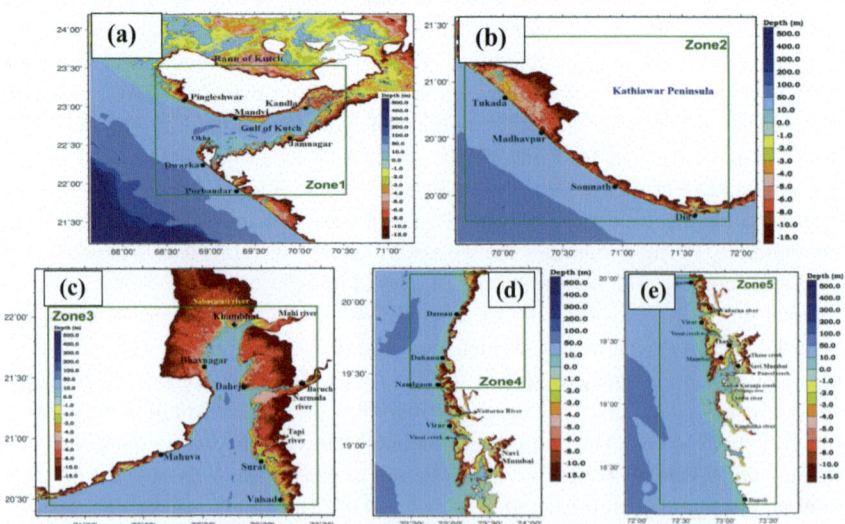

Fig. 11.14 Bathymetry (zoomed) derived from GEBCO and topography from SRTM of **a** Zone1, **b** Zone2, **c** Zone 3, **d** Zone 4, **e** Zone 5

Table 11.2 Maximum intensity of the cyclone considered for each zone

Zones	No. cyclone tracks	Max. pressure drop (ΔP)	Maximum wind speed (m/s)		
			Present scenario	7% Wind intensification	11% wind intensification
Zone 1	25	66 hPa	60 m/s	64 m/s	66.6 m/s
Zone2	13				
Zone3	13	50 hPa (Tracks 1–9)	52 m/s	56 m/s	58 m/s
		66 hPa (Tracks 10–13)	60 m/s	64 m/s	66.6 m/s
Zone4	9	40 hPa	45 m/s	49.2 m/s	51 m/s
Zone5	14				

45 m/s (40 hPa) to 60 m/s (66 hPa). The enhanced maximum wind speed for moderate and extreme scenarios has an increment of about 4–5 m/s and 6–7 m/s, respectively. The detailed bathymetry and topography of each zone derived from GEBCO and SRTM, respectively, is depicted in Fig. 11.14.

11.4.2 Results and Discussions

11.4.2.1 Zone1

Zone1 includes the area covering from north of Gujarat to Porbandar, including the Gulf of Kutch. The maximum simulated tidal range near Kandla, Mandvi and Porbandar is about 7 m, 3 m and 2 m, respectively. Total of 25 tracks are considered in this zone. The higher MWE in the region is mainly caused due to the funnelling shape of the Gulf, shallow offshore depths, high tidal range and the presence of small inlets. Figure 11.15a–f depicts the elevation along the coast and areas that could be affected by MWL during a cyclone for different CCS. The maximum extent of inundation is observed in the Rann of Kutch and adjoining low-lying areas of the innermost Gulf of Kutch for all the scenarios. The topography is within 5 m (Fig. 11.14). The inundated region is broadly classified based on the height of the water elevation. In this study, the areas of inundation are categorized into >7 m, >5 m, >3 m, >1 m and within 0.5 m to indicate the nature of vulnerability. The highest MWL of 9.4 m is noticed near the Kandla region for the present scenario. Simultaneously, it is further increased by 0.2 m and 0.6 m in moderate and extreme scenarios, respectively. There are many river inlets join the Gulf of Kutch near Kandla. Large amount of storm waters penetrate these inlets and spills out, resulting in high water levels and hence, the larger inundated area. This region falls into the most severe category of >7 m in all CCS.

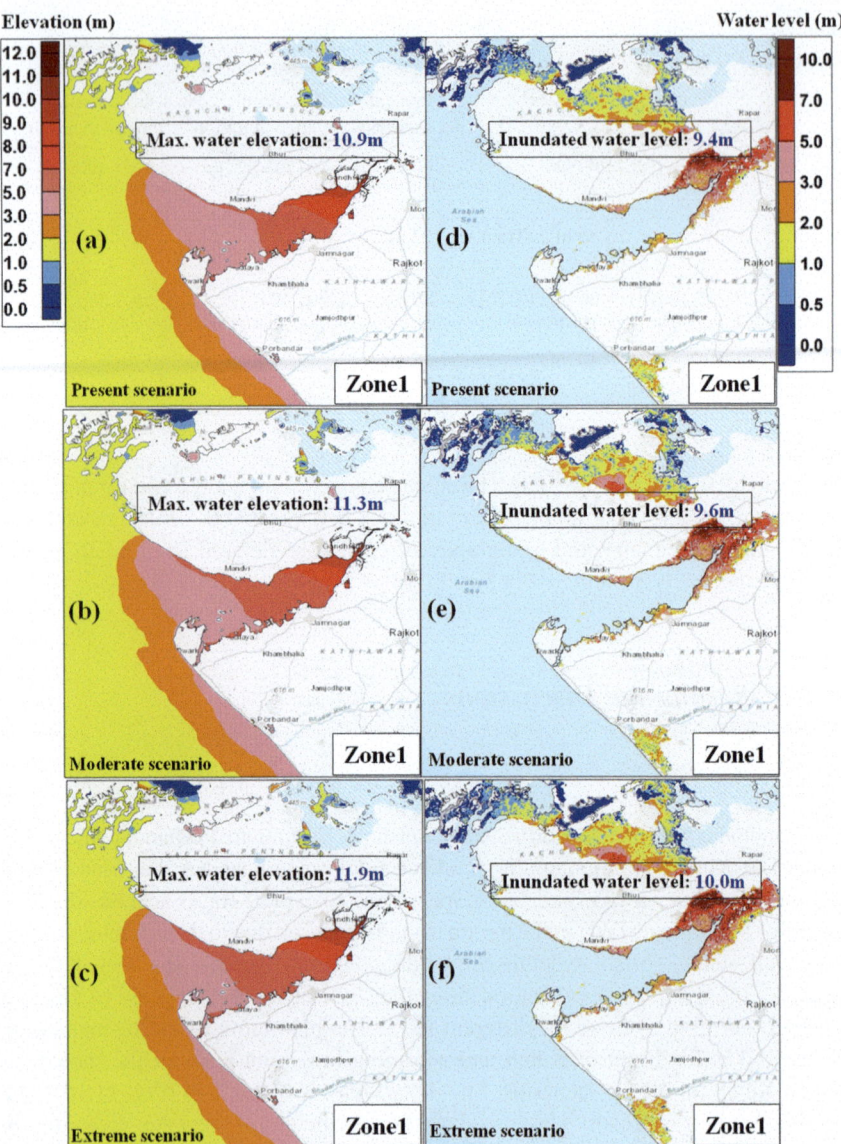

Fig. 11.15 Composite depiction of MWE generated due to the cyclone tracks of Zone1 for **a** present scenario, **b** moderate scenario **c** extreme scenario and MWL for **d** present scenario, **e** moderate scenario **f** extreme scenario

11.4.2.2   Zone 2

Zone 2 is comprised of the coast from Porbandar to Diu. A total of 13 synthetic tracks are used here for the model simulations. The continental shelf width of this zone is about 80–100 km and has a relatively straight coastline. Figure 11.16a–f depicts the elevation along the coast and areas that could be affected in terms of MWL during a cyclone for different CCS. The minimum topographic height is about 6 m along the coast (Fig. 11.14), which is higher than the elevation values and protects the coast from inland inundation. This area is not inundated even in the moderate

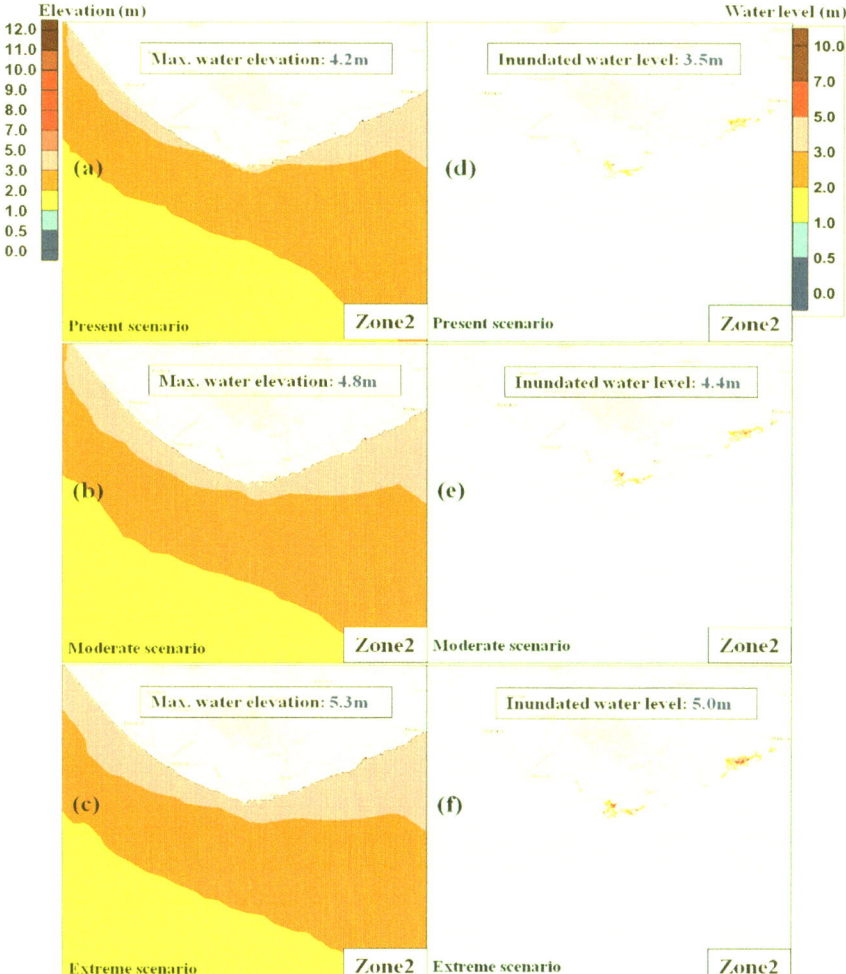

Fig. 11.16 Composite depiction of MWE generated due to the cyclone tracks of Zone 2 for **a** present scenario, **b** moderate scenario **c** extreme scenario and MWL for **d** present scenario, **e** moderate scenario **f** extreme scenario

and extreme scenarios (Fig. 11.16d–f. The cyclone, landfall close to Porbandar, has produced minimal extent of inundation in Zone2.

11.4.2.3   Zone 3

Zone3 covers the Gulf of Khambhat and the area is extended from Diu to Valsad. The simulated tidal range over this region increases from 3 m to 11–12 m from its mouth to the Gulf's interior. The tidal range near Diu is about 2.5 m, while it is about 5 m near Valsad, which is located on the right side of the Gulf (refer Fig. 11.13). A total of 13 synthetic cyclones are generated to simulate maximum elevation and inland flooding over this zone. These tracks are shifted and made landfall at every 10 km interval along this coast. Figure 11.17a–f depicts the elevation along the coast and areas that could be affected by MWL during a cyclone for different CCS. The highest MWL of 9 m, 9.6 m and 10 m are computed for the present, moderate and extreme scenarios, respectively, near Bhavnagar, located inside the Gulf (Fig. 11.17d–f). The presence of many small river systems in the Gulf of Khambhat's surroundings, like Sabarmati, Mahi, Narmada and Tapi and other inlets, enhance the elevation. The concave-shaped coast and shallow depth at the river's mouth and other inlets may also contribute to the higher elevation in the region. The adjoining area of Narmada will get inundated for the first time in the moderate scenario and the MWL reaches up to 1–3 m, which will increase to 5 m for the extreme scenario. The low-lying areas (<10 m topography) (refer Fig. 11.14), adjacent to Bhavnagar, are the most affected region of inundation and the MWL computed is about 9 m for the present scenario and is increased to 10 m for the extreme scenario.

11.4.2.4   Zone 4

This zone extends from Valsad (Gujarat) to Nandgaon (Maharashtra) and has the shortest coastline and the shelf width along this coast is about 330 km. The simulated maximum tidal range along this zone is~5 to 6 m (Fig. 11.5). For this zone, nine cyclone tracks are used for the model simulations. Figure 11.18a–f depicts the elevation along the coast and areas that could be affected by MWL during any cyclone for different CCS. The maximum simulated MWL is about 7.3 m in the present scenario, as shown in Fig. 11.18d. It is further increased to 8.2 m and 9.2 m for moderate and extreme scenarios, respectively (Fig. 11.18e, f). The maximum inundation is simulated along Virar and Navi Mumbai's coast, which falls in the neighbouring Zone5. In the present scenario, MWL around the location of Virar varies between 5 and 7 m and it is 2–5 m near the Navi Mumbai region. In the extreme scenario, MWL will reach up to 7–8 m and 3–7 m at these locations.

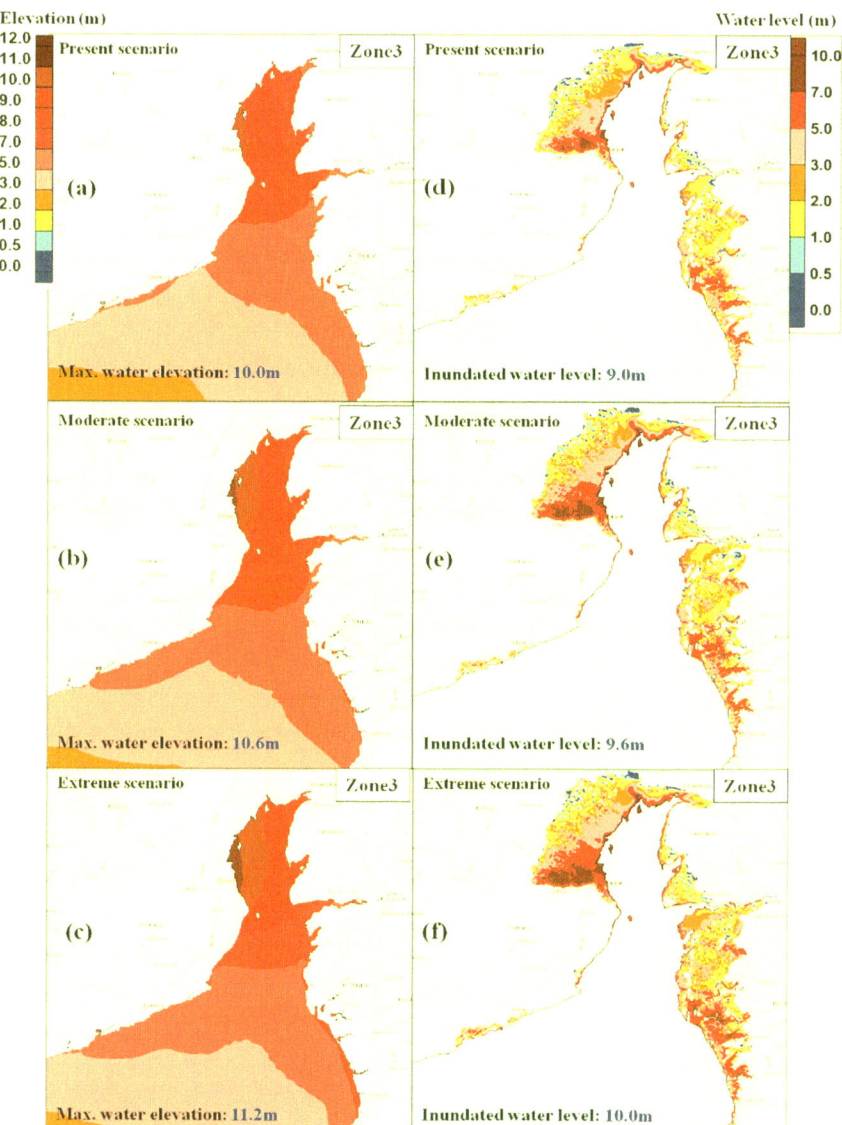

Fig. 11.17 Composite depiction of MWE generated due to the cyclone tracks of Zone3 for **a** present scenario, **b** moderate scenario **c** extreme scenario and MWL for **d** present scenario, **e** moderate scenario **f** extreme scenario

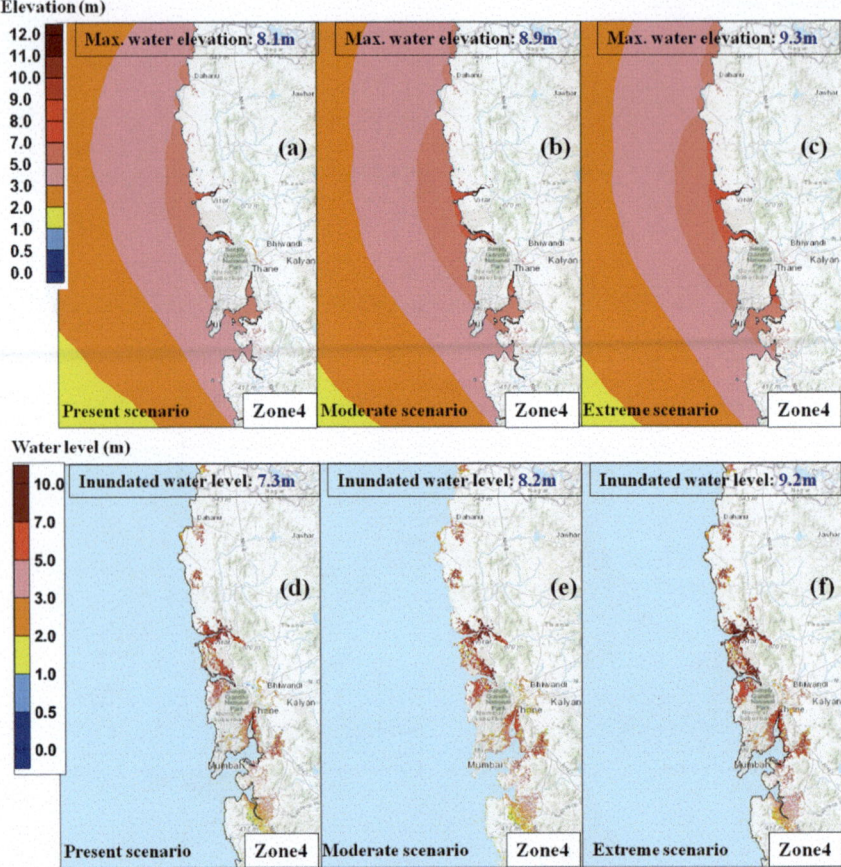

Fig. 11.18 Composite depiction of MWE generated due to the cyclone tracks of Zone4 for **a** present scenario, **b** moderate scenario **c** extreme scenario and MWL for **d** present scenario, **e** moderate scenario **f** extreme scenario

11.4.2.5 Zone 5

The coast from Nandgaon to Dapoli is covered in Zone5 and Mumbai region is also included in this zone. The complex coastal geometry and shallow coastal waters in the Navi Mumbai region generate higher tides. The presence of shallow waters and complex coastal geometry and the low-lying topography adjoining the Mumbai region increases the risk of coastal flooding. A total of 14 cyclone tracks are considered in this zone. For the present scenario, the simulated MWL is about 9 m, which resulted in more inundation across the adjacent low-lying regions of Navi Mumbai, Thane and the river banks (Fig. 11.19d). The average topography over this area is about 7 m. The MWL computed for the moderate and extreme cases are shown in Fig. 11.19e, f.

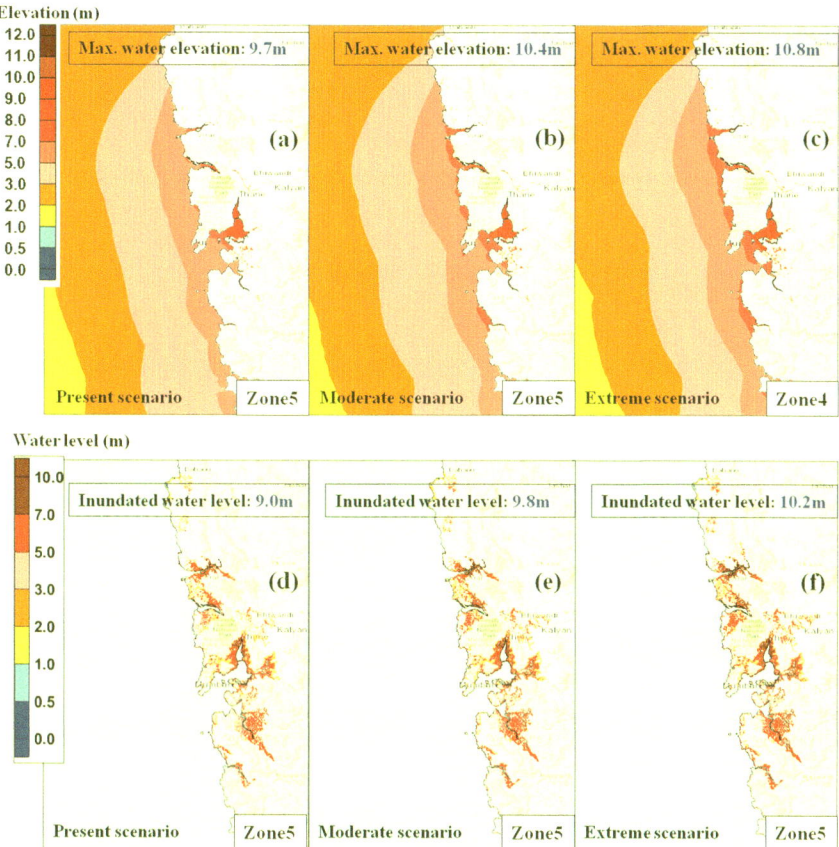

Fig. 11.19 Composite depiction of MWE due to the cyclone tracks of Zone5 for **a** present scenario, **b** moderate scenario **c** extreme scenario and MWL for **d** present scenario, **e** moderate scenario **f** extreme scenario

11.4.2.6   Maximum Extent of Coastal Inundation for Different Scenarios

The increase in MWE due to the moderate and extreme scenario above the present scenario is for all the five zones is also plotted to understand the impact of CCS (Fig. 11.20a). In Zone1, the maximum increase of 5 m is observed over a small region in the Rann of Kutch for the extreme scenario, which may be due to its low-lying topography. The extent of inundation is increased only by about 5% and 10% for the moderate and extreme scenarios, respectively. The impact of CCS on the extent of inundation is seen relatively less. However, an increase of 5 m in MWL for extreme scenarios is computed near the Gulf of Kutch and Rann of Kutch.

In contrast, the cyclones close to Diu in Zone2 generated MWL of about 5 m in the neighbouring Zone3 till Mahuva. The extent of inundation is observed up to 3 km

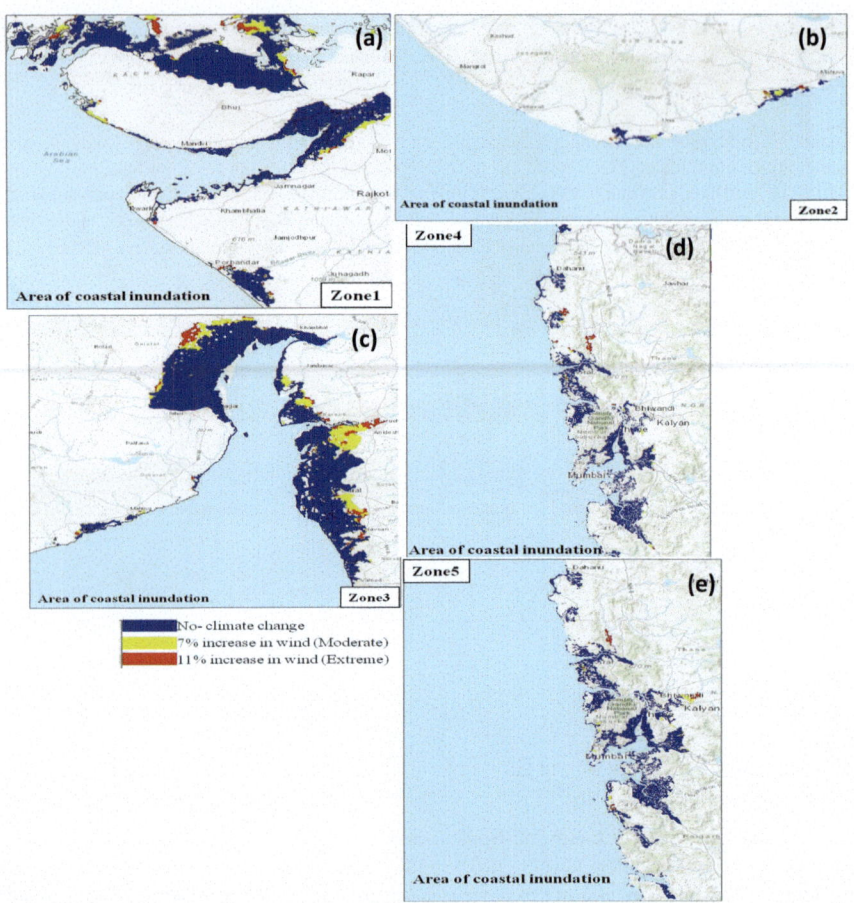

Fig. 11.20 Composite picture of maximum extent of coastal inundation for different scenarios **a** Zone1, **b** Zone2, **c** Zone3, **d** Zone4, **e** Zone5

with an MWL of 3.5 m along the coast from Diu to Mahuva for the present scenario (Fig. 11.20b). The maximum additional water levels of 1.0 m and 1.5 m are estimated above the present scenario in the same region for moderate and extreme cases. A highly elevated area of the Kathiawar Peninsula is restricting further intrusion of inundated waters landward. For moderate and extreme scenarios, ~20% and ~30% of extra area is inundated, respectively, compared to the no CCS. In Zone 3, about 10% additional area of total extent will be inundated with the moderate scenario and another 5% for the extreme scenario. The impact of CCS (Fig. 11.20c) is observed more near the river banks, both in the extent of inundation and water levels. However, the total extent of inundation due to cyclones in Zone 4 is comparatively less for any CCS, as shown in Fig. 11.20d. And the extra area of inundation in Zone 5 for moderate and extreme scenarios is negligible, which infers that this zone is unaffected by the

CCS for the area of inundation (Fig. 11.20e). Moreover, the maximum extent and height of inundation are observed along and across the river plains and creeks.

11.5   Conclusions

This chapter reviews the recent studies on storm tides and associated inland inundation along the Indian coasts. The simulations carried out with the state-of-the-art model, ADCIRC, are particularly discussed in this chapter. The discussion also covers numerical simulations performed by considering district-wise synthetic tracks, constructed based on past cyclone data to compute maximum possible elevations and corresponding coastal flooding for the present and climate change scenarios. The impact of climate change is seen by enhancing the present cyclonic wind intensity (no climate change or present scenario) in the model forcing by 7% (moderate scenario) and 11% (extreme scenario). Along the east coast, it is observed that river deltaic regions like Hooghly, Mahanadi, Brahmani, Baitarani, Krishna and Godavari are positively affected by higher water levels of flooding. In the present scenario, the northern districts of Odisha are most affected by coastal flooding, but during the CCS, the additional inundated area simulated is only 5–8%. In the extreme climate change scenario, the districts of Jagatsinghpur, Kendrapara and Bhadrak, along with the Odisha coast, will experience higher water levels of more than 11 m. In Tamil Nadu, the Kanyakumari district will witness an increase in the water levels by about 28% in the moderate scenario and 44% in the extreme scenario. In terms of the extent of coastal flooding, West Bengal is completely affected, with an inundated area of about 130 km. This is mainly caused due to the presence of the complex coastal geometry of the Hooghly estuary and its tributaries.

Mapping of coastal inundation is also presented for the coasts of north Maharashtra and Gujrat using climate change projections. The domain of this region is conveniently divided into five zones based on the past cyclone data. The coupled ADCIRC + SWAN model is used in this study for the simulation of maximum possible water elevations along the coast and subsequent flooding in the interior. In the case of no climate change scenario, the MWE of about 10 m is simulated inside the Gulf of Khambhat and Kutch, 7 m along Mumbai coast and the lowest levels of 4–5 m along the Porbandar to Diu coast. The additional water elevations of about 0.5–1.0 m and 1–2 m are simulated during moderate and extreme scenarios, respectively. The maximum extent of inundation is noticed in the Rann of Kutch, the Gulf of Khambhat and also near the Mumbai coast. However, the impact of climate change is small in the extent of inundation, with the maximum increase of only about 5 m in the inundation height (water level) along the Narmada and Tapi River banks and in the Great Rann of Kutch. The mapping of maximum elevations with coastal flooding along the Indian coasts is a crucial component that will assist the stakeholders at different levels for sustainable development plans and help in mitigation.

The chapter elaborates and highlights the interaction of storm surges, tides and wind waves in simulating maximum elevations along the coast and inland flooding due to cyclones. However, the only unresolved problem that needs to be given more attention for operational needs is the interaction of storm tides with river streams and the effect of cyclonic precipitation on the inundation. Therefore, it is essential to include river systems with accurate incorporation of river depths, freshwater discharge from the upstream end and rainfall during the cyclone period in the models to compute precise coastal flooding. This will significantly improve the accuracy in the computation of inundation, especially in the river deltaic regions.

Acknowledgements We are very thankful to the Department of Science and Technology for the financial support by awarding the project to IIT Delhi to carry out this work. We are also very grateful to the Indian Institute of Technology Delhi HPC facility and the Department of Science and Technology, Government of India, for giving financial support (DST-FIST 2014) for computational resources.

References

Antony, C., and A.S. Unnikrishnan. 2013. Observed characteristics of tide-surge interaction along the east coast of India and the head of Bay of Bengal. *Estuarine, Coastal and Shelf Science* 131: 6–11.

Antony, C., L. Testut, and A.S. Unnikrishnan. 2014. Observing storm surges in the Bay of Bengal from satellite altimetry. *Estuarine, Coastal and Shelf Science* 151: 131–140.

Antony, C., S. Langodan, H.P. Dasari, O. Knio, and I. Hoteit. 2021. Extreme water levels along the central Red Sea coast of Saudi Arabia: Processes and frequency analysis. *Natural Hazards* 105 (2): 1797–1814.

Antony, C., A.S. Unnikrishnan, Y. Krien, P.L.N. Murty, S.V. Samiksha, and A.K.M.S. Islam. 2020. Tide–surge interaction at the head of the Bay of Bengal during Cyclone Aila. *Regional Studies in Marine Science* 35: 101133.

Bacopoulos, P., Y. Tang, D. Wang, and S.C. Hagen. 2017. Integrated hydrologic-hydrodynamic modeling of estuarine-riverine flooding: 2008 Tropical Storm Fay. *Journal of Hydrologic Engineering* 22 (8): 04017022.

Bhaskaran, P.K., R. Gayathri, P.L.N. Murty, S. Bonthu, and D. Sen. 2014. A numerical study of coastal inundation and its validation for Thane cyclone in the Bay of Bengal. *Coastal Engineering* 83: 108–118.

Bhaskaran, P.K., S. Nayak, S.R. Bonthu, P.N. Murty, and D. Sen. 2013. Performance and validation of a coupled parallel ADCIRC–SWAN model for THANE cyclone in the Bay of Bengal. *Environmental Fluid Mechanics* 13 (6): 601–623.

BMTPC Vulnerability Atlas of India. 2006. Earthquake, windstorm and flood hazard maps and damage risk to housing. Building Materials and Technology Promotion Council, Ministry of Housing and Urban Poverty Alleviation, Government, of India, Delhi.

Carr, L.E., and R.L. Elsberry. 1994. Systematic and integrated approach to tropical cyclone track forecasting. *Part I: Description of the basic approach. Naval Postgraduate School Publ. NPSMR-002, Naval Postgraduate School, Monterey, CA.*

Carr, L.E., III., and R.L. Elsberry. 1995. Monsoonal interactions leading to sudden tropical cyclone track changes. *Monthly Weather Review* 123 (2): 265–290.

Church, J.A., J.R. Hunter, K.L. McInnes, and N.J. White. 2006. Sea-level rise around the Australian coastline and the changing frequency of extreme sea-level events. *Australian Meteorological Magazine* 55 (4): 253–260.

Cyriac, R., J.C. Dietrich, J.G. Fleming, B.O. Blanton, C. Kaiser, C.N. Dawson, and R.A. Luettich. 2018. Variability in coastal flooding predictions due to forecast errors during Hurricane Arthur. *Coastal Engineering* 137: 59–78.

Deo, A.A., and D.W. Ganer. 2014. Variability in tropical cyclone activity over Indian seas in changing climate. *International Journal of Science and Research* 4: 880–885.

Dietrich, J.C., M. Zijlema, J.J. Westerink, L.H. Holthuijsen, C. Dawson, R.A. Luettich Jr., R.E. Jensen, J.M. Smith, G.S. Stelling, and G.W. Stone. 2011. Modeling hurricane waves and storm surge using integrally-coupled, scalable computations. *Coastal Engineering* 58 (1): 45–65.

Dietrich, J.C., S. Tanaka, J.J. Westerink, C.N. Dawson, R.A. Luettich, M. Zijlema, L.H. Holthuijsen, J.M. Smith, L.G. Westerink, and H.J. Westerink. 2012. Performance of the unstructured-mesh, SWAN+ ADCIRC model in computing hurricane waves and surge. *Journal of Scientific Computing* 52 (2): 468–497.

Egbert, G.D., and S.Y. Erofeeva. 2002. Efficient inverse modeling of barotropic ocean tides. *Journal of Atmospheric and Oceanic Technology* 19 (2): 183–204.

Evan, A.T., J.P. Kossin, and V. Ramanathan. 2011. Arabian Sea tropical cyclones intensified by emissions of black carbon and other aerosols. *Nature* 479 (7371): 94–97.

Garratt, J.R. 1977. Review of drag coefficients over oceans and continents. *Monthly Weather Review* 105 (7): 915–929.

Gayathri, R., P.K. Bhaskaran, and P.L.N. Murty. 2019. River-tide-storm surge interaction characteristics for the Hooghly estuary, East coast of India. *ISH Journal of Hydraulic Engineering*, pp. 1–13.

Gayathri, R., P.L.N. Murty, P.K. Bhaskaran, and T.S. Kumar. 2016. A numerical study of hypothetical storm surge and coastal inundation for AILA cyclone in the Bay of Bengal. *Environmental Fluid Mechanics* 16 (2): 429–452.

Giardino, A., E. Elias, A. Arunakumar, and K. Karunakar. 2014. Tidal modelling in the Gulf of Khambhat based on a numerical and analytical approach.

Gray, W.M. 1985. Technical Document WMO/TD No. 72, WMO, Geneva, Switzerland, vol. I, pp. 3–19.

Gumbel, E.J. 1958. *Statistics of Extremes*, 375. New York: Columbia University Press.

Holland, G.J. 1983. Tropical cyclone motion: Environmental interaction plus a beta effect. *Journal of the Atmospheric Sciences* 40 (2): 328–342.

Holland, G.J. 1995. Scale interaction in the western Pacific monsoon. *Meteorology and Atmospheric Physics* 56 (1–2): 57–79.

Holthuijsen, L.H. 2010. *Waves in Oceanic and Coastal Waters*. Cambridge University Press.

Hoque, M.A.A., S. Phinn, C. Roelfsema, and I. Childs. 2018. Modelling tropical cyclone risks for present and future climate change scenarios using geospatial techniques. *International Journal of Digital Earth* 11 (3): 246–263.

Horsburgh, K.J., and C. Wilson. 2007. Tide-surge interaction and its role in the distribution of surge residuals in the North Sea. *Journal of Geophysical Research: Oceans*, 112(C8).

Ikeuchi, H., Y. Hirabayashi, D. Yamazaki, S. Muis, P.J. Ward, H.C. Winsemius, M. Verlaan, and S. Kanae. 2017. Compound simulation of fluvial floods and storm surges in a global coupled river-coast flood model: Model development and its application to 2007 Cyclone Sidr in Bangladesh. *Journal of Advances in Modeling Earth Systems* 9 (4): 1847–1862.

IPCC. 2014a. Stocker, T.F., D. Qin, G.-K. Plattner, M.M. Tignor, S.K. Allen, J. Boschung, A. Nauels, Y. Xia, V. Bex, and P.M. Midgley (eds.). 2014a. *Climate Change 2013: The Physical Science Basis. Contribution of Working Group I to the Fifth Assessment Report of IPCC the Intergovernmental Panel on Climate Change*. Cambridge: Cambridge University Press.https://doi.org/10.1017/CBO9781107415324.

IPCC. 2014b. Christensen, J.H., K.K. Kanikicharla, E. Aldrian, S.I. An, I.F. Albuquerque Cavalcanti, M. de Castro, W. Dong, P. Goswami, A. Hall, J.K. Kanyanga, A. Kitoh, J. Kossin, N.C. Lau, J.

Renwick, D.B. Stephenson, S.P. Xie, T. Zhou, L. Abraham, T. Ambrizzi, and L. Zou. 2013. Climate phenomena and their relevance for future regional climate change. In *Climate Change 2013 the Physical Science Basis: Working Group I Contribution to the Fifth Assessment Report of the Intergovernmental Panel on Climate Change* (Vol. 9781107057999, pp. 1217–1308). Cambridge: Cambridge University Press. https://doi.org/10.1017/CBO9781107415324.028.

Irish, J.L., and D.T. Resio. 2013. Method for estimating future hurricane flood probabilities and associated uncertainty. *Journal of Waterway, Port, Coastal, and Ocean Engineering* 139 (2): 126–134.

Jelesnianski, C.P., and A.D. Taylor. 1973. NOAA technical memorandum. *ERL, WMPO-3, 33.*

Jonkman, S.N., B. Maaskant, E. Boyd, and M.L. Levitan. 2009. Loss of life caused by the flooding of New Orleans after Hurricane Katrina: Analysis of the relationship between flood characteristics and mortality. *Risk Analysis: An International Journal* 29 (5): 676–698.

Karim, M.F., and N. Mimura. 2008. Impacts of climate change and sea-level rise on cyclonic storm surge floods in Bangladesh. *Global Environmental Change* 18 (3): 490–550.

Knutson, T.R., J.L. McBride, J. Chan, K. Emanuel, G. Holland, C. Landsea, I. Held, J.P. Kossin, A.K. Srivastava, and M. Sugi. 2010. Tropical cyclones and climate change. *Nature Geoscience* 3 (3): 157–163.

Knutson, T.R., J.J. Sirutis, G.A. Vecchi, S. Garner, M. Zhao, H.S. Kim, M. Bender, R.E. Tuleya, I.M. Held, and G. Villarini. 2013. Dynamical downscaling projections of twenty-first-century Atlantic hurricane activity: CMIP3 and CMIP5 model-based scenarios. *Journal of Climate* 26 (17): 6591–6617.

Le Provost, C., A.F. Bennett, and D.E. Cartwright. 1995. Ocean tides for and from TOPEX/POSEIDON. *Science* 267 (5198): 639–642.

Le Provost, C., F. Lyard, J.M. Molines, M.L. Genco, and F. Rabilloud. 1998. A hydrodynamic ocean tide model improved by assimilating a satellite altimeter-derived data set. *Journal of Geophysical Research: Oceans* 103 (C3): 5513–5529.

Lewis, M., P. Bates, K. Horsburgh, J. Neal, and G. Schumann. 2013. A storm surge inundation model of the northern Bay of Bengal using publicly available data. *Quarterly Journal of the Royal Meteorological Society* 139 (671): 358–369.

Lin, N., K. Emanuel, M. Oppenheimer, and E. Vanmarcke. 2012. Physically based assessment of hurricane surge threat under climate change. *Nature Climate Change* 2 (6): 462–467.

Longuet-Higgins, M.S., and R.W. Stewart. 1964. Radiation stresses in water waves; A physical discussion, with applications. *Deep-Sea Research* 11 (4): 529–562.

Luettich, R.A., and J.J. Westerink. 1999. Elemental wetting and drying in the ADCIRC hydrodynamic model: Upgrades and documentation for ADCIRC version 34. XX. *Contract Report, prepared for Headquarters, US Army Corps of Engineers, Vicksburg, MS.*

Luettich, R.A., J.J. Westerink, and N.W. Scheffner. 1992. ADCIRC: An advanced three-dimensional circulation model for shelves, coasts, and estuaries. Report 1, Theory and methodology of ADCIRC-2DD1 and ADCIRC-3DL.

Mandal, A.K., R. Ramakrishnan, S. Pandey, A.D. Rao, and P. Kumar. 2020. An early warning system for inundation forecast due to a tropical cyclone along the east coast of India. *Natural Hazards* 103: 2277–2293.

Mohapatra, M., and M. Sharma. 2015. Characteristics of surface wind structure of tropical cyclones over the north Indian Ocean. *Journal of Earth System Science* 124 (7): 1573–1598.

Murakami, H., G.A. Vecchi, and S. Underwood. 2017. Increasing frequency of extremely severe cyclonic storms over the Arabian Sea. *Nature Climate Change* 7: 885–889. https://doi.org/10.1038/s41558-017-0008-6.

Murty, P.L.N., K.G. Sandhya, P.K. Bhaskaran, F. Jose, R. Gayathri, T.B. Nair, T.S. Kumar, and S.S.C. Shenoi. 2014. A coupled hydrodynamic modeling system for PHAILIN cyclone in the Bay of Bengal. *Coastal Engineering* 93: 71–81.

Murty, P.L.N., J. Padmanabham, T.S. Kumar, N.K. Kumar, V.R. Chandra, S.S.C. Shenoi, and M. Mohapatra. 2017. Real-time storm surge and inundation forecast for very severe cyclonic storm 'Hudhud.' *Ocean Engineering* 131: 25–35.

Murty, P.L.N., A.D. Rao, K.S. Srinivas, E.P.R. Rao, and P.K. Bhaskaran. 2019. Effect of wave radiation stress in storm surge-induced inundation: A case study for the East Coast of India. *Pure and Applied Geophysics*, pp. 1–20.

Murty, T.S., R.A. Flather, and R.F. Henry. 1986. The storm surge problem in the Bay of Bengal. *Progress in Oceanography* 16 (4): 195–233.

Nayak, S., and P.K. Bhaskaran. 2014. Coastal vulnerability due to extreme waves at Kalpakkam based on historical tropical cyclones in the Bay of Bengal. *International Journal of Climatology* 34 (5): 1460–1471.

Nicholls, R.J., F.M. Hoozemans, and M. Marchand. 1999. Increasing flood risk and wetland losses due to global sea-level rise: Regional and global analyses. *Global Environmental Change* 9: S69–S87.

Olabarrieta, M., J.C. Warner, and N. Kumar. 2011. Wave-current interaction in Willapa Bay. *Journal of Geophysical Research: Oceans* 116(C12).

Pandey, S., and A.D. Rao. 2018. An improved cyclonic wind distribution for computation of storm surges. *Natural Hazards* 92 (1): 93–112.

Pandey, S., and A.D. Rao. 2019. Impact of approach angle of an impinging cyclone on generation of storm surges and its interaction with tides and wind waves. *Journal of Geophysical Research: Oceans* 124 (11): 7643–7660.

Pinheiro, J.P., C.L. Lopes, A.S. Ribeiro, M.C. Sousa, and J.M. Dias. 2020. Tide-surge interaction in Ria de Aveiro lagoon and its influence in local inundation patterns. *Continental Shelf Research* 200: 104132.

Poulose, J., A.D. Rao, and P.K. Bhaskaran. 2018. Role of continental shelf on nonlinear interaction of storm surges, tides and wind waves: An idealized study representing the west coast of India. *Estuarine, Coastal and Shelf Science* 207: 457–470.

Poulose, J., A.D. Rao, and S.K. Dube. 2020. Mapping of cyclone induced extreme water levels along Gujarat and Maharashtra coasts: A climate change perspective. *Climate Dynamics* 55 (11): 3565–3581.

Rao, A.D., P.L.N. Murty, I. Jain, R.S. Kankara, S.K. Dube, and T.S. Murty. 2013. Simulation of water levels and extent of coastal inundation due to a cyclonic storm along the east coast of India. *Natural Hazards* 66 (3): 1431–1441.

Rao, A.D., P. Upadhaya, S. Pandey, and J. Poulose. 2020a. Simulation of extreme water levels in response to tropical cyclones along the Indian coast: A climate change perspective. *Natural Hazards* 100 (1): 151–172.

Rao, A.D., P. Upadhaya, H. Ali, S. Pandey, and V. Warrier. 2020b. Coastal inundation due to tropical cyclones along the east coast of India: An influence of climate change impact. *Natural Hazards* 101 (1): 39–57.

Rego, J.L., and C. Li. 2010. Nonlinear terms in storm surge predictions: Effect of tide and shelf geometry with case study from Hurricane Rita. *Journal of Geophysical Research: Oceans* 115(C6).

Rupp, J.A., and M.A. Lander. 1996. A technique for estimating recurrence intervals of tropical cyclone-related high winds in the tropics: Results for Guam. *Journal of Applied Meteorology and Climatology* 35 (5): 627–637.

Sahoo, B., and P.K. Bhaskaran. 2016. Assessment on historical cyclone tracks in the Bay of Bengal, east coast of India. *International Journal of Climatology* 36 (1): 95–109.

Sahoo, B., and P.K. Bhaskaran. 2018. Multi-hazard risk assessment of coastal vulnerability from tropical cyclones—A GIS based approach for the Odisha coast. *Journal of Environmental Management* 206: 1166–1178.

Sindhu, B., and A.S. Unnikrishnan. 2013. Characteristics of tides in the Bay of Bengal. *Marine Geodesy* 36 (4): 377–407.

Singh, O.P., T.M.A. Khan, and M.S. Rahman. 2001. Has the frequency of intense tropical cyclones increased in the north Indian Ocean? *Current science*, pp. 575–580.

Srinivasa Kumar, T., P.L.N. Murty, M. Pradeep Kumar, M. Krishna Kumar, J. Padmanabham, N. Kiran Kumar, S.C. Shenoi, M. Mohapatra, S. Nayak, and P. Mohanty. 2015. Modeling storm surge and its associated inland inundation extent due to very severe cyclonic storm Phailin. *Marine Geodesy* 38 (4): 345–360.

Weber, D.D., and E.J. Englund. 1994. Evaluation and comparison of spatial interpolators II. *Mathematical Geology* 26 (5): 589–603.

Part IV
Earthquakes and Landslides

Chapter 12
Assessment of the Earthquake Risk Posed by Shale Gas Development in South Africa

Raymond J. Durrheim, Vunganai Midzi, Moctar Doucoure, and Musa S. D. Manzi

Abstract A strategic environmental assessment (SEA) of the economic, social, and environmental benefits and risks is an essential precursor to any major infrastructural and industrial project. The Karoo Basin of South Africa is thought to contain significant resources of shale gas. A SEA was commissioned by the Department of Environmental Affairs to help formulate regulations governing the exploration for shale gas, the extraction of any viable resources, and the decommissioning of any wells. Here, we report on the assessment of the risk posed by earthquakes that might by triggered by the injection of fluids. Natural earthquakes are rare within the Karoo, and the increase in earthquake risk posed by fracking is considered small. Nevertheless, a range of mitigation measures are recommended.

Keywords Shale gas development · Fracking · Earthquakes · Karoo Basin

12.1 Introduction

The governments of developing countries are usually eager to improve the standard of living of their citizens by building infrastructure such as roads, railways, power stations, electricity grids, dams, and pipelines; and by developing agricultural and

R. J. Durrheim (✉) · M. S. D. Manzi
University of the Witwatersrand, Private Bag 3, Wits 2050, South Africa
e-mail: Raymond.Durrheim@wits.ac.za

M. S. D. Manzi
e-mail: Musa.Manzi@wits.ac.za

V. Midzi
Council for Geosciences, Pretoria, South Africa
e-mail: vmidzi@geoscience.org.za

M. Doucoure
Nelson Mandela University, Port Elizabeth, South Africa
e-mail: Moctar.Doucoure@nmmu.ac.za

© The Centre for Science & Technol. of the, Non-aligned and Other
Devel. Countries 2022
A. S. Unnikrishnan et al. (eds.), *Extreme Natural Events*,
https://doi.org/10.1007/978-981-19-2511-5_12

mineral resources to stimulate the economy and create decent jobs. However, development can also have negative impacts on the environment and society. These costs and benefits should be carefully assessed, and measures to mitigate the undesirable effects should be formulated and taken into account when making project decisions.

The Karoo Basin of South Africa (Fig. 12.1) contains extensive shale formations that may hold significant resources of gas amenable to extraction using hydraulic fracturing technology ("fracking"), offering substantial economic and energy-security opportunities (Kuuskraa et al. 2011, 2013; Decker and Marot 2012). Fracking uses high-pressure fluids to create a network of fractures that extend for tens of metres from a borehole drilled into impermeable rocks (such as shale) at depths of several kilometres to provide pathways for the extraction of gas. The technology has greatly increased hydrocarbon production during the last decade, especially in the United States.

New jobs could be created in a region with high unemployment rates, but fracking and a gas industry could have environmental and social costs in a region where farming and ecotourism are important economic activities. Fracking uses large volumes of water, which is a scarce resource in the Karoo, and produces polluted "production water". There is a legitimate concern that ground and surface water, critical for farming and ecotourism activities and household use, could be over-exploited or contaminated. Furthermore, the injection of fluids into the Earth at pressures high and volumes great enough to fracture rocks will inevitably cause seismic events. Some events might be large enough to alarm residents and even damage vulnerable structures that are built from adobe (mud brick) or unreinforced masonry, including heritage buildings.

The development of shale gas using fracking has been presented to the South African public and decision-makers as a trade-off between economic opportunity and environmental protection. It became a highly divisive topic poorly informed by publically available evidence. To address this lack of critically evaluated information, a strategic environmental assessment (SEA) for shale gas development (SGD) was commissioned in February 2015 by the Department of Environmental Affairs with the support of the Departments of Energy, Mineral Resources, Water Affairs and Sanitation, Science and Technology, and Agriculture, Forestry and Fisheries; and the governments of the Eastern, Western, and Northern Cape provinces.

Rigorous and systematic risks assessments were carried out to address a broad range of issues, including the impact on energy planning and energy security; air quality and greenhouse gas emissions; occurrence of earthquakes; surface and underground water resources; waste planning and management; biodiversity and ecology; agriculture; tourism; the economy; social fabric; human health; sense of place values; visual, aesthetic and scenic resources; heritage resources; noise; electromagnetic interference with the Square Kilometre Array; and spatial and infrastructure planning. The final report was delivered in November 2016 (Scholes et al. 2016).

Here, we summarise the study of the earthquake risk posed by shale gas development (Durrheim et al. 2016), incorporating relevant information from complementary studies of the risks posed by geohazards on "energy corridors" (i.e. routes for a proposed gas pipeline network and an expansion of the national electricity grid, see

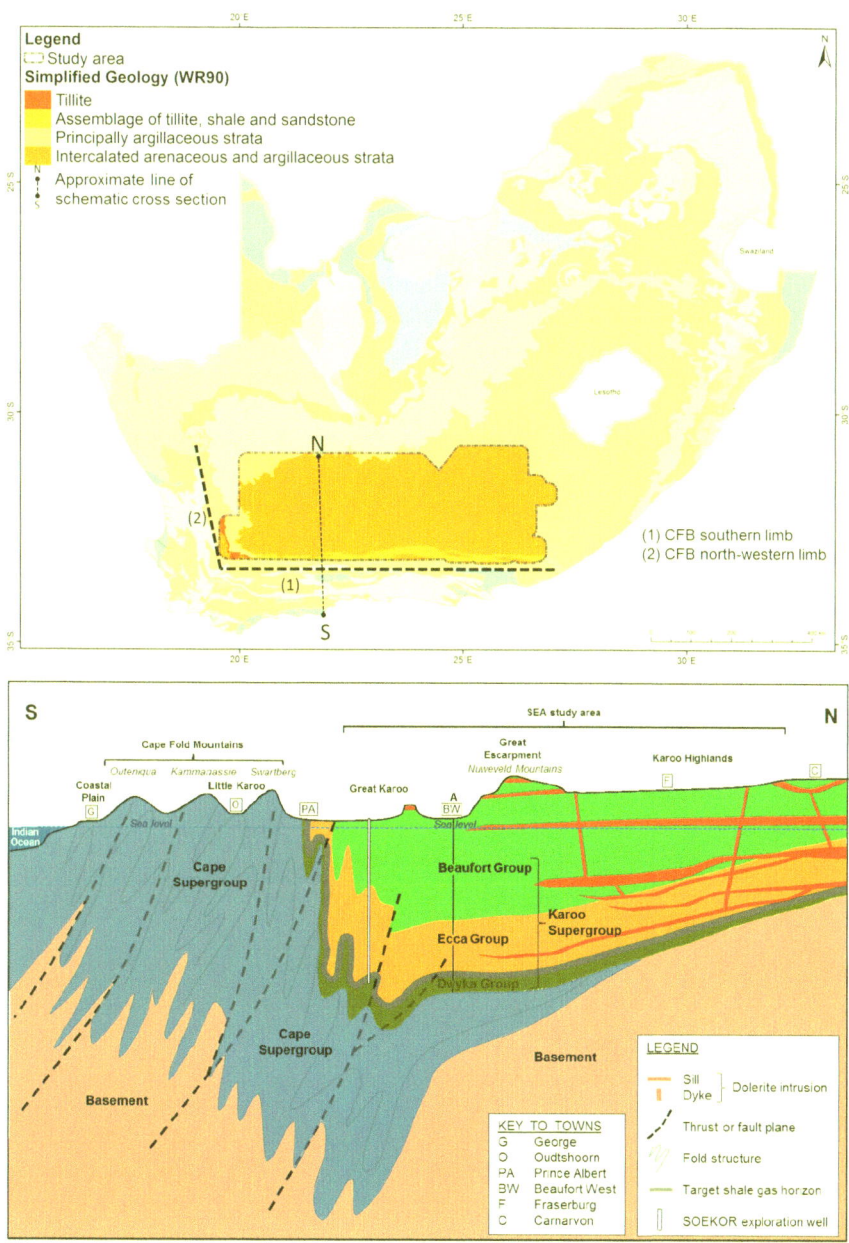

Fig. 12.1 Geological map and section showing the shale gas study area (Scholes et al. 2016). A 3-km-deep stratigraphic borehole (A) was drilled in 2021 near Beaufort West, reaching the Dwyka Formation

Durrheim and Manzunzu 2019a, b) and the seismic hazard assessment of proposed sites for the construction of nuclear power stations (Bommer et al. 2015).

12.2   Seismotectonics of the Karoo Basin and Environs

Southern Africa is, by global standards, a seismically quiet region as it is remote from the boundaries of tectonic plates, although natural earthquakes do take place from time to time (Fig. 12.2). Various tectonic forces, such as the spreading of the sea floor along the mid-Atlantic and mid-Indian ocean ridges, the propagation of the East African Rift System, and the response of the crust to erosion and uplift drive them. However, a low rate of seismicity does not mean that the maximum size of a future earthquake will be small; just that earthquakes are less frequent. Even a moderate-sized earthquake (such as the $M_L6.1$ that occurred near Cape Town in 1809, the $M_L6.3$ that occurred near Ceres in 1969, and the $M_W6.5$ that occurred in Botswana in 2017) can prove disastrous should it occur close to a town with many

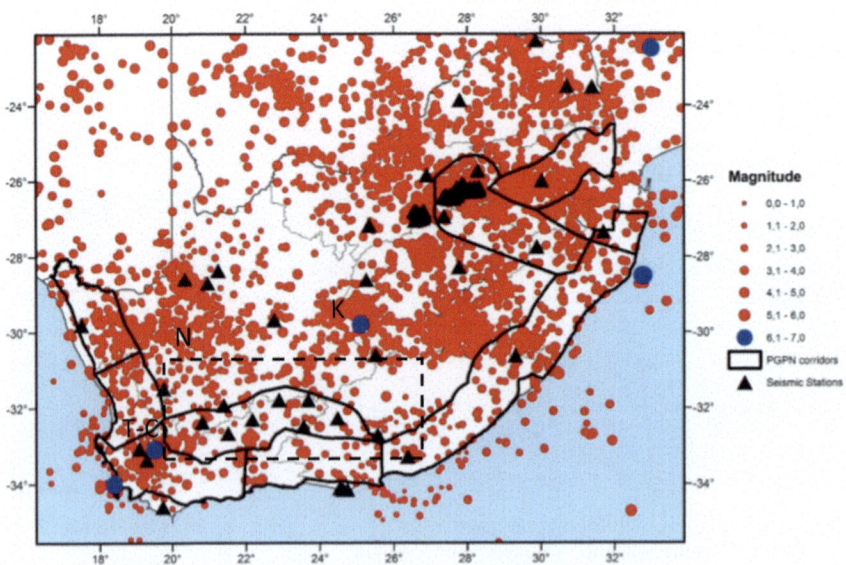

Fig. 12.2 Location of earthquakes in southern Africa from 1811 to 2014. Triangles indicate stations belonging to the South African Standard Seismograph Network (SANSN). T-C, N, and K indicate the Tulbagh-Ceres, Namaqualand and Koffiefontein earthquake clusters (*Source* Council for Geoscience, 2014). The shale gas development study area is indicated by a dashed rectangle; the other polygons represent "energy corridors" that were assessed for the construction of a phased gas pipeline network (PGPN) and the expansion of the national electricity grid (Durrheim and Manzunzu, 2019a, b)

vulnerable buildings, the terrain is steep and prone to landslides, or the soil is thick and prone to liquefaction.

The seismotectonics of southern Africa has been studied by Singh et al. (2011) and Manzunzu et al. (2019). Here, we summarise their findings in the regions that could have an impact on the shale gas development area.

Cape Fold Belt. The Cape Fold Belt lies to the west and south of the study area (Fig. 12.1). The largest instrumentally recorded earthquake to occur in South Africa was a M_L 6.3 event that struck the Ceres-Tulbagh region on 29 September 1969, causing widespread damage and claiming 12 lives. Modern concrete-frame buildings sustained relatively minor damage, some well-constructed brick houses were badly damaged, while many adobe-type buildings were completely destroyed. The maximum intensity on the Modified Mercalli Intensity (MMI)Scale was VIII (Van Wyk and Kent, 1974), indicating "slight damage to earthquake-resistant structures, considerable damage to solid buildings, and great damage to poorly built buildings". The event provides a useful reference for the vulnerability of typical Karoo farmsteads and heritage buildings.

The most significant recorded earthquakes along the southern limb of the Cape Fold Belt are events of magnitude M_L 5.2 (12 January 1968) and M_L 5.4 (11 September 1969) that had their epicentres near Willowmore and Calitzdorp, respectively. Palaeoseismic studies were conducted in this region as part of a seismic hazard assessment of proposed nuclear power station sites (Bommer et al. 2015). Goedhart and Booth (2016a) interpreted an 84-km-long scarp running parallel to the Kango Fault to be the surface expression of an earthquake. An 80-m-long, 6-m-deep, and 2.5-m-wide trench was dug across the fault, exposing twenty-one lithological units, six soil horizons, and nineteen faults strands. It was concluded that the scarp was produced by M_w 7.4 earthquake that occurred about 10,600 years ago (Goedhart and Booth 2016b).

Koffiefontein, Free State Province. A cluster of events is found near Koffiefontein in the southern Free State, to the northeast of the study area. An M_L 6.2 event that occurred on 20 February 1912 was felt over much of South Africa and assigned a maximum MMI of VIII (Brandt et al. 2005).

Namaqualand, Northern Cape. M_L 4 earthquakes occur from time to time in the region, which lies to the northeast of the study area.

Mining-related earthquakes in the Free State, North West, and Gauteng provinces. Most seismic activity in South Africa is induced by deep-level mining for gold and platinum. The mining regions are remote from the shale gas development area, but are mentioned as they provide useful examples of damage caused to surface structures by shallow induced earthquakes. The largest event to occur in a mining region to date was the M_L 5.5 earthquake that struck Orkney on 5 August 2014, causing damage to more than 600 dwellings, mostly constructed of unreinforced masonry (Midzi et al. 2015). MMI intensities of VII were experienced in several towns that lie within 20 km of the epicentre.

12.3 Relevant Legislation and Regulation

The Minister of Mineral Resources published the "Regulations for Petroleum Exploration and Production. Notice R466" under section 107 of the Mineral and Petroleum
Resources Development Act (2002) in the Government Gazette dated 3 June 2015.
These include regulations relevant to the risk posed by earthquakes induced by
hydraulic fracturing. The validity of these regulations were challenged in the high
court in April 2017 (Mathews 2017; Steyn 2017) and in the supreme court of appeal
in July 2019. At the time of writing (June 2021), the new Upstream Petroleum
Resources Development Bill had yet to be tabled in parliament.

12.4 Likely Impact of Fracking-Induced Ground Shaking

12.4.1 Triggering of Earthquakes by Fluid Injection

Beginning in the 1960s, efforts were made to enhance oil recovery by injecting high-
pressure fluids into reservoirs to "hydrofracture" the rock. About the same time,
technologies were developed to "steer" drilling bits so that targets could be reliably
hit. The technology advanced to the extent that the trajectory of a hole could be
deviated from the vertical to horizontal, enabling a far larger subsurface area to be
explored and exploited from a single drilling pad. Beginning in the 1990s, engineers
in the USA combined fracking and directional drilling to explore and exploit low
permeability source rocks directly on a large scale. This required a great deal of
technical development, including the use of a variety of chemical additives to enhance
the flow of oil and gas, and the introduction of sand grains to "prop open" the cracks.
Generally, unconventional reservoirs are at depths of several kilometres. In the case
of the Karoo Basin, the depth is likely to be in the range of 3 to 4 km.

The injection of fluids into the rock at pressures that exceed its tensile strength
will cause the intact rock to fracture, releasing some of the stored elastic energy
as vibrations. During fracking, this is done in a controlled manner. The density
and length of fractures are controlled by in situ conditions (e.g. stress field), rock
properties (e.g. rock strength), and the parameters of the fracking process itself (fluid
pressure, density with which the well casing is perforated, and the length of borehole
where the pressure is elevated). Generally, the desired length of fractures is of the
order of tens of metres, while the length of borehole that is fractured (or "stimulated")
at any one time is, at most, a few hundred metres. Rock fracturing inevitably releases
elastic energy stored in the intact rock, but generally the shaking is too weak to be
felt on the surface.

Earthquakes related to hydraulic fracturing are induced by at least three mechanisms:

i.   Cracking or rupturing of rocks in the vicinity of the wellbore that creates micro-earthquakes of very small magnitude, $M < 0$;
ii.   Interaction between fracking fractures and nearby faults, where the fracking fluid enters the fault zone. This may cause a change in pore fluid pressure that can trigger earthquakes of $0 < M < 3$, and rarely, but possibly, greater.
iii. Interaction between fracking fractures and nearby faults, through the transfer of stress through the rock. This may cause a change in the shear stress acting on the fault and trigger earthquakes of $0 < M < 3$, and rarely, but possibly, greater.

Many thousands of hydraulic fracture wells have been drilled worldwide (Warpinski 2013; Skoumal et al. 2018). Most only caused microseismic events ($M < 3$) imperceptible to humans, while, to the best of the author's knowledge, none of the few felt events have caused damage. It must be emphasised that the felt events ($M > 3$) associated with fluid injection are almost all associated with the injection of massive volumes of wastewater, and very rarely with the deliberate formation of fractures to liberate gas (i.e. fracking) (Ellsworth 2013). Several moderate earthquakes ($4 < M < 5$) that occurred near Fox Creek in Alberta have been associated with fracking operations (Atkinson et al. 2016). In this case, the formations that were being fracked were close to crystalline basement. It is postulated that the increase in pressure triggered slip on preexisting faults that extended into the basement. It should be noted that a rupture with an extent of the order of 1 km is required to produce a M4 earthquake. This is much greater than the length of fractures produced by fracking.

To date, all damaging events associated with fluid injection are associated with the disposal of large volumes of wastewater. The disposal of wastewater by injection into underground aquifers is forbidden by current South African legislation. Thus, hydraulic fracturing is considered very unlikely to induce a damaging event. However, we cannot entirely exclude the possibility that a shallow $M > 5$ event will be triggered. Fracking and fracking-triggered earthquakes could, however, cause damage and losses underground, even though the events might not be felt on the surface. For example, a well casing could rupture if a slipping fault intersects the borehole; and water resources could be contaminated should there be interaction between aquifers, hydraulic fractures, faults and leaking casings. However, there are no documented and verified cases of contamination of potable groundwater resources from the fracking process itself. Surface spills or faulty casing and poor well maintenance account for all proven instances of contamination.

12.5   Mitigation of Impacts

Several practical steps should be taken to mitigate the risk of fracking-induced earthquakes:

1. Seismicity should be monitored before, during and after fracking, ideally beginning at least two years before fracking starts and for three years after fracking ends. The Council for Geoscience database shows that moderately sized earthquakes (M > 4) have previously occurred in the region. National Seismograph Network stations have been installed in the region since 2016 to improve the sensitivity of monitoring (see Fig. 12.2).

2. Identify faults by mapping regional and local structures in the field, in boreholes and with geophysical methods. A 3-km-deep stratigraphic borehole was drilled near Beaufort West by the Council for Geoscience in 2021.

3. Measure stress using proximal boreholes and regional geodetic measurements.

4. Analyse seismicity, geological and stress data to identify pre-stressed and active/capable faults. Obtain orientations and slip tendency of identified active/capable faults.

5. Mitigate risks of fault movement by preventing fluids from flowing into pre-stressed faults by informed location of fracking wells.

6. Perform real-time microseismic monitoring during appraisal and production. Implement "traffic light" systems (i.e. feedback system) during fracking that will enable operators to respond quickly to induced earthquakes by either reducing the rate of fracking or stopping fracking altogether (Majer et al. 2007).

7. Determine the expected maximum magnitude of earthquakes, and the expected maximum ground motion at the fracking site and in the region.

8. Assess the building typologies in the region. Inspect buildings and structures prior to fracking to assess their condition. Reinforce vulnerable buildings and structures. Some simple measures may reduce the severity of earthquake damage. For example, buttress walls, strapping of hot water heaters (geysers) to rafters, stabilisation of towers carrying water tanks with anchor cables. Enforce building codes.

12.6   Risk Assessment Methodology

Hazard assessment is the process of determining the likelihood that a given event will take place. Probabilistic seismic hazard assessment (PSHA) is generally expressed in terms of the ground motion (for example, peak ground acceleration, PGA) that has a certain likelihood of exceedance (say 10%) in a given period (say 50 years). There are many PSHA schemes, but all require a catalogue of earthquakes (size, time, and location); the characterisation of seismically active faults and areas (usually in terms of the maximum credible magnitude and recurrence periods); and a prediction of variation in ground motion with distance from the epicentre. The longer the duration of the catalogue, the smaller the magnitude of completeness, and the better the zonation, the more reliable is the PSHA. The moment there is human interference (i.e. fluid injection), probabilistic hazard assessment techniques cannot be used. The most reliable approach is to consider analogous situations elsewhere in the world.

Risk assessment is an attempt to quantify the losses that could be caused by a particular hazard. It is calculated as follows:

Risk = likelihood of the hazard occurring × seriousness of consequences

The consequences of an earthquake depend on four main factors: the vulnerability of buildings to damage, the exposure of persons and other assets to harm, the cost of reconstruction, and the cost of lost economic production. Risk assessments are useful for raising awareness of possible disasters and motivating policies and actions to mitigate losses and avoid disasters. For example, vulnerable buildings may be reinforced, building codes enforced and insurance taken out to cover possible losses.

The outcomes of the risk assessment process are presented in a risk matrix (Fig. 12.3). The first step is to identify the phenomena that could cause damage and harm, e.g. strong ground motion, surface rupture, liquefaction, and landslides. The next step is to assess the likelihood that this will occur during the full-life cycle of shale gas development project. The duration of a shale gas development programme is likely to be of the order of 20 to 50 years (National Petroleum Council, 2011). The main stages are exploration (2–3 years), appraisal (2–3 years), development (3–5 years), production (10–30 years), and decommissioning (5–10 years). Earthquake hazard is routinely expressed in terms of the maximum ground motion that has a given probability (usually 10% and 2%) of being exceed in a certain period. A period of 50 years is generally used, which is similar to the maximum duration of a shale gas development programme. Ground motion is expressed in peak ground acceleration (expressed in m/s^2, or as a percentage of the acceleration of gravity, g); or the acceleration at a certain period of vibration, known as the "spectral acceleration". Spectral acceleration is useful in predicting the effect on structures, as buildings, dams, and bridges have a natural resonance period that depends on their dimensions.

The next step is to assess the consequence. This is usually expressed in terms of the loss of human life and economic loss on a scale that ranges from "slight" to "extreme". The adjudication of the categories must be discussed with interested and affected parties, considering other impacts of fracking (e.g. on water resources, air quality, ecotourism, and agriculture) and the severity of other risks that the population of the region is exposed to (e.g. floods, fires, vehicle crashes, and crime). The "consequence table" developed for the assessment of the risks posed by geohazards (ground shaking, surface displacement, soil liquefaction, and landslides caused by earthquakes; ground deformation caused by collapsing or swelling soils) to gas pipeline networks by Durrheim and Manzunzu (2019b) is shown in Fig. 12.4.

The consequences will obviously be influenced by the scale of shale gas development and the implementation of mitigating measures. In order to take this into account, a range of scenarios were developed (exploration only, "small" gas, and "big" gas; without and with mitigating measures) and compared to the base case where no shale development takes place. These are summarised in the "risk matrix" (Fig. 12.6).

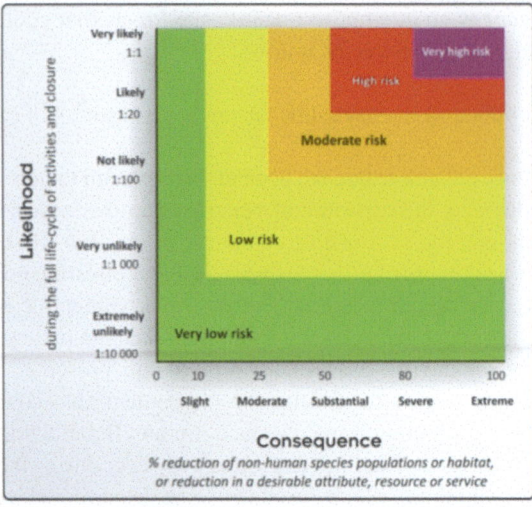

Fig. 12.3 Risk assessment diagram and table showing likelihood x consequence to determine risk (Scholes et al. 2016). Risk is qualitatively measured by multiplying the likelihood of an impact by the severity of the consequences to provide risk rating ranging from very low to very high

Risk category	Definition
No discernible risk	Any changes that may occur as a result of the impact either reduce the risk or do not change it in a way that can be differentiated from the mean risk experienced in the absence of the impact.
Very low risk	Extremely unlikely (<1 chance in 10 000 of having a consequence of any discernible magnitude); or if more likely than this, then the negative impact is noticeable but slight, i.e. although discernibly beyond the mean experienced in the absence of the impact, it is well within the tolerance or adaptive capacity of the receiving environment (for instance, within the range experienced naturally, or less than 10%); or is transient (< 1 year for near-full recovery).
Low risk	Very unlikely (<1 chance in 100 of having a more than moderate consequence); or if more likely than this, then the impact is of moderate consequence because of one or more of the following considerations: it is highly limited in extent (<1% of the area exposed to the hazard is affected); or short in duration (<3 years), or with low effect on resources or attributes (<25% reduction in species population, resource or attribute utility).
Moderate risk	Not unlikely (1:100 to 1:20 of having a moderate or greater consequence); or if more likely than this, then the consequences are substantial but less than severe, because although an important resource or attribute is impacted, the effect is well below the limit of acceptable change, or lasts for a duration of less than 3 years, or the affected resource or attribute has an equally acceptable and un-impacted substitute.
High risk	Greater than 1 in 20 chance of having a severe consequence (approaching the limit of acceptable change) that persists for >3 years, for a resource or attribute where there may be an affordable and accessible substitute, but which is less acceptable.
Very high risk	Greater than even (1:1) chance of having an extremely negative and very persistent consequence (lasting more than 30 years); greater than the limit of acceptable change, for an important resource or attribute for which there is no acceptable alternative.

12.6.1 Scenarios

There is no history of production of shale gas in South Africa, so this description of potential scenarios and related activities is necessarily hypothetical.

Consequence level	Slight	Moderate	Substantial	Severe	Extreme
Impact ↓					
Economic loss Repairs to property, loss of production, etc.	<R1 million	<R100 million	<R1 billion	<R100 billion (<2% of GDP)	>R100 billion (>2% of GDP)
Environmental loss Burning or pollution of grazing areas, orchards, plantations or forests	<1000 m²	<10,000 m² (1 hectare)	<1 km² 100 hectare	<100 km²	>100 km²
Human loss Injury and death	<10 injuries 0 deaths	<100 injuries <10 deaths	<1000 injuries <100 deaths	<10,000 injuries <1,000 deaths	≥10,000 injuries ≥1,000 deaths

Fig. 12.4 Consequence table used to assess risks posed by earthquakes to gas pipeline networks. South Africa's GDP in 2016 was about USD 300 billion (say R4500 billion). From Durrheim and Manzunzu (2019b)

Base Case: In the absence of shale gas exploration, natural events will occur from time to time. The important parameters are the maximum magnitude (M_{max}), the recurrence interval, and the likely ground motion. These parameters are difficult to determine in a region where the seismicity is low, the instrumental catalogue is of short duration, and there are no recordings of strong ground motion. The largest instrumentally recorded events in the region have magnitudes of $M_L 6.2$ (Koffie-fontein 1912) and $M_L 6.3$ (Ceres, 29 September 1969). In the absence of a lengthy and complete catalogue, it is standard practice to assume that the maximum credible magnitude is 0.5 units larger than the maximum observed event. Paleoseismic studies suggest that a $M_w 7.3$ event occurred along the Kango Fault (in the southern limb of the Cape FoldBelt) some 10,000 years ago (Goedhart and Booth 2016b).

A probabilistic hazard assessment for the region by Fernández and Du Plessis (1992) found that the PGA with a 10% probability of being exceeded in 50 years is less than 0.05 g in the scientific assessment area, rising to a value greater than 0.2 g in the Ceres region. The latest and most complete probabilistic seismic hazard assessment (PSHA) for South Africa was performed by the Council of Geoscience using an up-dated homogenised earthquake catalogue (Midzi et al. 2018; see Fig. 12.5), yielded much the same result. It is important to realise that the PSHA estimates are calculated on a relatively coarse grid ($0.5° \times 0.5°$). There is no quick and easy way to increase spatial resolution or reduce uncertainty in the PSHA calculations. This can only be done through decades or centuries of monitoring with a denser and more sensitive seismological network. Identification and mapping of palaeoseismic faults require extensive fieldwork.

Based on these studies, the occurrence of a felt earthquake (M > 3) within the scientific assessment area in a 50 year period is considered to be likely, while the occurrence of damaging earthquake (M > 6) is consider to be very unlikely. The

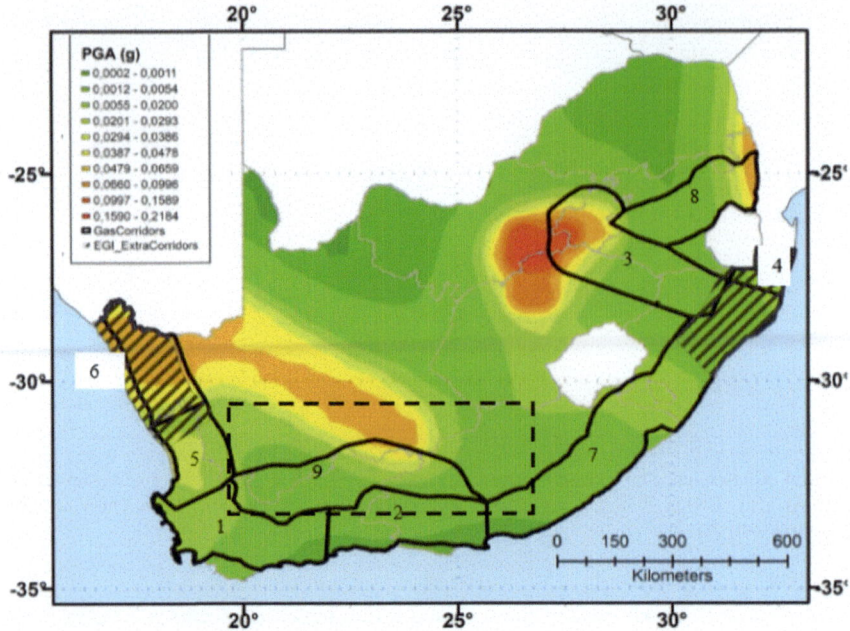

Fig. 12.5 Peak Ground Acceleration (PGA, in g) with 10% probability of exceedance in 50 years (Midzi et al., 2020). The shale gas development study area is indicated by a dashed rectangle; the other polygons represent "energy corridors" that were assessed for the construction of a gas pipeline network and the expansion of the national electricity grid (Durrheim and Manzunzu, 2019a, b)

Key Strategic Issue	Earthquakes		Without Mitigation			With Specified Mitigation		
Risk	**Scenario**	**Area**	**Likelihood**	**Consequence**	**Risk**	**Likelihood**	**Consequence**	**Risk**
DAMAGING earthquakes induced by hydraulic fracturing	Reference Case	Wells **WITHIN** **20 km** of towns	Very unlikely	Slight	Low	Very unlikely	Slight	Low
	Exploration Only		Very unlikely	Slight	Low	Very unlikely	Slight	Low
	Small Gas		Not likely	**Moderate**	Low	Not likely	Slight	Low
	Big Gas		Not likely	**Moderate - Substantial**	**Moderate**	Not likely	**Moderate**	**Moderate**
	Reference Case	Wells **BEYOND** **20 km** of towns	Very unlikely	Slight	Low	Very unlikely	Slight	Low
	Exploration Only		Very unlikely	Slight	Low	Very unlikely	Slight	Low
	Small Gas		Very unlikely	Slight	Low	Very unlikely	Slight	Low
	Big Gas		Unlikely	Slight	Low	Not likely	Slight	Low

Fig. 12.6 Earthquake risk matrix

area is sparsely populated, apart from the towns such as Beaufort West, Victoria West, Middelburg, Queenstown, Cradock, and Graaff-Reinet. Hence, the exposure is generally low and the consequences of an earthquake are generally considered to be slight. However, very few buildings in the region were constructed to withstand strong shaking. Most buildings, including important heritage buildings and dwellings, were built using unreinforced masonry and are thus vulnerable to damage similar to that experienced in the Ceres-Tulbagh region in 1969 and in Khuma Township near Orkney in 2014. Thus, the consequences of a shallow M > 5 earthquake occurring within 20 km of a town could be moderate or even substantial.

Scenario 1 (Exploration Only): Exploration activities include seismic surveys and the drilling of exploration and appraisal wells, and does not involve the large-scale injection of pressurised fluids although trial injection tests may be carried out at a few sites. The triggering of a felt earthquake is considered to be unlikely, and the triggering of a damaging event to be very unlikely. Thus, the risk posed by earthquakes in the scientific assessment area during the exploration phase is considered to be low and not significantly different from the base case.

Scenarios 2 (Small Gas Development): Should unconventional gas resources be developed in impermeable shale strata in the Karoo, fluid will be injected into boreholes at pressures that are high enough to cause the shale to fracture, thereby creating a network of pathways that enables the gas to be extracted. Under the "Small Gas" scenario, about 5 trillion cubic feet (Tcf) of gas are extracted. For comparison, the Mossel Bay offshore gas field will yield a total of about 1 Tcf. Downstream development in the Small Gas scenario results in a 1 000 MW combined cycle gas turbine (CCGT) power station located within 100 km of the production block. The triggering of a felt earthquake is considered to be likely, and the triggering of an event large enough to cause damage to surface structures very unlikely.

Scenarios 3 (Big Gas Development): The "Big Gas" scenario assumes a relatively large shale gas resource of 20 Tcf is developed. For comparison, the offshore conventional gas fields in Mozambique contain about 100 Tcf. Downstream development results in construction of two CCGT power stations, each of 2 000 MW generating capacity, and a gas-to-liquid (GTL) plant, located at the coast, with a refining capacity of 65 000 barrels (bbl) of liquid fuel per day. The triggering of a felt earthquake is considered to be likely, and the triggering of an event large enough to cause damage to surface structures very unlikely.

It is well known that humans can perceive ground vibration at levels as low as 0.8 mm/s, much lower than the level of vibration that will damage even the most fragile structures (about 6 mm/s). Experience gained from open pit mining shows that the main reason for complaints about ground vibration is not usually structural (or even cosmetic) damage, but the fear of damage and/or nuisance (Brovko et al. 2016). Good public relations and explanations will help to reduce anxiety and reduce complaints. Surveys of buildings and other structures (e.g. bridges and dams) should be carried out before, during and following any shale gas development activities, and occupants and owners reassured that they will be compensated for damage due to fracking.

The regulations gazetted by the Minister of Mineral Resources on 3 June 2015 provide sound guidelines for best practice and monitoring, although some aspects might require clarification, be unnecessarily stringent or prescriptive. Other mitigation measures to be considered should include:

i.   Establishment of "buffer zones" around towns (say 20 km in radius) where fracking operations are either prohibited or carried out under strict control,
ii.   Reinforcement of vulnerable buildings, such as farm and heritage buildings, schools, and hospitals,
iii.   Guarantees of compensation for any damage caused by fracking-induced earthquakes,
iv.   Enforcement of building regulations,
v.   Disaster insurance,
vi.   Training of emergency first responders, and
vii.   Earthquake drills in schools and work places (Drop, Cover, Hold On!).

At the time of original investigation (Durrheim et al. 2016), the Council for Geoscience operated only two seismograph stations within the scientific assessment area, and another four stations close to its perimeter. A further six stations have since been installed within the scientific assessment area, which should improve the threshold of completeness to M1 (V Midzi, pers. comm.)

The monitoring of seismicity at the well site is normally the responsibility of the holder of the production license. The location of small events should be accurately determined (say better than 100 m) in order to understand the dynamical processes that are taking place in the fracking area. Seismic arrays should be designed accordingly and advanced location algorithms used. Routine processing of seismic data should include an estimation of spectral parameters such as scalar seismic moment, seismic energy, and static stress drop, which will help to identify a stressed fault as is required by clause 89(1)(b) of the regulations.

The principal lack of information with regard to the assessment of the risk posed by earthquakes is the lack of baseline information on the regional stress field, seismicity, and active faults. It is clear from available information that there is some seismic activity in the region. However, given the sparse seismograph station distribution in the country, especially in the Karoo Scientific Assessment region, the available data is not adequate to identify and characterise the active structures. Improved monitoring by densifying the network would certainly assist. Detailed geological and geophysical studies of identified structures would also be necessary.

12.6.1.1 Risk Matrix

In order to illustrate the risk posed by earthquakes in the scientific assessment area, we considered a worst case scenario, using the Ceres-Tulbagh earthquake of 29 September 1969 (MMI VIII) and the Orkney earthquake of 5 August 2014 (MMI VII) as credible examples of natural and induced earthquakes, respectively. Should a $M > 6$ natural earthquake or a shallow $M > 5$ induced earthquake occur within 20 km of

Graaff-Reinet, dozens or even hundreds of heritage buildings and dwellings could be damaged, some severely. Dozens of people could lose their lives. Repair costs could average perhaps 20%-40% of the cost of the building stock, amounting to hundreds of millions of Rand, and the consequence would then be judged moderate to substantial. However, the likelihood of a natural event occurring is considered to be very unlikely, and the risk posed by this scenario is considered to be low, or at most moderate. Based on international experience, hydraulic fracturing is highly unlikely to induce an M > 5 earthquake, but this cannot be entirely excluded, and the consequences could be moderate or even substantial. The implementation of mitigating measures would decrease the likelihood and consequences to some extent, although this is difficult to quantify.

12.7   Conclusions and Recommendations

Certainty about what technically and economically extractable gas reserves may occur within the study area requires detailed exploration, including test fracking of the Karoo Basin shales. It is possible that no economically recoverable gas reserves exist at all in the study area. At the other extreme, the economically recoverable reserves may be greater than the Big Gas scenario considered here—in which case, associated impacts of shale gas development will be quantitatively larger, but not qualitatively different. Typically, only a small fraction of the total amount of gas present in a shale formation can actually be extracted at an affordable cost. The economic viability of shale gas development in the Central Karoo will depend on the gas price and production costs at the time when the gas is produced, which will be many years from now, if at all. In the interim, the extraction technology may advance, lowering the costs associated with production in future.

The natural occurrence of a damaging earthquake (M > 5) anywhere in the study area is considered to be very unlikely. The level of risk depends on the exposure of persons and vulnerable structures to the hazard. In the rural parts of the study area, the exposure is very low, the consequences of an earthquake are likely to be slight, and hence, the risk posed by earthquakes is considered to be low. While it is considered to be very unlikely that a damaging earthquake will occur within 20 km of a town, the consequences of such an event could be moderate or even substantial. Lives could be lost, and many buildings would need to be repaired. Hence, the risk in urban areas is considered to be moderate. Exploration activities associated with the exploration only scenario do not involve the large-scale injection of pressurised fluids, the risk posed by earthquakes in the study area during the exploration and appraisal phase is considered to be low and not significantly different to the base case.

Shale gas development (SGD) by fracking increases the likelihood of small earth tremors near the well bores. Only a few such tremors are likely to be strong enough to be felt by people on the surface. Many studies, in several parts of the world, demonstrate an increase in small earth tremors during fracking. The possibility that fracking will lead to damaging earthquakes through triggering movement on preexisting faults cannot be excluded, but the risk is assessed as low because the study area very rarely experiences tremors and quakes (Fig. 12.6). Damaging earthquakes associated with SGD elsewhere in the world are almost exclusively linked to the disposal of large volumes of waste water into geological formations (a practice forbidden by South African legislation); rather than the development of shale gas resources using fracking.

The risk to persons and assets close to fracking operations in rural areas, such as workers, farm buildings and renewable energy and Square Kilometre Array (SKA) radio telescope infrastructure, should be handled on a case-by-case basis. Vulnerable structures, including features of heritage importance, should be reinforced and arrangements made to insure or compensate for damage. Should particularly attractive shale gas resources be found close to towns, it is essential to inform local authorities and inhabitants of any planned fracking activities and the attendant risks; enter into agreements to repair or compensate for any damage; monitor the induced seismicity; and slow or stop fracking if felt earthquakes are triggered.

The possibility that an earthquake may be triggered by fracking cannot be excluded. The Earth's crust is heterogeneous and physical processes are complex. Rock properties and geodynamic stresses are not perfectly known, and the seismic history is incomplete. It is thus important that seismicity is monitored for several years prior to any fracking, and that a seismic hazard assessment is performed to provide a quantitative estimate of the expected ground motion. Monitoring should continue during SGD to investigate any causal link between SGD and earthquakes. Should any such link be established, procedures governing fluid injection practice must be reevaluated.

It is recommend that Council for Geoscience's seismic monitoring network be densified in the study area, and that vulnerability and damage surveys of buildings and other structures be carried out before, during and following any SGD activities. Other mitigation measures to be considered should include: monitoring of seismicity during SGD and the slowing or stopping of fracking if felt earthquakes are induced, schemes to guarantee compensation in the case of damage, disaster insurance, reinforcement of vulnerable buildings (especially farm and heritage buildings, schools, and hospitals), enforcement of building regulations, training and equipping of emergency first responders, and earthquake drills in schools and work places.

The Council for Geoscience installed an additional six seismograph stations since the 2016 assessment. In 2020 the Council began drilling a stratigraphic borehole near Beaufort West. At the time of writing (June 2021) the borehole had passed through the target Whitehill Formation and had entered the underlying Dwyka Formation, at a depth of about 2800 m below surface. A team from the University of the Witwatersrand, led by Professor Musa Manzi, conducted a series of active and passive seismic experiments near the borehole.

Acknowledgements Ray Durrheim acknowledges the support of the South African Research Chairs Initiative of the Department of Science and Technology and National Research Foundation.

References

Atkinson, G.M., D.W. Eaton, H. Ghofrani, D. Walker, B. Cheadle, R. Schultz, R. Shcherbakov, K. Tiampo, J. Gu, R.M. Harrington, and Y. Liu. 2016. Hydraulic fracturing and seismicity in the Western Canada Sedimentary Basin. *Seismological Research Letters* 87 (3): 631–647.

Bommer, J.J., K.J. Coppersmith, R.T. Coppersmith, K.L. Hanson, A. Mangongolo, J. Neveling, E.M. Rathje, A. Rodriguez-Marek, F. Scherbaum, R. Shelembe, and P.J. Stafford. 2015. A SSHAC level 3 probabilistic seismic hazard analysis for a new-build nuclear site in South Africa. *Earthquake Spectra* 31 (2): 661–698.

Brandt, M.B.C., M. Bejaichund, E.M. Kgaswane, E. Hattingh, and D.L. Roblin. 2005. *Seismic History of Southern Africa. Seismological Series 37.* Pretoria, South Africa: Council for Geoscience.

Brovko, F., T. Kgarume, N. Singh, A. Milev, B. Wekesa, R. Durrheim, T. Lumbwe, T. Pandelany, and M. Mwila. 2016. Development of a South African Minimum Standard on Ground Vibration, Noise, Air-blast and Flyrock near Surface Structures to be Protected. Project Final Report, SIM14-09-01. Mine Health and Safety Council.

Decker, J., and J. Marot. 2012. Resource assessment. Department of Mineral Resources Working Group, Report on hydraulic fracturing in the Karoo Basin of South Africa, Annexure A, 1–13.

Durrheim, R., M. Doucouré, and V. Midzi. 2016. Earthquakes. In *Shale Gas Development in the Central Karoo: A Scientific Assessment of the Opportunities and Risks*, ed. R. Scholes, P. Lochner, G. Schreiner, L. Snyman-Van der Walt, and M. De Jager. CSIR/IU/021MH/EXP/2016/003/A, ISBN 978-0-7988-5631-7. Pretoria: CSIR. Available at http://seasgd.csir.co.za/scientific-assessment-chapters/.

Durrheim, R.J., and B. Manzunzu. 2019a. Seismicity Assessment (Part 4.2.3, 9 pp) and Seismicity Specialist Report: Impacts of Earthquakes, Seismicity and Faults (Appendix C.3, 58 pp). In Department of Environment, Forestry and Fisheries, Strategic Environmental Assessment for the Expansion of Electricity Grid Infrastructure Corridors in South Africa. CSIR Report Number: CSIR/SPLA/EMS/ER/2019a/0076/B. ISBN Number: ISBN 978-0-7988-5648-5. Stellenbosch and Durban.

Durrheim, R.J., and B. Manzunzu. 2019b. Seismicity Assessment (Part 4.2.2, 10 pp) and Seismicity Specialist Report: Impacts of Earthquakes, Seismicity and Faults (Appendix C.2, 76 p). In Department of Environment, Forestry and Fisheries, Strategic Environmental Assessment for the Development of a Phased Gas Pipeline Network in South Africa. CSIR Report Number: CSIR/SPLA/EMS/ER/2019b/0077/B. ISBN Number: ISBN 978-0-7988-5649-2. Stellenbosch and Durban.

Ellsworth, W.L. 2013. Injection-induced earthquakes. *Science* 341: 1225942. https://doi.org/10.1126/science.1225942.

Fernández, L.M., and A. Du Plessis. 1992. *Seismic Hazard Maps of Southern Africa*. Geological Survey of South Africa, Pretoria.

Goedhart, M.L., and P.W.K. Booth. 2016a. A palaeoseismic trench investigation of early Holocene neotectonic faulting along the Kango Fault, southern Cape Fold Belt, South Africa-Part I: Stratigraphic and structural features. *South African Journal of Geology* 119: 545–568.

Goedhart, M.L., and P.W.K. Booth. 2016b. A palaeoseismic trench investigation of early Holocene neotectonic faulting along the Kango Fault, southern Cape Fold Belt, South Africa-Part II: Earthquake parameters. *South African Journal of Geology* 119: 569–582.

Kuuskraa, V., S. Stevens, T. Van Leeuwen, and K. Moodhe. 2011. World Shale Gas Resources: An initial assessment of 14 regions outside the United States. US Energy Information Administration, 1–365. Prepared by: Advanced Resources International, Inc. Available at: http://www.eia.gov/analysis/studies/worldshalegas/pdf/fullreport.pdf.

Kuuskraa, V., S. Stevens, T. Van Leeuwen, and K. Moodhe. 2013. World shale gas and shale oil resource assessment. Prepared for United States Energy Information Administration, 2013. Technically recoverable shale oil and shale gas resources: An assessment of 137 shale formations outside the United States. Available at: http://www.eia.gov/analysis/studies/worldshalegas/pdf/fullreport.pdf.

Majer, E.L., R. Baria, M. Stark, S. Oates, J. Bommer, B. Smith, and H. Asanuma. 2007. Induced seismicity associated with Enhanced Geothermal Systems. *Geothermics* 36 (3): 185–222.

Manzunzu, B., V. Midzi, T.F. Mulabisana, B. Zulu, T. Pule, S. Myendeki, and G.W. Rathod. 2019. Seismotectonics of South Africa. *Journal of African Earth Sciences* 149: 271–279.

Mathews, C. 2017. Court moves may halt fracking exploration: Sunday Times, 9 April 2017. https://www.pressreader.com/south-africa/sunday-times/20170409/282505773456226.

Midzi, V., B. Manzunzu, T. Mulabisana, B.S. Zulu, T. Pule, and S. Myendeki. 2020. Probabilistic seismic hazard maps for South Africa. *Journal of African Earth Sciences* 162: 103689.

Midzi, V., B. Zulu, B. Manzunzu, T. Mulabisana, T. Pule, S. Myendeki, and W. Gubela. 2015. Macroseismic survey of the M_L5.5, 2014 Orkney earthquake. *Journal of Seismology* 19(3): 741–751.

National Petroleum Council. 2011. Life Cycle of Onshore Oil and Gas Operations. Prepared by the Onshore Operations Subgroup of the Operations & Environment Task Group, Working Document of the NPC North American Resource Development Study.

Scholes, R., P. Lochner, G. Schreiner, L. Snyman-Van der Walt, and M. de Jager. (eds). 2016. *Shale Gas Development in the Central Karoo: A Scientific Assessment of the Opportunities and Risks.* CSIR/IU/021MH/EXP/2016/003/A, ISBN 978-0-7988-5631-7. Pretoria: CSIR. Available at http://seasgd.csir.co.za/scientific-assessment-chapters/.

Singh, M., A. Kijko, and R.J. Durrheim. 2011. First-order regional seismotectonic model for South Africa. *Natural Hazards* 59 (1): 383–400.

Skoumal, R.J., R. Ries, M.R. Brudzinski, A.J. Barbour, and B.S. Currie. 2018. Earthquakes induced by hydraulic fracturing are pervasive in Oklahoma. *Journal of Geophysical Research, Solid Earth* 123: 10,918–10,935. https://doi.org/10.1029/2018JB016790.

Steyn, L. 2017. Red tape puts brake on fracking: Mail and Guardian, 21 April 2017, Available at https://www.pressreader.com/south-africa/mail-guardian/20170421/281981787468471.

Van Wyk, W.L., and L.E. Kent (eds.). 1974. *The Earthquake of 29 September 1969 in the South-western Cape Province, South Africa.* Seismologic Series 4, Geological Survey of South Africa, South Africa.

Warpinski, N.R. 2013. Understanding hydraulic fracture growth, effectiveness, and safety through microseismic monitoring. In *ISRM International Conference for Effective and Sustainable Hydraulic Fracturing.* 20 May 2013. International Society for Rock Mechanics, 123–135. Available: http://cdn.intechopen.com/pdfs-wm/44586.pdf. 2016, May 31.

Chapter 13
Living Safely with Earthquakes in Asia

R. K. Chadha

Abstract On the scale of "abnormality" extreme events occupy the top position. The term is commonly used to refer to the phenomena of nature that have disastrous consequences like the cyclones, storms, rainfall, flooding, extreme high and low temperatures, snowfall, wild fires, droughts, earthquakes, and tsunami, these events have shown an increase in their frequency in recent times. Barring earthquakes and tsunami, which are caused by tectonic forces inside the earth, all others are attributed to global climate change. Severity of earthquakes increases with magnitude where earthquakes of M > 6 can cause a lot of damage in densely populated areas, earthquakes of M > 8 can totally destroy the communities near the epicenter. Data in the last 120 years has shown that while there is no change in decadal numbers of earthquakes in the magnitude range between M7.5 and M8.4, and earthquakes exceeding $M \geq 8.5$ have clustered in a 10–15 year time window. The first cluster comprising six earthquakes was observed during 1905–1920, followed by the second and third clusters that had seven and six earthquakes each during 1950–1965 and 2004–2012. Only five earthquakes exceeding $M \geq 9.0$, which can be termed as extreme events, also occurred in these clusters. Earthquakes and tsunami are low-probability high-risk phenomena and are responsible for more than 50% of the casualties caused by natural catastrophes because of their suddenness and absence of prior warnings. The Asia–Pacific region experiences more natural disasters than any other region in the world with greater impact on people and infrastructure, owing to growing population living in poverty and low quality houses. This paper discusses our understanding of earthquakes and tsunami in Asia and the challenges faced by people and suggests possible strategies to minimize its impact. Extreme events have occurred in the past and will continue to occur in the future. Let us plan and manage these natural hazards and learn to live safely with them.

Keywords Earthquakes · Tsunami · Seismic hazard · Asia · Subduction zones · Mitigation

R. K. Chadha (✉)
Raja Ramanna Fellow, CSIR-National Geophysical Research Institute, Hyderabad, Telangana 500007, India
e-mail: rajen0555@gmail.com

© The Centre for Science & Technol. of the, Non-aligned and Other Devel. Countries 2022
A. S. Unnikrishnan et al. (eds.), *Extreme Natural Events*,
https://doi.org/10.1007/978-981-19-2511-5_13

13.1 Introduction

Extreme events generally refer to weather- or climate-related phenomena like cyclones, extreme high and low temperatures, drought, rainfall, and flooding, which historically, have dominated disaster deaths. This global climate events have global impact, but in the twenty-first century two shock events related to a tsunami in the Indian Ocean in 2004 and an earthquake in 2010 in Haiti (www.usgs.gov), collectively killed more than half a million people in Asia and in the Caribbean. The damage caused to buildings and infrastructure exceeded economic losses of billions of US dollars.

Although earthquakes are a global phenomenon, it has been observed that developing countries, especially in the Asia–Pacific Region, are more vulnerable to earthquake and tsunami hazard in comparison with developed countries in Europe and the west in terms of loss of human lives. The last 120 years of global earthquake data had shown that there is no appreciable change in the decadal average frequency of damaging earthquakes of $M \geq 6.0$, but the loss of human lives related to this phenomenon have increased considerably in the recent times. However, earthquakes of $M \geq 9.0$, considered extreme events, have shown clustering in time in a 10–15 year time window since the last century (Shearer and Stark 2012). Three clusters of six to seven earthquakes of $M \geq 8.5$ in each time window have occurred during 1905–1920, 1950–1965, and 2004–2012. So, in case of earthquakes and tsunami, an event need not be extreme to kill people and cause colossal economic losses.

Since earthquakes occur a few kilometers below the Earth's surface and are not open to direct observations (unlike climate phenomena), predicting them in terms of place, timing and magnitude are beyond the realm of the present state of knowledge. At best one can make indirect measurements of some known precursors, which have been found to be associated with earthquakes. But these studies are fraught with danger of providing misleading information, which could be counter-productive. Now and then, one finds statements in newspapers claiming successful prediction of an earthquake after it has occurred. None of these claims withstand to scientific scrutiny, as they are not based on sound scientific basis understood as of now. Whatever knowledge we have gained today about the earthquake phenomenon is through inferences drawn from the recordings of the seismic waves at different locations or laboratory experiments on rock failure mechanism. So, while predicting earthquakes is not possible, attempts are being made in the direction of saving human lives as well as to minimize economic losses. In spite of achieving great success in providing successful warnings for tsunami in the recent past, there are several instances when the phenomenon has turned into an extreme event and has claimed several hundreds of lives in Asia.

Earthquakes are natural phenomena, which are often turned into mega-disasters due to unplanned human activities and lack of awareness about the hazards associated with them. A seismic hazard is a physical attribute of an earthquake like displacement along faults, strong ground shaking, landslides, ground failure, liquefaction, lateral spreading, and tsunami generation if the earthquake occurs under sea. The vulnerability of the population or infrastructures to these seismic hazards decides the size of a disaster. Quantitatively, a seismic risk can be defined as

$$\text{Seismic Risk} = \text{Seismic Hazard} \times \text{Vulnerability}$$

where seismic hazard is the physical attribute of an earthquake and vulnerability is the potential for damage to the built-in environment, which is the primary cause of casualties. The seismic risk becomes zero if an earthquake occurs in a remote region with no built environment and hence, no vulnerability and no disaster.

A lot of progress has been made to understand the genesis of earthquakes, their sources and the processes. The advancement of knowledge in the field of earthquake engineering and its application in the real world has helped in managing earthquake hazard to a great extent, but still people die in large numbers in developing countries even when a moderate size earthquake strikes. The challenges faced by people and other stakeholders in mitigating the earthquake hazard in Asia and other developing countries are discussed.

13.2 Earthquake Hazard

Earthquakes do not occur in a haphazard way; they need distinct fault lines along which the displacement takes place when an earthquake occurs. The global seismicity patterns bring out two clearly defined linear features along which the earthquakes concentrate (Fig. 13.1). This observation laid the foundation for the "Plate Tectonic

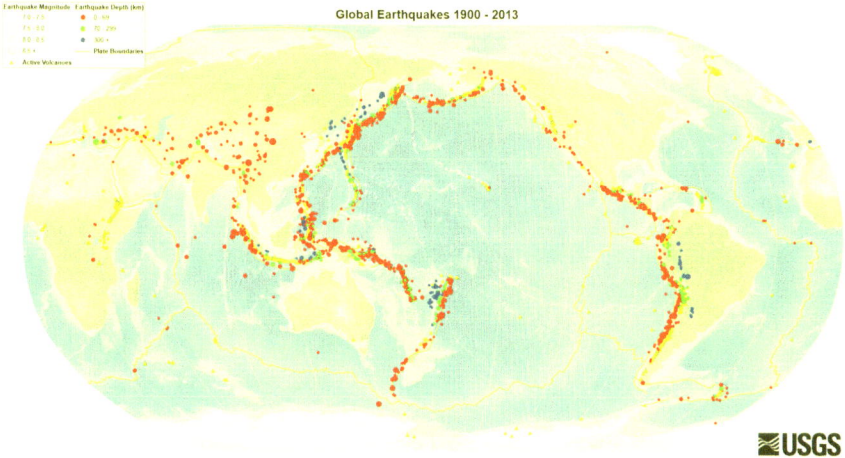

Fig. 13.1 Earthquakes of M ≥ 7.0 during 1900–2013 are shown in different colors according to their depths. Major lithospheric plates defined by mid-oceanic ridges and subduction boundaries are shown (*Source* www.usgs.gov); https://en.wikipedia.org/wiki/Ring_of_Fire)

Theory" during the mid-twentieth century. The linear feature observed in the Atlantic and Indian Oceans are mid-oceanic ridges and the other along the entire circumference of the Pacific Ocean and parts of Indian Ocean along Sunda Arc from Sumatra to Java in Indonesia are subduction zones. The region bounded by these features defines a lithospheric plate that is relatively less active with micro to moderate size earthquake activity. These plates are both continental and oceanic. While, the mid-oceanic ridges represent a divergent plate boundary between the two plates, which move apart, laterally, the subduction zones are the convergent or collision plate boundary along which a plate subducts below the other plate. Mostly, earthquakes of $M \leq 7.0$ dominates spreading ridges, with few exceptions of higher magnitudes, whereas the subduction zones are primary locales of $M \geq 8.0$ earthquakes including the extreme events of M9.0 or greater. There are two distinct collision boundaries in the world. Firstly, the Circum-Pacific Belt, which is also referred to as the "Ring of Fire", which represents interaction between the oceanic Pacific plate with other continental and oceanic microplates. The other collision boundary is the Alpine-Himalayan Belt, which represents a continent–continent collision between the Indian and Eurasian plates. These subduction zones are also the regions of tsunami generation, which occur when one of the subducting plates is oceanic and earthquakes occur on thrust faults with vertical movement. Mid-oceanic ridges do not generate tsunami as earthquakes occur along strike-slip faults with lateral movement, unless a tsunami wave is created due to secondary effect like large-scale-triggered submarine landslides.

An interesting observation, which needs to be pointed out here, is that the pattern of earthquake distribution will not deviate much when plotted for any other time window since the inception of earthquake monitoring using global seismic networks in the last several decades.

13.2.1 Earthquake Hazard in Asia

Since the twentieth century more than 90 earthquakes of $M \geq 8.0$ have occurred, globally (Fig. 13.2). Except for a few, most of the earthquakes occurred along three convergent plate boundaries, (i) the Pacific plate and other microplates in the Asia–Pacific region, (ii) the Indian plate subducting below the Burmese plate along Sunda Arc from Sumatra to Java in the Indian Ocean, and (iii) continent–continent collision boundary along Alpine-Himalayan Belt.

The juxtaposition of Asian countries next to tectonic collision boundaries exposes them to greater earthquake and tsunami hazard relative to African and European continents. Although the major part of the north and south American continents is free of major earthquake zones, their western margins (being on the Circum-Pacific Tectonic Belt) are locales of great earthquakes and tsunami generation. The Australian continent is among the least in terms of earthquake hazard, though tsunami waves generated away from the continent are sometimes observed on its shores as small ripples.

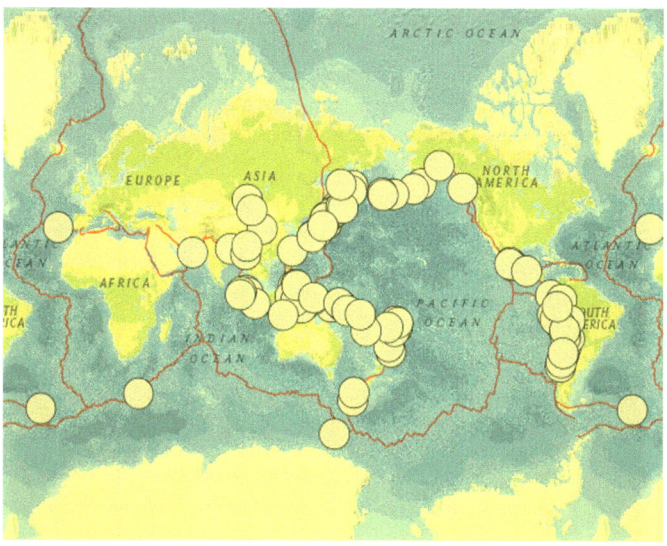

Fig. 13.2 Shows global earthquakes of M ≥ 8.0, globally since 1900 (*Source* www.usgs.gov)

Home to more than 4.5 billion people, Asia is the most densely populated continent with 60% of the world's population. The continent covers an area of about 17,212,000 square miles with a population density of 246.11 per square mile and an average annual growth of about 1%. Such a huge population is faced with the highest risk of earthquakes and tsunami. Two extreme tsunami events in 2004 in the Indian Ocean and in 2011 off the coast of Japan highlight the severe earthquake related hazard in the Asia–Pacific region.

With rapid urbanization of mega-cities, especially in Asia, assessment of earthquake hazard has become a critical parameter to be taken into account in urban planning. Different strategies have to be adopted for urban centers located along and away from shorelines. Standard operating procedures (SOPs) need to be developed while planning a city and its infrastructure, especially for critical structures like nuclear power plants or irrigation dams. Globally, there has been a growing concern about the vulnerability of human lives and structures to seismic hazard. In the United Nations proclaimed "International Year of the Planet Earth", in 2008 out of the ten themes "Hazards—minimizing risk and maximizing awareness" was identified as one of the important themes (Chadha 2010). In India, several initiatives on seismic hazard studies have been undertaken by several research organizations, the Ministry of Science and Technology, and universities. A new Ministry of Earth Sciences has been created in India to address the issues of Earth related research. It is commonly agreed that, in the absence of a reliable earthquake prediction model at present, more stress has to be laid on the assessment and mitigation of seismic hazard.

13.2.2　Earthquake Hazard in India

The Seismic Zoning map of India (Fig. 13.3) shows earthquake hazard in the country (BIS 2002). In the last three decades, increased earthquake threat has been observed in the country where more than 50,000 lives have been lost due to earthquakes and tsunami.

Earthquake sources in the Himalaya: Most of Himalayan belt and the adjoining regions come under the Seismic Zone V and IV. Earthquakes of M ≥ 7.5 have occurred in the Himalaya since historic times in Shillong, Himachal Pradesh, Bihar-Nepal Border, Arunachal Pradesh and parts of Pakistan in the west (Fig. 13.4). The entire Himalayan belt is considered to be one of the most seismically active regions of the world because of the collision tectonics. The recent earthquake of M_w7.9 in 2015,

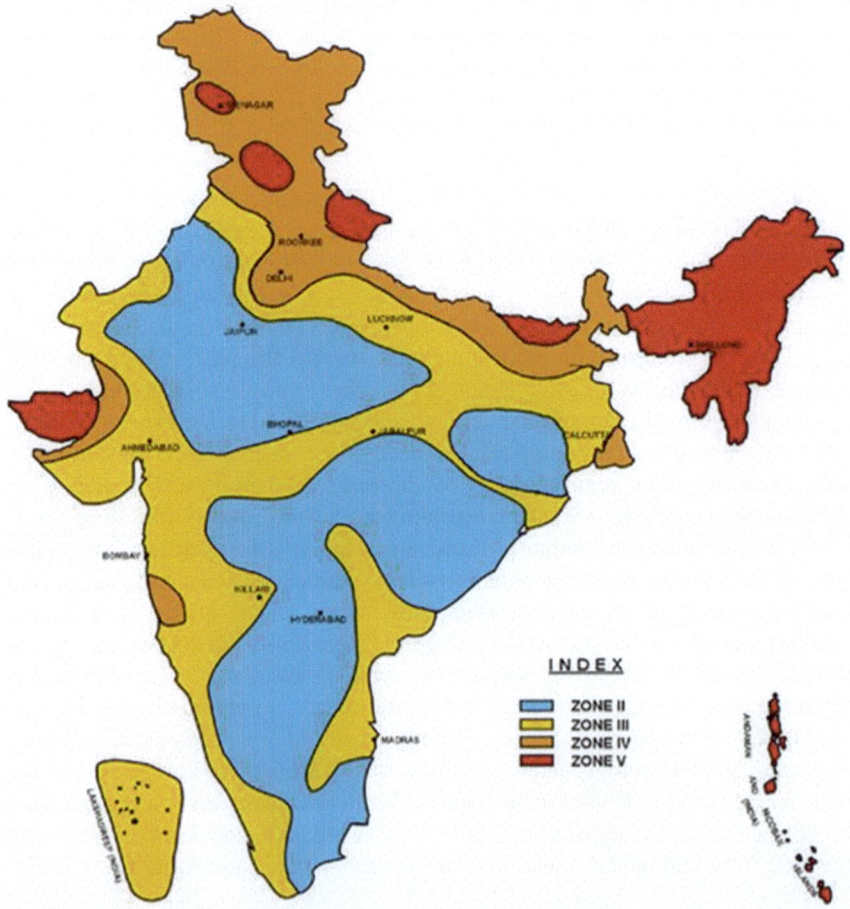

Fig. 13.3　Seismic Zoning Map of India (BIS 2002) with Zone V as the maximum hazard

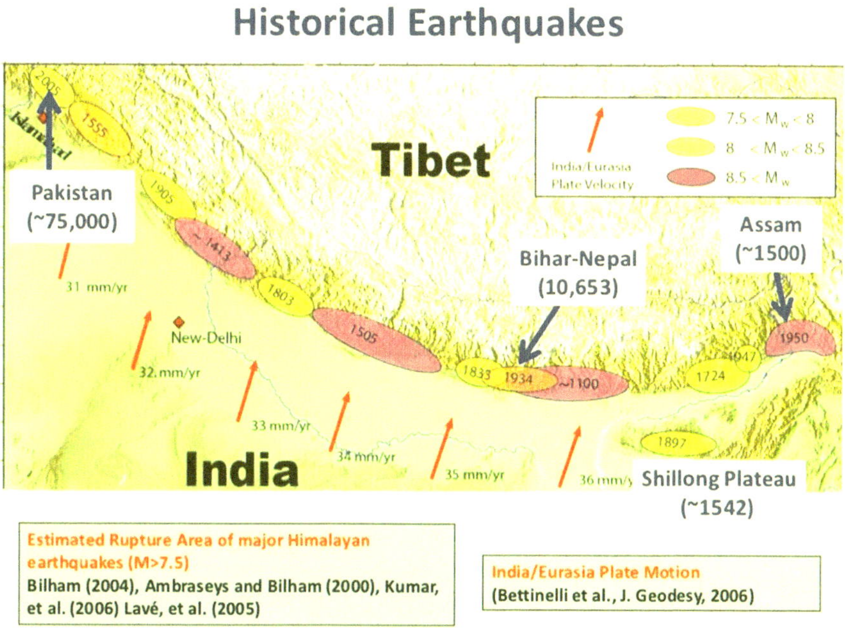

Historical Earthquakes

Fig. 13.4 Shows major earthquakes in the Himalaya from west to Northeast India, since historic times. Figures in bracket are the number of casualties in an around the epicentral and adjoining region (*Source* https://www.slideshare.net/GRFDavos/disaster-risk-reduction-in-the-hindu-kush-himalayan-region-14117839)

with a largest aftershock of $M_w 7.2$, also occurred in the Nepal Himalaya (Fig. 13.6). Several of the recent devastating earthquakes of moderate magnitudes above M6.0 occurred in the Uttarkashi and Chamoli regions in the Garhwal Himalaya (Gupta et al. 2020).

Earthquake sources in the Indian Peninsular Shield: The Peninsular shield of India falls mostly in Zone II and III, which has not experienced earthquakes of M > 8.0, but still several thousands of lives were lost during Koyna earthquake in 1967, Latur in 1993, Jabalpur 1997 and Bhuj in 2001 (Gupta 1993; Mandal et al. 2000; Gupta et al., 2001). Moderate size earthquakes in the shield have occurred along the known weak zones and lineaments like the E-W trending Narmada-Son and West Coast faults and the Godavari Rift. Several micro-earthquake swarms, with the highest magnitude less than 4.5, were also reported along with few scattered events in the entire shield region (Fig. 13.5). This pattern of seismicity is typical of stable continental region (SCR) elsewhere in the world (Gupta and Johnston 1998).

So, it is not a simple relationship between magnitudes of earthquakes and loss of human lives or damage to infrastructures. It depends on several other variables like local site conditions, accelerations of ground motions, distance from the earthquake

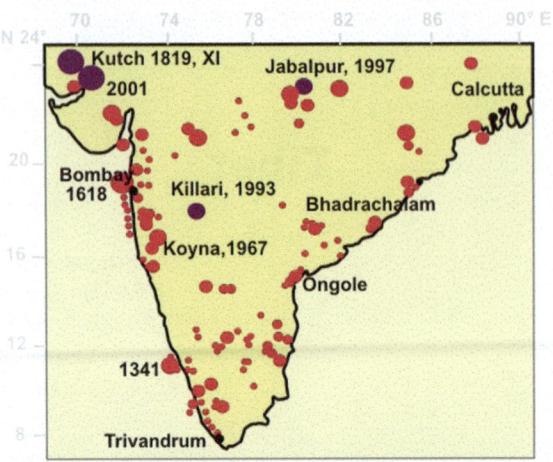

Fig. 13.5 Shows moderate earthquake activity of the Indian Shield with earthquake of M ≤ 6.5 (Kusala Rajendran, Centre for Earth Science, IISc, Bangalore, India, 2008)

source, quality of the built environment, soil types, etc. Earthquake damages to structures start usually from earthquakes exceeding M ≥ 5.0. That is why, the statement that "Earthquakes do not kill people, buildings do" holds true.

Geneses of earthquakes in India: The earthquakes in the Indian plate are caused by the release of elastic strain energy created and replenished by the stresses along weak zones resulting from the continent–continent collision of the Indian plate with the Eurasian plate along the Himalayan boundary. The Indian plate is bounded by the spreading center in the southwest in the Indian Ocean, the collision zone along the Himalaya and transforms faults along the oceanic ridges in the west and east (Fig. 13.1). Earthquakes caused by the processes of continent–continent plate collision have been in progress for several tens of millions of years. The longevity of this process indicates the average conditions of elastic strain accumulation and release within the Indian shield and its boundaries. This understanding of the processes is based on a relatively short period of instrumental record that is not adequate to quantify the processes to characterize earthquake generation with cyclic durations exceeding a few hundred years. It is uncertain whether the earthquakes of the past century are typical of long-term recurrence rates or magnitudes, because the short time window is insufficient to average random fluctuations in slip on the plate boundary, or strain adjustments within the Indian craton. Long-term monitoring with dense GPS measurements may offer valuable clues and will help in quantification of the processes (Bilham et al. 2001).

Reasons for earthquake damage and casualties: Post earthquake studies have shown that damage to structures are caused due to several reasons, most important being, (i) magnitude and depth of the earthquake, (ii) nearness to the epicenter, (iii) shaking, (iv) design of the structure, (v) local site conditions like thickness of the soil layer,

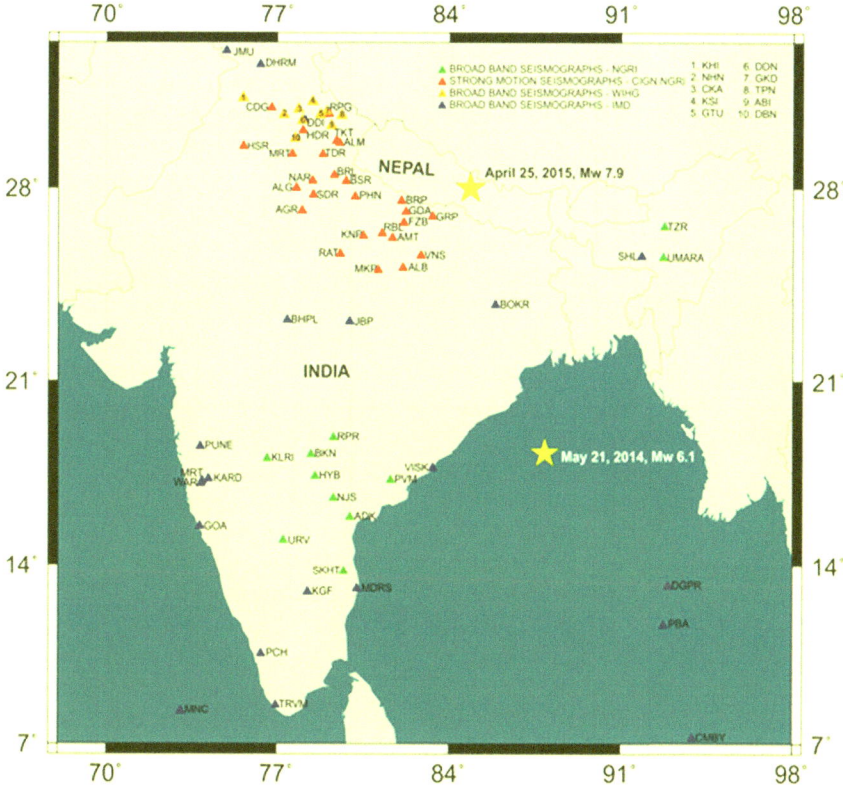

Fig. 13.6 Earthquakes of Mw6.1 in the Bay of Bengal and Mw7.9 earthquake in Nepal are shown as stars. Triangles are broadband seismic stations, which recorded these earthquakes in India. Different color triangles represent different organizations operating these seismic stations (Chadha et al. 2016)

and (vi) quality of construction. While the first three are beyond the human control, the damage due to the other three factors can be countered by a conscious human effort.

Moderate size earthquakes in the Indian shield have killed more than 30,000 people in the last three decades. Studies after the Latur earthquake in 1993, Jabalpur 1997 and Bhuj 2001 have shown that the primary cause of mortality was building collapse and majority of deaths occurred indoors at home. Local construction practices and materials were associated with increased mortality risk, which includes unreinforced masonry, mud and stonewalls, concrete panel, and wood construction. Other factors associated with high mortality were local site conditions, intensity and distance to earthquake epicenter, low socioeconomic status and to some extent, response to rescue (Jain 2005).

Implication of far field effect: In the last few decades, it has been observed that large earthquakes can cause damage to structures even in the far field in addition to the near field of the epicentral region, claiming human lives and causing heavy economic losses. While static displacement and strong shaking cause damages to structures in the near field, far field damages are essentially due to strong ground motions. Ground motions in far field are known to dramatically amplify or attenuate depending on the medium properties through which seismic waves traverse. During the M_w 7.7 Bhuj earthquake in 2001 in Gujarat, structures suffered damages at distances of more than 350 km from the epicenter of the earthquake (Mandal and Chadha 2008). The recent M_w 7.9 Nepal earthquake in 2015 also induced strong shaking due to transient seismic waves and claimed several hundred lives in parts of Bihar and Uttar Pradesh in India at distances of more than 150–200 km (Singh et al. 2017). Earlier, this phenomenon was also noticed during the 1985 Mexican earthquake where Mexico City suffered heavy damage due to an earthquake that occurred at more than 350 km distance in the Pacific Ocean. The reason for such damage was found to be the presence of soft sediments in the Mexico City, which amplified the seismic waves at local sites inducing severe shaking causing collapse of buildings.

In India, the Indo-Gangetic Plains (IGP) present a similar scenario where a pile of sediments reaching few thousands of meters in thickness rests on the basement rocks. The IGP runs almost parallel to the arcuate-shaped Himalayan mountain, which are seismically very active with earthquakes of $M > 7.5$ (Srinagesh et al. 2011). The basin is filled with extensive tracts of alluvium deposited by the major rivers like Ganga, Yamuna, and Indus, and their tributaries on to the flexed part of the Indian plate in front of the gigantic Himalayan orogen. The IGP comprises several sub-basins deepening from south to north. The sediment fill in the Ganga basin constitutes an asymmetrical sediment wedge that is a few tens of meters in thickness in the south, progressively increasing in thickness up to 5 km in the northernmost part (Srinivas et al. 2013). Home to cities like Chandigarh, Amritsar, Dehradun, Delhi, Lucknow, Kanpur, Allahabad, Patna, and Kolkata, which have large and dense populations, the earthquake risk is the highest. Also, the presence of several critical structures like nuclear power plants and large dams in this region demands highest attention and should be a priority on the list of earthquake hazard assessment studies in the country, especially after the tragedy of Fukushima Nuclear reactors in Japan after the 2011 earthquake. Similarly, the city of Guwahati and other urban centers in the northeast India located in the sediment-filled Brahmaputra valley possess equally severe hazard from Himalayan earthquakes.

On May 21, 2014, an M_w 6.1 earthquake occurred in the Bay of Bengal and was felt strongly over a large area in India, in cities as far away as Delhi and Jaipur, approximately a distance of about 1600 km, which was unusual because of the oceanic location of the earthquake and its moderate magnitude. The earthquake was an intraplate, strike-slip event located far from the plate boundaries (Fig. 13.6). The depth of earthquake was estimated to be in the range of 60–80 km by modeling of direct P waves together with the depth phases of pP and sP. The unusually large felt area is attributed to path effect, a consequence of the relatively large depth of the source. A 26 broadband seismic station network in the Indo-Gangetic Plains (CIGN), installed to study the strong ground motions in the Ganga basin, recorded this earthquake. Other broadband seismic stations located in different tectonic environments like the Andaman subduction zone, Himalayan collision zone, eastern Dharwar craton, Shillong plateau, and south west corner of the Deccan Volcanic province to name a few, also recorded this earthquake (Chadha et al. 2016).

Figure 13.7 shows the traces of records at 25 seismic stations of CIGN and other networks as a function of distance from the earthquake epicenter. The first arrivals

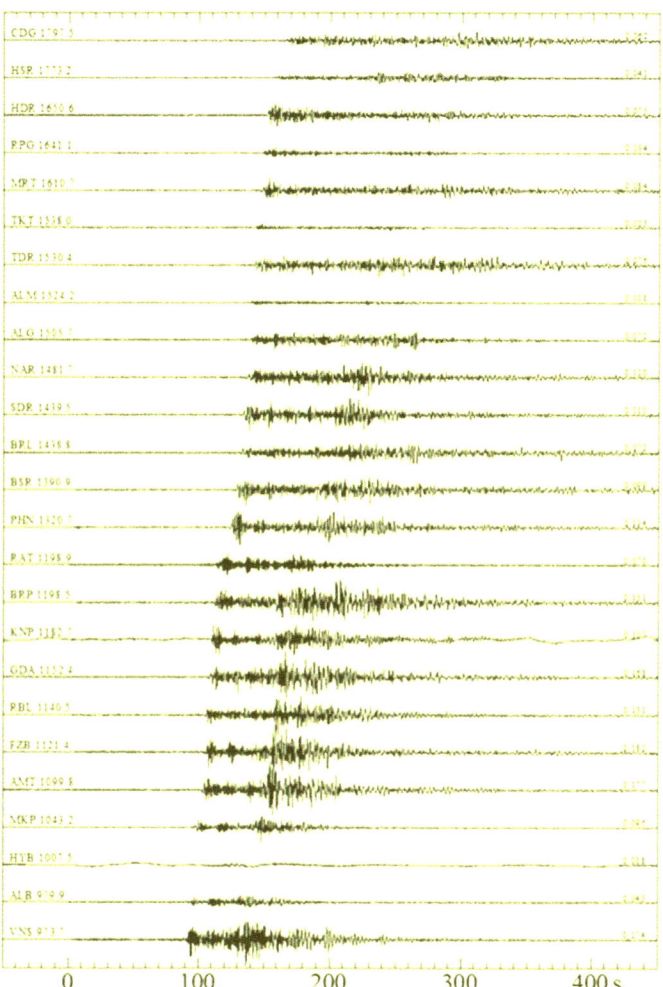

Fig. 13.7 North–south component velocity seismograms of the Bay of Bengal earthquake of May 21, 2014, (M_w 6.1) recorded by the strong motion seismographs of the CIGN, plotted in ascending order of epicentral distance from bottom to top. Traces begin–50 s from the P-wave arrival. The number following each station code indicates the epicentral distance. The number shown on right side of the Y-axis are counts/volts recorded by the seismometer and gives the values of acceleration after multiplying with the sensitivity of the instrument

are the p-wave group followed by larger amplitude of surface waves. It is clearly seen that that seismic stations located on hard sites show attenuation and stations on soft soil amplify the seismic waves, especially the surface waves, which are the main cause of strong shaking in far field even at distances of more than 1500 km (Chadha et al. 2016). The analysis of seismograms further showed that Peak Ground Acceleration (PGA) values, a parameter primarily used in earthquake hazard assessment, for seismic stations located on soft sites in the Indo-Gangetic plains are found to be systematically higher in comparison with the stations located in other geological terrains in India for the same epicentral distances. A good correlation was also found between sediment thickness obtained in the IGP and the amplification of seismic wave propagation, highlighting the importance of local site effects while assessing earthquake hazard in a region. Initial analysis of the other earthquake in Nepal in 2015 also indicated similar results.

Tsunami hazard along the Indian coast: The 2004 Great Indian Ocean tsunami provided an opportunity to initiate tsunami research in the country and led to the establishment of a Tsunami Early Warning Centre in 2007 in Hyderabad, India. This was the deadliest ocean-related disaster which claimed close to 300,000 lives while propagating through the oceans on the earth with devastating effects on the Indian Ocean rim countries like Indonesia, Thailand, Malaysia, Myanmar, Bangladesh, India, Sri Lanka, Maldives, and Africa. This event qualifies to be an "extreme event" related to earthquake hazard.

Tsunami is water waves generated by disturbance caused by large submarine earthquakes in the subduction zones or trenches, submarine landslides and explosive volcanism or a rare possibility of a meteorite impact. Major factors for tsunami generation are: (i) magnitude, (ii) depth, (iii) the nature of faulting, and (iv) the rupture of ocean floor. The Indian Ocean is dominated by the presence of mid-oceanic ridges where the Indo-Australian and African plates are moving away from the Antarctic plate along these ridges. The earthquakes associated with these ridges are along strike-slip faults where the dominant movement is horizontal along transform faults and hence, will not generate tsunami. In the east, Indian plate subducts along Andaman–Sumatra region below the Burmese plate and most of the earthquakes occur along thrust faults where the movement is in the vertical direction generating tsunami waves. In the west, there is another subduction zone along Makran coast of Pakistan where earthquakes can generate tsunami affecting the Indian coastline.

Impact of Tsunami on the Indian coast: Tsunami impact on shore is studied based on run-up heights and inundation distance. The velocity of the tsunami wave depends on water depth in the ocean. For waves with 200–2000s periods, the velocities are of the order of 700–900 km/hour in the open sea. Tsunami gain heights near the shore where the ocean depth becomes shallower and hence, bathymetry plays a major role in developing high tsunami waves.

The average distance of the eastern Indian shoreline from tsunami sources in the Andaman-Sumatra region varies between 1500 and 1800 km, and hence, the time required for a tsunami wave to travel will be approximately 2 h or more. In 2004, the tsunami took 150 min to reach the east coast of India. The worst affected was the coastline along Tamil Nadu from Chennai to Nagapattinam in the south. A maximum run-up height of over 5 m at Nagapattinam was reported where the maximum inundation was about 800 m (Chadha et al. 2005; Chadha and Rajendran 2005).

There are a few other factors, which are crucial in the assessment of tsunami impact on shorelines. Firstly, strike direction of the fault plays an important role in the distribution of tsunami energy in the ocean. The maximum energy is focused perpendicular to the strike of the fault and decreases in intensity along the strike direction. Secondly, energy dissipation occurs whenever there are obstructions to the propagating tsunami. Thirdly, small differences in local run-up and coastal topography can result in large differences in tsunami inundation and associated loss of life and damage to structures (Yeh et al. 2006).

13.3 Results and Discussion

The disposition of the Asia Pacific region *vis-à-vis* the tectonics makes Asian countries more vulnerable to earthquakes and tsunami hazard (Chadha et al. 2007). Knowing this reality, country-specific strategies have to be adopted in such a way that the impact of these hazards is not turned into disasters for the people of Asia. Development cannot be stopped, but care should be taken that the development is sustainable in the face of vagaries of nature. A few of the strategies already under implementation in India and other countries are discussed here.

A disaster cycle can be divided into two, viz., (i) pre-disaster activities and (ii) post-disaster actions. Since, the disaster phenomenon is assumed as cyclic, both the pre- and post-stages overlap in terms of responsibilities and activities. As the saying goes "prevention is better than cure" implies that the better the pre-disaster preparation, the less the impact of the disaster. Since disasters are inevitable and we are a long way from preventing them in totality, let us start with post-disaster actions and activities.

Post-disaster actions: The executive mostly handles the post-disaster actions. It involves three phases. Firstly, how good is the **response** of the executive to a disaster? The speed of rescue operations decides the fate of survivors followed by their immediate rehabilitation with proper shelter, food, clean water, and medical care so that post disaster deaths are minimized. The role of NGOs had been highlighted in several disasters where they performed yeoman service to the needy and affected people. Secondly, in the **recovery** phase, when people are left to their mercy, the role of compensation and trained psychiatrists is very important to reduce the trauma of the disaster where people having lost their kith and kin and belongings are in disarray and looking for honorable resettlement. Finally, the third phase is of **development** after the disaster, which decides how well you prepare for the next disaster. This phase requires a variety of expertise in different fields to build up a strong and resilient community, which become better and better after each disaster.

Pre-disaster activities: Prevention is the first phase of the pre-disaster cycle, which overlaps with the development phase of the post-disaster phase where scientists and engineers start playing their roles. Primarily, this phase requires domain knowledge in scientific and technology fields, which provide basic foundation toward reaching a disaster-free ideal situation. While earth scientists provide information on seismic genesis and sources, civil and structural engineers lay the foundation for earthquake-resistant structures. This is the most important phase of the disaster cycle and is described in detail.

In India, several initiatives are taken to systematically study the earthquake hazard assessment, both qualitatively and quantitatively, with a focus to save human lives and damage to infrastructure and to reduce the economic losses during earthquakes (Chadha and Rajendran 2005). One of them is the preparation of a seismic hazard map for the India and adjoining countries like Pakistan, Afghanistan, Nepal, parts of China, Burma, and parts of Sumatra under a GSHAP program (Bhatia et al. 1999) and another is the microzonation maps of major cities in India like Jabalpur, Dehradun, Delhi, Chandigarh, Lucknow, Guwahati, and Bangalore. Several others are being pursued. These maps take inputs from seismology and other geophysical studies, geological, and geotechnical data in their preparation. The majority of seismic codes in the world accepts structural damages to buildings during an earthquake, given that there is no human loss. Undeniably, many such damages have occurred due to earthquakes in the past. Many new constructions were unaffected by the improvements of codes, but the earthquake safety of existing buildings is under question.

The Indian subcontinent faces serious earthquake threat due to rapid growth of urban population, wherein nearly 60% of landmass in India lies in moderate-to-severe earthquake prone areas. However, it faces serious earthquake threat due to rapid urbanization, wherein over 80% of the population is living in this 60% landmass. In last two decades, India has witnessed damaging earthquakes in 2001 Bhuj, Gujarat, 2004 Sumatra and Indian Ocean tsunami, 2005 Kashmir, 2009 Andaman islands, 2011 Sikkim and another 24 moderate-to-severe earthquakes. More than 400,000 human lives were lost and economic losses worth billions of US dollars were suffered. However, similar intensity earthquakes in the US and Japan did not lead to such an enormous loss of lives, as the structures in these countries are earthquake resistant. During the 2001 Bhuj earthquake, it was estimated that about 370,000 houses and huts were completely destroyed, while another 931,000 were partially destroyed. In the 2005 Kashmir earthquake, 400,153 houses and 6,298 schools were damaged completely. These failures require immediate attention to assess seismic vulnerability of the existing buildings and suggest possible solutions to retrofit them (Pradeep and Murty 2014). Since the detailed assessment of buildings is a complex and expensive task, it cannot be performed on all the buildings in an area. This can be achieved in three steps: (i) rapid visual screening, (ii) preliminary assessment, and (iii) detailed evaluation.

To start with, all Asian countries should undertake at least the first step of rapid visual screening (RVS) of different housing typologies like reinforced concrete, brick masonry, stone masonry, rammed earth and hybrid in their countries to calculate RVS scores, which will provide a quantitative estimate of the earthquake risk in urban cities. An RVS survey is basically a "sidewalk survey" in which an experienced screener visually examines a building to identify features such as the building type, seismic zone, soil conditions, horizontal and vertical irregularities, apparent quality in buildings and short column, etc., that affect the seismic performance of the building. This sidewalk survey is carried out based on the checklists provided in a Performa for all different typologies of the buildings. Other important data regarding the building including the occupancy of the building and the presence of nonstructural falling hazards is also gathered during the screening. A performance score corresponding to these features is calculated for the building based on numerical values on the RVS form. The performance score is compared to a "cut-off" score to determine whether a building has potential vulnerability and whether an experienced engineer should evaluate it further.

Several developed countries facing earthquake threat follow this RVS method for their urban areas. For example, the Federal Emergency Management Agency (FEMA, 1998) developed a number of guidelines in the USA for seismic risk assessment and rehabilitation of buildings (Sucuoglu et al. 2007). These guidelines have undergone several revisions with added new knowledge and experience, e.g., FEMA 310 includes a process for seismic evaluation of existing buildings, along with the introduction of an analysis procedure for screening, preliminary evaluation, and detailed evaluation. In Greece, RVS procedure was developed based on 102 buildings, affected by 1999 Athens earthquake and used for future earthquake events using fuzzy logic to categorize buildings into five different damage grades. In Canada, National Research Council, Canada (NRCC, 1993) suggested the RVS method based on a seismic priority index, which accounts for both structural and nonstructural factors including soil conditions, building occupancy, building importance, falling hazards, occupied density, and the duration of occupancy. In Japan, the procedure is based on a seismic index (SI) for total earthquake-resisting capacity of a story, which is estimated as the product of the basic seismic index based on strength and ductility indices, irregularity index, and time index (TI). The New Zealand code (NZSEE, 2006) recommends a two-stage seismic performance evaluation of buildings. The initial evaluation procedure (IEP) involves making an initial assessment of performance of existing buildings against the standards required for a new building. A standard of 33% or less of the new building standards imply that the building is assessed as "potentially earthquake prone" in terms of the building act and that a more detailed evaluation will be required. Haseeb et al. (2011) provided seismic codes after damaging earthquakes in Muzaffarabad region in Pakistan. In Turkey, the RVS method is based on the ratio of roof displacement capacity to roof displacement demand determined for life safety performance criteria and collapse prevention performance criteria by Bogadici University and Istanbul Technical University. Later, this method was improved on the basis of 454 reinforced concrete buildings surveyed after the 1999 Duzce earthquake and classified into four damage grades.

In India, there have been some efforts toward developing RVS methods. Methodologies for RVS have been proposed for ten different types of buildings. The procedure requires identification of the primary structural load carrying system and the building attributes that are expected to modify the expected seismic performance for the lateral load resisting system under consideration. Building types have been grouped into six vulnerable classes based on European Macro Seismic scale (EMS) recommendations. Damage to structures has been categorized in different grades, depending on their impact on the seismic strength of buildings and the damage levels sourced from EMS.

Several vulnerability studies of some cities in India have been taken up like in Dehradun where loss estimation is calculated based on buildings and population. In Kanpur, preliminary evaluation was carried out on 30 multistoried RC buildings. The study revealed that large opening, horizontal and vertical projections, presence of soft and weak stories, and short column effects are major weaknesses in the buildings at Kanpur from seismic safety point of view. In Gujarat, a RVS was conducted on around 20,000 buildings in Gandhidham and Adipur cities (Srikanth et al. 2010). Though there is a large variation in construction practices, about 26% of buildings were RC buildings and 74% were of brick masonry. In Nanded, Maharashtra a RVS revealed that there are wide variety of construction practices, however, predominantly buildings are classified as per material used, i.e., reinforced concrete, stone and brick masonry, tin shade and other buildings, about 70% are reinforced concrete buildings, 26% are stone and brick masonry, 2% are tin shade, and 1% are other buildings. A detailed study has been done for buildings regarding the structural aspect and effect of earthquakes.

The second phase of pre-disaster cycle is the **mitigation,** which overlaps with the prevention phase. One of the reasons for high number of deaths in the developing Asian countries, which is now emerging as crucial, is the lack of proper education, awareness, and communication of earthquake risk to the general public. After the 2004 Indian Ocean tsunami, a very profound statement was made by Lori Dengler of Humboldt State University in the American Scientist Magazine (Jan 15, 2005) that *"Even without a warning system, even in places where they did not feel the earthquake, if people had simply understood that when you see the water go down, when you hear a rumble from the coast, you don't go down to investigate, you grab your babies and run for your life, many lives would have been saved"*.

This statement underpins two instances in India and Sri Lanka after the 2004 tsunami. In Vishakhapatnam on the east coast of India more than 100 lives were lost when people went inside the ocean when it receded more than a kilometer and later they could not out run the impending tsunami wave which followed. On the contrary, in Sri Lanka, a teacher Victor Desosa saved the village of Galbokka because he knew what to do when the water in ocean receded. Only one person was killed in his village, whereas in the nearby villages casualty rates were 70–90%.

Since, earthquakes are low frequency—high-risk phenomenon, there is greater challenge to effectively connect with people to reduce the disaster impact. Scientists and Engineers have to develop effective communication skills to deal with general public directly using visual media, as their word is more likely to be believed. But, the communication of earthquake risk to public is fraught with great danger and should be handled carefully. While addressing the general public they have to be extra cautious to convey the scientific information in a simple language understood by a layman. Failing to do so will lead to repercussions beyond imagination. The recent incident associated with L'Aquila earthquake in 2009 in Italy is an eye opener to the entire scientific community about the seriousness of the issue of communication with pubic.

In October 2012, six scientists of the Italian National Commission for the Forecast and Prevention of Major Risks were convicted and sentenced to six years' imprisonment on charges of multiple manslaughter for downplaying the likelihood of a major earthquake six days before its occurrence in April 2009 in L'Aquila in Italy. This verdict by an Italian court shook the entire scientific community, world over and created a fear psychosis among the earth scientists of facing legal action over statements that are inherently uncertain. It was a strange verdict because how can one downplay occurrence of a major earthquake without even understanding the phenomenon. This was definitely a case of either miscommunication on the part of the scientists or misunderstanding on the part of the judiciary. The scientists were accused of giving inexact, incomplete and contradictory information about the danger of tremors, which occurred prior to the main quake on April 6, 2009 and claimed 309 lives. The scientists were later acquitted in higher court verdict. Although, this stray incident, condemned by the entire earth science community was seen as an error in judgment, it brought out the importance of spreading awareness of the earthquake risk in the right perspective and its limitations across the cross section of society through sound and innovative communication channels.

The last phase in the disaster cycle is the **preparedness,** which is a condition prior to the occurrence of a disaster. This phase depends on the efficiency of the executive in keeping the equipment updated, conducting awareness with periodical disaster drills and well-trained manpower machinery with clear-cut assignment of duties in case of a disaster.

13.4   Conclusions

(1)   Earthquake and tsunami hazard in the Asian countries are well known. Since, these hazards are highly unpredictable in nature stress should be on prevention and preparedness where human lives could be saved and economic losses minimized.

(2)   In today's highly globalized economy, when a disaster occurs it can create unpredictable turmoil not just in the affected area but, all over the world. Countermeasures against these large-scale disasters are crucial for sustainable development of the global economy to ensure human security. It is time now that international frameworks are activated and collaborations encouraged which will help gathering information on disaster mitigation in the Asia–Pacific Region, including Japan.

(3)   Disaster preparedness and mitigation efforts are resource and funds intensive and require collective wisdom of the executives, scientists, engineers, and other stakeholders, if human losses are to be avoided. A strong leadership at the

highest level is required to convey the conviction and will to stand up to the vagaries of nature and coexist with them.

Acknowledgements The author gratefully acknowledges and thanks all the researchers whose work has been taken while preparing this manuscript. The views expressed in the manuscript are based on author's own research, the contents available in published literature and other information available on Internet and are duly acknowledged. I sincerely thank Dr. Amitava Bandopadhyay who provided an opportunity to express my views on this very important issue.

References

Ambraseys, N., and R. Bilham. 2000. A note on the Kangra Ms = 7.8 earthquake of 4 April 1905. *Current Science* 79: 101–106.

Bettinelli, P., J.P. Avouac, M. Flouzat, F. Jouanne, L. Bollinger, P. Willis, and G.R. Chitrakar. 2006. Plate motion of India and interseismic strain in the Nepal Himalaya from GPS and DORIS measurements. *Journal of Geodesy* 80: 567–589.

Bhatia, S.C., M.R. Kumar, and H.K. Gupta. 1999. A probabilistic seismic hazard map of India and adjoining region. *Annals of Geophysics* 42 (6): 1153–1164.

Bilham, R., V.K. Gaur, and P. Molnar. 2001. Himalayan seismic hazard. *Science* 293: 1442–1444.

Bilham, R. 2004. Earthquakes in India and the Himalaya: Tectonics, geodesy and history. *Annals of Geophysics* 47: 839–858.

BIS. 2002. IS 1893-2002 (part 1): Indian standard criteria for earthquake resistant design of structures, Part 1—General provisions and buildings. Bureau of Indian Standards, New Delhi.

Chadha, R.K., G. Latha, Harry Yeh, Curt Peterson and Toshitama Katada. 2005. The tsunami of the great Sumatra earthquake of M 9.0 on 26 December 2004 – Impact on the east coast of India. *Current Science* 88(8): 1297–1300.

Chadha, R.K., and Kusala Rajendran. 2005. Coping with disasters in future: Some Indian Initiatives. *Current Science* 89 (1): 15.

Chadha, R.K., G.A. Papadopoulos, and A.N. Karanci. 2007. Disasters due to natural hazards. *Natural Hazards* 40 (3): 501–502.

Chadha, R.K. 2010. Seismic hazard in India—Practical aspects and initiatives during IYPE. In *Geophysical Hazards*, ed. Tom Beer, 151–160. Berlin: Springer.

Chadha, R.K., D. Srinagesh, D. Srinivas, G. Suresh, A. Sateesh, S.K. Singh, X. Perez-Campos, G. Suresh, K. Koketsu, T. Masuda, K. Domen, and T. Ito. 2016. CIGN: A high resolution seismic network in Central Indo-Gangetic plains, foothills of Himalayas: First Results. *Seismological Research Letters* 87 (1): 37–46.

FEMA 310, 1998. Handbook for the Seismic Evaluation of Buildings: A Pre-standard - A Handbook, Federal Emergency Management Agency, Washington D.C.

Gupta, H.K. 1993. The deadly Latur earthquake. *Science* 262: 1666–1667.

Gupta, H.K., and A.C. Johnston. 1998. Stable continental region are more vulnerable to earthquakes than once thought. *Eos, Transactions American Geophysical Union* 79(27): 319–321.

Gupta, H.K., N.P. Rao, B.K. Rastogi, and D. Sarkar. 2001. The deadliest intraplate earthquake. *Science* 291 (5511): 2101–2102.

Gupta, H.K., K.A. Sabnis, R. Duarah, R.S. Saxena, and S. Baruah. 2020. Himalayan earthquakes and developing an earthquake resilient society. *Journal of the Geological Society of India* 96 (5): 433–446.

Haseeb, M., Aneesa B. Xinhailu, Z.K. Jahan, A. Iftikhar, and M. Rizwan. 2011. Construction of earthquake resistant buildings and infrastructure implementing seismic design and building code in Northern Pakistan 2005 earthquake affected area. *International Journal of Business and Social Science* 2(4): 168–177.

Jain, S.K. 2005. The Indian earthquake problem. *Current Science* 89 (9): 1464–1466.

Kumar, S., S.G. Wesnousky, T.K. Rockwell, R.W. Briggs, V.C. Thakur, and R. Jayangondaperumal. 2006. Paleoseismic evidence of great surface rupture earthquakes along the Indian Himalaya. *Journal of Geophysical Research (solid Earth)* 111: B03304.

Lave, J., D. Yule, S.N. Sapkota, K. Basant, C. Madden, M. Attal, and R. Pandey. 2005. Evidence for a great medieval earthquake (c. 1100 A.D.) in the central Himalayas, Nepal. *Science* 141: 1302–1305.

Mandal, P., and R.K. Chadha. 2008. Three dimensional velocity imaging of the Kachchh seismic zone Gujarat, India. *Tectonophysics* 452: 1–16.

Mandal, P., B.K. Rastogi, and H.K. Gupta. 2000. Recent Indian earthquakes. *Current Science* 79(9): 1334–1346.

NRCC. 1993. Manual for screening of buildings for seismic investigation by Institute for Research in Construction, National Research Council Canada, Ottawa, United States of America.

Pradeep, K.R., and C.V.R. Murty. 2014. Earthquake safety of houses in India: Understanding the bottlenecks in implementation. *Indian Concrete Journal* 88 (9): 51–63.

Shearer, P.M., and P.B. Stark. 2012. Global risk of big earthquakes has not recently increased. *PNAS* 109 (3): 717–721.

Singh, S.K., D. Srinagesh, D. Srinivas, D. Arroyo, X. Pérez-Campos, R.K. Chadha, G. Suresh, and G. Suresh. 2017. Strong ground motion in the Indo-Gangetic plains during the 2015 Gorkha, Nepal, earthquake sequence and its prediction during future earthquakes. *Bulletin of the Seismological Society of America* 107(3). https://doi.org/10.1785/0120160222

Srikanth, T., K.R. Pradeep, P.S. Ajay, B.K. Rastogi, and K. Santosh. 2010. Earthquake vulnerability assessment of existing buildings in Gandhidham and Adipur Cities, Kachchh, Gujarat (India). *European Journal of Scientific Research* 41 (3): 336–353.

Srinagesh, D., S.K. Singh, R.K. Chadha, A. Paul, G. Suresh, M. Ordaz, and R.S. Dattatrayam. 2011. Amplification of seismic waves in the central Indo-gangetic Basin, India. *Bulletin of the Seismological Society of America* 101(5): 2231–2242.

Srinivas, D., D. Srinagesh, R.K. Chadha, and M. Ravi Kumar. 2013. Sedimentary thickness variations in the Indo-Gangetic foredeep from inversion of receiver functions. *Bulletin Seismological Society of America* 103 (4): 2257–2265.

Sucuoglu, H., U. Yazgan, and A. Yakut. 2007. A screening procedure for seismic risk assessment in urban building stocks. *Earthquake Spectra* 23 (2): 441–458.

Yeh, Harry, R.K. Chadha, M. Francis, T. Katada, G. Latha, C. Peterson, G. Raghuraman, and J.P. Singh. 2006. Tsunami runup survey along the southeast Indian Coast. *Earthquake Spectra* 22: S173–S186.

Chapter 14
The 24 June 2020 Earthquake in Southern Ghana

Paulina Amponsah, Vunganai Midzi, Prince Amoah, and Andrew Tetteh-Cofie Tetteh

Abstract An earthquake of local magnitude 4.2 occurred in Southern Ghana on 24 June 2020 and was felt widely in the Greater Accra region and some parts of the Central Region. The Ghana Geological Survey Authority (GGSA) recorded and located the epicenter offshore Dansoman, near Accra. In response to this event, the National Data Centre (NDC), Ghana, embarked on a project to conduct studies on the earthquake and the earthquake effects. The study involved the NDC soliciting information from the general public, especially residents of the area through questionnaire surveys and interviews. All the data obtained were sorted, analyzed, and evaluated. Analysis of the responses from the survey, along with observations made by the team, and personal inquiries made, showed the distribution of the earthquake intensities with a maximum intensity value of IV (Modified Mercalli Intensity scale) observed close to the epicenter. In general, most parts of Accra and Tema experienced Intensity III, while Intensity II was experienced in areas like Adenta, Oyarifa, Abokobi, Prampram, Dodowa, Oyibi, Somanya, and Aburi. Even though the 24 June 2020 earthquake did not cause any damage to property and resulted in fatalities, the increase in seismic events in southern Ghana in recent years, and the high concentration of strategic assets and human population in the area increases the seismic risk. Severe damage is expected in the event of a large earthquake in the future.

Keywords Ghana · Earthquakes · Seismic risk · Mitigation

P. Amponsah (✉) · A. T.-C. Tetteh
National Nuclear Research Institute, National Data Centre, Ghana Atomic Energy Commission, P.O. Box LG 1055, Legon-Accra, Ghana
e-mail: pekua2@yahoo.com

V. Midzi
Council for Geoscience, Engineering and Geohazards Unit, Silverton, Pretoria, South Africa
e-mail: vmidzi@geoscience.org.za

P. Amoah
Nuclear Regulatory Authority, Legon-Accra, Ghana

© The Centre for Science & Technol. of the, Non-aligned and Other Devel. Countries 2022
A. S. Unnikrishnan et al. (eds.), *Extreme Natural Events*,
https://doi.org/10.1007/978-981-19-2511-5_14

14.1 Introduction

Ghana is characterized by low to moderate seismic activity as it is situated within the intraplate region of West Africa, which is deemed to be fairly typical of a "stable" region. However, many earthquakes have occurred in southern Ghana, primarily in Accra and its vicinity, with major ones in 1862, 1906, and 1939 (Junner 1941). Bacon and Banson (1979) and Amponsah (2002) attributed the seismicity of the south-eastern portion of Ghana to the Akwapim fault zone. Burke (1969) also noted that seismic activity in the coastal areas of the West African region like Accra may be associated with the interaction of the continental margin with the offshore chain transform zone.

On 24 June 2020, three earthquakes of local magnitude ranging from 3.5 to 4.2 hit the southern part of the country within ten minutes, causing residents to panic. The first event of local magnitude (M_L) 3.7 occurred at about 10:47 pm local time and was followed by the main tremor of M_L 4.2 at approximately 10:53 pm (local time). It was felt mostly in the Greater Accra region for about 10–20 s, but with no damage reported. The Ghana Geological Survey Authority (GGSA) recorded all the events and located the epicenter of the main event at the coast near the town of Dansoman (Fig. 14.1). The intensity of the shock was greatest in Dansoman, Ablekuma, Awoshie, Gbawe, Mallam, and also around Weija, Kasoa, and Winneba where the intensity reached IV on the Modified Mercalli Intensity (MMI) scale. The third M_L 3.5 event occurred at about 10:57 pm (local time).

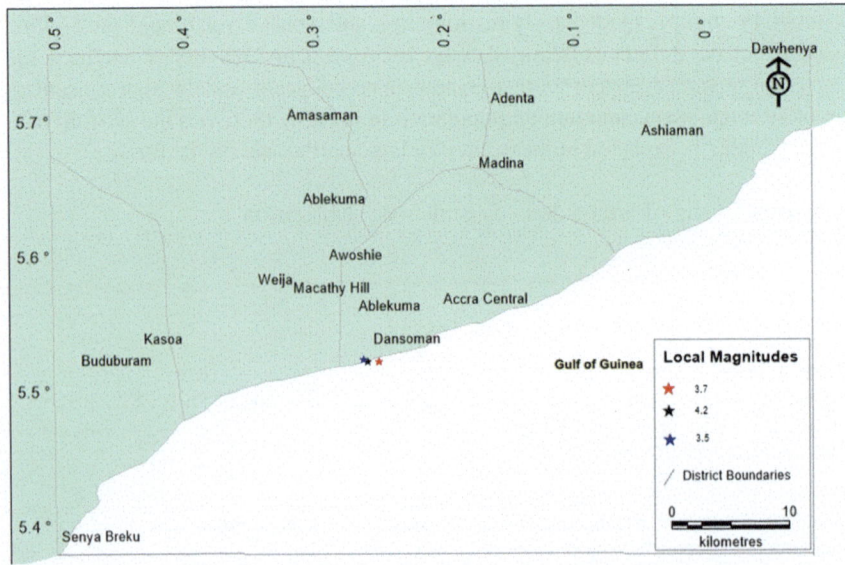

Fig. 14.1 Map showing the location of the 24 June 2020 earthquakes

The National Data Centre (NDC) of the Ghana Atomic Commission catalogs seismic events in Africa and provides information on significant earthquakes in Ghana. A survey was therefore undertaken by a team of scientists from the NDC to determine the impact of the main event on the populace. This paper presents results of the studies conducted on the earthquake and the earthquake effects.

14.2   Tectonic Framework

Zones with distinct tectonic elements define the tectonic structure of SE-Ghana and its offshore region (Fig. 14.2). These are the Akwapim fault zone and the Coastal boundary fault. The occurrence of earthquakes in Ghana seems to be influenced by these two fault zones.

The Akwapim fault zone includes a fault system running northeast-southwest (Fig. 14.2) outlined by the Western Boundary Fault (WBF) and the Eastern Boundary Fault (EBF), which are overthrusts of Neo-proterozoic age (Ahmed et al. 1977). Muff and Efa (2006) indicated that at a later stage the Akwapim Togo belt was subjected

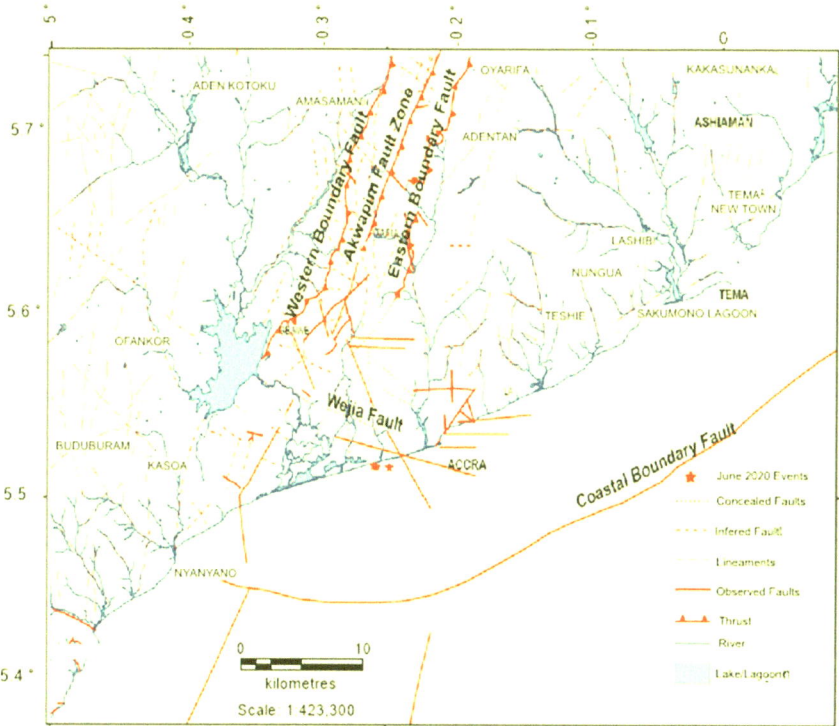

Fig. 14.2 Fault Map of the Study Area (modified from Muff and Efa 2006)

to a block-tectonic style of deformation and several normal local faults developed in recent years. Most of the seismic events in Ghana are associated with the Akwapim fault zone.

Faults in the coastal area and near the continental shelf include the Weija fault and the Coastal Boundary Fault (CBF), which is the most prominent. The Coastal boundary fault strikes approximately N60° to 70° E at a distance of 3–5 km from the coast. West of Accra, the fault bends to strike E-W and intersects the Akwapim fault zone. The Coastal boundary fault forms the northern margin of the Keta basin (Blundell 1976) and was probably active during the entire deposition period.

14.3   Review of Past Earthquakes

Ghana has experienced several significant earthquakes in the last two centuries. The earliest known destructive earthquake in the country was a local magnitude (M_L) 5.7 earthquake that occurred in Axim on 18 December 1636 at about 2 pm. Junner (1941) reported that buildings and underground workings of the Portuguese gold mine at Aboasi, located near the river Duma, north-east of Axim, collapsed and many of the workers were buried alive. The walls of Axim's Fort St. Antonio were cracked.

Since the 1636 Axim earthquake, two other severe earthquakes, both of magnitude M_L6.5 occurred in 1862 and 1939. The 1862 earthquake took place along the coast of Accra and caused significant damage. Every stone building in Accra was destroyed and the forts in Accra and the Christiansburg Castle were rendered uninhabitable. It is said that the African part of the city was almost entirely demolished as the epicenter was near Accra resulting in greater intensity in the city. There was noticeably lower intensity of shaking inland than along the coast. The intensity also decreased to the west as no damage was recorded at Cape Coast (Junner 1941). The earthquake resulted in three fatalities in Accra. The damage and fatalities could have been much worse but for the lightweight type of construction at the time (Ambraseys and Adams 1986).

Similarly, the earthquake of 22 June 1939, the most recent devastating earthquake in Ghana, was felt by individuals over an area of about 770,000 km², and more than 800 km from Accra in most areas. The shock resulted in 17 fatalities, 1350 injured, and over 1 million pounds sterling worth of property damage. The loss of life was lower than might have been expected from an earthquake of that magnitude and the density of the population of Accra. According to Junner (1941), this was because the shock occurred at a time when everyone was awake and many were outdoors, and also because there were no fire outbreaks, which is typical of severe earthquakes.

Many other smaller earthquakes have occurred recently in Ghana, as reported by Amponsah et al. (2012) and Amponsah et al. (2020). Most of the events were of magnitude less than 5 (Table 14.1). These include the three earthquakes of 24 June 2020 with local magnitude ranging from 3.5 to 4.2 that hit the southern part of the country in about 10 min. The main event was of magnitude 4.2.

Table 14.1 Notable earthquakes in Ghana

Date	Magnitude (M_L)	Intensity	Location	Effects
1615	–	–	Elmina, Central region	Fortress at Elmina was destroyed
18 Dec 1636	5.7	IX	Axim, Western region	Workers buried in a Goldmine
10 July 1862	6.5	IX	Accra, Greater Accra region	Stone buildings in Accra were razed to the ground. The Christiansburg castle was uninhabitable. Three people died
1872	4.9	VII	Accra, Greater Accra region	
20 Nov 1906	5.0	VIII	Ho, Volta region	Felt in the north-eastern part of Ghana. Government buildings were damaged
1914–1933	–	–	Southern Ghana	Several shocks were recorded by the seismograph
22 June 1939	6.5	IX	Offshore	17 People died, 1350 injured, and over 1 million pounds sterling worth of property damage
1964	4.5	–	Akosombo, Eastern region	Felt around Akosombo area
1969	4.7	–	Tema, Greater Accra region	Felt around Tema
6 March 1997	4.8	–	Accra, Greater Accra Region Ghana	Cracks in buildings in Accra
18 May 2003	4.0	–	Accra, Greater Accra region	Slight cracks in some buildings
24 March 2018	4.0	–	Accra, Greater Accra region	No damage reported

14.4 Methodology

In response to the main 24 June 2020 earthquake ((M_L 4.2 event), the staff of the NDC embarked on a survey to collect information about the event. Open-ended questionnaires were administered randomly to the public, especially residents in the Greater Accra Metropolitan Area and other neighboring communities to share their experiences. Due to the Covid-19 Pandemic and lockdown, Google forms were used in delivering questionnaires to the general public. Individuals who were not familiar with the approach were assisted via telephone calls. A few interviews were also conducted in person. Responses were then sorted and analyzed to produce intensity data points (IDPs). This was achieved following the procedure by Musson and Cecic (2002) and also implemented by Midzi et al. (2013). Intensities were assigned to locations that were broad enough to contain a sufficient amount of data, as well as fine enough not to group observations made under very different local conditions together. The Modified Mercalli Intensity Scale, MMI-56 (Richter 1958) was used to determine which description of the different intensity levels is best suited to the sum of the data for a particular location under consideration, thus creating intensity data points (IDPs). IDPs were in turn used to prepare an isoseismal map of the study area.

14.5   Results and Discussion

In total, 160 responses were received from the questionnaire administered, and interviews conducted. About 98% of the respondents verified the occurrence of an earthquake on 24 June 2020. The time range reported by the respondents was between 10:00 pm and 11:15 pm local time. This is within the time range as recorded by the Seismic Network in Ghana. Some respondents commented on having felt two earthquakes. However, care was taken to ensure that only observations associated with the main event were collected.

From the data received, over 90% of the respondents were in Accra and neighboring towns (Fig. 14.3). A few respondents reported observing objects swing or heard rattling noises during the earthquake. No cracks in walls were reported by any of the respondents. Although there was a series of earthquakes within a short period, the magnitude recorded by the seismic network in Ghana was in the range of 3.5–4.2, which may account for almost no damage being reported.

The responses from the questionnaires indicated that most of the buildings are situated on loose soils and sloppy surfaces which can result in amplification of ground motion and thus increase the risk of a disastrous earthquake should an event of large magnitude occur. Also, from Fig. 14.4, only 20% of the respondents indicated that geotechnical or seismic parameters were considered before their buildings were constructed. The remaining 80% of the respondents either did not know whether geotechnical or seismic investigations were conducted on the land before their buildings were erected or no geotechnical or seismic parameters were considered. This calls for effective and stringent enforcement of building codes or regulations in the country.

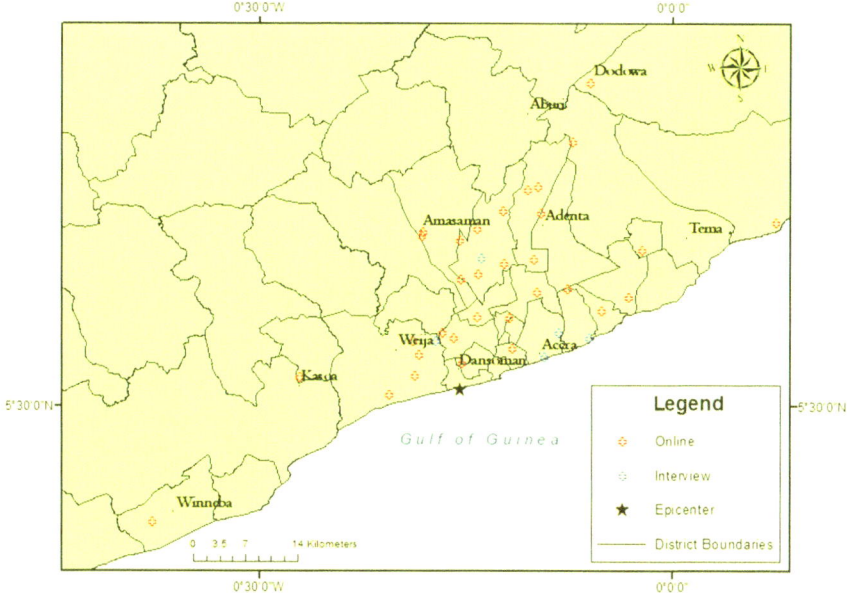

Fig. 14.3 The geographical location of respondents

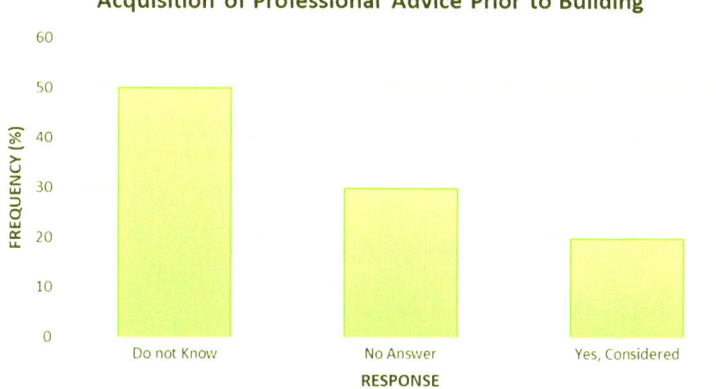

Fig. 14.4 Acquisition of professional advice before building

14.6   Isoseismal Map

Following the procedure outlined in the methodology section, a total of 48 IDPs were created and their spatial distribution is displayed in Fig. 14.5. An evaluation of the IDPs revealed that 27 IDPs were generated using information from one or two questionnaires, while the rest of the IDPs were created using multiple questionnaires.

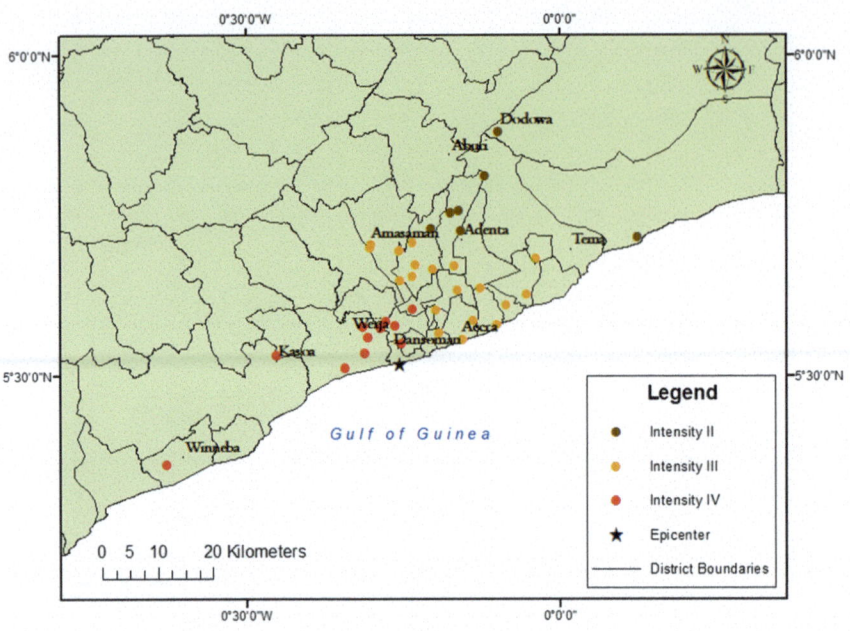

Fig. 14.5 Spatial distribution of intensity data points (IDPs)

On average, information from three questionnaires was used in creating each IDP. It is clear from Fig. 14.5 that observed intensity values decreased north-eastwards. This is seen in the isoseismal map prepared using the IDPs (Fig. 14.6). From Figs. 14.5 and 14.6, it can be seen that the severity of the earthquake only reached Intensity IV, which occurred in towns and communities close to the epicenter (offshore Dansoman). It can also be noted that most parts of Accra and Tema experienced Intensity III, while Intensity II was experienced in areas like Adenta, Oyarifa, Abokobi, Prampram, Dodowa, Oyibi, Kpone Bawaleshie, Somanya, and Aburi.

Isoseismal IV: This intensity was observed along the coast up to the towns of Dansoman, Gbawe, Mallam, Ablekuma, Awoshie, Odorkor, Tesano, Accra-central, Weija, Kasoa, and Winneba. These towns situated on granitods suffered moderate shaking accompanied by a rumbling sound "like a moving truck" which resulted in doors, windows, and beds rattling, as well as the falling of picture frames, hangers, paintings, and other artifacts.

Isoseismal III: Towns a bit further inland and also along the coast but to the northeast of the epicenter, felt shaking of Intensity III. These include Madina, La, Teshie, East Legon, Roman Ridge, North Ridge, Nungua, Haatso, Atomic, Dome, Taifa, Kwabenya, Ashongman, Tema, Sakumono, Lashibi, Spintex, Pokuase, and Amasaman. Most of these towns are underlain by the Accraian and Dahomeyan rocks. The shock in some places like La and Teshie was marginally higher but hardly enough to get them within isoseismal IV. The shock was slightly felt by several

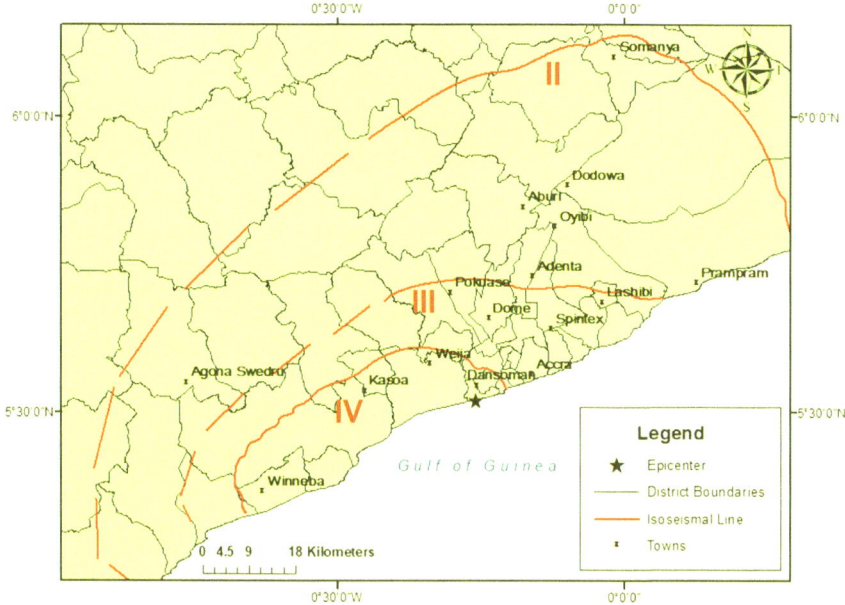

Fig. 14.6 Isoseismal map created using prepared IDPs. The Black Star Represents the Location of the Mainshock (Offshore Dansoman)

residents without causing any alarm, and most residents did not recognize it was an earthquake.

Isoseismal II: The biggest towns within this isoseismal are Adenta, Oyarifa, Abokobi, Danfa, Prampram, Dodowa, Somanya, and Oyibi. Within the isoseismal, the shock was very slight and felt by only a few people.

14.7   Conclusions

Earthquakes in Ghana are concentrated in the southern part of the country, where the Akwapim fault zone intersects the Coastal boundary fault. It is also important to note that some of the epicenters have been located offshore and may be related to the activity of the Coastal boundary fault. The 24 June 2020 earthquake might have been caused by the activity of the Weija fault and other observed faults which are closer to the epicenter as well as the Coastal boundary fault (Fig. 14.2). Even though the main earthquake (M_L 4.2) did not result in damage to property and loss of human lives, the increase of seismic events as observed in southern Ghana in recent years, and the high concentration of strategic assets and human population in the area increases the seismic risk of the region. Responses solicited from residents of Accra and environs after the 24 June 2020 earthquakes indicated high ground shaking in communities along the coast. The higher ground shaking in this region is likely due to the proximity of the area to the earthquake source as well as possible amplification due to site effects. Given this scenario, it is unfortunate that very few people seek professional seismic and geotechnical advice before putting up their buildings. A systematic framework for refocusing National Earthquake Preparedness and Response must be established.

Acknowledgments The authors are grateful to the staff, and National Service Personnel of the National Data Center of the Ghana Atomic Energy Commission. We appreciate with gratitude the immense assistance from the Ghana Geological Survey Authority. Also grateful to the National Nuclear Research Institute for their assistance. We would like to thank all residents who willingly participated in the survey. This research is prepared in the frame of the IGCP-659 (UNESCO) project on Seismic Hazard and Risk in Africa.

References

Ahmed, S.M., P.K. Blay, S.B. Castor, and G.J. Coakley. 1977. The geology of field sheets No. 33 Winneba NE 59, 61 and 62. Accra. SW, NW and NE. *Ghana Geological Survey, Bulletin* 32: 1–47.

Ambraseys, N.N., and R.D. Adams. 1986. Seismicity of West Africa. *Annales Geophysicae* 4: 679–702.

Amponsah, P.E. 2002. Seismic activity in relation to fault systems in southern Ghana. *Journal of African Earth Sciences* 35: 227–234.

Amponsah, P.E., G. Leydecker, and R. Muff. 2012. Earthquake catalogue of Ghana for the time period 1615–2003 with special reference to the tectono-structural evolution of South-East Ghana. *Journal of African Earth Sciences* 75: 1–13.

Amponsah, P., I. Opoku Ntim, and G. Nortey. 2020. Earthquake risk in Ghana: Efforts and challenges. *Arabian Journal of Geosciences* 13. https://doi.org/10.1007/s12517-020-05665-4.

Bacon, M., and J.K.A. Banson. 1979. Recent seismicity of southeastern Ghana. *Earth and Planetary Science Letters* 44: 43–46.

Blundell, D.J. 1976. Active faults in West Africa. *Earth and Planetary Science Letters* 31: 287–290.

Burke, K. 1969. The Akwapim fault, a recent fault in Ghana and related faults of the Guinea Coast. *Journal of Mining and Geology* 4: 29–38.

Junner, N.R. 1941. The Accra earthquake of June 1939. *Gold Coast Geological Survey Bulletin* 13: 3–41.

Midzi, V., J.J. Bommer, F.O. Strasser, P. Albini, B.S. Zulu, K. Prasad, and N.S. Flint. 2013. An intensity database for earthquakes in South Africa from 1912 to 2011. *Journal of Seismology* 17: 1183–1205.

Muff, R., and E. Efa. 2006. Explanatory notes for the geological map for urban planning 1:50,000 of Greater Accra Metropolitan Area. *Ghana Geological Survey*.

Musson, R.M.W., and I. Cecic. 2002. Macroseismology. In *International Handbook of Earthquake and Engineering Seismology*, ed. W.H.K. Lee, H. Kanamori, P.C. Jennings and C. Kisslinger, 81A, 807–822.

Richter, C.F. 1958. *Elementary Seismology*. San Francisco: Freeman.

Chapter 15
Landslides and Slope Instability in Mussoorie and Nainital Townships (Uttarakhand) in Present Climate—Change Scenario

Vikram Gupta, Kalachand Sain, and Ruchika Sharma Tandon

Abstract Landside and slope instability are common in the Himalaya and have been noted to increase in the present climate change scenario. Here, we document the landslides and slope instability conditions in the hilly townships of Mussoorie and Nainital located at an elevations of ~2000 m above mean sea level in the Lesser Himalaya of Garhwal and Kumaon region of Uttarakhand, respectively. It has been observed that there is great lateral variation in the spatial distribution of rainfall in these townships, but in general, Mussoorie township has witnessed consistent rainfall of around 2020 mm per annum, with exceptionally high rainfall in the year 1998 and 2020, whereas the rainfall in Nainital township is more erratic with numerous extreme rainfall events. It had witnessed continuously higher annual rainfall between 2010 and 2017. Besides, there is a significant increase in the intensity of rainfall (i.e. amount of rainfall during a number of days) in both the townships, during the monsoon. However, the increase in intensity is more conspicuous in Nainital township. This change in the climatic pattern in the form of concentrated rainfall had adversely affected the slope instability as well as the occurrences of landslides in both the hilly townships.

15.1   Introduction

Landslides and related mass movement activities are common and widespread in the mountainous region impacting the landform development. These are part of a normal geomorphic cycle and are regulated by geological, geomorphological and geotechnical characteristics of the terrain. In general, lithology, structure, tectonics, terrain morphology, hydrological conditions and the anthropogenic pressure on the

V. Gupta (✉) · K. Sain
Wadia Institute of Himalayan Geology, Dehradun, India
e-mail: vgupta_wihg@yahoo.com; vgupta@wihg.res.in

R. S. Tandon
Graphic Era Deemed to be University, Dehradun, India

© The Centre for Science & Technol. of the, Non-aligned and Other Devel. Countries 2022
A. S. Unnikrishnan et al. (eds.), *Extreme Natural Events*,
https://doi.org/10.1007/978-981-19-2511-5_15

slope are the conditioning factors for the development of landslides and the slope instability, whereas earthquake and rainfall are mainly the triggering factors.

The Himalaya is infested with both earthquake- and rain-triggered landslides. Earthquake-triggered landslides are contemporaneous with the occurrence of earthquakes (Gupta et al. 2015; Rosser et al. 2021). The repeated micro-seismicity in an area may also induce a slope towards the instability, which in future may be triggered by any agent including rainfall (Gupta et al. 2021). Rain-induced landslides are prevalent in the Himalayan terrain and occur every year, particularly during the monsoon, which usually starts in mid-June and ends in September. During this period, besides the occurrence of disastrous landslides, phenomena in the form of development of cracks, subsidence, small-scale debris wash, erosional features, etc. occur at many places and serve as primary indicators of slope instability and that may intensify and be sites of landslides in near future. Therefore, it is essential to map the areas of active landslides as well as slope instability for the disaster management strategy of a region.

Uttarakhand Himalaya is known for the occurrence of landslides, triggered by both earthquakes and rainfall. Some of the disastrous examples of landslides are 1977 Tawaghat landslide (Kali valley), 1978 Gangnani landslide (Bhagirathi valley), 1979 Kontha landslide (Mandakini valley), 1980 Gyansu landslide (Bhagirathi valley), 1990 Neelkanth Mahadev landslide (Rishikesh), 1998 Okhimath landslide (Madhmaheshwar valley), 1999 Malpa landslide (Kali valley), 2001 Phata-Byung landslide (Mandakini valley), 2002 Budhakedar landslide (Balganga valley) and 2003 Varunavat Parvat landslide (Bhagirathi valley) causing great loss of lives and properties in the region (Pande 1990; Paul et al. 2000; Sah et al. 2003; Gupta and Bist 2004; Chaudhary et al. 2010; Martha and Kumar 2013; Maikhuri et al. 2017). During recent times, higher concentration of extreme climatic scenarios in the form of concentrated rainfall has been observed at many places causing loss of lives and damage to private and public properties (Dobhal et al. 2013). These concentrated rainfall events may or may not qualify to be called as 'cloudburst' asper the definition of cloudburst being the 'rainfall > 10 cm occurring in an hour over a larger area' (Ashrit 2010). One of the disastrous examples of extreme climatic events in Uttarakhand Himalaya was the 2013 incessant rainfall, popularly known as the '*Himalayan Tsunami*' or the '*Kedarnath Disaster*'. This event had generated > 10,000 small- and large-scale landslides in the entire Uttarakhand Himalaya (Martha et al. 2015).

These rain-induced landslides vary in dimension ranging from few tens of cubic metres to millions of cubic metres and maybe shallower or deep-seated. These may be single one-time episodic instantaneous movement or may even reoccur every year during the rainy season. These landslides have contributed to the release and creep of natural slopes in the region (Bhasin et al. 2002). Some of the typical rain-induced landslides, which took a heavy toll of lives as well as a huge loss of properties and environment in the Uttarakhand Himalaya, are summarized in Table 15.1. Apart from these, small-scale landslides, taking place by the side of drainage, may also block the flow of streams. This causes damming and makes disasters for the people living downstream (Gupta and Sah 2008). Nevertheless, many of the landslides go unreported in the Higher Himalaya, particularly when these do not interfere with

Table 15.1 Typical examples of rain-induced landslides in the state of Uttarakhand

S. no	Location/name of landslide	Date of occurrence of landslide	Causes and consequences
1	Nainital	19 September 1880	Heavy rainfall claimed 151 lives
2	Birehi landslide (Birehi Gad, Alaknanda river)	September 1893	A massive landslide blocked Birehi river, and artificial Guna Tal was formed, and subsequently, on August 5, 1894, lake was breached causing destruction in the downstream Chamoli, Karanprayag, Ruderaprayag and Srinagar region
3	Kailakhan landslide (Nainital)	7 August 1898	Heavy rainfall killed 29 people in the region
4	Garbyang landslide (Pithoragarh District)	1937	Incessant rainfall caused huge landslide in the Garbyang village of Pithoragarh district and killed many people
5	Mandakini and Madhyamaheshwar	1962	Heavy rainfall caused landslide that blocked Mandani river for many days, and subsequent breaching caused disaster in whole of Madhyamaheshwar and Mandakini rivers killing many people. This was the worst flood in the Madhyamaheshwar and Mandakini rivers
6	Belakuchi landslide (Alaknanda valley)	20 July 1970	Landslide and flash floods in Alaknanda valley caused considerable loss of life and property, and the entire village was washed away killing 70 people
7	Dobata landslide (Pithoragarh district)	19 July 1971	Excessive rainfall and cloudburst in Dobata village killed 12 people and buried 35 houses
8	Tawaghat landslide (Kali valley)	15 August 1977	Heavy rainfall caused landslide that killed 44 people, damaged 100 houses and 2 km stretch at many places. Life was disrupted in an area of 50 km^2

(continued)

Table 15.1 (continued)

S. no	Location/name of landslide	Date of occurrence of landslide	Causes and consequences
9	Kanldiya Gad landslide, (Bhagirathi valley)	6 August 1978	Breaching of lake in Kanldiya Gad due to excessive rainfall caused flash flood in Bhagirathi valley. Debrani village was washed away killing 25 people, and Manari-Bhali hydroelectric project was damaged
10	Kontha landslide (Mandakini valley)	August 1979	Kontha landslide due to incessant rainfall in the Mandakani valley killed 50 people and 100 cattle and destroyed about 150 houses and affected an area of 10 km^2
11	Gyansu landslide (Bhagirathi river)	23 June 1980	High intensity rainfall caused Gyanshu landslide near Uttarkashi township that killed 45 people
12	Kanodiya Gad landslide (Bhagirathi valley)	9 September 1980	High intensity rainfall killed 15 people, and Uttarkashi–Gangotri highway was damaged
13	Neelkanth Mahadev root landslide (Rishikesh)	1990	Rainfall caused huge landslide in Neelkanth Mahadev root near Laxmanjhula, in Rishikesh that left more than 100 pilgrims dead
14	Bhimgoda landslide (Haridwar township)	23 August 1994	Rainfall-triggered landslide killed one child and destroyed 2 houses and 150 m long rail-track. The rail traffic was disrupted for 21 days
15	Tiloth Nala landslide (Bhagirathi valley)	20 August 1995	This was caused by cloud burst followed by debris flow along Tiloth Nala near Uttarkashi (Bhagirathi valley). The slide damaged 200 m road section and 18 buildings

(continued)

human development and occur in uninhabited areas. Such phenomena are not called disasters.

Many of the townships in the Lesser Himalayan are situated at higher elevation ranging between 2000 and 2500 m above mean sea level, where the monsoon rain is pronounced. In Uttarakhand, Mussoorie and Nainital are two such hilly townships,

Table 15.1 (continued)

S. no	Location/name of landslide	Date of occurrence of landslide	Causes and consequences
16	Okhimath landslide (Mandakini valley)	11–12 and 18–19 August 1998	Cloud burst generated numerous landslides including Okhimath landslide, and breaching of the landslide dam caused flash floods in Madhyamaheshwar and Kaliganga valleys killing 101 people, 422 cattle and washed away 820 houses and 411 ha agricultural land
17	Malpa landslide (Kali valley)	August 1997	Heavy rainfall generated rockfall and debris flow along Malpa Gad in the Kali valley killing 211 people
18	Kailash Mansarovar landslide	17 August 1998	Intense rainfall caused massive landslide at the root of Kailash Mansarovar that buried 169 pilgrims
19	Phata landslide (Mandakini valley)	16 July 2001	Heavy rainfall/cloud burst generated numerous landslide and debris flow in Phata and Byung Gad affecting around 14 villages and killing 27 people and 53 livestock. 154 houses were damage, and > 43 ha agricultural land was washed away
20	Khanara landslide (YamunaValley)	30 August 2001	Heavy rainfall generated Khanara landslide in the Yamuna valley blocking the continuous flow of the Yamuna river, and its subsequent breaching damaged 100 m road section
21	Guna landslide (Alaknanda valley)	30 August 2001	Cloud burst in late hours in Guna village of Ghanshyali tehsil killed 7 people. Forty six houses were damaged along with damage to the agricultural land

(continued)

Table 15.1 (continued)

S. no	Location/name of landslide	Date of occurrence of landslide	Causes and consequences
22	Budha Kedar landslide (Balganga valley)	10 August 2002	Cloud burst and debris flow in in the Medh and Dharm Ganga valleys, Budha Kedar area, killed 29 people and injured 31 people in the adjoining area. 16 houses were completely damaged along with the loss to live-stocks, agricultural land and bridges. Micro-hydroproject in Budha Kedar was also damaged
23	Varunavat Parvat landslide (BhagarathiValley)	24 September 2003	Heavy rain activated the Tambakhani landslide in Varunavat Parvat along the right bank of Bhagarathi river at Uttarkashi township. Nine hundred houses are damaged, and some of the four-storey buildings at the toe portion of slide are completely buried under sliding mass. About 5000 people were affected
24	Isolated landslides in Uttarakhand	2009	Heavy rainfall and cloudbursts left > 70 people dead in separate incidents all across the state of Uttarakhand
25	Kedarnath Disaster	June 2013	High intensity rainfall and flash flood in the upper reaches of Uttarakhand caused series of landslides in the entire state of Uttarakhand and killed more than 5000 people
26	Balia Nala landslide (Nainital township)	September 2014	High intensity rainfall in the caused landslide long the right bank of the Balia Nala damaging the entire Rais hotel colony locality and the footpath. More than hundred families living in the area were displaced

(continued)

Table 15.1 (continued)

S. no	Location/name of landslide	Date of occurrence of landslide	Causes and consequences
27	Stone lay area compound landslide (Sher-ka-Danda hill) (Nainital township)	July 2015	Heavy rainfall in the area damaged the link road connecting Birla Vidya Mandir School with the Nainital Town Mall Road. Huge debris were brought down on the Mall Road as well into the Lake causing serious environmental issues
28	Lower Mall Road landslide, Nainital	August 2018	Lower Mall Road of Nainital township collapsed causing serious environmental issues as well as disruption of traffic movement in the town
29	Naina (China) Peak landslide	29 Jan 2020	Heavy snowfall and rainfall triggered landslide at the peak posing threat to the people and building of high court at the base. The portion of the hill also fell in 1990s causing damage to the high court buildings
30	Landslides in Mussoorie township and Dehradun—Mussoorie road	July–August 2020	Heavy rainfall in the area damaged Dehradun—Mussoorie link road

which are situated at high altitudes and famous for tourism activities. The landslide hazard potential in these two townships in the present climate change scenario has been presented. Since various infrastructural development projects like the construction of tunnels and ropeways in Mussoorie are being planned, this study may be utilized by the planners and the decision-makers for mitigating disasters that may be caused by future landslide hazards and planning for disaster management.

15.2 Study Area

The study area, located in the state of Uttarakhand, is broadly divisible into Garhwal and Kumaon region. Mussoorie is located in the Garhwal region, whereas Nainital is in the Kumaon region. The study area around Mussoorie lies between longitude 77°59′59″E to 78°07′46″E and latitude 30°25′58″N to 30°29′08″N, whereas the area around Nainital lies between longitude 79°25′35″E to 79°28′32″E and latitude 29°24′28″N to 29°20′05″ (Fig. 15.1). Both the townships are famous hill stations

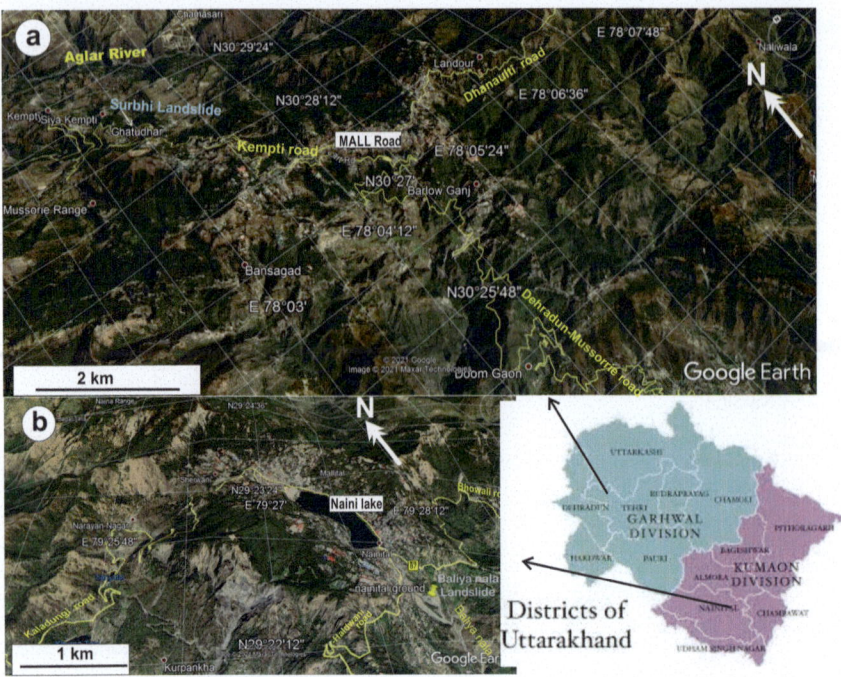

Fig. 15.1 Location map of the Mussoorie and Nainital township in the Garhwal and Kumaun division of Uttarakhand

and thus are occupied by a highly variable floating population during the peak tourist season in summer and winter. The elevation in the Mussoorie region varies between 900 and 2290 m, whereas in Nainital, it is between 1380 and 2542 m above mean sea level (msl). Mussoorie town is located on a flat and rugged ridge with southern slopes that are steeper and much dissected as compared to northern slopes, which are relatively gentle and are covered with thick vegetation. Other geomorphic features present in the area are broader valleys, flat ground, fluvial terraces, gentlysloping spurs, high relief hills, narrow elongated valleys, rocky cliffs, rounded hills and rounded hills with knobs. The habitation in the area is denser in the central part near Mall Road, whereas sparse at other localities. The highest elevation (~2290 m above msl) is noted at Lal Tibba, whereas the minimum elevation of ~800 m is at the Aglar river, a tributary of the Yamuna river. There are many waterfalls in the area like Kempty falls, Shekhar falls, Jharipani falls, Mussoorie falls, Bhatta falls, etc. Nainital town is located at an elevation of ~2000 m above msl. Its surrounding areas are characterized by steep slopes, very high relief and rugged topography. It is densely populated around Nainital Lake. The lake is bounded by the high and steep Naina peak on the north-west side, by the Tiffin top to the south-west and snow-view peaks on the north (Fig. 15.1).

Geologically, both the townships are located in the Outer Lesser Himalaya to the north of the Main Boundary Thrust (MBT) that separates it from the Siwaliks. In general, the rocks of the Lesser Himalaya are highly folded, faulted and thrusted meta-sediments. The geological and structural settings of both the townships are depicted in Fig. 15.2.

The rocks in the Mussoorie area and its environs form part of the western extension of the NW–SE trending Mussoorie Syncline (Auden 1934). These rocks are

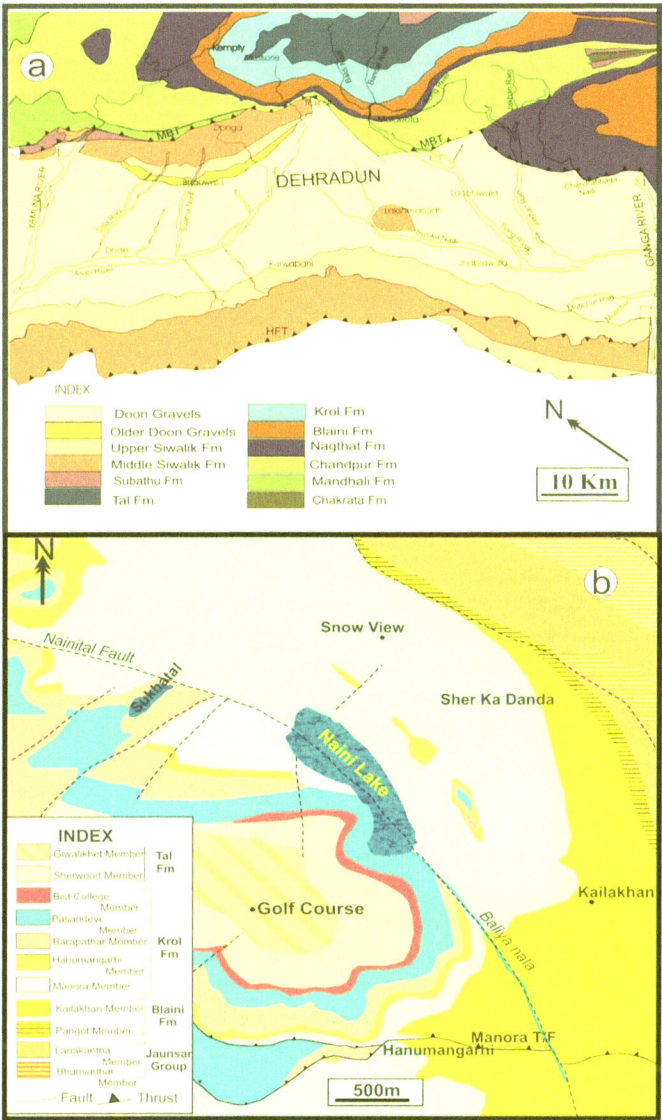

Fig. 15.2 Regional geological map of the Mussoorie and Nainital

invariably classified into Chandpur Formation, Nagthat Formation, Blaini Formation, Krol Formation and Tal Formation that were exposed along various road cut
sections. Towards the south, the MBT brings the Chandpur Formation in juxtaposition with Doon gravels. The MBT is and well exposed at Sahansai Ashram. Chandpur
Formation mainly constitutes phyllite, slates, siltstone and greywacke and is overlain by the quartzite and slate belonging to the Nagthat Formation, which, in turn,
is overlain by the Blaini Formation constituting conglomerate, siltstone, greywacke,
slate and sandstone. However, a greater part of the study area consists of limestone, dolomitic limestone and dolomite belonging to the Krol Formation (Tewari
and Qureshy 1985). The Krol Formation is overlain by Tal Formation, which occu-
pies the core of the Mussoorie syncline. The Tal Formation is further divisible into
lower and upper Tal. The lower Tal constitutes four distinct members: Chert Member,
Argillaceous Member, Arenaceous Member and Calcareous Member, whereas the
upper Tal is represented by the Quartzite Member. The geological setup of the area has
been studied in detail by Auden (1934), Shanker (1971), Panikkar and Subramanyan
(1997) and Banerjee et al. (1997).

Similar to Mussoorie town, the rocks in Nainital township constitute limestone,
shale and slate belonging to the Blaini, Infrakrol, Krol and Tal Formations of the
Lesser Himalaya. These formations have further been divisible into various members
and are shown in Fig. 15.2. The Blaini Formation, dominant in the north-eastern
and eastern side of the Nainital Lake, mainly comprises conglomerates, associated
with purple slate, quartzitic and dolomitic limestone. The Krol Formation mainly
consists of argillaceous limestone, grey and blue dolomitic limestone, dolomite and
tuffaceous limestone, including red and purple ferruginous shale, different coloured
calcareous slate, greywacke, siltstone and fine-grained muddy sandstone. The Tal
Formation is represented by carbonaceous shale with subordinate dolomitic limestone characterized by nodules, lamina and stringers of phosphatic materials with
purple green shale intercalated with muddy fine-grained sandstone and siltstone. It
is best developed around southwestern side of the Nainital Lake. A number of folds,
faults and thrusts traverse through the area. A NW–SE trending Nainital Fault, also
referred as Lake Fault (Valdiya, 1988), traverses through the Nainital Lake. The fault
extends towards the south-eastern end of the lake and follows the Balia Nala, and
here, it is referred to as 'Balia Nala Fault'. There are numerous offshoots from this
fault, the signatures of which have been observed in the premises of the Birla Vidya
Mandir School, Kailakhan area and the snow-view locality. Other notable faults in
the area are NE-SW trending Giwalikhet Fault, observed near the golf course area
and E-W trending Hanumangarhi—Krishnapur Fault or Manora Fault, observed near
Durgapur (Fig. 15.2). The geological setup of the area has been studied in detail by
Middlemiss, (1890), Holland (1897), Auden (1934) and Heim and Gansser (1939).
Both the townships are located in zone IV of seismic zonation map of India (BIS
2002).

15.3 Engineering Geological Conditions of Slope Forming Material

Geo-engineering properties of rocks constituting the slopes around Mussoorie and Nainital have been extracted using direct as well as indirect methods. The soil samples collected from the landslide area from both the areas were tested for cohesion and friction angle using direct shear apparatus. A series of unconsolidated–undrained direct shear tests were performed. The tested samples were predominantly non-plastic and sandy silt in both regions. It has been observed that the cohesion of the soil is negligible, reaching a maximum up to 1.95 kPa, whereas friction angle varies between 27° and 36.4°, indicating low shear strength of the soil. In order to assess the condition of rock mass in both the area, the rock mass rating (RMR) of rocks proposed by Bieniawski (1973) was measured at different sections covering the entire townships. These mainly include the measurement of the strength of the rocks and the orientation conditions of the various discontinuities present in the rock mass. The strength of the rocks has been assessed using the Schmidt hammer rebound (R-) value, whereas orientation and conditions of all the discontinuities present in the rock mass were assessed in the field. The R-value of rocks around the Mussoorie and Nainital townships is variable between 10 and 15 with an average value of 12 for limestone, while for shale and slate, it is variable between 14 and 35 with an average R-value of about 22. Based on the correlation between R-value and compressive strength of rocks, it has been suggested that the compressive strength of these rocks is less than 20 MPa, thus classified as a weak rock (Barton and Choubey 1977; Tandon and Gupta 2015). The calculated RMR as per the standard procedure also indicates poor quality of rock mass around the Mussoorie and Nainital hills.

15.4 Climate Pattern

Both the regions have a tropical climate with pleasant summers and cold winters. The average summer temperature is around 25 °C, and the winter temperature drops to ~0 °C. During winter, both the townships often experience snowfall. Daily rainfall data of both the townships for the years (1995–2020) have been analysed, and it has been noted that the average annual precipitation for Mussoorie ranges between 846–6734 mm, and 1302–4773 mm for Nainital with monsoon months, i.e. June–September receiving > 80% rainfall (Fig. 15.3).

Though both the townships are located at an elevation of ~2000 m above msl, still there is a large lateral variation in the spatial distribution of rainfall. The average yearly rainfall in Mussoorie is ~2020 mm; however, in the years 1998 and 2020, the area witnessed exceptionally high rainfall which is ~2.5 and 3.3 times the average yearly rainfall, respectively. The monsoon season also recorded 2.8 times and 3.7 times the average rainfall. However, the rainfall pattern in Nainital is erratic with numerous extreme rainfall events and with continuously higher annual rainfall

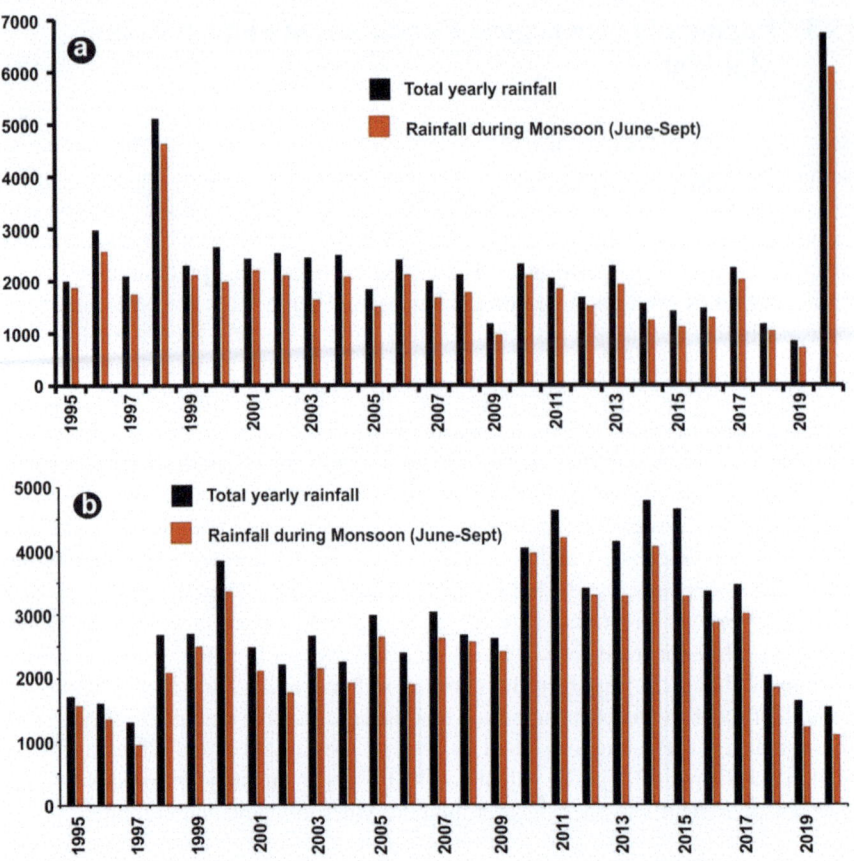

Fig. 15.3 Bar diagram of average annual rainfall and the rainfall during monsoon months in the **a** Mussoorie and **b** Nainital townships

between 2010 and 2017. Besides, there is a significant increase in the intensity of rainfall (i.e. amount of rainfall during a number of days) in both the townships, during the monsoon. However, the increase in intensity is more conspicuous in Nainital. This change in the climatic pattern in the form of concentrated rainfall had adversely affected the slope instability, as well as the occurrences of landslides in both the hilly townships. These are described in the succeeding sections.

15.5 Landslide Hazard Potential

Landslides in both the townships are known to occur since the geological past. However, the landslides and slope instability in both the townships have increased manyfold, primarily due to increased extreme rainfall events and also due to the

Fig. 15.4 a Surabhi resort landslide on the Mussoorie–Kempty road that was initiated in August 1998 because of extreme climatic rainfall event; **b** reactivation of the paleo-landslide deposit after the 1998

excessive anthropogenic pressure on the slope. The landslide activities and slope instability in both the townships are briefly described below:

a.   Mussoorie

Mussoorie township and its surroundings have a history of hazardous landslides and signature of slope instability in the past. Earlier landslides in the area were reported mainly due to wrongful practices of limestone mining, resulting in cave-ins and slumping. However, in order to prevent the environmental degradation further, mining activities in the entire area were banned in 1961. Recently the climate change and more specifically with the change in the rainfall pattern in the form of higher extreme rainfall events led to occurrences of landslides and slope instability in the region. One example of such landslides is the Surabhi Resort landslide on the Mussoorie–Kempty road that was initiated in August 1998 due to excessive rainfall (Gupta and Ahmed 2007) (Fig. 15.4a). This had occurred on the slope, which was occupied by the palaeo-landslide deposits (Fig. 15.4b), and caused loss of life and damage to properties. Besides, there are many slopes in the region that had started sliding, including the slopes near the Kempty fall (Uniyal and Rautela, 2005).

The entire Mussoorie and surrounding regions also received extreme rainfall during the monsoon of 2020 (Fig. 15.3), leading to the occurrence of more than 40 landslides in the Mussoorie township including the landslides on the Dehradun—Mussoorie road. During the intense precipitation in August 2020, an area with 50 m depth in the Dehradun—Mussoorie road near Kolhu Khet was washed away (Fig. 15.5a), causing hardship to the locals and tourists. In another landslide event in the same year, the intense rainfall on 31st August 2020 (Fig. 15.5b) resulted into the accumulation of debris on the Mussoorie—Dehradun road, which blocked the movement of vehicles for manyhours.

Considering various causal factors of landslides like lithology, landuse-landcover (LULC), slope, aspect, curvature, elevation, road cut, drainage and lineament, a

Fig. 15.5 a Damaged part of the road due to landslide; **b** landslide on the Dehradun—Mussoorie road after intense precipitation in August 2020

landslide susceptibility map (LSM) of the area, based on the bivariate statistical Yule coefficient (YC) method, has been prepared (Ram et al. 2020) (Fig. 15.6). The results show that ~44% of the study area falls under very high, high and moderate landslide susceptible zones, and ~56% of the area falls into the low and very low landslide susceptible zones. The dominant part of the area falling under high and moderate landslide susceptible zones lies in the area covered by highly fractured Krol limestone exhibiting slopes ranging between 65° and 77°. Further, the central part of the township encompassing Nagar Palika Parishad ward, Library ward and

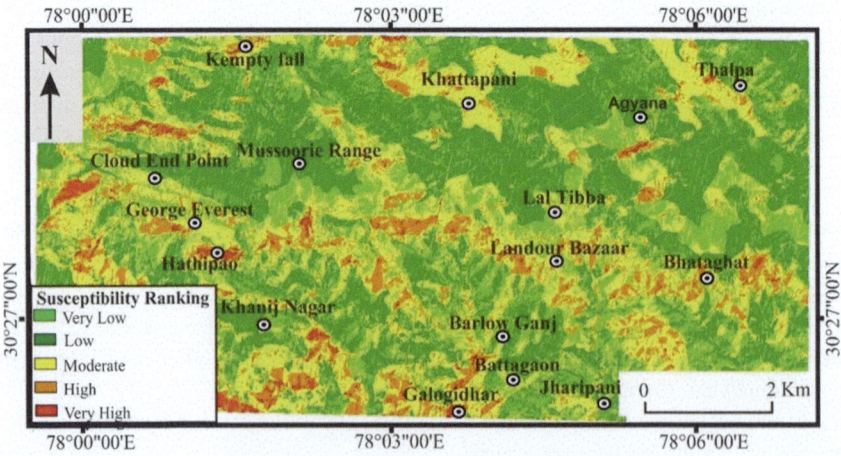

Fig. 15.6 Landslide susceptibility map of the Mussoorie and its environs prepared using the bivariate method

Bhadraj ward falls dominantly in the high and very high landslide hazard zones (Ram and Gupta 2021). ~23% of total buildings (~1604 buildings) with habitation of ~8000 persons in the township are prone to high and very high landslide risk (Ram and Gupta 2021).

Therefore, it is essential to monitor the area from time to time and the results of the present study may be utilized as a reference by the planners and decision-makers for further planning and development of the area.

b.   Nainital

Nainital is known to have occurrences of landslides in the past, and about half of the area of the Nainital is covered with debris generated by landslides (Valdiya 1988). The earliest record of the landslides in the area dates back to 1867 and 1880 when Naini peak-Sher-ka-Danda spur failed following a heavy rainfall. The area again witnessed disastrous landslides in 1893, 1898, 1924, 1989 and 1998, and more recently after 2009 due to increased and concentrated rainfall (DMMC 2011; Gupta et al. 2017). Some of the characteristics of landslides and slope instability in the area post 2009 are described as below:

i.   *Rais Hotel Locality—September 2014 Balia Nala Landslide*

Balia Nala (drainage) located on the southern side of the Nainital Lake drains its excess water. The first record of landslides on the right side of Balia Nala dates back to 17 August 1898 following 102 cm of incessant rainfall for 8 days (Middlemiss 1898). The slide surged across the valley damaging the right bank as well, and since then, both sides of the Balia Nala are continually progressively expanding. A planar landslide had occurred in the Rais hotel locality on the right bank of Balia Nala on September 10, 2014 after the continuous rainfall (Gupta et al. 2016) (Fig. 15.7a, b). The year 2014 recorded the highest total annual rainfall (4773 mm) as well as the second highest rainfall during the monsoon season (4063 mm) in the last twenty six years (1995–2020). Preceding the occurrence of the landslide on 10 September, 465 mm rainfall had fallen during September 1–9 and 33 mm on the day of occurrence

Fig. 15.7 a The crown portion of the Balia Nala landslide damaging the GIC–Krishnapur footpath and the retaining wall; **b** Panoramic view of the Balia Nala landslide exhibiting endangered habitations

Fig. 15.8 **a** The crown portion of the landslide located on the Sher-ka-Danda hill slope damaging the narrow link road connecting Birla Vidya Mandir School with the town Mall Road; **b** downslope view of the landslide exhibiting the threat to the habitations

of landslide. Further, the area also recorded the highest intensity of 68 mm of rainfall per day during June–September 2014 (Gupta et al. 2016) and this was the main trigger for the occurrence of this landslide. The slide had completely damaged the ~20 m stretch of the concrete footpath connecting GIC–Krishnapur and ~360m^2 retaining wall on the slope (Fig. 15.7a) and partially damaged ~20 houses in the vicinity and endangered all otherlocal houses (Fig. 15.7b). Prior to the occurrence of landslides in the area, cracks in the houses and on the ground have been observed, clearly indicating that the slopes were moving slowly (Gupta et al. 2016). Cracks had also been observed in the building of the Govt. Inter College, located on the crown portion of the landslide.

ii.   *Sher-ka-Danda Hill–July 2015 Stone Lay Area Compound Landslide*

A landslide in the form of debris flow had occurred on the south-west facing slope of the Sher-ka-Danda hill on July 5, 2015 after the incessant rainfall. The slide had damaged the narrow link road connecting Birla Vidya Mandir School with the town Mall Road (Fig. 15.8a) The rainfall has mobilized the debris cover on the bedrock generating a huge volume of debris, exposing the bedrock. The year 2015 had recorded the second highest annual rainfall (4641 mm) during 1995–2021, of which about 71% had fallen during the monsoon season (Fig. 15.3). Though, on the day of occurrence of landslide, five mm of rainfall had been recorded, but 416 mm rain had fallen during July 1–4, preceding the day, of which just 300 mm fell on July 01, 2015. These made the soils in the slopes saturated with water.

The slope wash material was transported downslope towards the lake (Fig. 15.8b). Since there is no roper drainage to drain off the water from the hill, and the lower part of the slope is occupied with commercial activities, the debris material finds its way through a narrow passage of about 2 m wide, between two buildings, namely Everest Hotel and India Hotel, and accumulated on the Mall Road and partly finds its way into the Nainital Lake. It was estimated that about 1500 m^3 of debris has been deposited on the road and part has been transported into the lake (Gupta et al. 2017).

Fig. 15.9 View of the collapsed portion of the lower Mall Road

15.5.1 Land Subsidence on Lower Mall Road in Nainital—August 2018

A portion of the lower Mall Road, close to Mallital (between Grand Hotel and HDFC Bank), subsided and consequently washed away on 18th August 2018 and again on 25th August 2018 (Fig. 15.9). The collapse was observed to have occurred on the concave portion of the road. Prior to the collapse, numerous curvilinear cracks, with concavity towards the lake, were observed both on the lower Mall Road and on either side of the collapsed portion of the road indicating the progressive development of the slope instability. The total affected portion of the road was ~115 m, of which 25 m stretch had collapsed. It was reported that the appearance of the cracks on the Mall Road, particularly in the proximity of the subsided section, was a routine phenomenon observed every year post 2010.

iii. *Naina (China) Peak Landslide—2020*

The Naina Peak hill is located to the north of Sukha Tal in Nainital city. This is a NW–SE trending hill, the highest point of which is at ~2611 m, whereas the base is at ~2000 m. The slopes in the upper and middle part of the hill are about near vertical and ~70°, respectively, and are made up of highly shattered and jointed shale and limestone. The area was quiescent for more than 30 years. However, with the recent climate change pattern in the region, the area started sliding in January 2020 (Fig. 15.10a). To the west of this landslide, numerous ~30–40 m long tension cracks with 50–100 cm wide openings were observed in a zone of about 100 m indicating the active nature of slopes in the region (Fig. 15.10b).

15.6 Discussion and Conclusion

Landslides and slope instability in the hilly terrain are very common. In general, the main factors responsible for the higher susceptibility of landslides in both the

Fig. 15.10 a The crown portion of the landslide located on the Naina peak; **b** The signature of the instability on the slope of the Naina peak

Mussoorie and Nainital townships are weak geological settings and numerous active tectonic discontinuities. The dominant part of both the townships comprises limestone, dolomitic limestone, shale, slate and siltstone belonging to the Krol Formation. The characteristic features of the rock types of this formation are that the rocks are highly jointed, fractured and micro-fractured, and thus, their unconfined compressive strength of the rock mass is very low as evidenced by the lower Schmidt hammer rebound (R-) value. The rocks in the area are further shattered due to the presence of faults. A major Nainital Lake Fault or the Balia Nala in Nainital and their offshoots have also been observed on the Naini peak—Sher-ka-Danda spurs, near the Nainital ropeway locality, in the premise of the Birla Vidya Mandir school and in the Kailakhan area. These faults have also contributed to the shattering of the rocks. Thus, there is a higher probability of landslides in the area. The continuous movement along these faults has been observed and evidenced in the continuous growth of cracks in the buildings located in the vicinity (Gupta et al. 2016). Kotlia et al. (2009) also recorded considerable movement per year along the Balia Nala Fault using a network of global positioning systems (GPS).

Large-scale landslide susceptibility as well as hazard, vulnerability and risk assessment indicating the different landslide hazard and risk zones of the Mussoorie township has been carried out at the level of municipality ward (Ram et al. 2020; Ram and Gupta 2021). Studies confirmed that about one fourth of the total buildings with habitation of ~8000 are at risk due to landslide. Further with the increased extreme rainfall incidences, like in 2020, these number may increase. It is pertinent to mention that year 2020 led to > 45 number of landslides at different places in Mussoorie. These landslides continue to pose serious threat to the habitations and also damaged the Dehradun—Mussoorie link road causing inconvenience to the locals and tourists.

In general, the slopes after failure stabilize themselves after attaining the angle of repose, provided there is not any change in environmental conditions. However, throughout the northwestern Himalaya, there is pronounced visibility of climate change in the form of (i) more area under the influence of rainfall, particularly in the inner Lesser Himalaya and the Higher Himalaya, (ii) higher concentrated rainfall in the Lesser Himalaya and (iii) increased frequency of concentrated rainfall over a small area, many times leading to cloudburst (Gupta and Sah 2008). Due to this climate change, the cloudbursts are not restricted to the Lesser Himalaya in the monsoonal dominant zone, but also have been reported from cold desert regions such as from Leh region in August 2010 (Ashrit 2010). This is one of the pronounced examples of visibility of climate change in the Himalayan region.

This climate change has the major implication on the slope instability in the area. The slopes, which had acquired an angle of repose and were meta-stable under a particular set of prevalent climatic regime quite for a long time, are now moving so as to adjust itself under a new climatic regime. It is envisaged that the increased rainfall patterns and/or the concentrated rainfall in an area will cause more landsliding, and the slopes will continue to move until these acquire the angle of repose.

In summary, it has been concluded that slopes in and around Mussoorie and Nainital townships are geologically unstable. The instability has been observed at many places in the form of opening up of cracks, ground fissure and development of new scars on the hill, such observed along both sides of the Balia Nala and on the Naina peak in Nainital. The propensity for instability is mainly due to inherent geological and structural setup of the area in the form of highly weathered rocks traversed by weak structural lineament, like Balia Nala Fault along with the presence of low strength thin veneer of debris on slopes. The conditions of slope instability have further worsened due to climate change, and more so in Nainital township as during 2009–2014, an increase in annual rainfall of about 70% has been recorded (Fig. 15.3).

Landslide susceptibility maps (LSM) for many areas in the Himalaya, including Mussoorie, have been prepared (Kumar et al. 2021; Ram et al. 2020; Ram and Gupta 2021). With the present scenario, it is necessary that these maps must be updated under a new climate change scenario or climate change must be considered for the preparation of these susceptibility maps. In the meantime, it is of utmost important to take immediate prevention measures in the form of construction of a proper drainage network on the slopes so that ingress of water into the slope is minimized. This will help arrest the movement of the slope.

References

Ashrit, R. 2010. *Investigating the Leh 'cloudburst'*. National Centre for Medium Range Weather Forecasting, Ministry of Earth Sciences, India.

Auden, J.B. 1934. The geology of the Krol belt. *Geological Survey of India* 67 (4): 357–454.

Banerjee, D.M., M. Schidlowski, F. Siebert and M.D. Brasier. 1997. Geochemical changes across the Proterozoic–Cambrian transition in the Durmala phosphorite mine section, Mussoorie Hills, Garhwal Himalaya, India. *Palaeogeography Palaeoclimatology* 132 (1–4): 183–194.

Barton, N., and V. Choubey. 1977. The shear strength of rock joints in theory and practice. *Rock Mechanics* 10 (1): 1–54.

Bhasin, R., E. Grimstad, J.O. Larsen, A.K. Dhawan, R. Singh, S.K. Verma, and K. Venkatachalam. 2002. Landslide hazards and mitigation measures at Gangtok Sikkim Himalaya. *Engineering Geology* 64 (4): 351–368.

Bieniawski, Z.T. 1973. Engineering classification of jointed rock masses. *Transaction of the South African Institutions of Civil Engineers* 15: 335–344.

BIS Code 1893. 2002. *Earthquake hazard zoning map of India.* www.bis.org.in.

Chaudhary, S., V. Gupta and Y.P. Sundriyal. 2010. Surface and sub-surface characterization of Byung landslide in Mandakini valley, Garhwal Himalaya. *Himalayan Geology* 31 (2): 125–132.

DMMC. 2011. Slope instability and geo-environment issues of the area around Nainital; Disaster Mitigation and Management Centre, Dehradun, India, 92 p. http://dmmc.uk.gov.in/files/pdf/Nai nital_Enviornmental_Degradation.pdf.

Dobhal, D.P., A.K. Gupta, M. Manish, and D.D. Khandelwal. 2013. Kedarnath disaster: Facts and plausible causes. *Current Science* 105 (2): 171–174.

Gupta, V., and I. Ahmed. 2007. Geotechnical characteristics of Surabhi Resort landslide in Mussoorie, Garhwal Himalaya, India. *Himalayan Geology* 28(2): 21–32.

Gupta, V., and K.S. Bist. 2004. The 23 September 2003 Varunavat Parvat landslide in Uttarkashi township, Uttaranchal. Current Science, 1600–1605.

Gupta, V., and M.P. Sah. 2008. Impact of the trans-Himalayan landslide lake outburst flood (LLOF) in the Satluj catchment, Himachal Pradesh. *India. Natural Hazards* 45 (3): 379–390.

Gupta, V., R.K. Bhasin, A.M. Kaynia, R.S. Tandon, and B. Venkateshwarlu. 2016. Landslide hazard in the Nainital township, Kumaun Himalaya, India: The case of September 2014 Balia Nala landslide. *Natural Hazards* 80 (2): 863–877.

Gupta, V., A.K. Mahajan, and V.C. Thakur. 2015. A study on landslides triggered during Sikkim earthquake of September 18, 2011. *Himalayan Geology* 36: 81–90.

Gupta, V., R.S. Tandon, B. Venkateshwarlu, R.K. Bhasin, A.M. Kaynia. 2017. Accelerated mass movement activities due to increased rainfall in the Nainital township, Kumaun Lesser Himalaya, India. *Zeitschrift für Geomorphologie* 61(1): 29–42.

Gupta V., A. Paul, S. Kumar, and B. Dash. 2021. Spatial distribution of landslides vis-à-vis epicentral distribution of earthquakes in the vicinity of Main Central Thrust (MCT) zone, Uttarakhand Himalaya, India. *Current Science* 120(12): 1927–1932.

Heim, A., and A. Gansser. 1939. Central Himalaya. *Memoirs of the Society Natural Sciences* 73: 1–245.

Holland, T.H. 1897. *Report on the geological structure and stability of hill slopes around Nainital, Geological Survey of India.* Report: Unpubl.

Kotlia, B.S., L.M. Joshi, R.K. Dumka, and K. Kumar. 2009. Vulnerability of the Balia Nala landslide at Nainital: Preliminary GPS analysis. Natural Resource Conservation in Uttarakhand (Ed B.L. Sah), 136–150.

Kumar, S., V. Gupta, P. Kumar, and Y.P. Sundriyal. 2021. *Coseismic landslide hazard assessment for the future scenario earthquakes in the Kumaun Himalaya.* India: Bulletin of Engineering Geology and the Environment. https://doi.org/10.1007/s10064-021-02267-6.

Maikhuri, R.K., A. Nautiyal, N.K. Jha, L.S. Rawat, A. Maletha, P.C. Phondani, and G.C. Bhatt. 2017. Socio-ecological vulnerability: Assessment and coping strategy to environmental disaster in Kedarnath valley, Uttarakhand, Indian Himalayan Region. *International Journal of Disaster Risk Reduction* 25: 111–124.

Martha, T.R., and K.V. Kumar. 2013. September 2012 landslide events in Okhimath, India—an assessment of landslide consequences using very high resolution satellite data. *Landslides* 10 (4): 469–479.

Martha, T.R., P. Roy, K.B. Govindharaj, K.V. Kumar, P.G. Diwakar, and V.K. Dadhwal. 2015. Landslides triggered by the June 2013 extreme rainfall event in parts of Uttarakhand state, India. *Landslides* 12(1): 135–146.

Middlemiss, C.S. 1890. Physical geology of the Sub-Himalaya of Garhwál and Kumaun. *Memoirs of the Geological Survey of India* 24 (2): 59–200.

Middlemiss, C.S. 1898. *Report on Kailakhan landslip near Nainital of 17th August, 1898.* Calcutta: Government Press.

Pande, R.K. 1990. Tawaghat landslide of Kumaun Himalaya, India. *Indonesian Journal of Geography* 20.

Panikkar, S.V., and V. Subramanyan. 1997. Landslide hazard analysis of the area around DehraDun and Mussoorie, Uttar Pradesh. *Current Science,* 1117–1123.

Paul, S.K., S.K. Bartarya, P. Rautela, and A.K. Mahajan. 2000. Catastrophic mass movement of 1998 monsoons at Malpa in Kali Valley, Kumaun Himalaya (India). *Geomorphology* 35 (3–4): 169–180.

Ram, P., and V. Gupta. 2021. Landslide hazard, vulnerability, and risk assessment (HVRA), Mussoorie township, lesser Himalaya, India. *Environment, Development and Sustainability,* 1–29.

Ram, P., V. Gupta, M. Devi, and N. Vishwakarma. 2020. Landslide susceptibility mapping using bivariate statistical method for the hilly township of Mussoorie and its surrounding areas, Uttarakhand Himalaya. *Journal of Earth System Science* 129 (1): 1–18.

Rosser, N., M. Kincey, K. Oven, A. Densmore, and T. Robinson, D.S. Pujara, and M.R. Dhital. 2021. Changing significance of landslide hazard and risk after the 2015 Mw 7.8 Gorkha, Nepal Earthquake. *Progress in Disaster Science* 10: 100159.

Sah, M.P., A.K.L. Asthana, and B.S. Rawat. 2003. Cloud burst of August 10, 2002 and related landslides and debris flows around Budha Kedar (Thati Kathur) in Balganga valley, district Tehri. *Himalayan Geology* 24: 87–101.

Shanker, R. 1971. Stratigraphy and sedimentation of Tal Formation, Mussoorie Syncline, Uttar Pradesh. *Journal of the Palaeontological Society of India* 16: 1–15.

Tandon, R.S., and V. Gupta. 2015. Estimation of strength characteristics of different Himalayan rocks from Schmidt hammer rebound, point load index, and compressional wave velocity. *Bulletin of Engineering Geology and the Environment* 74 (2): 521–533.

Tewari, V.C., and M.F. Qureshy. 1985. Algal structures from the Upper Krol-Lower Tal formations of Garhwal and Mussoorie synclines and their palaeo-environmental significance. *Journal of Geological Society of India* 26: 111–117.

Uniyal, A., and P. Rautela. 2005. Disaster management strategy for avoiding landslide induced losses to the villages in the vicinity of the Himalayan township of Mussoorie in Uttaranchal (India). *Disaster Prevention and Management* 14 (3): 378–387.

Valdiya, K.S. 1988. Tectonics and evolution of the central sector of the Himalaya. *Philosophical Transactions of the Royal Society of London. Series A, Mathematical and Physical Sciences* 326 (1589): 151–175.

Part V
Impact Assessment

Chapter 16
Impact Assessment and Adaptation Options for Climatic Change in Paddy Cultivation: A Case Study in Ampara District, Sri Lanka

A. Narmilan, A. M. M. Asmath, and N. Puvanitha

Abstract Agriculture is one of the most vulnerable sectors of the economy due to climate change impacts. The key objective of this study was to identify futuristic mitigation practices of paddy production in Ampara District, Sri Lanka, and to assess the risks of climate change. The climate change consequences such as increasing sea level would make rice production vulnerable to climate change. With the sea-level increase, the effect of salinity could permeate, resulting in lower fertility of agricultural land. Moreover, rice diseases and rodents are much threatened by climate change in the study area. The results of this study showed the losses in agricultural production. The highest crop losses appeared to occur in the Ampara District in the last decade. Furthermore, the significant threats for agricultural crop damage were seen from extreme events such as droughts (52.2%), floods (38.9%), and severe wind events (4.2%). However, paddy farmers are practicing various adaptive strategies to buffer the impacts to cope with the situation. These adaptation measures need further improvement to experience reduced implications of climate change. Therefore, the study recommends that the farmers need to practice a variety of techniques such as mixed cropping, intercropping, construction of structural flood control measures, precision agriculture, and cultivating-tolerant paddy varieties to observe minimum damaging impacts.

Keywords Ampara · Climate change · Paddy production · Flood · Drought

A. Narmilan (✉) · A. M. M. Asmath
Department of Biosystems Technology, Faculty of Technology, South Eastern University of Sri Lanka, Oluvil, Sri Lanka
e-mail: narmilan@seu.ac.lk

A. M. M. Asmath
e-mail: mohamedasmath@seu.ac.lk

N. Puvanitha
Department of Agriculture, Hardy, Sri Lanka Institute of Advanced Technological Education, Colombo, Sri Lanka

© The Centre for Science & Technol. of the, Non-aligned and Other Devel. Countries 2022
A. S. Unnikrishnan et al. (eds.), *Extreme Natural Events*,
https://doi.org/10.1007/978-981-19-2511-5_16

16.1 Introduction

Climate change is a persistent and long-term phenomenon involving constant time and commitment. Global past climates have alternated between ice ages and warmer periods than today over relatively long-time scales (millions to tens of millions of years) due to natural causes such as carbon dioxide concentrations in the atmosphere, continent positions, ocean circulation, solar output, and the Earth's orbit. However, since the mid-twentieth century, human-caused greenhouse gas emissions have been the primary driver of global climate change, trapping additional heat in the atmosphere (Licker 2020). It is a significant global environmental challenge and a developmental issue of great importance to Sri Lanka (IPCC 2007). Being an island, Sri Lanka is particularly vulnerable to all projected impacts of climate change, including an increase in land and sea surface temperatures, changes in the amount of precipitation and pattern, extreme weather events, and sea-level rise (Eriyagama et al. 2010). Agriculture makes up a large part of the economy. Therefore, it is vulnerable to climate change. Herath and Dharmakeerthi (2010) warn that climate change will have a detrimental effect on agriculture because it will cause temperatures to rise, decrease rainfall, and increase carbon emissions. Like climate change, 90% of respondents are concerned about the annual crop production. Johnston et al. (2010) found that warmer temperatures would decrease crop yields by preventing flowers' pollination. In such a case, rice yields might be recorded with the minimum temperature increase in the growing season.

Only by adopting proper adaptation practices can we cope with any detrimental impacts of climate change. Also, preparation and assessment should be performed in conjunction with the appropriate survey, planning, architecture, and execution. The implications of climate change have to be defined, quantified, and necessary to combat its consequences. Ampara is home to the country's leading agricultural producer, rice, coconuts, and vegetables. However, recently, it has been reported that climate change will be very mild on the Ampara District's agriculture sector and food security by the severe flood, drought, earthquake, hurricane, and high salinity water intrusion (Marambe et al. 2015).

On the other hand, in Ampara, there is still a scarcity of data on climate change. The shortage of data in Sri Lanka raises strategic planning, policy formulation, and prioritization challenges. Therefore, the primary objective of this study was to evaluate and analyze climate change risks and adaptation strategies to minimize the impact of climate change in paddy cultivation in Ampara District, Sri Lanka.

16.2 Methodology

16.2.1 Description of the Study Area

The geographic coordinates of Ampara are 7.231759 latitudes, 81.647344 longitudes, and 36.88 m elevation. The topography within 3.21 km of Ampara features only minor elevation changes, with an overall elevation of 39.01 m and an average above sea level of 31.69 m. Inside 16.09 km, there are only modest differences in elevation (602.89 m). Within 80.46 km contains significant variations in elevation of 2043.98 m (Weather Spark 2021).

16.2.1.1 Land Use Pattern

The area within 3.21 km of Ampara is covered by cropland (66%), trees (14%), and shrubs (11%), within 16.09 km by cropland (54%) and shrubs (23%), and within 80.46 km by water (54%) and cropland (21%) (Weather Spark 2021).

16.2.1.2 Climatic Conditions

In Ampara, the summers are hot and overcast; the winters are short, warm, wet, and mostly cloudy; and it is oppressive year-round. Over the year, the temperature typically varies from 23.9 °C to 33.9 °C and is rarely below 22.2 °C or above 35.6 °C. The wetter season lasts 2.5 months, from October 3 to December 21, with a greater than 41% chance of a given day being wet. The possibility of a wet day peak at 62% on November 7. The drier season lasts 9.5 months, from December 21 to October 3. The slightest chance of a wet day is 20% on June 28. The most rain falls during the 31 days centered around November 19, with an average total of 8.1 inches. The average hourly wind speed in Ampara experiences significant seasonal variation over the year. The windier part of the year lasts for 3.8 months, from May 20 to September 15, with average wind speeds of more than 12.39 km per hour. The windiest day of the year is June 18, with an average hourly wind speed of 15.39 km per hour (Weather Spark 2021).

16.3 Results and Discussion

16.3.1 Paddy Cultivation in Ampara District

Agriculture is an important sector of the national economy, adding to the country's growth. The agriculture sector contributes around 7% of the overall GDP. (Central

Bank of Sri Lanka 2018). Demand for paddy will rise by 28% in 2050, and it has been the primary crop for food production, jobs, and rural livelihood for centuries in Sri Lanka. Furthermore, rice is the main crop cultivated in Sri Lanka, occupying 34% of the cultivated area. During the *Maha* season (September to March), on average, 560,000 ha area is cultivated, and in the *Yala* season (May to the end of August), on average, 310,000 ha area is developed. Approximately, 1.8 million farm families are engaged in paddy cultivation island-wide (Rice Cultivation 2021). Ampara District plays an essential role in the rice production of Sri Lanka, and the majority of the population in the community is farming rice for their livelihood. In 2014 and 2015, the total area sown was 83,133 ha. In 2015, the area harvested was 65,793 ha. The district is expected 617,000 metric tons yield of paddy in both seasons, and Ampara is expected to produce around 20% of the national requirement of paddy (Central Bank of Sri Lanka 2018).

16.3.1.1 Impact of Climate Change on Paddy Cultivation

Climate change is an irreversible phenomenon seen worldwide in various forms, such as rising temperatures, rising sea levels, droughts, floods, hurricanes, and landslides. It is caused directly or indirectly by human action that changes the composition of the global environment in addition to natural climate fluctuations observed over comparable periods. (Esham and Garforth 2013). Climate change due to global warming induces precipitation changes, increasing sea levels and temperatures that will impact paddy cultivation and producing these goods (Watanabe and Kume, 2009). Rice thrives in wet environments where other crops fail to grow. It is vital to alleviate flooding because rice can only survive underwater for brief periods. Sea-level rises and the increased strength of tropical hurricanes with climate change are projected to reduce rice yields. Various predictions about future climate conditions, carbon dioxide levels, and humidity changes make it challenging to forecast future crop yields. IRRI research indicates that with a 1 °C increase in nighttime temperatures, rice yields can fall by about 10%.

16.3.1.2 Statistical Data for Impact on Drought and Flood with the Unique References to Ampara District

Drought and floods are the most influencing parameters for climate change in Sri Lanka. Since 2011, irregular rainfall marked by flood/drought cycles during the northeast monsoon has contributed to a rise in the number of people impacted by disasters in Sri Lanka. The prevailing drought has impacted up to 951,597 people in many districts. 95,334 Eastern Province individuals, 428,181 Northern Province individuals. According to Fig. 16.1, agricultural sectors are affected mainly by floods (39.37%) and drought (52.89%) from 1974 to 2008in Sri Lanka (Disaster Information Management system in Sri Lanka 2021) (Figs. 16.2 and 16.3).

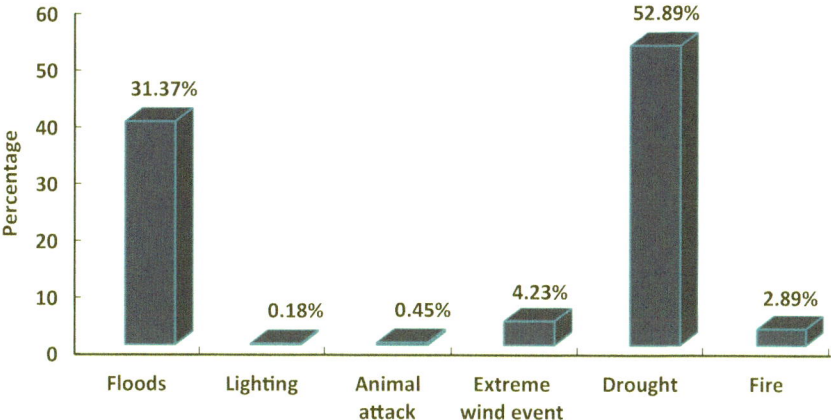

Fig. 16.1 Agricultural loss due to disasters in Sri Lanka

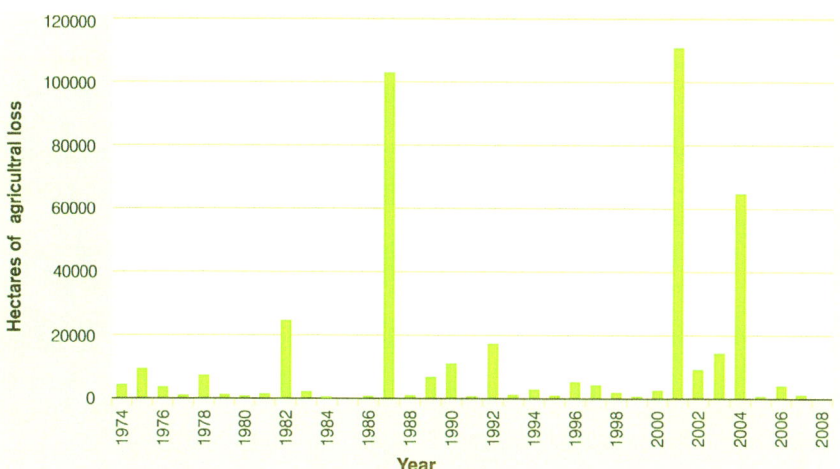

Fig. 16.2 Agricultural loss due to drought in Sri Lanka

The spatial distribution indicates (Fig. 16.4) that the Kurunegala and Ampara districts tend to have the most considerable crop losses. The significant loss of crops was due to droughts (52.2%), flooding (38.9%), and intense wind incidents (4.2%). According to data review (Fig. 16.5), the population of Gampaha and Batticaloa (dark brown) regions was most affected due to natural disasters during 2013. The other districts in which large numbers of people were affected were Ampara (red color), Nuwaraeliya, and Kalutara (orange color) (DIMS 2021). Flooding has hit Kalmunai, Alayadiwembu, and Ninthavur the hardest in Ampara. Flooding can be controlled by improving water construction methods and implementing suitable preparedness strategies.

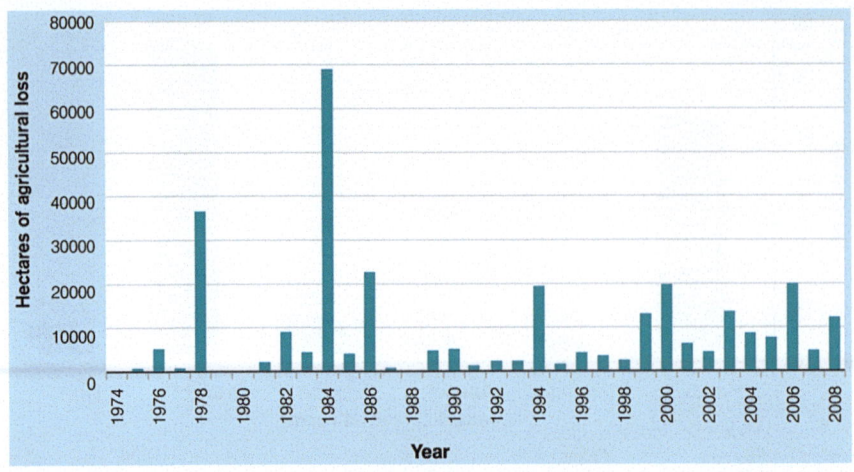

Fig. 16.3 Agricultural loss due to flood in Sri Lanka

Figure 16.6 shows the agricultural loss due to drought. It demonstrates that from 3530 to 10,984.14 hectares area of agrarian land was impacted by lack from 1974 to 2008 in the district of Ampara. Figure 16.7 shows the geographical spread and the density of people affected by the drought during 1974–2008. In 1974, 90,550 people were affected, while in 2008, 236,230 individuals were involved in the district of Ampara due to drought (DIMS 2021).

Figure 16.8 indicates that during 1974–2008, more than 36,583.51 hectares of land were damaged in the districts of Ampara, most of which belonged to agricultural land. Figure 16.9 shows the number of people affected by flooding across the country. In 1974, 724,025 people were affected by drought, while in 2008, 2,394,704 people were impacted by drought in the Ampara District (DIMS 2021). Table 16.1 shows the cost of losses and damages in Batticaloa, Polonnaruwa, Anuradhapura, and Ampara districts due to floods in 2011 (Disaster Profile of Sri Lanka 2020).

Figures 16.10 and 16.11 indicate the vulnerability level of paddy production due to floods and drought in all the Sri Lankan districts (MOE 2011).

16.3.1.3   Possible Adaptation Techniques for the Impact of Drought

Agricultural drought occurs when the supply of soil moisture to plants has decreased beyond the required level, impacting crop yields and agricultural production (Bhandari and Panthi, 2014). Climate change is causing a decline in the water supply in Sri Lanka (Williams and Carrico 2017). Losing access to clean water has adverse effects on farmers' profits and contributes to food insecurity. Scientists have been studying semidwarf, hybrid, and new form varieties to achieve better yields. Enhanced crop yield is possible with modern techniques such as marker-aided breeding and genetically modified crops (Maclean 1997). Planting drought and heat-resistant varieties

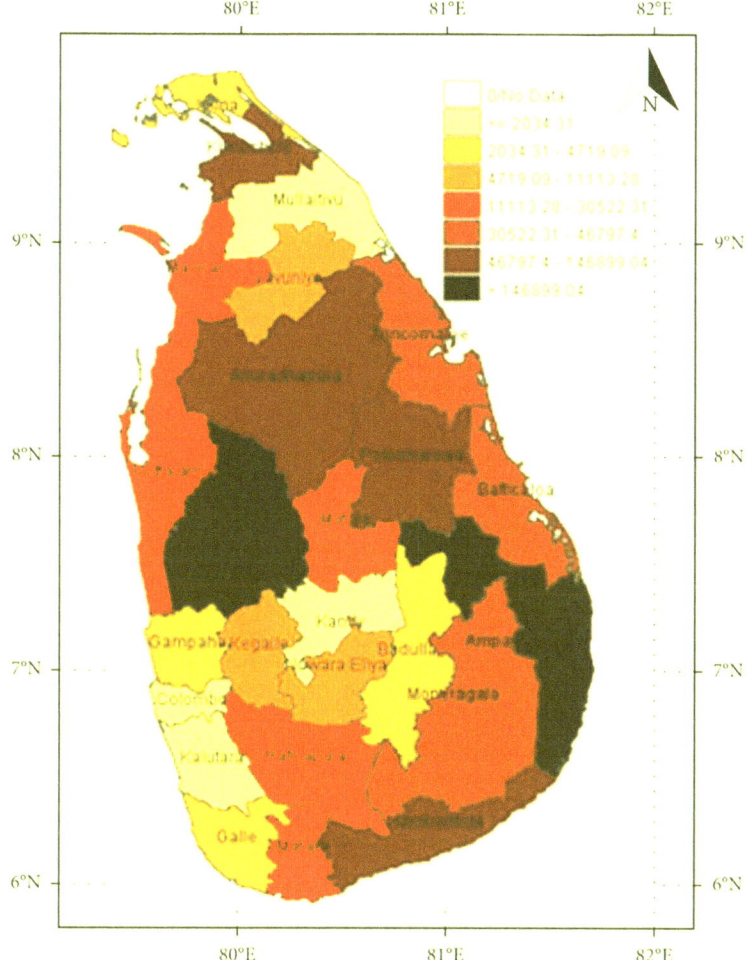

Fig. 16.4 Agricultural loss due to disasters (1974–2008)

are one of the most successful adaptation strategies in many areas of the world experiencing drought stress. Scientists have established and mapped genes for drought and heat tolerance. They are developing novel varieties, essential for all rice production in China due to year-to-year variation in rainfall (Maclean 1997). To counter the effects of climate change, farmers need to implement new methods of water production. In areas with insufficient irrigation water or drought, water-efficient technology such as alternate wetting and drying, land leveling, increased tillage, package preparation, and direct seeding will increase crop yields (Maclean 1997). Farmers are also encouraged to plant 'other field crops' during drought cycles that need less water than rice paddy, such as millet, maize, or soy in lowlands traditionally used for rice cultivation. Farmers with good ties with agricultural extension programs are more

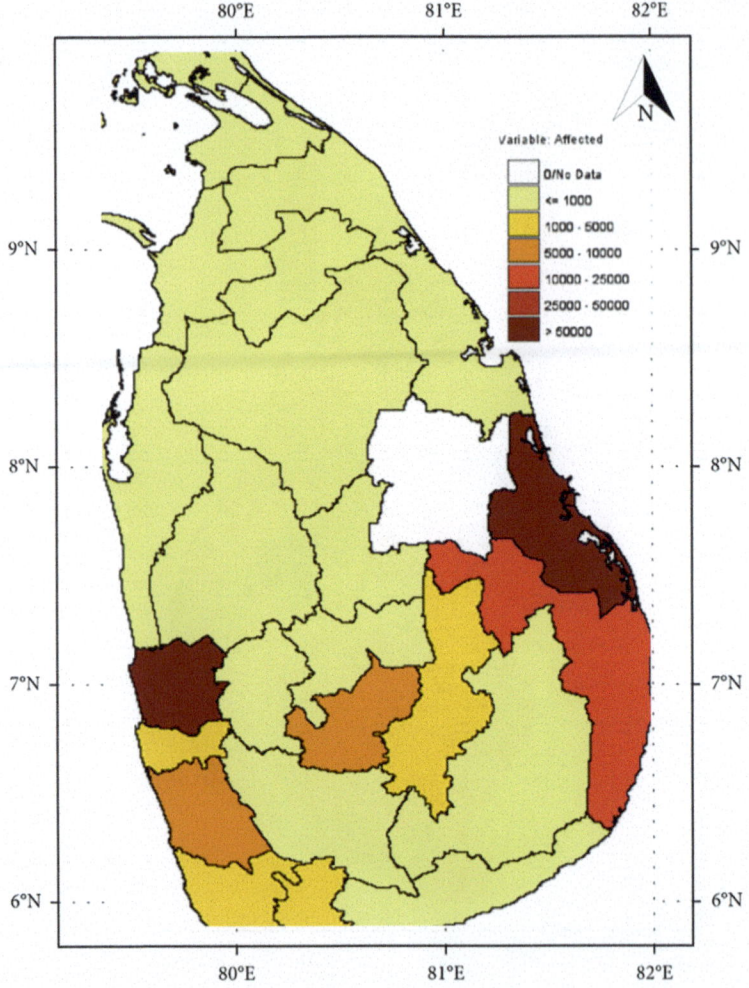

Fig. 16.5 Natural disaster-affected population (1974–2008)

likely to use the strategies promoted by these programs, i.e., planting non-rice crops in their paddy land programs (Williams and Carrico 2017).

To encourage a range of various techniques, the government and farmers' organizations are in collaboration. These techniques include—(i) *kakulama*, a local dry seeding method historically used during the dry season; (ii) the 'parachute technique' of transmitting nursery-raised seedlings into muddy fields; (iii) recycling irrigation water by pumping it back to the head of a local system; and (iv) breeding and distributing hybrid seeds, especially those that mature faster than traditionally planted varieties (Wassmann et al. 2009). Farmers are planting short-term seed varieties resistant to drought to buffer against the effects of drought (Williams and Carrico 2017).

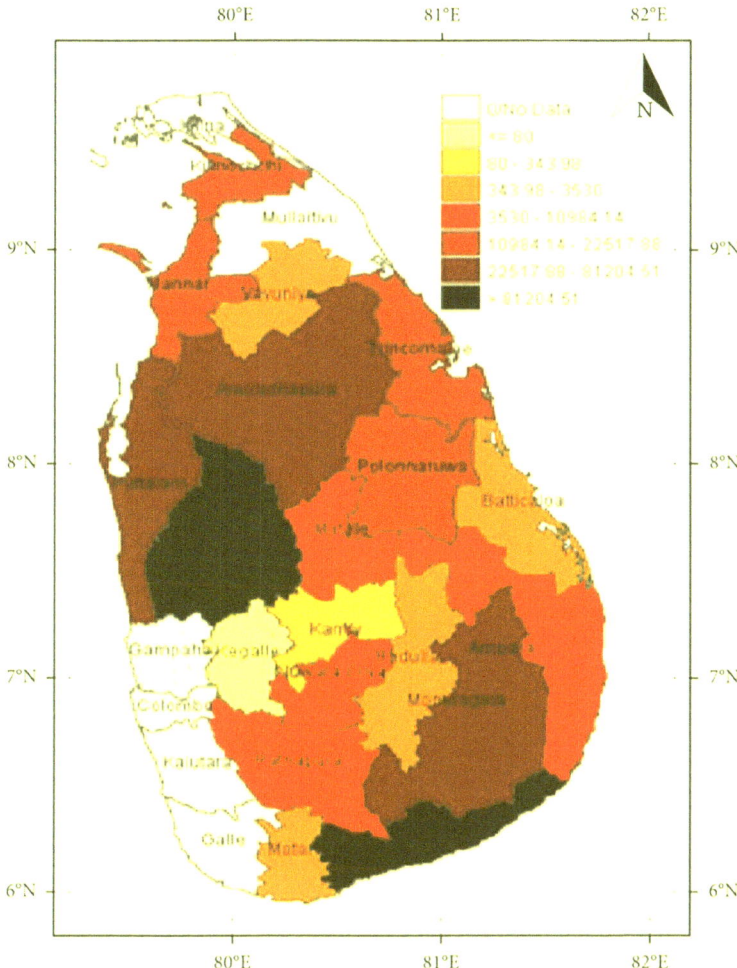

Fig. 16.6 Agricultural loss due to drought (1974–2008)

By enhancing the quality of preparations and events and improving water conservation, the Internet of Things will help farmers tackle problems such as shortage and drought (Narmilan and Niroash 2020). In reaction to more dynamic and evolving weather, implement new water harvesting methods and introduce educational and outreach programs to boost the acceptance of water harvesting activities.

16.3.1.4   Possible Adaptation Techniques for the Impact of Floods

Flood is a natural phenomenon with geomorphological, physical, social, and ecological ramifications with numerous direct and indirect effects. It accounts for around

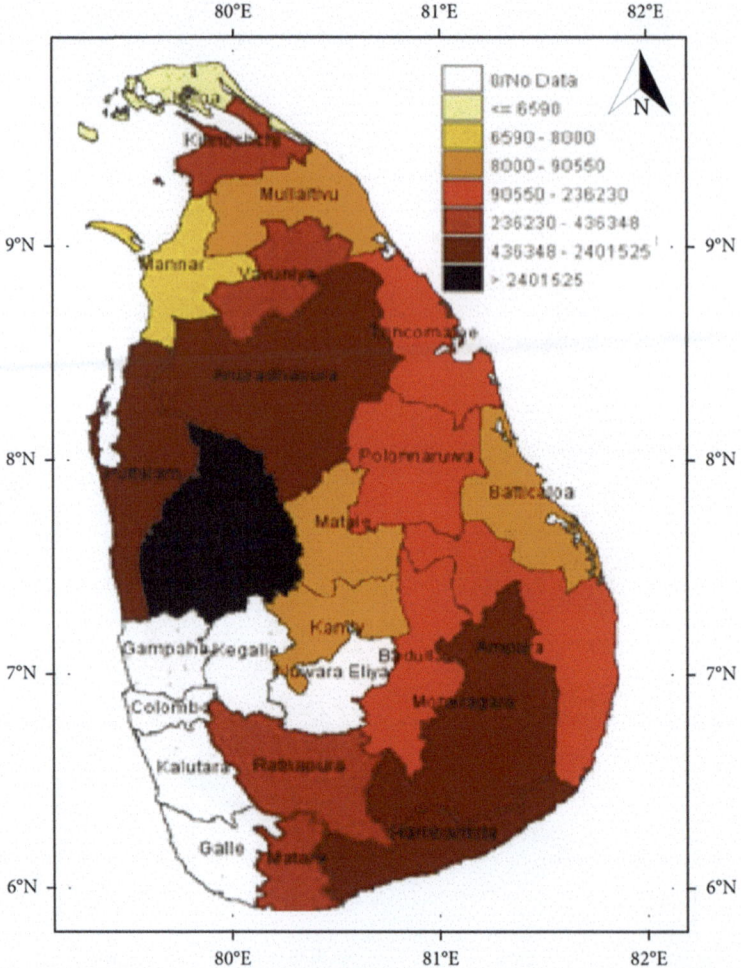

Fig. 16.7 People affected due to drought (1974–2008)

one-third (34%) of the overall number of worldwide natural disasters that have occurred. Farmers are implementing various adaptive strategies to mitigate the consequences of climate change. One of the adaptive practices to minimize flood stress is mixed cropping or intercropping, as co-growing plants can also improve their nearby microclimate by engaging with the rhizosphere (Brooker 2006). Recent research has shown that rice (Oryza spp.) can increase the resistance of co-growing cereal crops to flood pressure by releasing O_2 into the rhizosphere of its dryland (Kamwele 2017). Flood protection entails systemic flood control mechanisms such as dam or river dike construction and non-structural measures such as flood monitoring and warning, flood risk and risk assessment, public participation, administrative arrangements, and so on (Tingsanchali 2011).

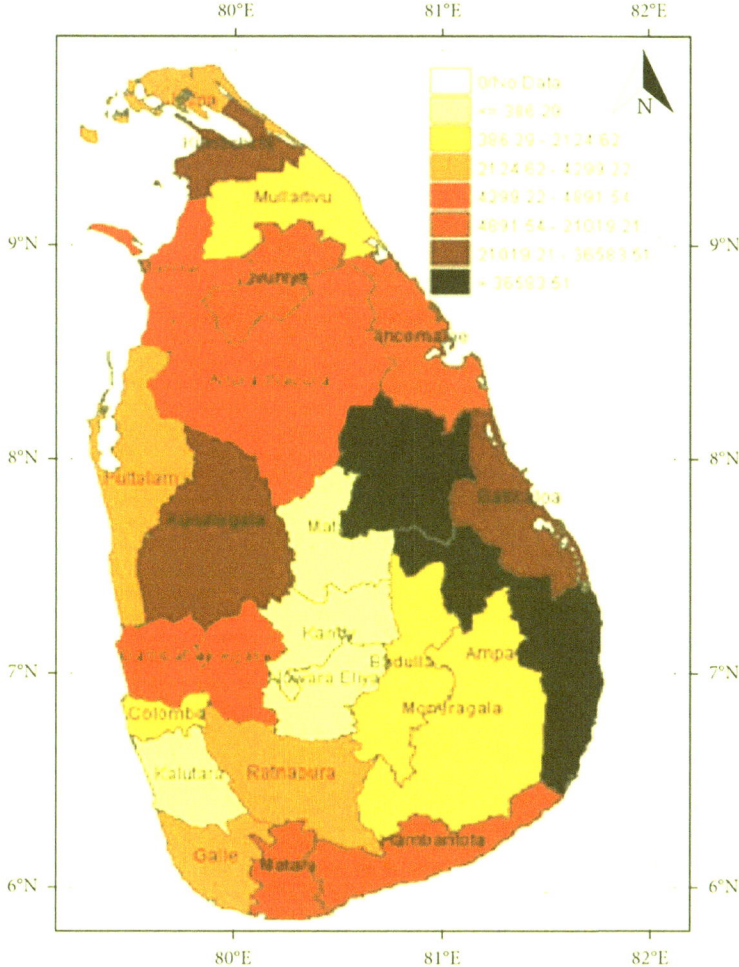

Fig. 16.8 Agricultural loss due to floods (1974–2008)

According to the environment charity, avoiding erosion and wetland runoff, refor-esting upstream regions, and rebuilding damaged wetlands would significantly miti-gate climate change effects on flooding. Dikes are fencing or walls built to defend the ground from damage to the river. It is possible to make them with gravel, bricks, rocks, sandbags, or wood. Properly constructed, dikes will shield crops from floods. The dike can delay water flow or divert water, or direct water flows to specific crops or other areas, such as the irrigation canal (Narmilan 2018). Improved rice varieties will reduce the impact of floods by planting flood tolerances. Healthy peat bogs can also mitigate erosion by slowing the influx of surface and subsurface water due to rugged surface vegetation from heavy rainfall events (Narmilan 2018). Protected bogs, irrigation raises the peat surface's soil moisture deficit, thus improving soil

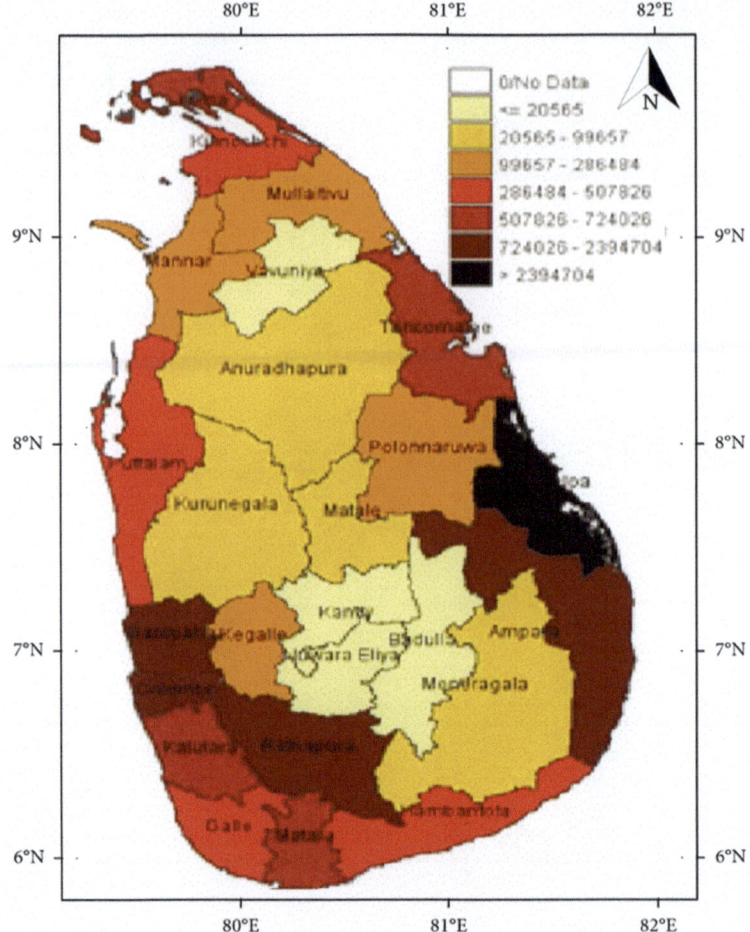

Fig. 16.9 People affected due to flood (1974–2006)

Table 16.1 Losses and damages in Batticaloa, Polonnaruwa, Anuradhapura, and Ampara districts due to flood in 2011	Sector	Cost of damage and losses (million LKR)
	Housing	7575
	Agriculture	15,070
	Irrigation	3000
	Road	48,916
	Livestock	1914
	Total	77,475

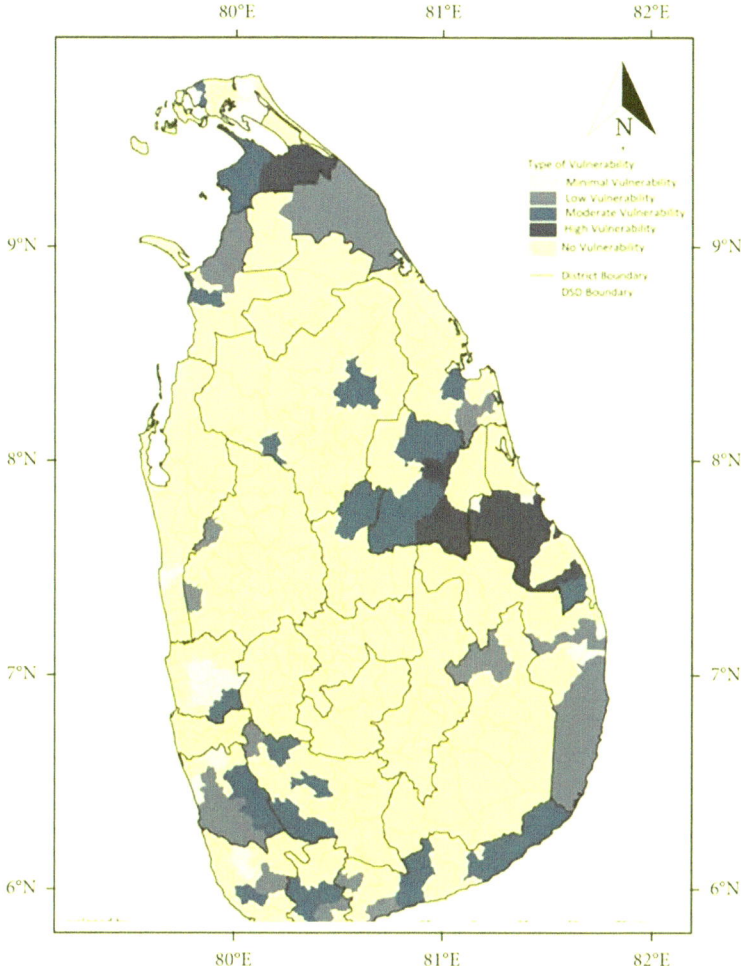

Fig. 16.10 Paddy sector vulnerability to flood exposure

water's storage potential. However, the drains still increase the influx of water to the canals, exacerbating the floods (Scottish EPA 2015).

Improving surface conditions is one of the flood controls schemes, and animal hooves will compress the soil so that the water runs out quickly instead of storing, holding, and gradually letting go of the moisture. Well-drained soil can absorb vast volumes of rainwater, preventing it from flooding into rivers. In addition to these adaptive practices, planting trees and hedges around fields to mitigate erosion, constructing dikes and irrigation canals/pipes to manage the flow of paddy water, enhancing drainage through raised beds, ridges or mounds, planting early ripening paddy to avoid flood seasons, planting hybrid varieties that are flood-tolerant will reduce the effect of floods. (Narmilan 2018). Farmers can use the Internet of things

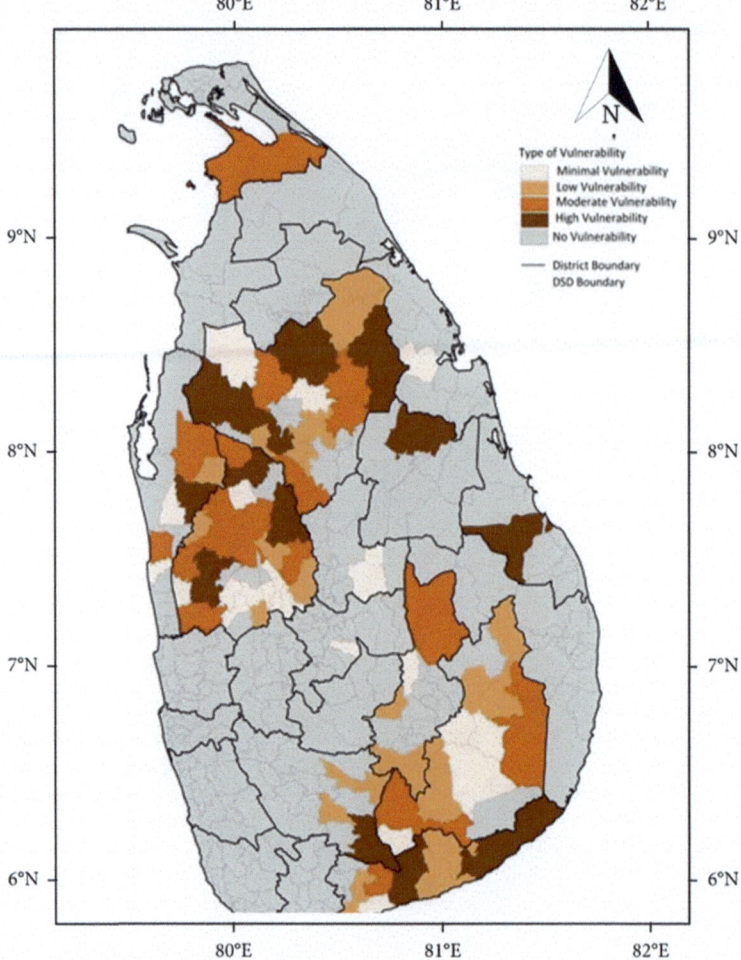

Fig. 16.11 Paddy sector vulnerability to drought exposure

to alleviate flooding problems and expand their plans and activities (Narmilan and Niroash 2020).

16.4   Conclusion and Recommendations

Adverse weather and climate change, resource scarcity, high input prices, access to credit difficulties, and low production quality affect agricultural productivity. They ultimately affect farmers' incomes in the district of Ampara. Climate change primarily affects grain and other agricultural production. Drought and flood conditions are significant climate changes that directly affect agricultural production and raise the poverty rate among poor rural farmers. For farmers to respond successfully to fight change, the lack of awareness about recent climate change and its impacts on production among the people living in the study area needs to be remedied. Farmers are practicing agricultural adaptation, including agronomic adaptation and modern enhanced crop varieties, but the number of farmers practicing agronomic adaptation is declining. To increase crop production, it is preferable to use alternative irrigation methods such as sprinklers and drips in highland regions.

However, appropriate and specific information on the effects of climate change is usually lacking in the Ampara District. Therefore, to sustain agricultural crop production in a shifting environment, management methods and cultivars would have to be changed. The problems mentioned above can be solved by encouraging local and international research partnerships between researchers in the same area, sharing and applying the information generated to tackle related issues. The development of information technologies in the context of the Internet, GIS, remote sensing, satellite communication, etc., will aid in the preparation and application of danger mitigation.

References

Bhandari, G., and B.B. Panthi. 2014. Analysis of agricultural drought and its effects on productivity at the different district of Nepal. *Journal of Institute of Science and Technology* 19 (1): 106–110.

Brooker, R.W. 2006. Plant–plant interactions and environmental change. *New Phytologist* 171 (2): 271–284.

Central Bank of Sri Lanka. 2018. *Annual report 2009*. Colombo, Sri Lanka.

Disaster Information Management system in Sri Lanka. Desinventar.lk. 2021. Retrieved 30 January 2021, from http://www.desinventar.lk/.

Eriyagama, N., V. Smakhtin, L. Chandrapala, and K. Fernando. 2010. *Impacts of Climate Change on Water Resources and Agriculture in Sri Lanka: A Review and Preliminary vulnerability Mapping*, 1–56. Colombo, Sri Lanka: International Water Management Institute.

Esham, M., and C. Garforth. 2013. Agricultural adaptation to climate change: Insights from a farming community in Sri Lanka. *Mitigation and Adaptation Strategies for Global Change* 18 (5): 535–549.

Herath, A.U.K., and L.K.S.U. Dharmakeerthi. 2010. Country paper: Sri Lanka regional workshop on strategic assessment for climate Change adaptation in natural resource management Colombo, Sri Lanka.

IPCC. 2007. Climate change: Impacts, adaptation, and vulnerability. In *Contribution of Working Group II to the Fourth Assessment Report of the Intergovernmental Panel on Climate Change*, ed. M.L. Parry, O.F. Canziani, and J.P. Palutikof, 1–131. Cambridge: Cambridge University Press.

Johnston, R.M., C.T. Hoanh, G. Lacombe, A.N. Noble, V. Smakhtin, D. Suhardiman, S.P. Kam, and P.S. Choo. 2010. Rethinking agriculture in the Greater Mekong Subregion: How to sustainably meet food needs, enhance ecosystem services and cope with climate change. *Colombo Sri Lanka* 26: 207.

Kamwele, A.S. 2017. *Mitigation of Flood Stress for Semi-arid Cereals by the Mixed-Seedling with Rice (Oryza sativa)*, 1–109.

Licker, R. 2020. Global climate change. *AccessScience*. Retrieved October 28, 2021, from https://doi.org/10.1036/1097-8542.757541.

Maclean, J. 1997. *Rice almanac*. China: International Rice Research Institute.

Marambe, B., R. Punyawardena, P. Silva, S. Premalal, V. Rathnabharathie, B. Kekulandala, U. Nidumolu, and M. Howden. 2015. Climate, climate risk, and food security in Sri Lanka: the need for strengthening adaptation strategies. In *Handbook of Climate Change Adaptation*, 1759–1789.

Ministry of Environment, Sri Lanka. 2011. *Climate change vulnerability data book*. Sri Lanka. Retrieved from http://www.climatechange.lk/adaptation/Files/Final_Climate_Change_Vulnerability_Databook.pdf.

Narmilan, A. 2018. *Structural measures for flood risk mitigation in the agricultural field: a review*. 6th International Symposium on Advances in Civil and Environmental Engineering.

Narmilan, A., and G. Niroash. 2020. Reduction techniques for consequences of climate change by Internet of Things (IoT) with an emphasis on the agricultural production: A review. *International Journal of Science, Technology, Engineering and Management-A VTU Publication* 2 (3): 6–13.

Practices for Sustainable Development (ACEPS-2018). 185–190.

Rice Cultivation. Doa.gov.lk. 2021. Retrieved 30 January 2021, from https://doa.gov.lk/rrdi/index.php?option=com_sppagebuilder&view=page&id=42&lang=en#:~:text=Rice%20cultivation-,Rice,rice%20to%20about%20870%2C000%20ha.

Scottish Environmental Protection Agency. 2015. Natural flood management handbook.

Tingsanchali, T. 2011. Urban flood disaster management. *Procedia Engineering* 32 (2012), 25–37. Available online at www.sciencedirect.com.

Wassmann, R., S.V.K. Jagadish, S. Heuer, A. Ismail, E. Redona, R. Serraj, R.K. Singh, and G. Howell. 2009. Climate change affecting rice production: The physiological and agronomic basis for possible adaptation strategies. *Advances in Agronomy* 101: 59–122. https://doi.org/10.1016/S0065-2113(08)00802-X.

Watanabe, T., and T. Kume. 2009. A general adaptation strategy for climate change impacts on paddy cultivation: Special reference to the Japanese context. *Paddy and Water Environment* 7 (4): 313.

Weatherspark.com. 2021. Average weather in Ampara, Sri Lanka, year-round—Weather Spark. Retrieved from https://weatherspark.com/y/110322/Average-Weather-in-Ampara-Sri-Lanka-Year-Round#:~:text=In%20Ampara%2C%20the%20summers%20are,or%20above%2096%25C2%B0F.

Williams, N.E., and A. Carrico. 2017. Examining adaptations to water stress among farming households in Sri Lanka's dry zone. *Ambio* 46: 532–542. https://doi.org/10.1007/s13280-017-0904-z.

Part VI
Integrated Disaster Risk Reduction

Chapter 17
Contributions to a Comprehensive Strategy Design for Disaster Risk Reduction Related to Extreme Hydroclimatic Events in Latin America and the Caribbean

Alejandro Linayo

Abstract This article summarizes the experience and the main results of a regional comparative research carried out during 2017 at the request of the Inter-American Development Bank. This study was carried out by the Research Center for Disaster Risk Reduction (CIGIR) with the purpose of identifying financial strategies for reducing the risk of disasters associated with the occurrence of floods and debris flows in the particular context of Latin America and the Caribbean. The article starts with a description of the high levels of disaster risk associated with extreme hydroclimatic events that have been identified for the region and with a very brief description of the impact that some disasters of this type have left in several countries. Subsequently, the conceptual foundations of the approach to the comprehensive management of socio-natural risks—with which most of the countries of the region have been trying to advance in their efforts to reduce the impact of disasters on their development processes—are described. The results of a comprehensive and regionally relevant strategy for disaster risk management associated with torrential avalanches and floods are subsequently presented. A strategy was designed based on the consultation of about 50 experts from 14 countries to highlight the need to work simultaneously in five large areas of intervention, where a total of 25 specific priority actions are identified as actions that must be attended in order to reduce the regional impact of this type of disaster. Finally, the results of an assessment of the levels of comprehensiveness of the efforts being made by the various countries of the region in order to reduce the risk of hydroclimatic disasters are shown. The starting point of this final analysis is the detailed review of the individual efforts that regional countries have been reporting in their national reports of advances to climate change adaptation (NDC) and that has been presented within the framework of the Paris agreement COP21. The results of this analysis suggest the importance of continuing to work on the design and adequate implementation of comprehensive agendas to reduce the risk of hydroclimatic disasters in societies such as those in

A. Linayo (✉)
Research Center On Disaster Risk Reduction CIGIR, Merida, Venezuela
e-mail: alejandrolinayo@gmail.com

© The Centre for Science & Technol. of the, Non-aligned and Other Devel. Countries 2022
A. S. Unnikrishnan et al. (eds.), *Extreme Natural Events*,
https://doi.org/10.1007/978-981-19-2511-5_17

Latin America. Societies in which their particular political, institutional, and socio-cultural practices define so strongly the patterns of occupation and the sustainability of the development efforts are promoted in their territories.

Keywords Urban risk management · Hydroclimatic disasters · Climate change · Disaster risk reduction

17.1 Introduction

Reports of the International Disasters Database EM-DAT of the Centre for Research on the Epidemiology of Disasters (EM-DATA, 2021) state that hydroclimatic extreme events triggered 71 percent of disasters that have occurred worldwide since 1990. These reports also mention that flooding has become the most common disaster worldwide since 1990. In fact, from 1990 to 2019, a total of 9,924 disasters have been registered globally, of which 42 percent were triggered by floods, and 30 percent were triggered by storms, including cyclones, hurricanes, tornadoes, blizzards, and dust storms (Fig. 17.1).

The same trends described about the alarming growing frequency and global impact generated by disasters associated with extreme hydroclimatic events can be found—with practically no variations—in each one of the sub-regions and countries that integrate the broad Latin American and Caribbean context (Fig. 17.2).

One of the causes that could lay behind the growing impact of disasters triggered by hydroclimatic extreme events in this region is the change in global climate conditions that are consistently warming water temperatures in the Pacific and Atlantic Oceans and that could be promoting a higher frequency and impact of extreme rainfalls, and also harder manifestations of "La Niña" and "El Niño" climatic oscillations

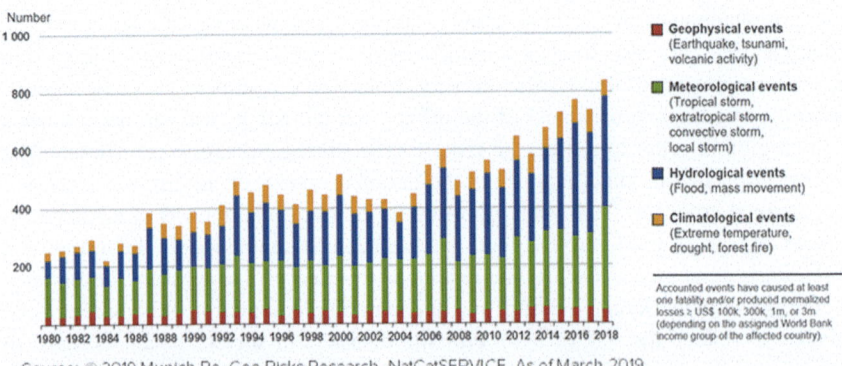

Source: © 2019 Munich Re, Geo Risks Research, NatCatSERVICE. As of March 2019.

Fig. 17.1 Total annual losses related to disasters triggered by meteorological and hydrological events. *Source* Munich Re Geo Risks Research cited on the website of the Insurance Information Institute III-2020

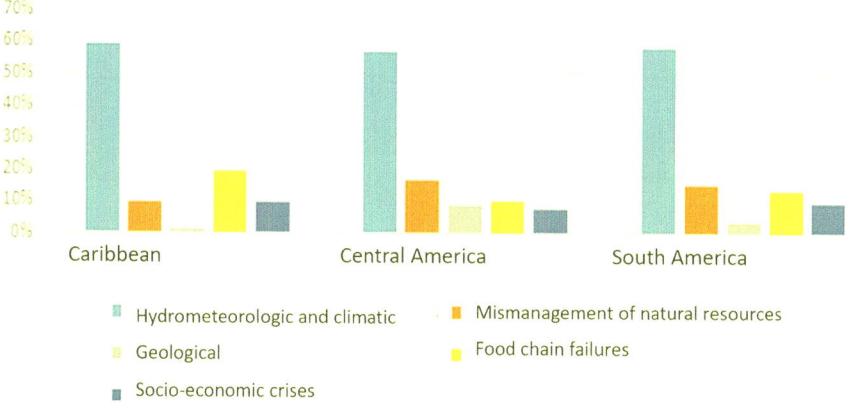

Fig. 17.2 Main disasters types in Latin America and the Caribbean region. *Source* Food and Agriculture Organization FAO (2018)

phenomena and its well-known effects in the intensity and frequency of droughts, storms, and floods in different countries (Wilches-Chaux 2007).

It has been also mentioned (Liñayo 2002; Wilches-Chaux 2007) that—beyond any kind of climatic changing consideration—the main cause that is behind the increasing frequency and impact of hydroclimatic disasters is the unsustainability level of the models of territorial occupation and development that characterize most of countries, and the permanent and exacerbated construction of urban and rural vulnerability scenarios that today characterize most Latin American and Caribbean regions.

The fact is that, no matter the causes, disasters triggered by extreme hydroclimatic events have been continuously striking Latin American countries during last decades. Some examples of these destructive events that could be mentioned are the hurricane Mitch that hit several Central American countries in 1998, killing 11,000 persons and producing losses of 6 billion USD dollars; the floods and landslides of Rio de Janeiro (Brazil) that killed nearly 1000 persons in 2011; the floods and debris flows registered during "El Niño" phenomenon in 1997–1998 that killed 600 persons and left losses over 7.5 billion USD dollars; and the disaster of Vargas that stroke the north coast of Venezuela in December 1999 killing around 10.000 deaths (casualties moves between 1000 and 30,000 victims).

To have a better idea about the real impact of each one of these disasters, we could mention about the last-mentioned event that the disaster of Vargas that hit Venezuela in December of 1999 is recognized as the worst disasters in the last decades that struck this country. During this event, torrential rains and severe mudslides killed thousands of people, destroying thousands of homes and collapsing almost all the infrastructure (potable water, roads, energy lines, phone lines, etc.) of the most affected areas. As a result of this situation, several coastal towns (Los Corales, Macuto, Mamo, El Piache, etc.) were buried under 1 to 3 m (3.2 to 9.8 feet) of mud and a high percentage of homes were simply swept away to the ocean. Estimated damages calculated for this

Fig. 17.3 Example of the destruction left after the disaster of Vargas (Venezuela—December 1999)

disaster were around 4 billion USD dollars. It is estimated that more than 8,000 homes were destroyed, 75,000 people were displaced, and more than 60 km (37 miles) of the coastline was significantly altered (Fig. 17.3).

Painfully, since the occurrence of those terrible experiences, the regional level of exposition to this kind of disasters—far from being reduced–has increase during the last years, and multiple studies suggest (IADB 2021a;FAO 2018; EM-DAT 2021) that Latin America and the Caribbean are regions of the world where climate-related disasters will hit hardest the development efforts if compressive agendas for disaster risk reductions are not well designed and implemented in the near future. (Fig. 17.4).

The urgency of that calls and the growing impact of climate-related disasters in Latin America has led to the emergence of a large number of initiatives focused to an adequately characterizing and comprehensively intervening of the complex conditions that usually precede the impact of this type of calamities. In most cases, these efforts have been originated within the national institutional structures of regional countries and usually have been totally or partially supported by projects financed by the wide network of development cooperation agencies that currently work in the region.

As the various protagonists and promoters of initiatives to reduce the impact of disasters have advance in the understanding of the multifactorial and highly complex nature underlying the causes of the aggravation of this problem in the region, a new and more comprehensive disaster reduction paradigm has been gaining ground throughout: The Integrated Management of Socio-natural Disaster Risks approach.

Although a complete description of the fundamentals that characterize this particular way to comprehensively reduce the impact of disasters cannot be addressed in this article, it is important to mention that this approach invites us to understand disasters—rather than as unexpected and unfortunate events—as undesired but inevitable manifestations of the unsustainability of the models of territorial occupation and development that characterize most of Latin American countries. In this sense, it is

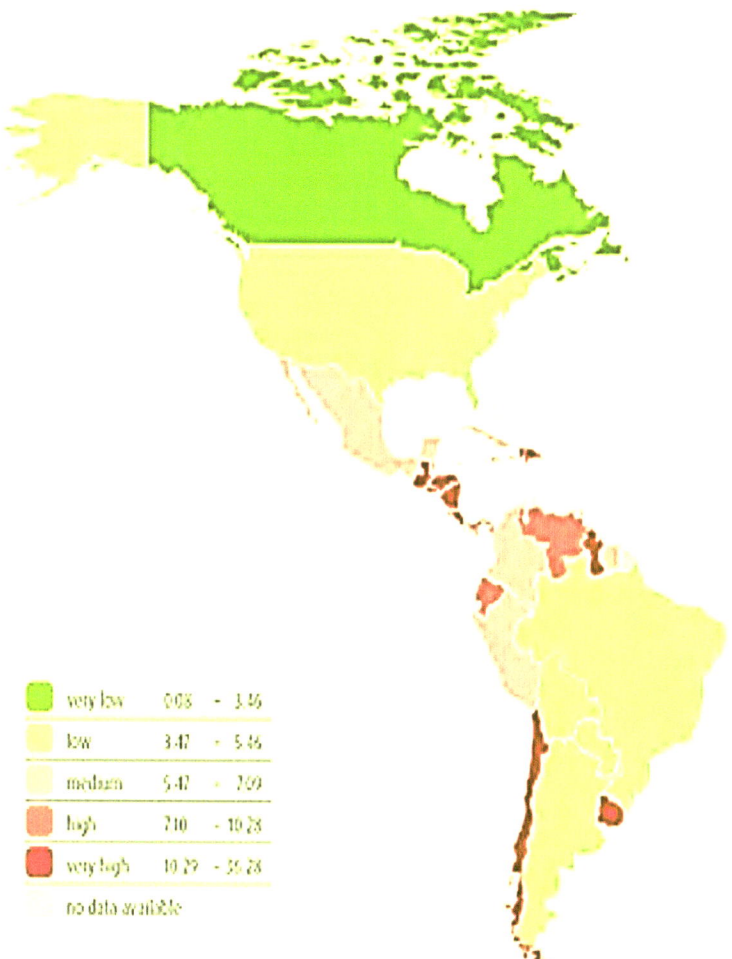

Fig. 17.4 Regional map of disaster risk indices (2019). *Source* Institute for Law of Peace and Armed Conflict at Ruhr University Bochum World Risk Report 2019—Global map of natural disaster risk

essential to understand that it will be very difficult to reduce the regional impact of disasters as long as we fail to transform the social and institutional processes that allow—and also enhance—the permanent and exacerbated construction of urban and rural vulnerability scenarios that today characterize most Latin American and Caribbean countries.

From a perspective of this nature, it is logical to promote a treatment of disaster problematic that should no longer be limited to merely preparing societies to respond adequately to future calamities, since this effort only addresses the "symptomatic treatment" of a disease that, as we mentioned above, is generated by the unsustainability of the social and institutional development practices that prevail in the region.

Socio-natural (because disasters clearly are not "naturals") disaster risk management approach invites us to prioritize the identification, intervention, and transformation of the social and institutional processes that allow the construction of risk scenarios, and in this sense, it strives to identify better ways to avoid both that human settlements continue becoming a threat to the dynamics of the natural environment and the territory and that the territory dynamics continue becoming a threat to human settlements (Wilches-Chaux 2007).

The convincing power of above declarations has led various regional governments to assume a strong commitment to work toward strengthening their socio-natural disaster risk management capacities. A commitment that must be understood as a whole transformation of planning and implementation of national development processes, and that in no case can be addressed just as another task that can be assigned to national emergency management and/or disaster preparedness institutions. What socio-natural risk management promotes is an integral commitment to sustainability that needs to be addressed by each and every one of the sectorial actors of a nation's development. In this sense, various regional countries have even conceptualized disaster risk management as the capacity of society and its officials to avoid the conditions that generate disasters by acting on the causes that produce them. In this sense, this national effort should be understood as a necessary characteristic of public management and a condition for the sustainability of any sectorial development effort.

Unfortunately, and without failing to recognize the conceptual and normative advances that have been made in this sense in different Latin American countries, the poor impact in the reduction of impacts of disasters that these political commitments has had, at least so far, shows that a much more accentuated and concrete effort is needed to enable disaster risk management to move beyond the merely discursive and normative plane and can be translated into much more concrete and tangible agendas of sectorial action that could be really capable of stopping and reversing the unacceptable levels of vulnerability that can be found in most of Latin American and Caribbean countries.

The concrete effort mentioned above seems to demand important regional agreements of conceptual and methodological type, and in order to advance in this direction, the Inter-American Development Bank promoted during the second half of 2017, a regional comparative study that could help in the definition of a comprehensive strategy to strengthen national disaster risk capacities in case of events triggered by debris flows and floods.

17.2 Materials and Methods

The project in which this research is framed was entitled "Identification of Financial Strategies for Climate Change Mitigation in the Water and Sanitation Sector in Latin America and the Caribbean." For this project, four main objectives were defined: (1) to analyze expected regional climate scenarios in the short and medium term; (2) to identify financial strategies for adaptation to climate change in the water and sanitation sector; (3) to identify financial strategies for climate change adaptation in solid waste management sectors; and (4) to identify comprehensive regional strategies for reducing the risks of torrential avalanches and floods.

In relation to this last component, an effort was made—between June and December 2017, in which 45 regional experts from 14 countries participated—to diagnose and design consensus-based priority action and strategies, and in which the impacts associated with this type of disasters were analyzed, including the trend scenarios of frequency and associated impacts, the national legal frameworks, policies and commitments developed by each country on the subject, and various case studies considered to be illustrative of the impact and illustrative ways to potentially reduce this kind of disasters.

In terms of the methodological approach that supported this study, it could be mentioned very briefly that a mixed approach was prioritized, combining qualitative and ethnographic[1] methods—research approaches that are frequently used when dealing with organizational and/or social phenomena that are not strongly structured—with quantitative and statistical research methods that were particularly used in several statistical analysis and for design and use of indicators. Both efforts were also supported by a solid component of inquiry that was executed following the methodological principles that rule the documentary research.

Finally, a major part of the process of collecting and discussing data and results about the multiple areas of knowledge involved in this project was developed through the use of the V-RED online training platform currently available at the Center for Integrated Risk Management Research (CIGIR), which, thanks to its advantages, made it possible to hold the 32 working meetings and 18 regional workshops that were required to achieve the objectives of this research.

17.3   Results and Discussion

From the process of regional comparative inquiry, diagnosis and consensual design of strategies, and priority actions developed within the framework of this research, it was possible to obtain an important set of products; however, given the space limitations defined for this editorial project, we will limit ourselves to describe in this article two central results: The first of these will be focused on summarizing the strategy that was proposed to promote integrated risk management for torrential avalanches and floods in Latin America and the Caribbean.

The second result that we will summarize in this document shows the results of the efforts done to assess the comprehensiveness of the national commitments to reduce the risk of disasters associated to debris flows and floods, and that were assumed until 2017 by the countries of Latin America and the Caribbean, in the framework

---

[1] Some methodologies used were stake holder analysis, interviews, comparative documental research, and consensus strategies as Snow Ball Sampling and Delphi analysis.

of their National Commitments (NDCs) considered as part of the Paris Agreement for the reduction of Climate Change COPXXI.

17.3.1 Comprehensive Strategy for Action

The strategy for debris flows and floods disaster reduction that was obtained after this effort promotes an agenda of 24 key actions that are distributed in 5 action areas. This scheme is largely based on the conceptual contributions suggested by Lavell & Maskrey (2014) as a general guide for the implementation of efforts aimed at the comprehensive management of socio-natural risks.

In this sense, the original scheme suggested by Lavell shows that comprehensive efforts to reduce disaster risk scenarios demand simultaneous and articulated execution of at least four basic groups of actions:

17.3.1.1 Actions for the Identification and Characterization of Risk

Focused in this particular case in efforts aimed at knowing in as much detail as possible, the levels of threat of torrential avalanches and floods that could exist in each territorial space of interest, as well as the levels of vulnerability of the infrastructure that could be exposed to such events.

17.3.1.2 Actions for Prospective Risk Management

Focused on the design and implementation of measures to avoid future exposure to the risk of debris flows/floods in the processes of development and territorial occupation that are currently being carried out in each country. This effort has been identified as the most efficient and clever that could be promoted to reduce socio-natural risks of this nature; however, its efficient and sustained application in weakly structured socio-institutional contexts such as those of Latin America is always very complex.

17.3.1.3 Actions for Corrective/Compensatory Risk Management

In this case, the efforts are directed to the prioritization, design, and development of investments in infrastructure and engineering works tending to mitigate/reduce the potential impact of debris flows and/or flood risks that already exist and have been adequately identified in specific urban and/or rural settlements. This group of actions includes mitigation measures that several experts refer as "structural measures" and whose high costs make it difficult for their widespread use in practically all the countries of the region.

17.3.1.4   Actions for Reactive Disaster Response Management (Consummated Risks)

This effort is essentially focused on strengthening institutional and social capacities that could guarantee a rapid and efficient response in case of disaster scenario. This type of effort constitutes the dominant paradigm of the regional agenda for risk reduction, and it is always important to point out that although its actions rarely prevent losses in development infrastructure exposed to torrential landslides and floods, they have proven to have the capacity to markedly reduce morbidity and mortality associated with this type of disasters.

The original proposal based on these four areas of action was reviewed by various Latin American specialists, and, as a result of this, it was suggested that a fifth area of action be included. This fifth group of actions was termed cross-cutting regional priority actions, and inside three actions were suggested to maximize the efficiency and sustainability of the efforts that could be promoted within the framework of a comprehensive effort to reduce the impact of disasters associated with torrential avalanches or floods.

The schematic version of this comprehensive and particularly regionally adapted strategy for debris flows and flood risk reduction is presented in Fig. 17.5, and in the next paragraphs, we will briefly describe both the main characteristics of each group of efforts suggested and the general orientation of each defined action.

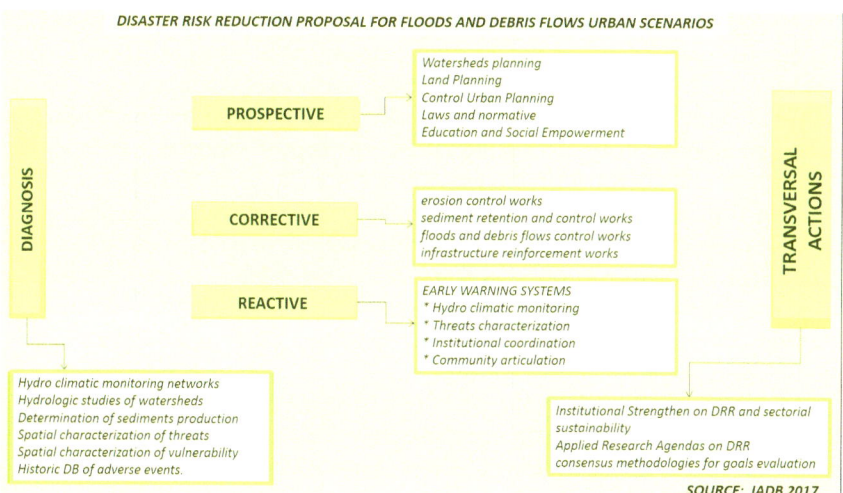

Fig. 17.5 Proposed strategy for integrated risk management of disasters associated with debris flows and floods in Latin America and the Caribbean region (IADB 2021b)

17.3.2   Actions for Disaster Risk Diagnostic and Characterization

In this first group, six actions were identified aimed at knowing in as much detail as possible, the levels of threat associated to debris flows, floods, mass movements, etc., to which the watersheds considered of interest could be exposed, as well as the different types of vulnerability that could exist in the human settlements located within them.

One of the most important products that should be obtained as part of this effort is the identification of exposure spots to various extreme events—generated on the basis of two-dimensional hydrological simulation models—since these diagnoses are very useful for estimating losses associated with extreme events and for assessing the cost/efficiency of potential investments in mitigation projects (Fig. 17.6).

The six priority actions suggested in this group are summarized in the Table 17.1.

17.3.3   Actions for Prospective Disaster Risk Management

In this second group, five actions were prioritized, and all of them focused on the design and implementation of measures that promote the effective control of territorial occupation processes of areas that could be exposed in the future to the risk of floods/debris flows. This type of effort has traditionally been identified as the most efficient and clever way to reduce socio-natural risks of this kind; however, its application in regional contexts of Latin America is always very complex.

Most of the actions proposed in this regard are usually aimed at strengthening the instruments and processes for planning and management of watersheds, territory, and urban growth, and this usually represents a major articulation and coordination effort due to the different levels of institutional competence that are usually involved in this type of task. In any case, the various actions suggested in this group always tend to strengthen the capacities of the organizations responsible for territorial planning and management processes, in order to ensure that disaster risk prevention could be adequately incorporated as a fundamental conditioning factor of the occupation and use of the territory.

A very important and relevant aspect of the regional reality that was highlighted in this regard is the importance that these efforts could be linked with community inclusion and empowerment initiatives that were considered fundamentals to ensure a successful implementation. In this sense, education and community work—when well understood and not limited to teaching the population what to do "before, during, and after" an adverse event—can become a very useful strategy to prevent the construction of future disaster risk scenarios.

The intervention and empowerment of social actors in these planning processes should be based on approaches aimed at reversing the practices and social perceptions that promote the construction of highly vulnerable local contexts. In this

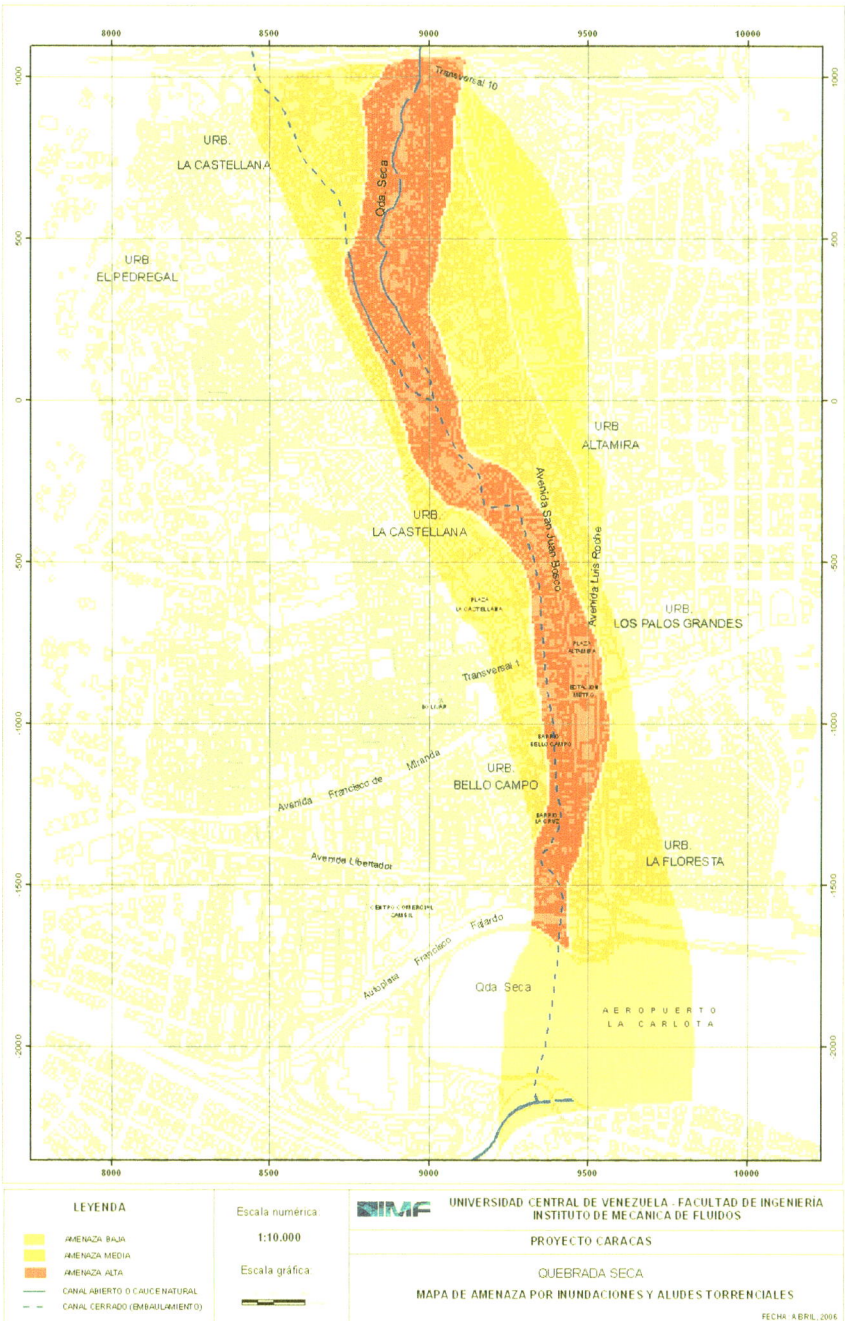

Fig. 17.6 Example of debris flow hazard maps (2D simulation) prepared by the Institute of Mechanic of Fluids of the Central University of Venezuela (IMF-UCV cited on IADB 2021b)

Table 17.1 Priority actions associated with the diagnosis of risk scenarios

Item	Purpose	Action plans
Baseline data inventory and production	Includes the collection/production of cartographic base (GIS), digital elevation models, urban maps, inventory of actors/regulations, etc. of the area	For the regional context, everything suggests that there is abundant information available in each country and that the effort should be focused on collection and systematization
Strengthen hydrometeorological monitoring	This involves evaluating/strengthening the existing climate monitoring capacity for that locality	This effort should focus on the national organization responsible for hydroclimatic monitoring
Hydrological study/modeling of the basin	Determine flow characteristics (hydrographs) of estimated floods at different points and for different return periods	The priority use of these studies is the proper management of water resources, but their results are of interest for hydroclimatic risk reduction
Hydrological study/modeling of the basin	Know the volume of sediments carried by normal and extraordinary flows and their granulometric characteristics	Of great interest for the design of mitigation projects. Must be accompanied by geomorphological characterization of the basins
Hazard 2D mapping	Mapping of extent and hazard of areas exposed to floods/torrential landslides	Based on mathematical simulation of flow (MMF) in digital elevation models (DTM) that allow estimation of depths/flow velocity
Vulnerability characterization	Identify both the degree of exposure of infrastructure and the potential impact that this could have in the event of an event	As part of this effort, assessments of community risk perception levels and diagnoses of social vulnerability should be promoted

sense, these processes should be oriented to provide the population the capacity to identify and promote more sustainable occupation practices, incorporating effective disaster prevention and mitigation criteria, without neglecting aspects of disaster preparedness, response, and recovery.

The five priority actions suggested for the prospective approach to disaster risk management of extreme hydroclimatic events DRM-HC are summarized in the following Table 17.2.

17.3.4 Actions for Corrective Disaster Risk Management

As mentioned above, this effort essentially aims the design and development of engineering works capable of mitigating/reducing damages that could be produced

Table 17.2 Priority actions to promote prospective risk management

Item	Purpose	Action plans
Consolidate basin management plan	Diagnose and strengthen plans, programs, and efforts that could be carried out for basin management	This effort should give priority to the correction of inadequate intervention processes identified in the upper and middle areas of the basin
Strengthen land management plans	Diagnose and strengthen the land management plans that could govern the basin in which we are working	This effort again requires a solid coordination with various entities and efforts in each country that take on this type of task
Strengthen urban management plans	Diagnose and strengthen the urban management plans that could govern the urban areas of the basin	Here, the articulation should be prioritized with the sectorial entities of housing/urbanism and with the departmental, district, and township levels of government
Strengthen occupancy regulations	Inventory and strengthen the entire regulatory and legal framework that governs occupancy processes in areas that could be exposed to HC risk scenarios	Here again, it is important to coordinate with departmental, district, and township governments
Education and community intervention	Consolidate locally relevant efforts to teach citizens—more than how to act before, during, and after an adverse hydroclimatic event—how to identify and better coexist with the natural dynamics of the watershed they occupy	In the case of education, these efforts should focus on the existing school infrastructure. In the community issue, they demand initiatives that guarantee the citizen's "right to know" and subsequently the negotiation of possible forms of action and community commitment to the sustainability of each locality

in those debris flows and/or floods risks scenarios that already exist and that have been adequately identified.

Unlike what was mentioned in the previous group of actions, corrective measures for hydroclimatic disaster risks are usually costly and tend to translate into investment projects that commonly require special funds that are not always available in the ordinary operating institutional budgets of most of the countries studied. Hence, the widespread use of these measures is often very difficult.

In order to exemplify and classify the types of mitigation works that are most frequently used for the corrective management of disaster risk associated with torrential landslides and floods in our region, a very simplified classification scheme is summarized in the following Table 17.3.

Table 17.3 Objectives, goals, and structural measures for corrective management (IADB 2017)

Objective	Target	Measure
Reduce erosion	Decrease surface erosion in watersheds Increase stability of slopes and carcaves Decrease vertical and lateral erosion in streambeds	• Watershed management • Reforestation and soil bioengineering • Drainage control • Terracing • Stabilization of unstable slopes and gullies • Bottom control by means of sleepers or dams • Control of flows by means of sleepers or stepped dams • River bank protection
Debris flow control	Interception and retention of sediments Flow detour to adjacent areas Flow conveyance to safe areas	• Open and closed dams • Sedimentation ponds • Flow diversion and channelization works • Channelization of the waterway downstream of the last retention dam
River flood control	Stabilize processes/spaces susceptible to generate mass movements	• Longitudinal berms and dikes • Storage ponds/reservoirs for flood cushioning • Increasing channel conveyance capacity (cutting of meanders, widening, or deepening of the channel, flow diversion)
Geotechnical stabilization works	Estabilizar procesos/espacios susceptibles generar movimientos de masas	• Cable-stayed screens • Retaining walls • Control of surface/served waters

The high cost of these efforts usually demands a solid economic justification on the part of decision makers, and in this sense, it is once again important to highlight the contribution that 2D mapping of urban hazard areas can provide for this purpose. It should be remembered that these diagnoses—when are created with good mathematical simulation models of expected flows (MMF) and supported with solid digital elevation models (DEM)—make it easier to compare the estimated economic damages that could be recorded in anurban area if the basin were not intervened, with the damages that would be recorded if a set of mitigation works were developed (Fig. 17.7). From the contrast of both scenarios, it is easy to estimate the savings that could be made, and this can become a solid justification for the investments that could be required.

One last but important aspect that is not often considered when this type of projects is undertaken is that when a single mitigation initiative is promoted, it is always important to plan integrally and with a long-term view how that specific mitigation

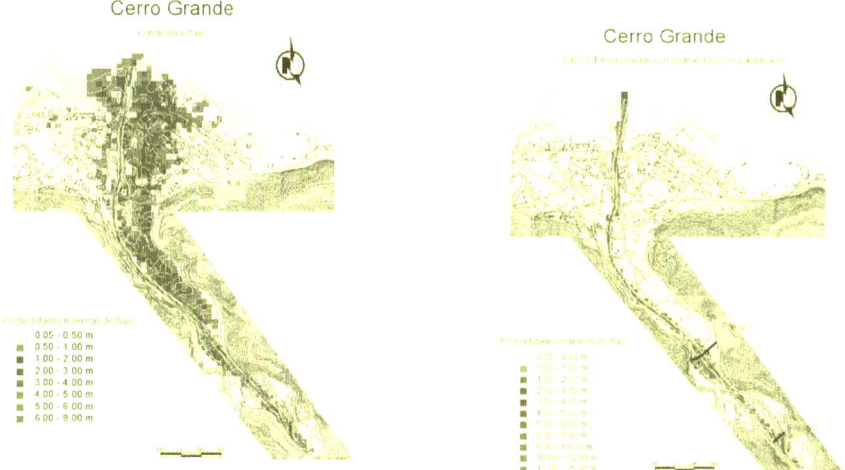

Fig. 17.7 Simulations of debris flow damages in an urban watershed (Cerro Grande—Vargas/Venezuela) with and without mitigation works (Bello & López 2010 cited on IADB 2021b)

project should be integrated with the rest of mitigation works that may be required in a basin to guarantee an integral intervention. Regional experience suggests that it is important to avoid isolated initiatives that tend to ignore the fact that these works must eventually operate in coordination with other complementary works—even those not yet built—in order to maximize their efficiency in the long term.

The six priority actions suggested for the corrective approach to disaster risk management of extreme hydroclimatic eventsDRM-HC are summarized in Table 17.4.

17.3.5   Actions for Preparation and Response (Reactive Disaster Risk Management)

Reactive risk management measures constitute efforts that is focus essentially on strengthening institutional and community coordination capacities that guarantee a rapid and efficient response in case of consummation of hydroclimatic risk scenarios in disaster (Fig. 17.8).

The design and implementation of Early Warning Systems—also known by their acronym EWS—are currently the most widespread strategy to promote social and institutional preparedness and response capacities in case of disasters associated with the occurrence of extreme hydroclimatic events, and the worldwide interest and increasingly frequent use of EWS have given rise to the establishment of a significant number of guiding principles, methodological guidelines, and even checklists

Table 17.4 Priority actions associated with corrective disaster risk management

Item	Purpose	Action plans
Integrated mitigation working plans	Generate preliminary plans that identify and articulate the different mitigation works that could be undertaken in a watershed	This effort serves to avoid common mistakes in the region that are associated with ad hoc interventions with no overall or long-term vision
Sediment control projects	Develop works to reduce surface erosion on slopes, increase stability of slopes and gullies, and/or reduce vertical and lateral erosion in river beds	They include reforestation and soil bioengineering, drainage control, terracing, stabilization at the foot of unstable slopes and gullies, control of the riverbed by means of sleepers or stepped dams, and bank protection
Flood control projects	Develop mitigation works to reduce water levels and river flood spots	Includes the use of berms and longitudinal dykes, storage ponds/flood buffers, increasing channel conveyance capacity (meander cutting, widening, deepening, diversion, etc.)
Debris flow control projects	Develop works for the interception and retention of materials, diversion of flows to adjacent areas, or their conveyance to safe areas	Includes the use of open, semi-closed, and closed dams, retention ponds, diversion and channelization work, installation of screens for material retention, etc.
Slope stabilization projects	Develop works for stabilization of land susceptible to mass movements (landslides, slides, or flows)	Aimed at reducing risks associated with geomorphological processes and include the use of cable-stayed screens, retaining walls, terracing, etc.
Mitigation works for the protection of strategic areas	Special works for the reinforcement and protection of strategic points	These are always very specific interventions that include reinforcement, retaining walls, etc.

that have been developed (ISDRR-UNO 2005) by various specialized international organizations to guide the effective use of this kind of effort.

A mistake that is frequently made when proposing Early Warning Systems for extreme hydroclimatic events capable of generating floods, torrential avalanches, etc. is the assumption that such efforts must be focused only on the implementation of telemetric detection networks and monitoring stations for meteorological and/or hydrological variables; this mistake promotes that these initiatives are usually assigned to experts in areas such as meteorology, hydraulics, electronics, instrumentation, remote sensing, and telecommunications.

Fig. 17.8 Examples of typical torrential landslide risk mitigation works. The upper figures show metal barriers for protection against torrential landslides built in the mountains of Chosica in Lima (Peru). *Source* GEOBURG cited on IADB (2021c). The figures in the middle and bottom of the page show different types of mitigation works built in the watersheds of Vargas state. *Source* López (2011) cited on IADB (2021b)

This erroneous conception has been pointed out by various international organizations and even in various international forums specialized. In these spaces, global standards and guidelines have been defined which establish that EWS efforts imply a set of actions and investments in the following four components or subsystems: monitoring subsystem, hazard characterization subsystem, institutional articulation and coordination subsystem, and the social and public community awareness and articulation subsystem. On the basis of these four subsystems suggested by the UN International Strategy for Disaster Reduction, the priority actions were suggested for the reactive treatment of the risk of floods and torrential landslides in our study were defined.

Table 17.5 Priority actions associated with preparative and response to disasters

Item	Purpose	Action plans
Inventory/evaluation of EWS Early Warning Systems SAT	Inventory, evaluate and strengthen existing early warning initiatives and/or define new EWS that should be consolidated in the basin	It is assumed that the most accepted mechanism for HC disaster preparedness/action is the EWS, and in this sense, efforts to strengthen its 4 constituent subsystems are promoted
Monitoring subsystem	Design, installation, and maintenance of devices to control hydroclimatic parameters that could activate response protocols	It has been noted that large national synoptic climate monitoring networks do not always provide the intensity or scale of monitoring that an EWS requires. (ISDRR-EWC)
Threat characterization subsystem	Identify geo-spatially the levels of exposure to floods that could occur in the local space in which an EWS system is installed and operates	This effort can be covered with the geospatial hazard characterization efforts mentioned in the diagnostic actions
Institutional articulation subsystem	Definition, dissemination, and validity of the inter-institutional coordination protocols that should govern the operation of the EWS	This defines where the data arrive, who processes them, and what decisions and inter-institutional action should be taken in the event of an alert declaration
Social articulation subsystem	Ensuring that the population is aware of the existence of the EWS and is articulated and empowered in its management and operation	It has been demonstrated in international comparative studies (EWCII 2003) that this is one of the weakest links and that it has had the greatest impact on the failure of these systems in the world

Some characteristics associated with each suggested action are presented in Table 17.5.

17.3.6 Cross-Cutting Priority Actions

The inclusion of this fifth group of actions was essentially due to the recommendations of various regional specialists who considered that it was very important to include some concrete efforts and could be convenient to maximize the efficiency and sustainability of a comprehensive agenda for debris flows and flood risk reduction in a context such as the Latin American one.

As part of the concrete actions of this last group, it was suggested to include efforts considered as baseline and cross-cutting conditions that could maximize the probability of success of the whole process. The first type of actions suggested here is related to the design and implementation of applied research agendas that promote the existence of permanent spaces for discussion, diagnosis, and design of new proposals. In this sense, it was considered essential that the design of this effort be based on processes of broad consensus among experts from different disciplines.

The second and third set of actions are aimed at establishing mechanisms to promote the implementation of efforts at local and community levels, and in this sense, the design of instruments for management control and evaluation of efforts at local government levels is proposed. Finally, a permanent effort of strengthening of community participation and empowerment to promote governance for local management of risk scenarios is suggested (Table 17.6).

Table 17.6 Cross-cutting priority actions

Item	Purpose	Action plans
Applied research agenda	Consolidate an applied research and technology development agenda to permanently support hydroclimatic risk management efforts	Efforts of this nature could be promoted by the entities promoting scientific and technological development policies in each country
Local DRM-HC evaluation/monitoring	Consensually define a set of indicators of risk reduction performance levels that characterize local levels of government	In this sense, it is suggested to promote the review and adaptation of regional efforts that have allowed the construction and use of DRR indexes for local government levels, instead of the direct application of these tools
Actions for community governance and empowerment	Establish principles, mechanisms, and forms of action and articulation that promote participation, empowerment, and community co-management of the different initiatives to be implemented for local risk management	This is one of the most complex challenges that must be addressed, and for this, it is convenient to review the serious regional efforts that have been developed to characterize/intervene the social perception of risk and social vulnerability and improve the socialization of information

17.3.7 Assessment of the Comprehensiveness of Country Commitments

The last product that we would like to refer to in this document is linked to the comparative assessment of the level of comprehensiveness of the efforts to reduce the risk of debris flows and floods that was promoted by the countries of Latin America and the Caribbean. To achieve this, we proceeded to construct an indicator that we called Sectorial Index of Country Commitment to Flood and Torrential Flood Risk Management (ICPGRIAT), which could facilitate—based on the systematic review of the documents of national progress in the fulfillment of commitments acquired by nations as part of the Paris Agreement COP XXI (NDCs)—the quantification of the level of comprehensiveness of the efforts to reduce the risk of this type of disasters in each country of the region.

The rating scale defined for this valuation instrument varies in a range from 0—which represents no reference in the country's NDCs to initiatives for Flood and Landslide Risk Management—to 4—which represents strong and explicit support for Flood and Landslide Risk Management in the country's NDCs. The results of this exercise allowed us to define that the regional average value (μ) of the use of this indicator in Latin America and the Caribbean countries was 1.60, with a maximum score found in Argentina (2.62) and a minimum in Suriname (0.42). It should be noted that this assessment made it possible to point out that for the moment of the evaluation (2017), most of the countries of the region shows ICPGRIAT below 2, and that was the minimum suggested baseline that—in the opinion of regional experts—should be accepted (Fig. 17.9).

Additionally, to the "global" assessment of the comprehensiveness of national efforts for the reduction of debris flows and flood disasters that was developed through the application of the ICPGRIAT, assessments of the comprehensiveness of the partial efforts that were developed by the countries in each of the groups of actions that were suggested in the regional strategy to promote the comprehensive reduction of the risk

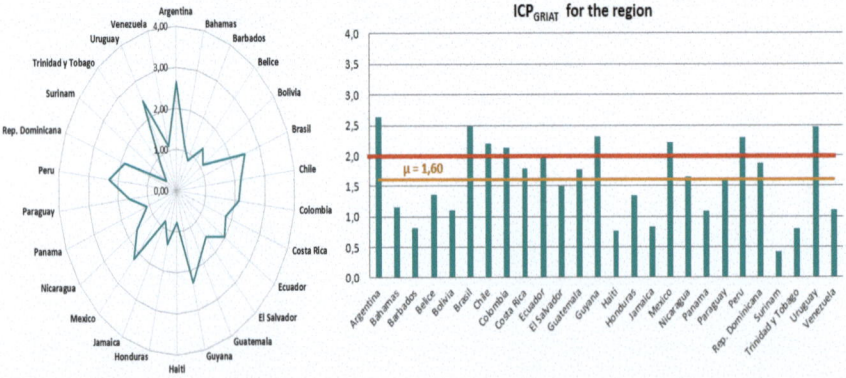

Fig. 17.9 ICPGRIAT values obtained for Latin America and the Caribbean (IADB 2021c)

of debris flows and floods were also measured as part of this effort. The results of these diagnoses are shown in the following figure (Fig. 17.10).

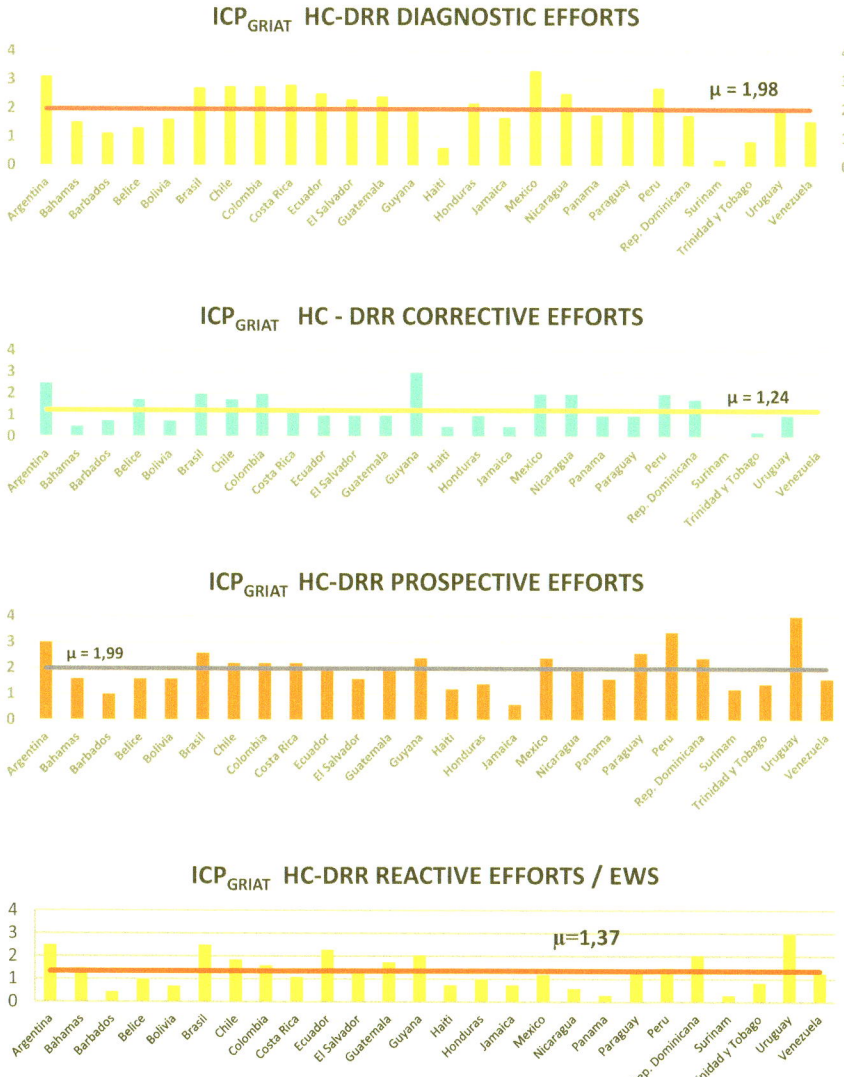

Fig. 17.10 ICPGRIAT values obtained for each of the first 4 groups of actions defined in the strategy for integrated management of torrential landslide and flood risks in Latin America and the Caribbean (IADB 2021b)

17.4   Conclusions

The multifactorial and highly complex nature underlying the causes of the worsening impact of disasters—particularly in contexts as complex as those prevailing in Latin America and the Caribbean—demands multiple efforts to unveil and try to intervene the reasons for the growing impact of disasters. As a result of this effort, it is essential to understand that it will be difficult to reduce the impact of socio-natural disasters in the countries of this region of the planet as long as we fail to transform the processes that currently enhance the levels of vulnerability and unsustainability that characterize most of our models of development and land occupation.

Along this article has been summarized the experience and main results of a regional comparative research carried out during 2017 by the Research Center for Disaster Risk Reduction (CIGIR) with the purpose of identifying financial strategies for reducing the risk of disasters associated with the occurrence of floods and debris flows in the particular context of Latin America and the Caribbean, a region of the world that is recognized by its high levels of disaster risk associated with extreme hydroclimatic events.

It has been briefly described in this article how the lessons left by last regional disasters have been promoting during last few decades important advances toward an approach to a comprehensive management of socio-natural disaster risks reduction, but more and better specific action agendas still are required for a better advance in the goal to reduce the impact of floods, debris flows, and similar extreme hydroclimatic events.

To cooperate in the definition of a comprehensive strategy for risk reduction of this kind of disasters, it has been proposed here a comprehensive action agenda that contemplates five large areas of intervention, that include a total of 25 specific priority actions that must be attended. This action agenda allowed us to assess of the levels of comprehensiveness of the efforts that has being made by the countries of Latin America and Caribbean in order to reduce their risks of hydroclimatic disasters.

As a result of this research, it has been possible to provide to a better understanding of the complexity of actions related with the goal to reduce the impact of disasters in a regional context characterized by very complex scenarios of inadequate land occupation, poverty, unequal rich distribution, and where hydroclimatic disasters essentially should be understood as a symptomatic manifestation of a long chain of unsustainable development processes which tend to build permanently the risk scenarios that will precede the impact of future calamities.

References

EM-DATA. 2021. International Disasters Database EM-DAT. Centre for Research on the Epidemiology of Disasters. School of Public Health of the Université Catholique de Louvain. Brussels, available in https://www.emdat.be/.

EWC. 2003. Integrating early warning into relevant policies; early warning conference II, Bonn - Germany. Available in: https://www.unisdr.org/2006/ppew/info-resources/ewc2/.

FAO. 2018. *The impact of disasters on agriculture and food security 2017*. Global Report of the Food and Agriculture. Organization. United Nations, New York, 2019. Chapter 1, p. 25. http://www.fao.org/3/I8656EN/i8656en.pdf.

IADB. 2017. Riesgo climático y definición de estrategias financieras para su mitigación en el sector agua y saneamiento en ALC: Inundaciones, Technical Document prepared to the Inter American Development Bank. http://dx.doi.org/10.18235/0003085.

IADB. 2021a. *Estrategias Financieras para la Adaptación al Riesgo Climático en el Sector Agua y Saneamiento de América Latina y El Caribe: Aludes Torrenciales e Inundaciones*, vol. I. Washington: Banco Interamericano de Desarrollo http://dx.doi.org/10.18235/0003086.

IADB. 2021b. *Estrategias Financieras para la Adaptación al Riesgo Climático en el Sector Agua y Saneamiento de América Latina y El Caribe: Aludes Torrenciales e Inundaciones*, vol. II. Washington: Banco Interamericano de Desarrollo. http://dx.doi.org/10.18235/0003087.

IADB. 2021b. *Estrategias Financieras para la Adaptación al Riesgo Climático en el Sector Agua y Saneamiento de América Latina y El Caribe: Aludes Torrenciales e Inundaciones*, vol. III. Washington: Banco Interamericano de Desarrollo. http://dx.doi.org/10.18235/0003088.

IFHV. 2019. *World risk report 2019 fOCUS: Water supply*. RUHR Universitat—Germany 2019; Risk Indices Dashboard. https://www.maplecroft.com/risk-indices/.

III. 2020. *Munich Re Geo risks graphic report on total annual losses related to disasters*. Insurance Information Institute. https://www.iii.org/graph-archive/96424.

IPCC. 2012. Managing the risks of extreme events and disasters to advance climate change adaptation. A special report of working groups I and II of the intergovernmental panel on climate change. In Field, C.B., V. Barros, T.F.Stocker, D. Qin, D.J. Dokken, K.L. Ebi, M.D. Mastrandrea, K.J. Mach, G.-K. Plattner, S.K. Allen, M. Tignor, and P.M. Midgley (eds.)], 582 pp. Cambridge, United Kingdom and New York, NY, USA: Cambridge University Press.

ISDRR-UNO. 2005. Memories 3rd world conference on early warning systems. EWC—Bonn, Germany. Conclusions, 7–18.

Lavell, A., and A. Maskrey. 2014. The future of disaster risk management. *Environmental Hazards* 13 (4): 267–280. https://doi.org/10.1080/17477891.2014.935282.

Liñayo, A. 2006. *National strategies for prevention and mitigation of disasters: Educative aspects.* Technical Document, Ministry of Planning—Caracas, 78–96.

Liñayo, A., and R. Estevez. 2003. *Bases para la formulación de Politicas Científicas en Reducción de Desastres.* Ministry of Science and Technology—Caracas, 75–92.

Liñayo, A. 2002. *A systemic approach to Disasters Administration in Latin America*, 87–114. Merida: Soft Systems Research Center, University of Los Andes.

Maskrey, A. 1993. Los Desastres NO son Naturales. LaRED—FLACSO. Editorial Tercer Mundo Bogotá.

Wilches-Chaux, G. 2007. "¿Qué nos pasa?", Red de estudios Sociales en Prevención de Desastres, OXFAN, ARFO Editores.